Solutions to Advanced Problems in Organic Chemistry

for

Competitive Examinations

Akshay Choudhary

Mandakini Choudhary

Pearson

ISBN 978-93-325-7831-9
ISBN 978-93-325-7836-4 (Combo Pack)

First Impression, 2017
Sixth Impression, 2024
Seventh Impression, 2025

Published by Pearson India Education Services Pvt. Ltd, CIN: U72200TN2005PTC057128,

Head Office: 1st Floor, Berger Tower, Plot No. C-001A/2, Sector16B, Noida 201301 U.P., India.

Registered Office: Featherlite, 'The Address' 5th Floor, Survey No 203/10B, 200 Ft MMRD Road, Zamin Pallavaram, Chennai - 600044.
Website: in.pearson.com, Email: companysecretary.india@pearson.com

Compositor: SRS Technologies, Puducherry
Printed in India at Sai Printo Pack Pvt Ltd

Contents

CHAPTER 1

General Organic Chemistry

LEVEL 1

1. (i)

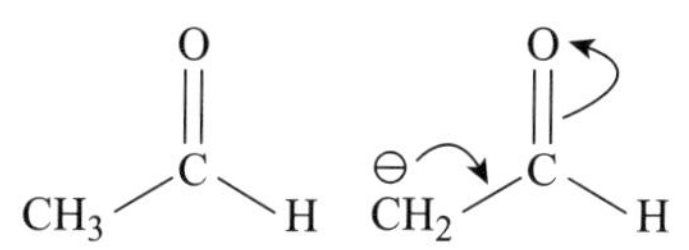

(ii)

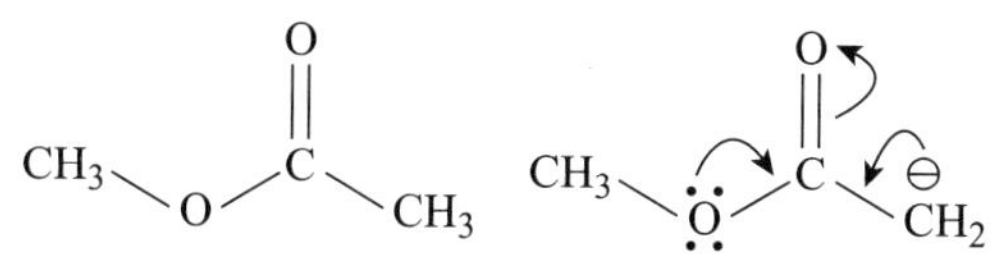

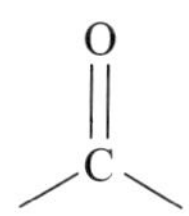 is in cross-conjugation

(iii)

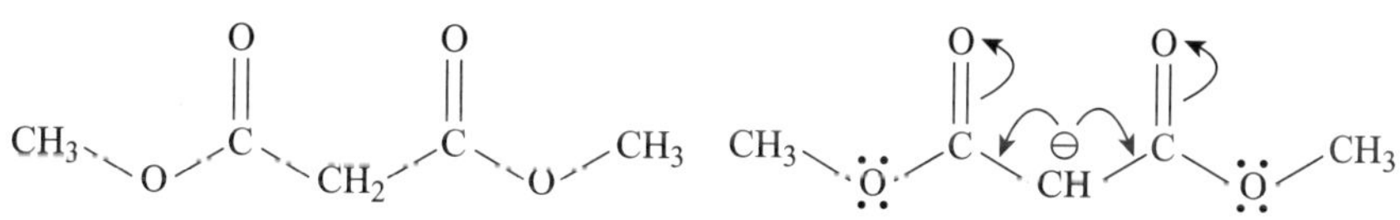

Negative charge is dispensed on two

–C=O

Ans: (d)

2. (i) (ii) (iii)

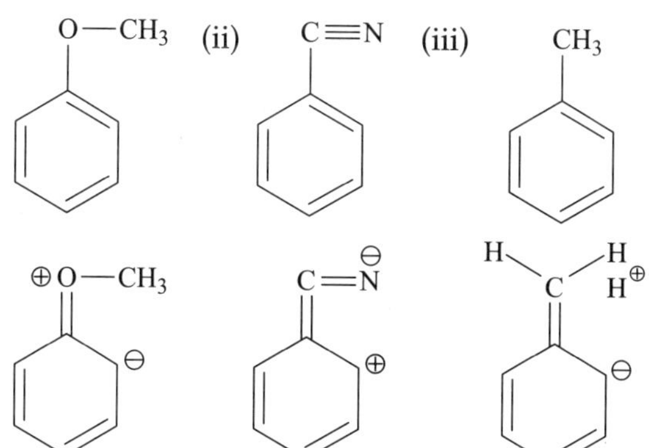

(+R effect) (– R) (+ H)

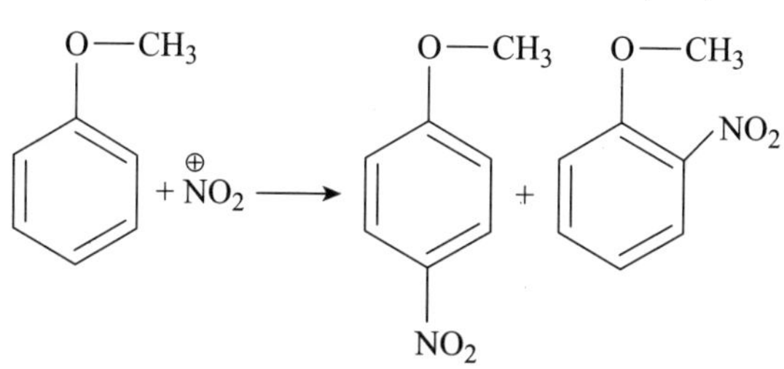

Ans: (c)

3. 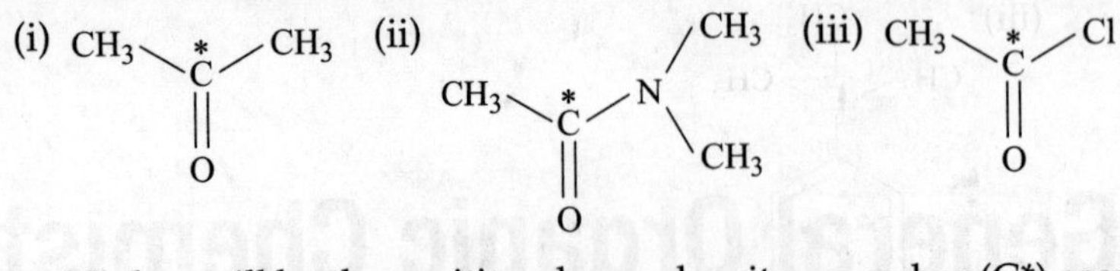

Higher will be the positive charge density on carbon(C*) easy will be the hydrolysis. **Ans:** (d)

4. 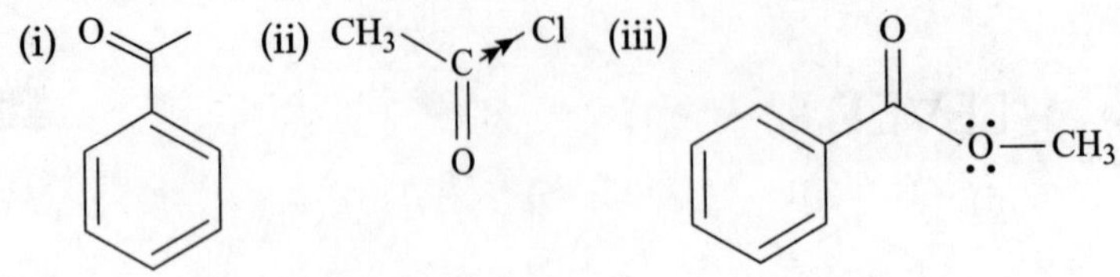

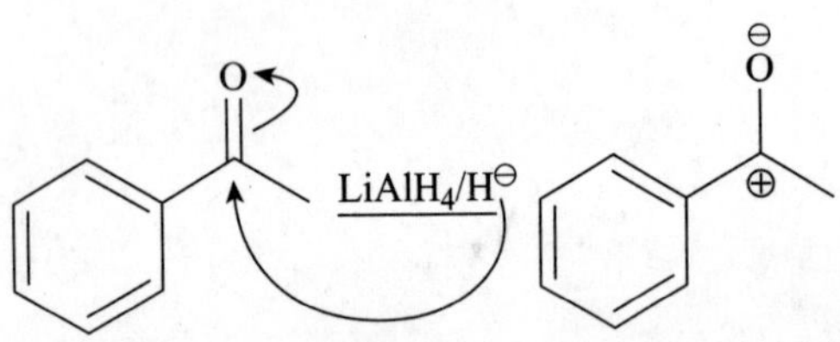

Higher will be the positive charge density easy will be the attack of $H^{\ominus}$

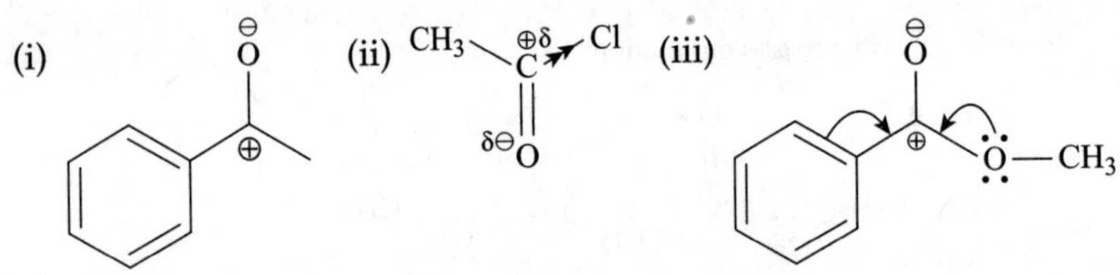

Ans: (a)

5. 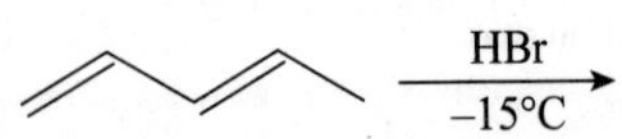

Ans: (a)

7. (i) $CH_2{=}CH{-}C({=}O){-}CH_3$ (ii) $CH_2{=}CH{-}CH_3$ (iii) $CH_2{=}CH{-}\ddot{O}{-}CH_3$

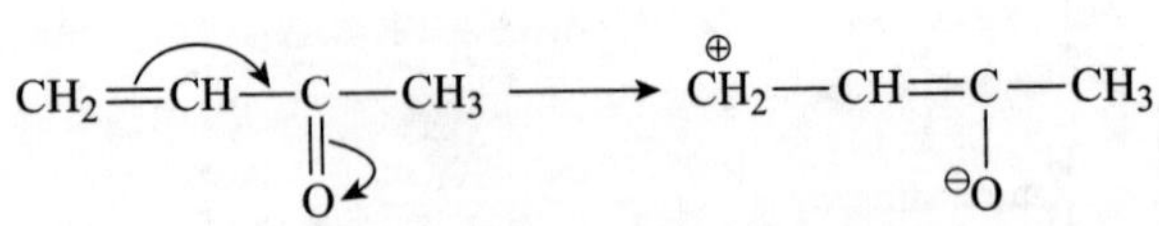

Hence less reactive towards electrophilic addition

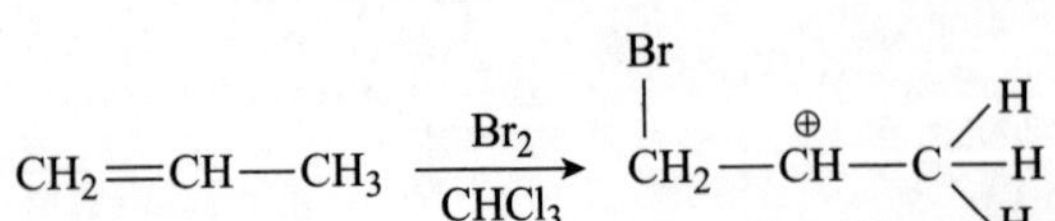

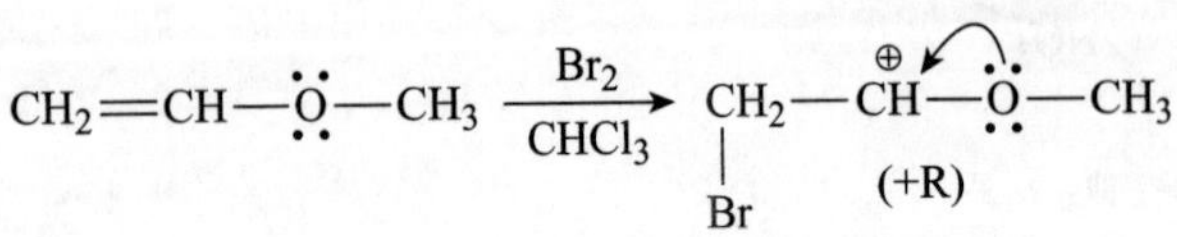

Ans: (b)

8. (i) (ii) (iii)

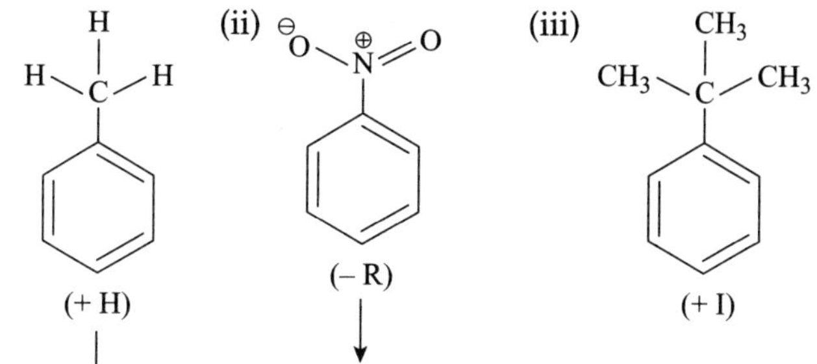

(+ H) → Activated ring; (– R) → Deactivated ring; (+ I)

Ans: (c)

9. (i) (ii) (iii)

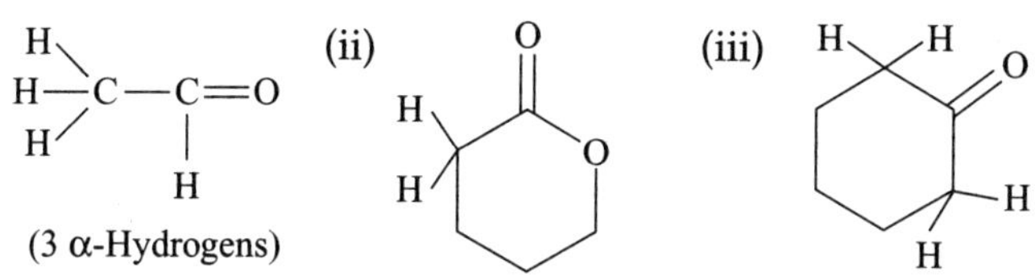

(3 α-Hydrogens) (2 α-Hydrogens) (4 α-Hydrogens)

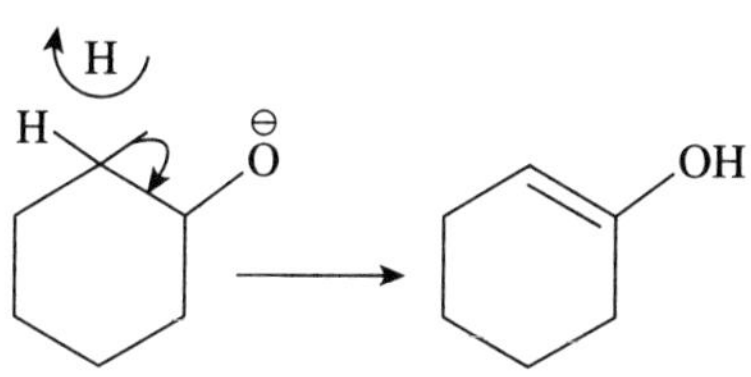

Ans: (d)

11. (i) (ii) (iii)

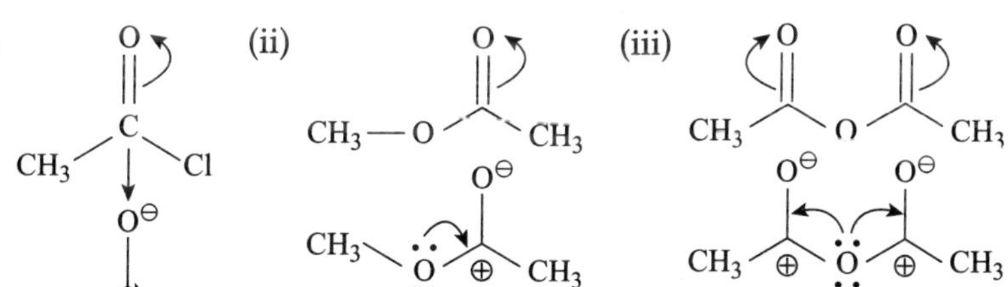

(+R) (+R cross conjugation)

Ans: (c)

12. (i)

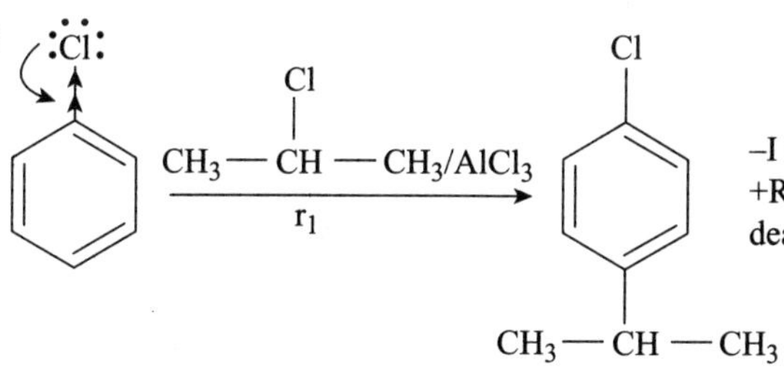

–I effect dominates +R, So in net ring deactivating

(ii)

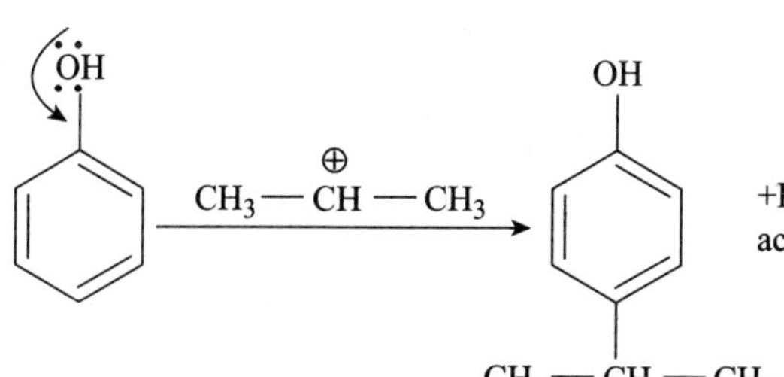

+R effect of –OH activates positive ring

(iii)

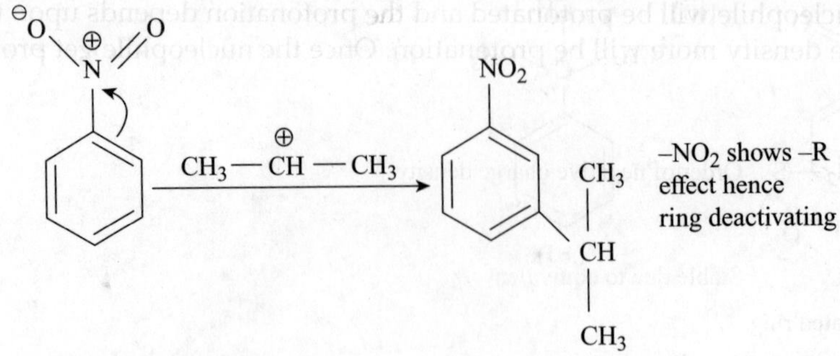

Ans: (d)

13. (i)

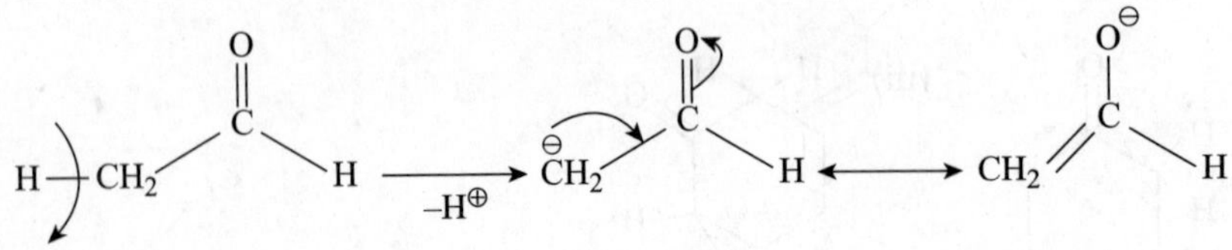

(ii)

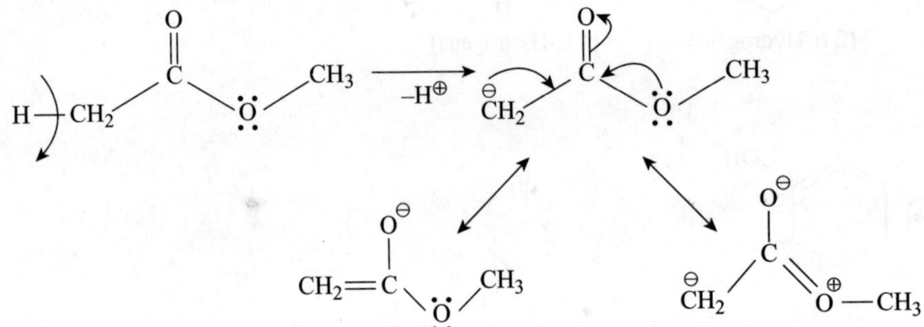

(iii)

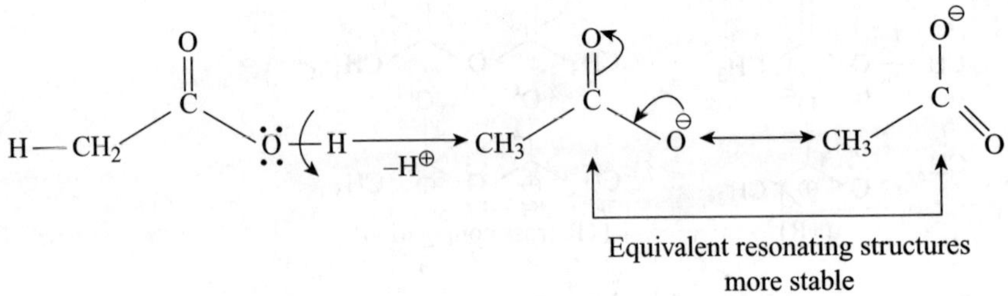

Ans: (d)

14. In MeMgBr the attacking nucleophile is $-Me^{\ominus}$

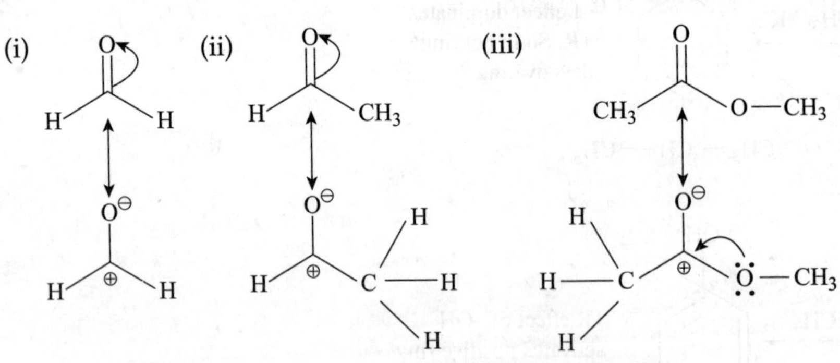

Ans: (a)

15. (i) $CH_3-CH_2-\overset{\ominus}{S}$ (ii) $CH_3-CH_2-\overset{\ominus}{O}$ (iii) $CH_3-C(=O)-\overset{\ominus}{O}$

In polar protic solvent the nucleophile will be protonated and the protonation depends upon the charge density, higher will be charge density more will be protonation. Once the nucleophile get protonated, it becomes weak nucleophile.

$CH_3—CH_2—\overset{\ominus}{O}$ $CH_3—CH_2—\overset{\ominus}{S}$ Order of negative charge density

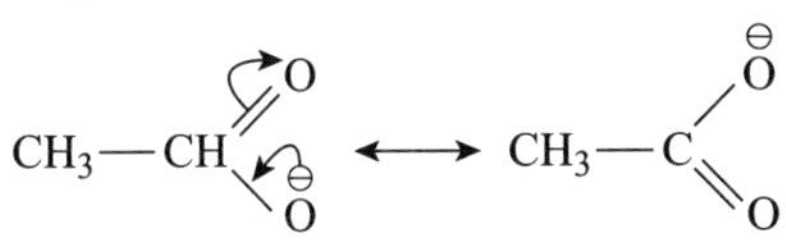

Stable due to equivalent R.S

Ans: (a)

16. 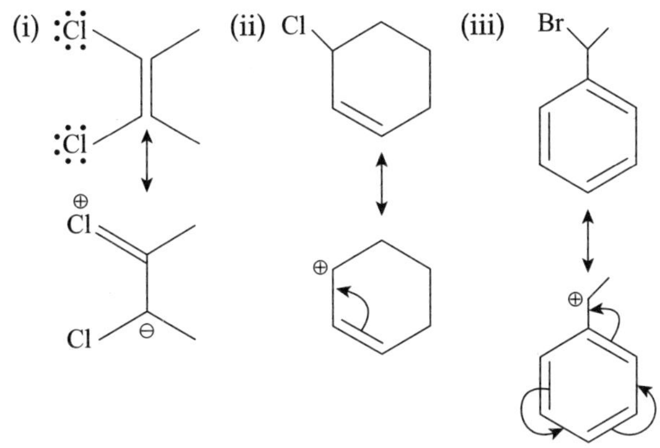

Ans: (b)

17. 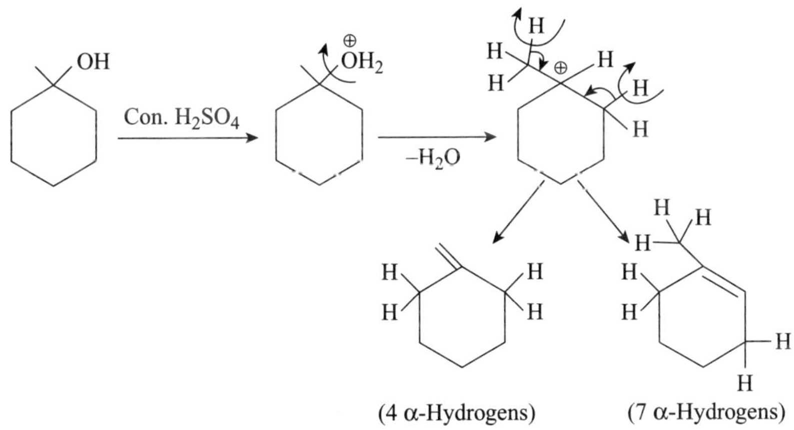

Stability of alkenes depends upon number of α-Hydrogens.

Ans: (c)

18. 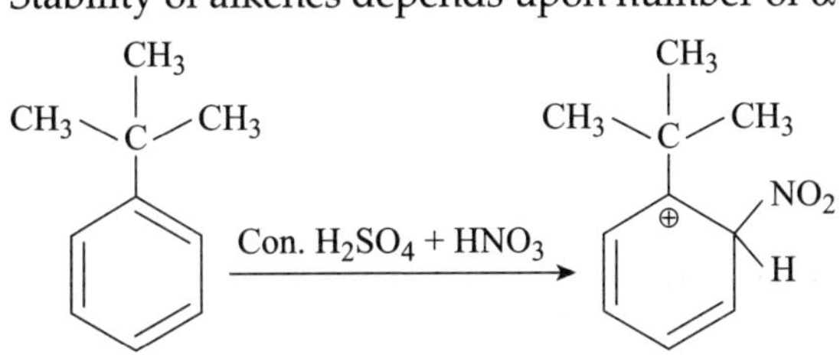

This can be decided by the stablility of a complex yield or stability of intermediate

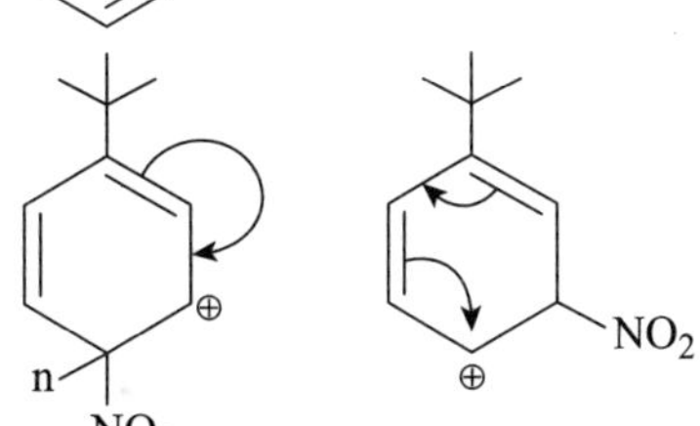

Ans: (d)

19.

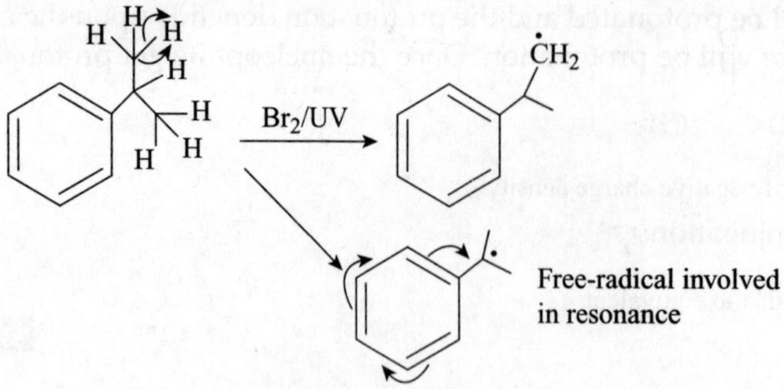

During this reaction free-radical intermediate is formed and relative yield depends upon the stability of the intermidiate (free-radical). This free-radical doesn't involve in resonance.

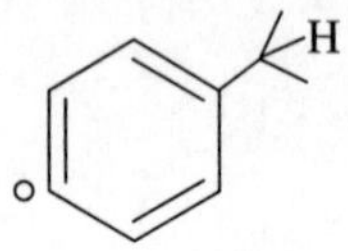
This free-radical doesn't involve in resonance.

Ans: (b)

21.

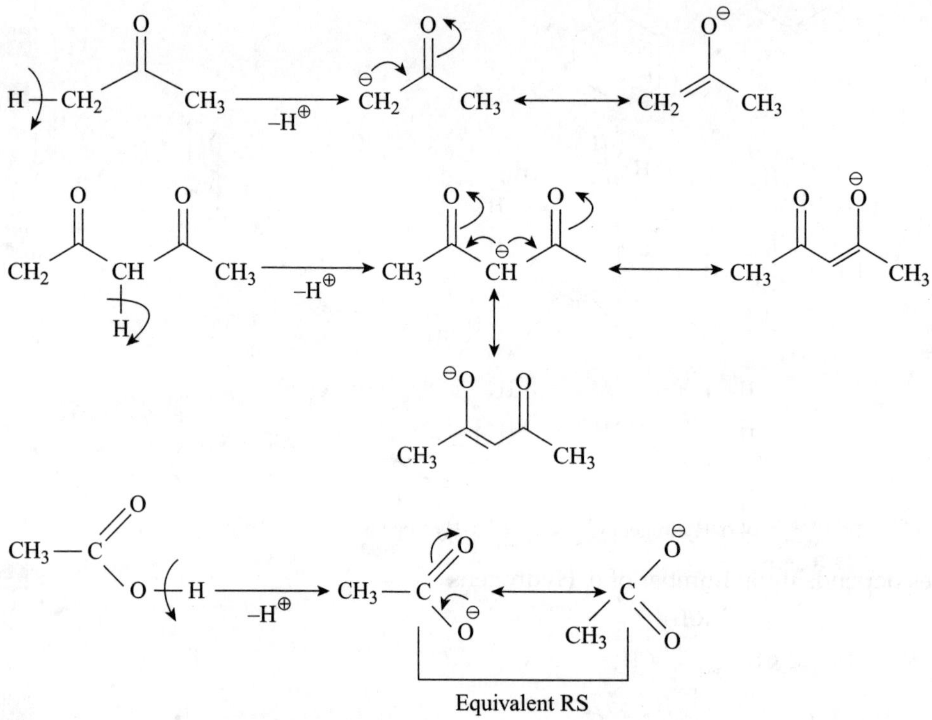

Ans: (a)

22. In protic medium protonation of the nucleophile depends upon charge density order of protonation in polar protic medium.

$CH_3—OH > CH_3—NH_2 > CH_3—CH$

∴ Nucleophilicity order

$CH_3—\ddot{C}H > CH_3—\ddot{N}H_2 > CH_3—\ddot{O}H$

Ans: (b)

23.

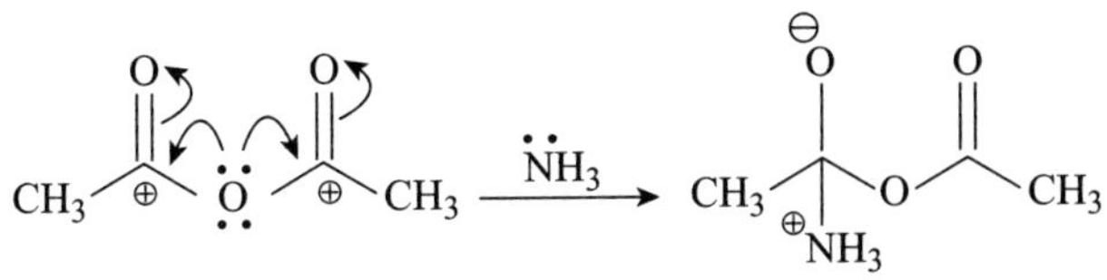

Note: Lone pairs of oxygen are in cross conjugation.

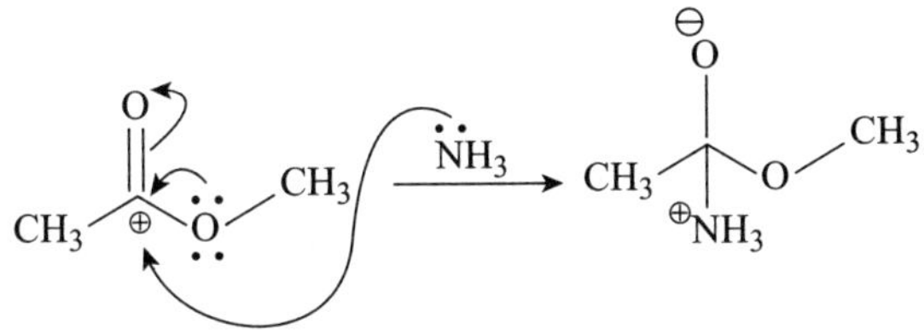

Here lone pair of oxygen is in 100% resonance with only single –C=O bond.

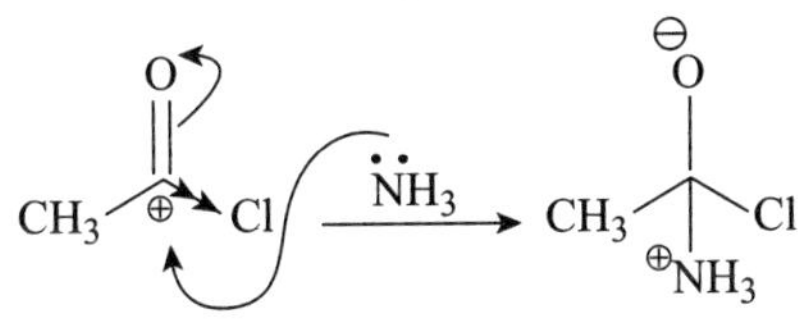

Ans: (d)

24.

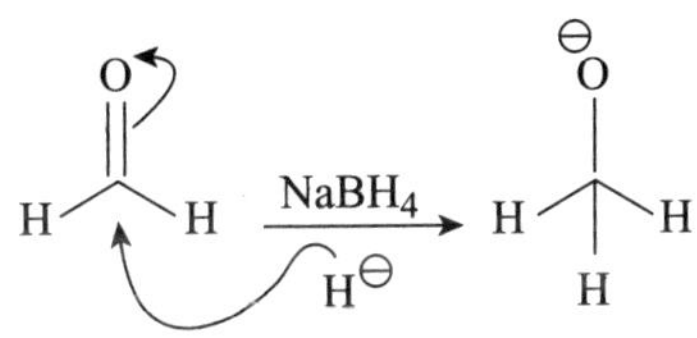

Ans: (a)

25.

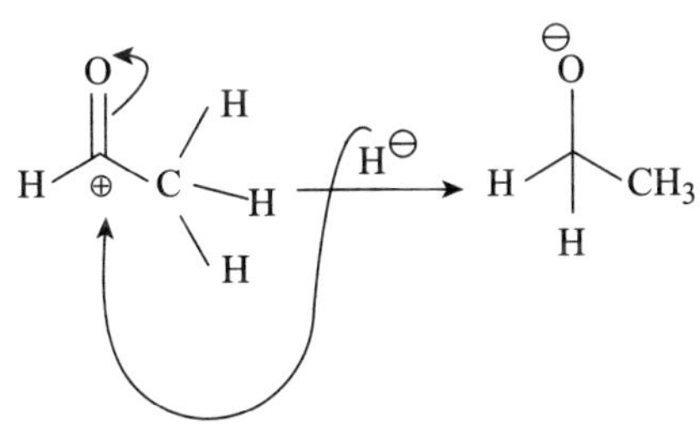

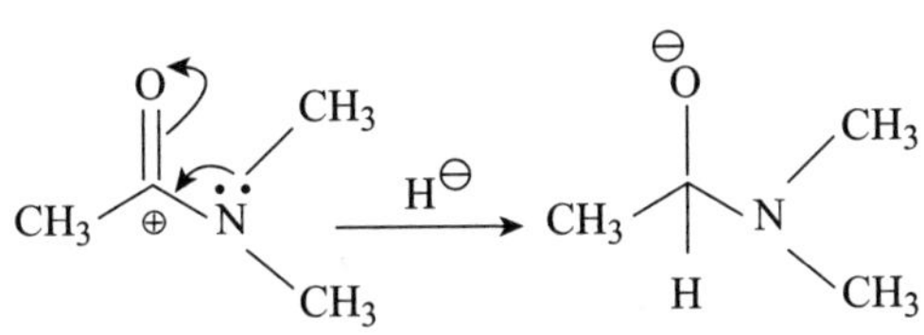

Ans: (d)

27. (i) $CH_3—CH{=}CH—\dot{C}H_2 \longleftrightarrow CH_3—\dot{C}H—CH{=}CH_2$

(ii) $\dot{C}H{=}CH—CH_3$ (least stable)

(iii)

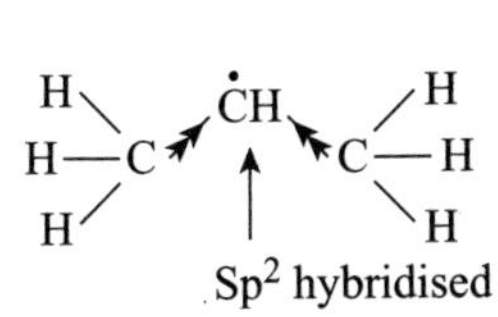

Stabilised by hyperconjugation

Ans: (c)

29. (i) 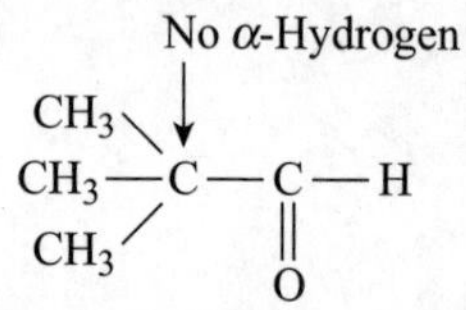

(ii) 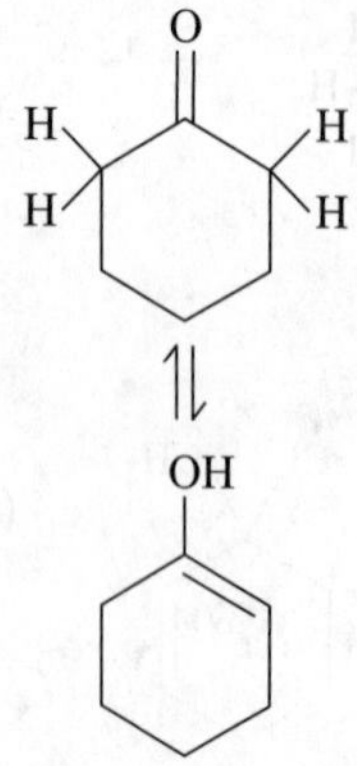
(iii) 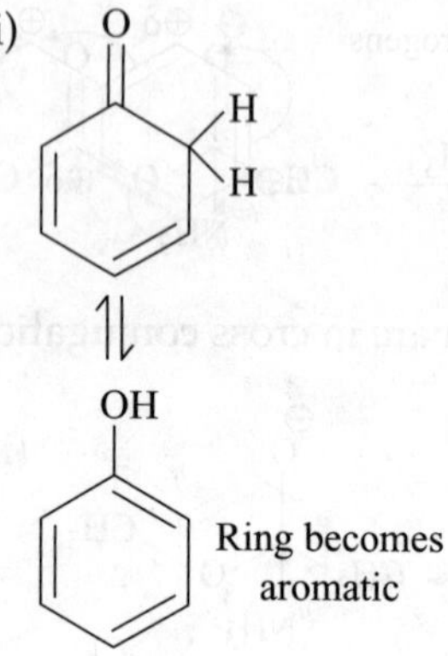

Ans: (d)

31. (i) 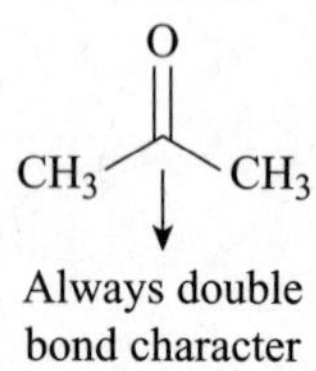

(ii) 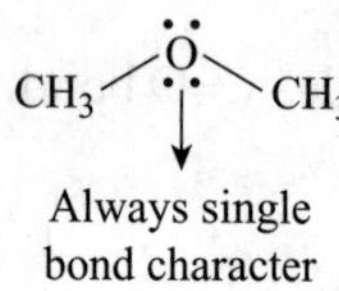

(iii) 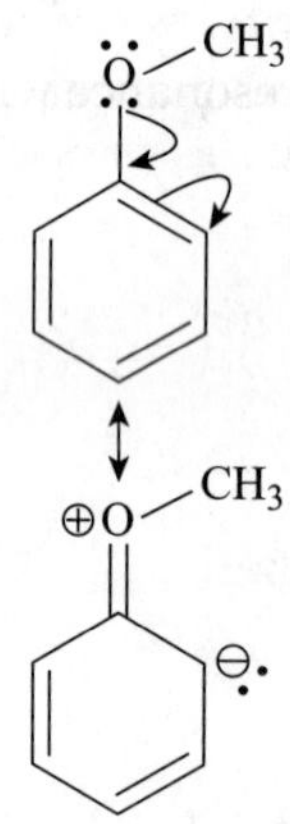

Single and double bond character due to resonance

Ans: (b)

32. (i) 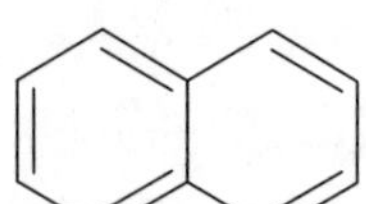 (ii) 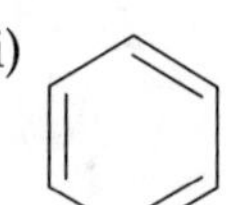 (iii) 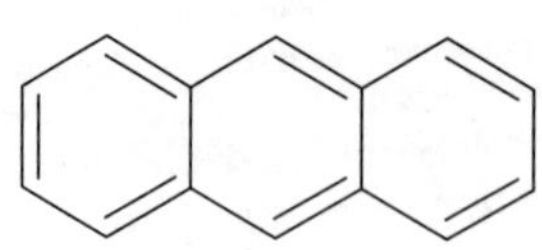

Hint: Resonance energy is directly proportional to number of conjugations.

Ans: (d)

33. (i) 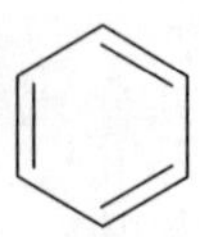 (ii) 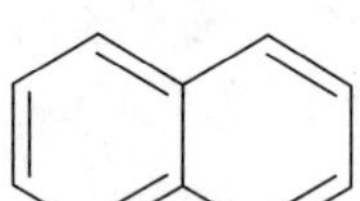 (iii) 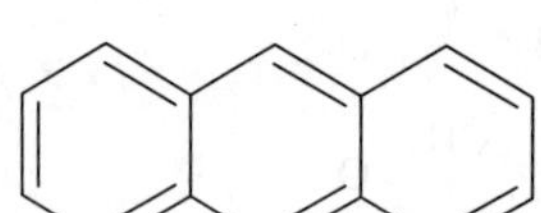

In this case resonance energy depends upon number of perfect benzene rings in the given compound.

Ans: (a)

34. With increase in carbon numbers, boiling point increases with same number of carbon atom, with branching boiling point increases.

Ans: (b)

35. (i) 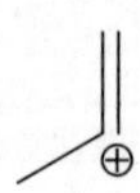 (ii) 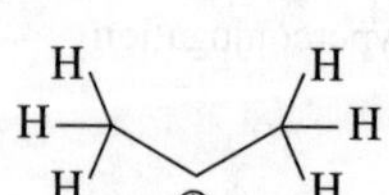 (iii) 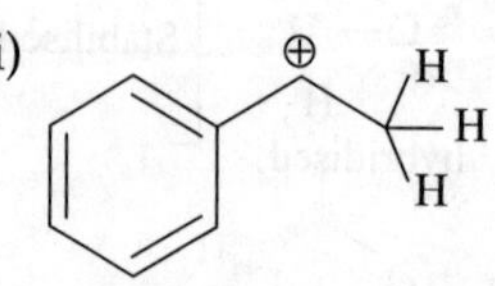

Positive charge on double bond and triple bond is less stable

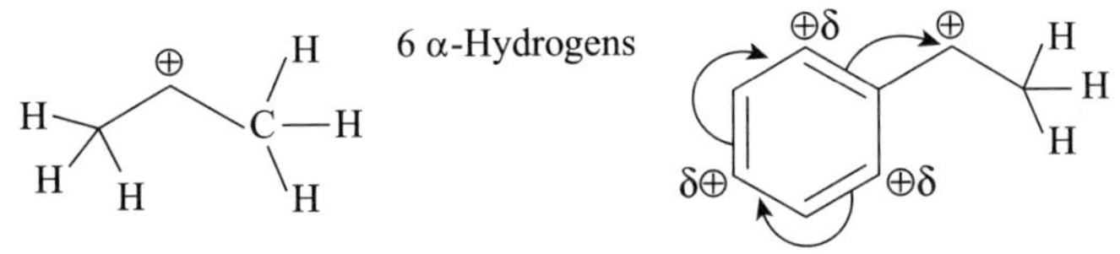

Order of stability

Ans: (d)

36. (i)

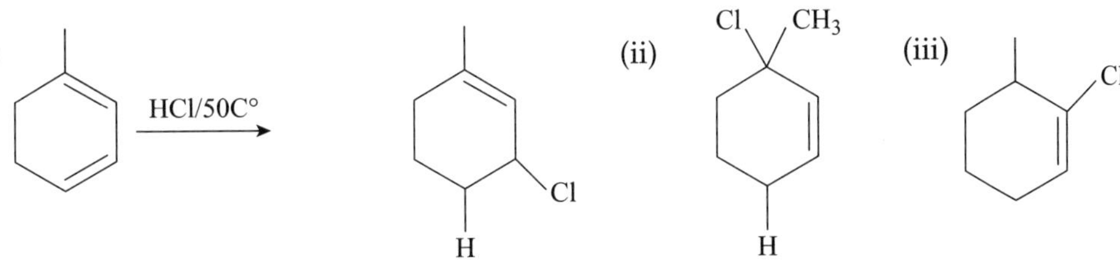

[A+ high temp –HCl addition is thermodinamically favourable]

Ans: (b)

37.

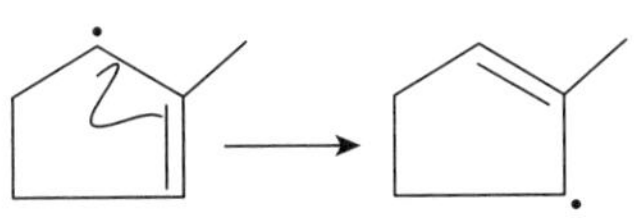

Ans: (a)

38.

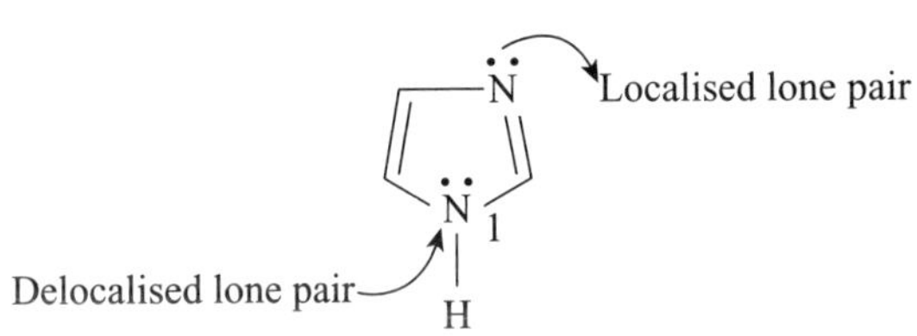

Hint: Lone pair of $–N^1$ is involved in resonance while the lone pair of $–N^3$ is localised.

Ans: (c)

39.

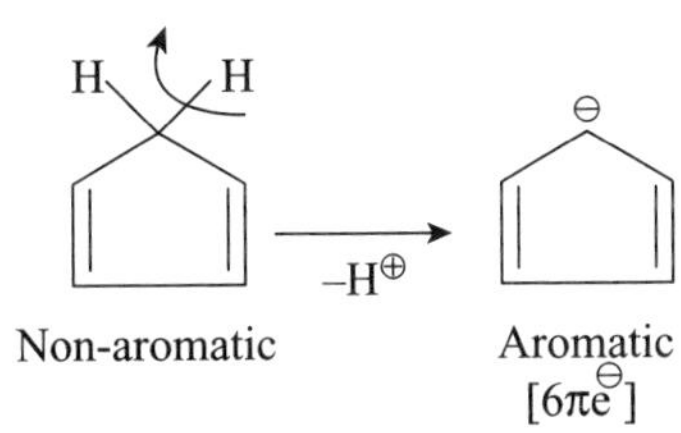

Ans: (d)

40.

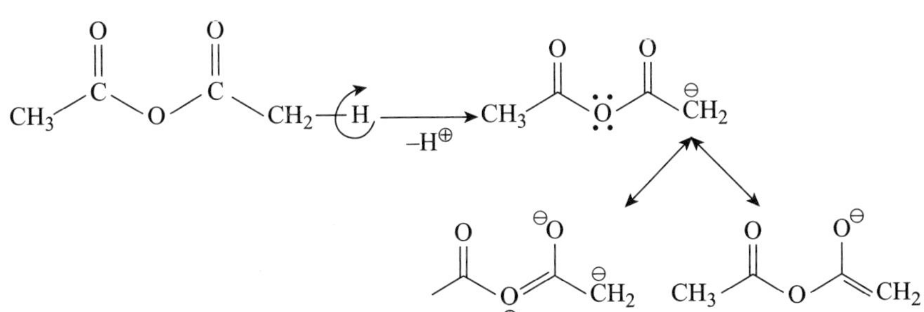

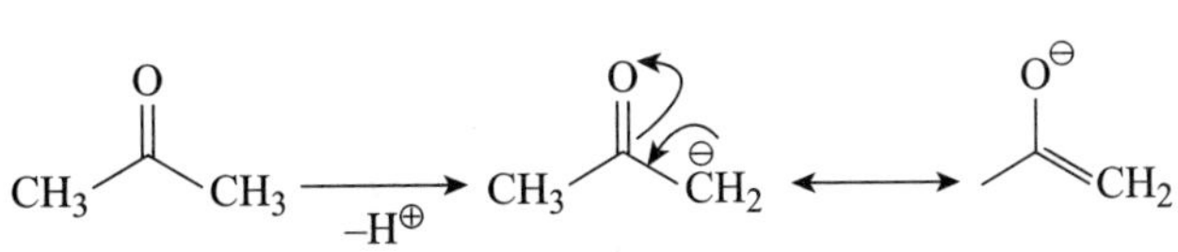

Ans: (c)

42. (i) (ii) (iii)

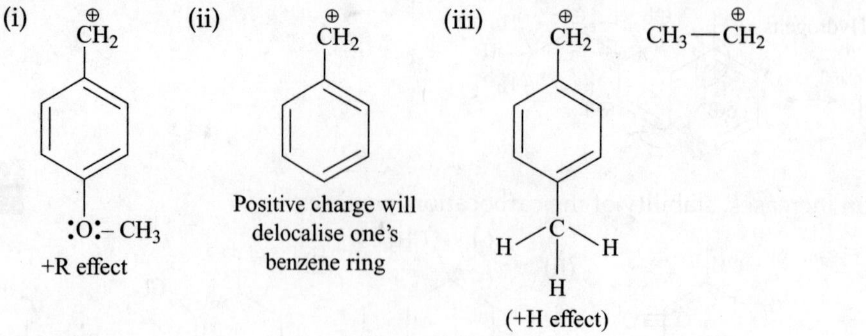

Ans: (c)

43. (i)

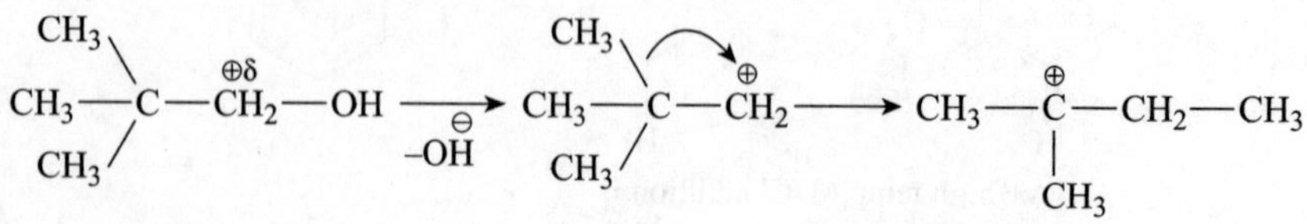

(ii)

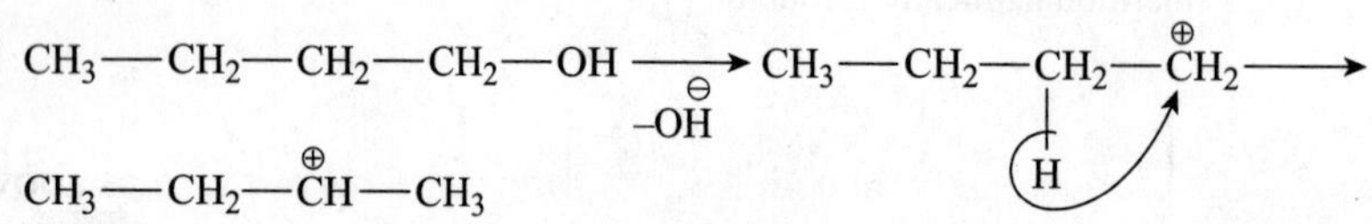

(iii)

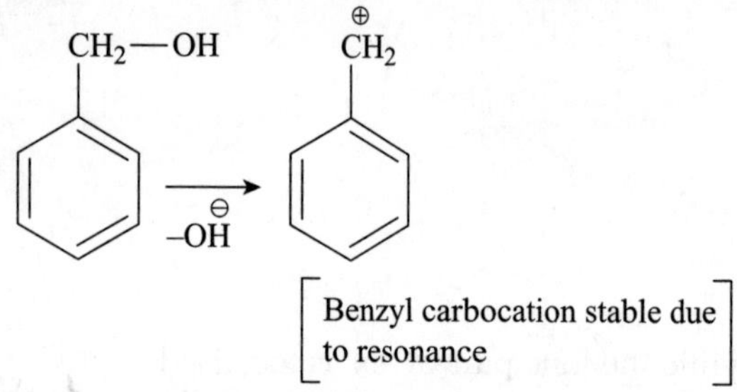

(iv)

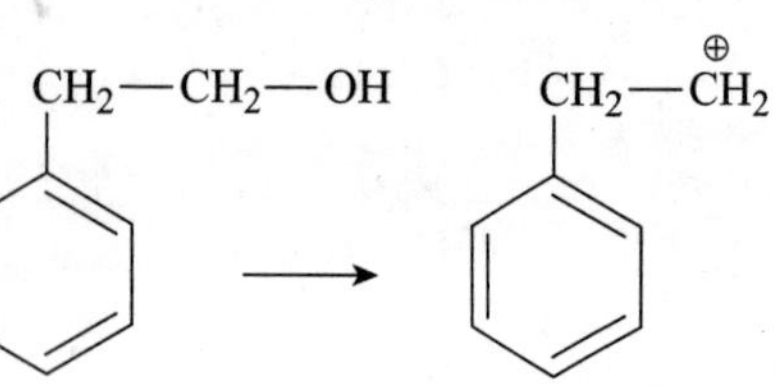

Ans: (c)

44.

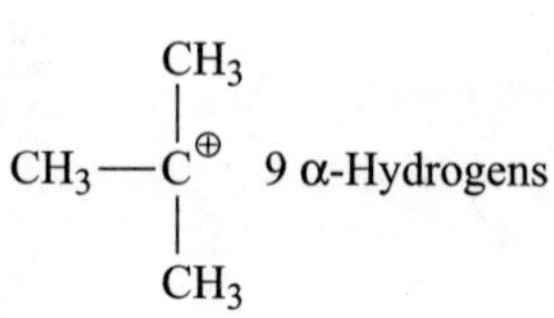

Ans: (a)

45. (i)

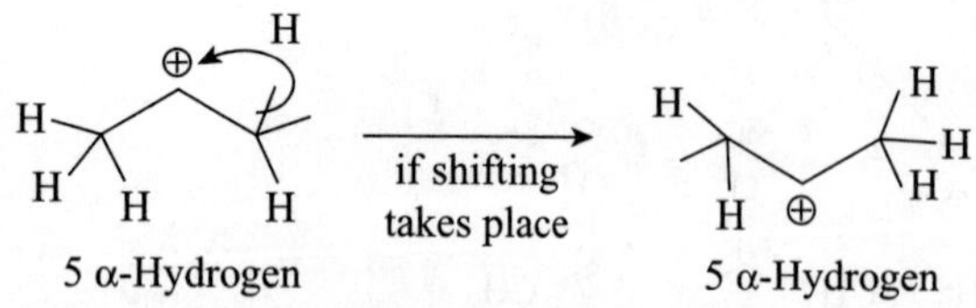

Hence shifting doesn't take place.

(ii)

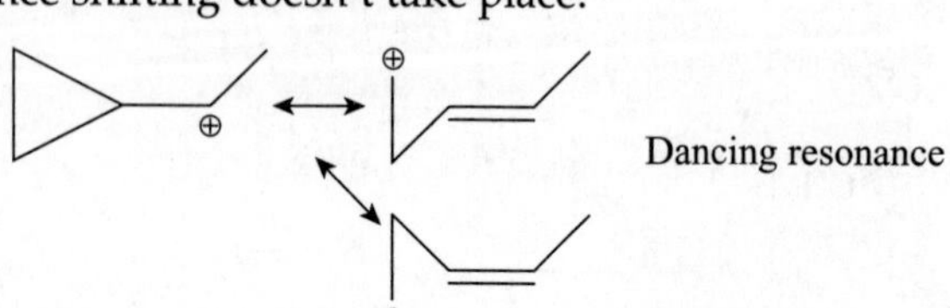

Ans: (b) and (c)

46.

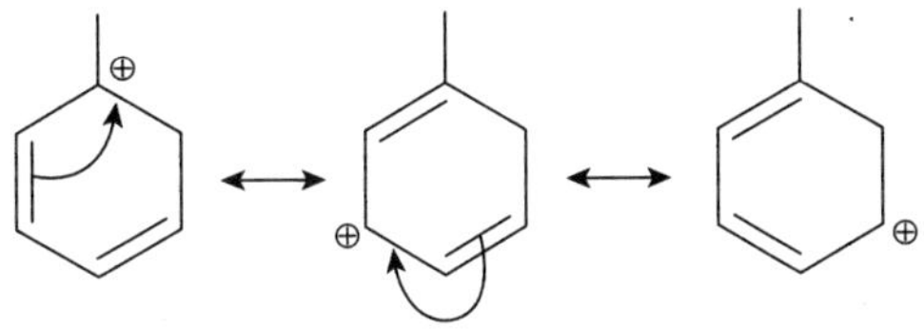

If number of conjugation increases, stability of the carbocation increases.

Ans: (d)

47.

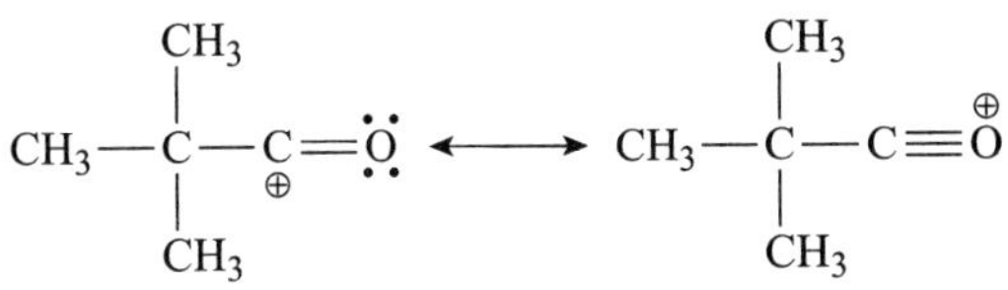

Species with complete octate

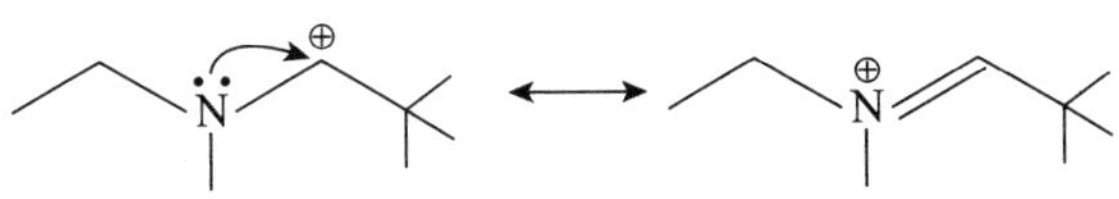

Species with complete octate

Whenever carbocation is formed next to lone pair then sharing of lone pairs takes place as above.

Ans: (a) and (c)

48.

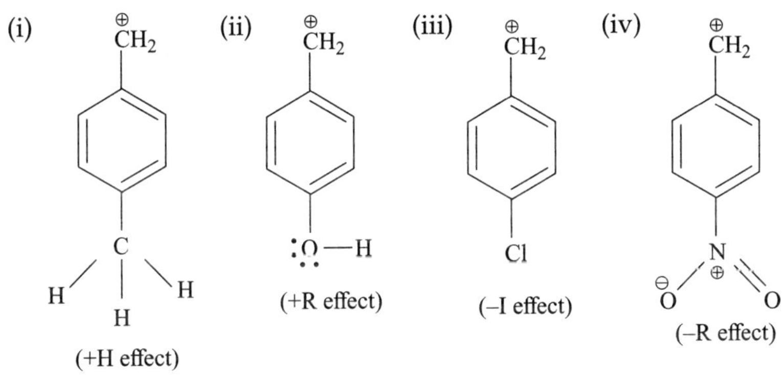

(i) (+H effect) (ii) (+R effect) (iii) (–I effect) (iv) (–R effect)

Ans: (b)

49.

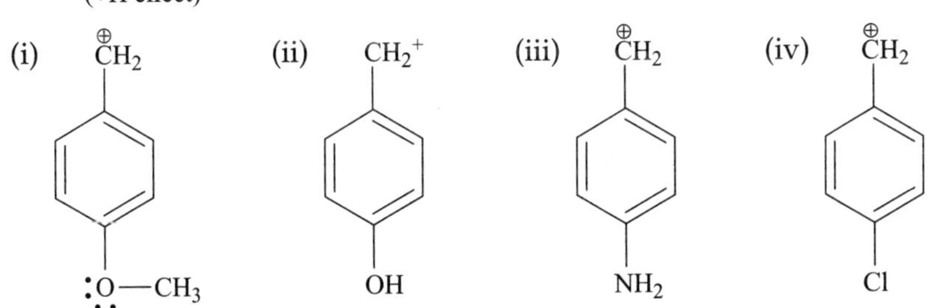

(a) (b) (c) (d)

Stability depends upon the order of +R effect

$-\ddot{N}H_2 > -\ddot{O}-CH_3 > -\ddot{O}H$

Ans: (c)

50. (i)

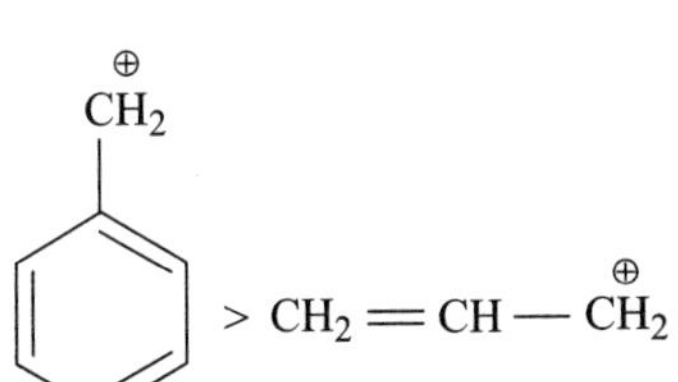

with increase in conjugation stability increases.

(ii) 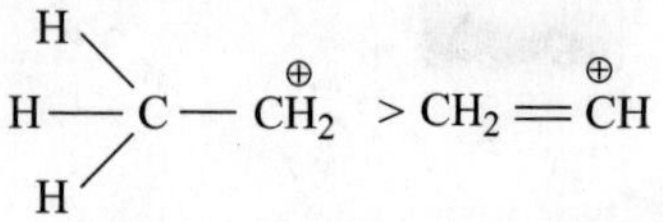

(+H effect)

∴ Positive charge on double and triple bond is less stable.

(iii) 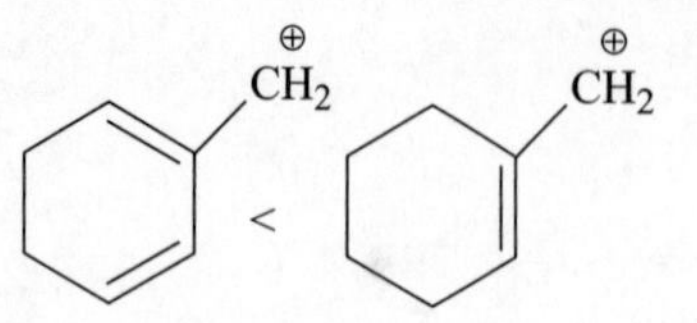

Double bond in conjugation with +ve charge.

(iv) 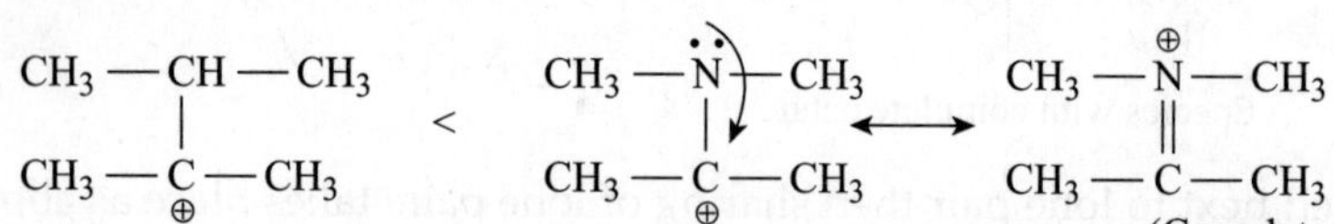

(Complete octate)

Ans: (a)

51. (i) 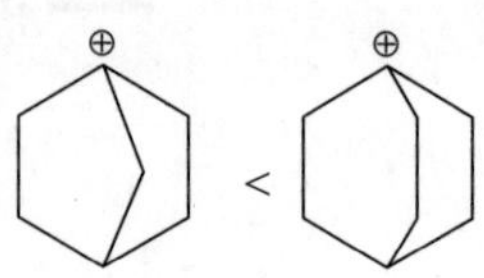

[with increase in number of carbon +I effect increases and due to increase in ringsize strain ↓ Se]

(ii) 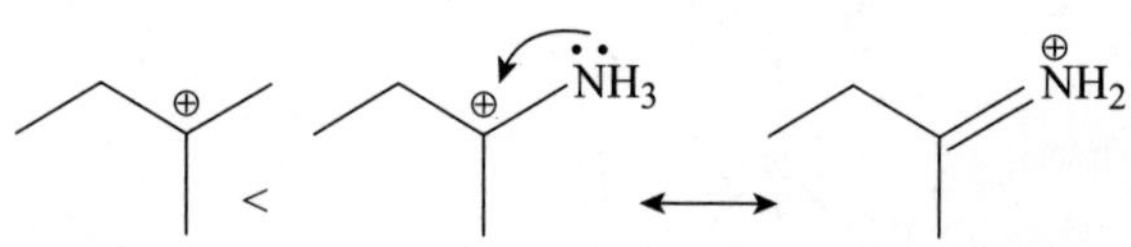

Species with complete octate are more stable

Ans: (a)

52. (i) 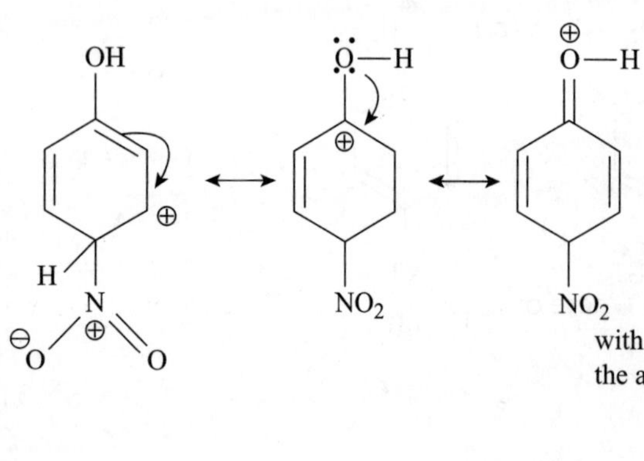

with complete octate of all the atoms

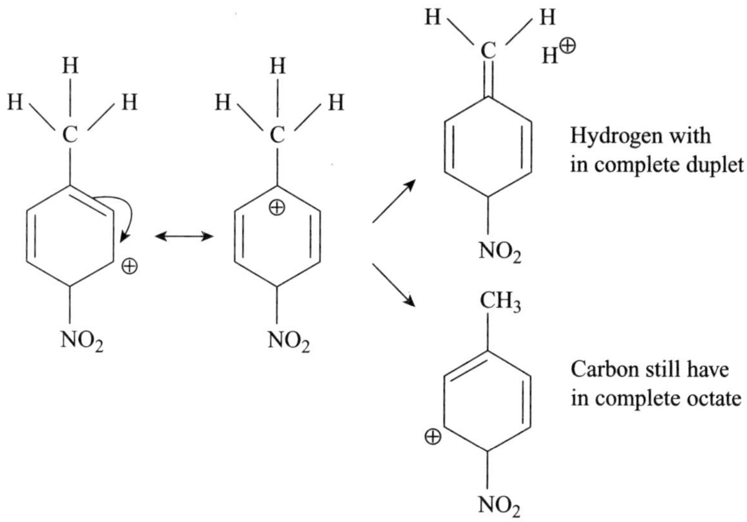

(ii)

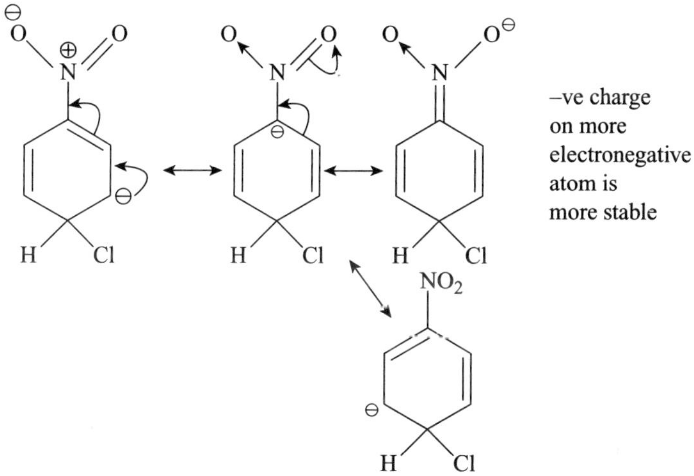

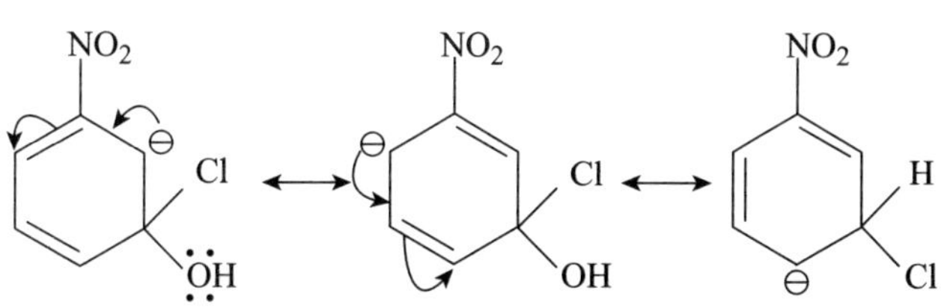

In this negative charge does not go in $-NO_2$ group where oxygen holds the negative charge more easily.

Ans: (a)

53. (i)

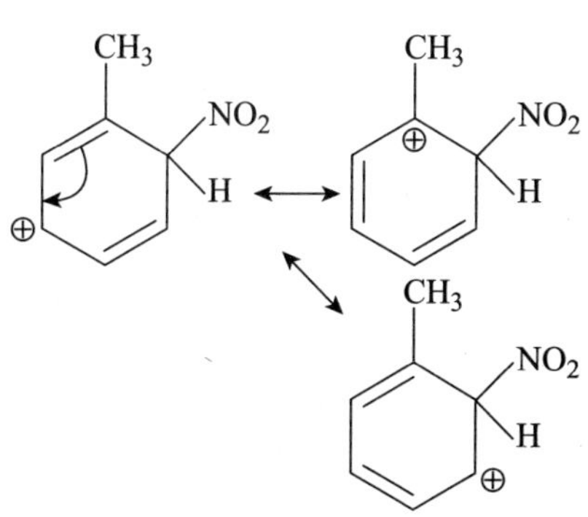

(ii)

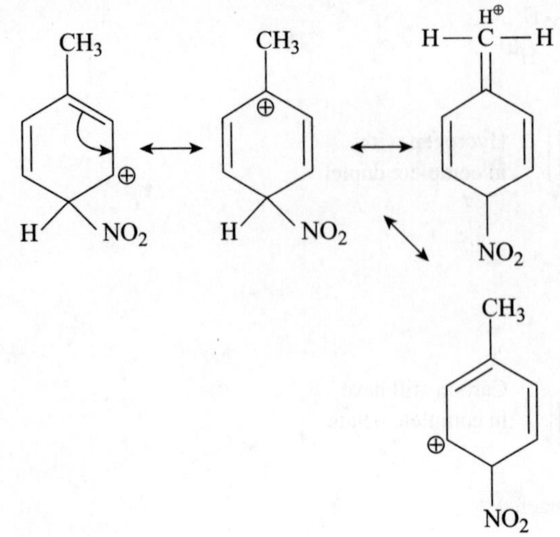

(iii)

(c)

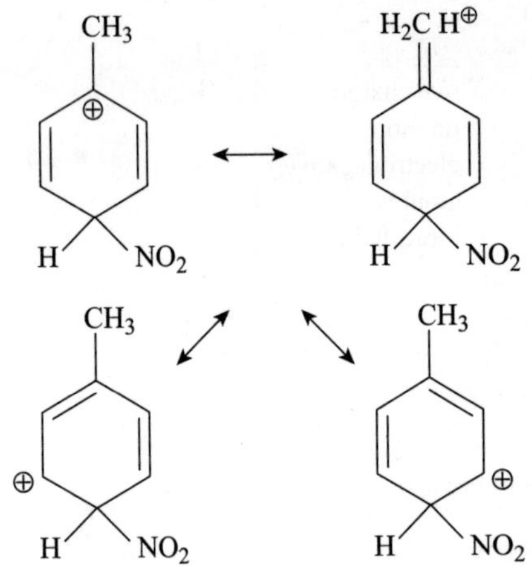

['b' and 'c' are same cation but resonating structure 'c' is more stable]

(iv)

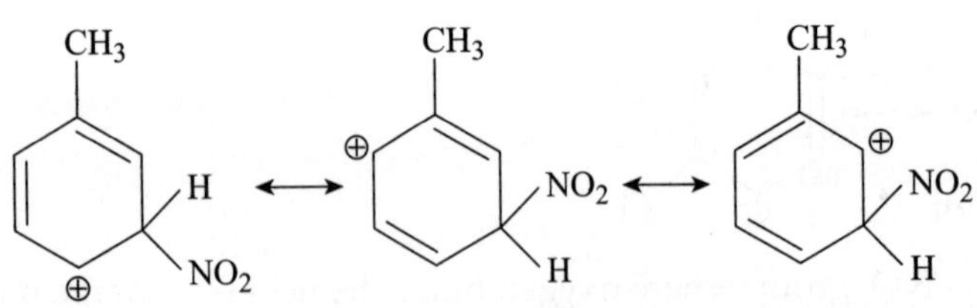

Simply

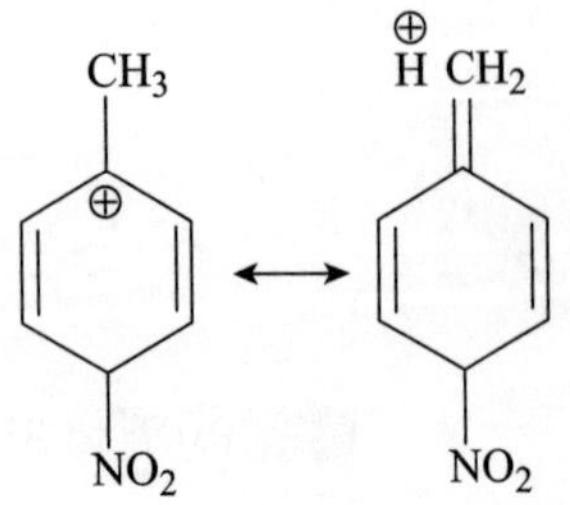

Most stable

3° Carbocation

Ans: (c)

54.

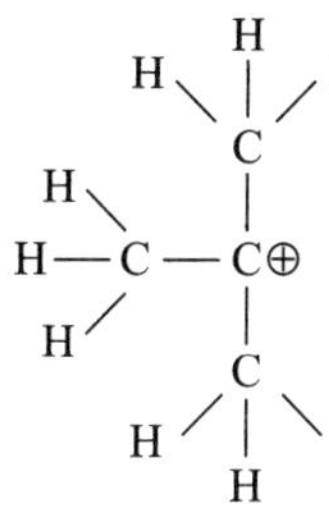

More stable
order of H-effect
$-CH_3 > -CD_3 > -CT_3$

Ans: (b)

55. (i)

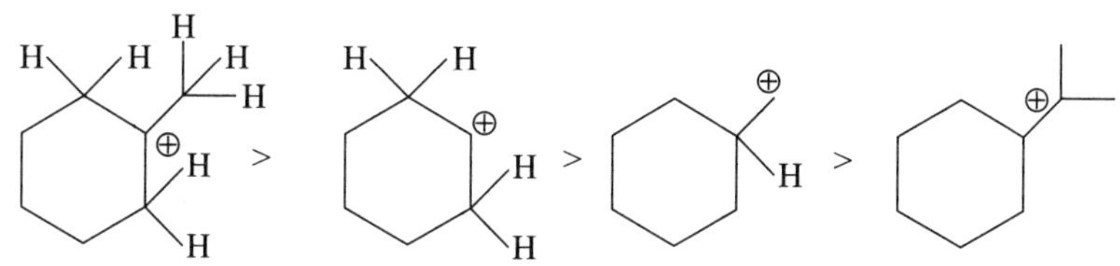

(7 α-Hydrogens) (4 α-Hydrogens) (1α-Hydrogens)

Ans: (b)

56. Under thermodynamic control more stable alkens are formed. First of all check the number of α-Hydrogen in the given product if number of α-Hydrogen are equal then check stereostability (trans-form more stable then cis).

Hence

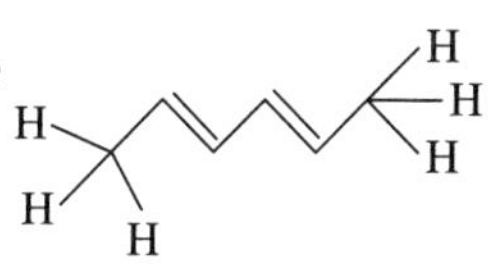

more stable

Ans: (c)

57. (i)

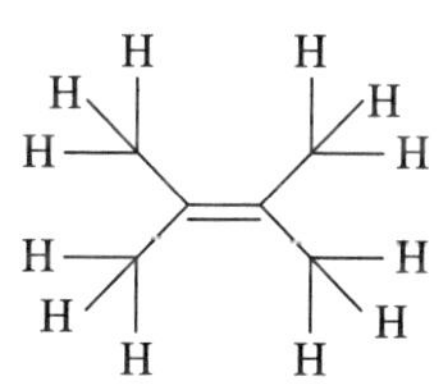

(9 α-Hydrogens)

(ii)

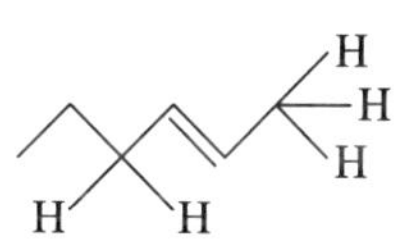

(5 α-Hydrogens)

(iii)

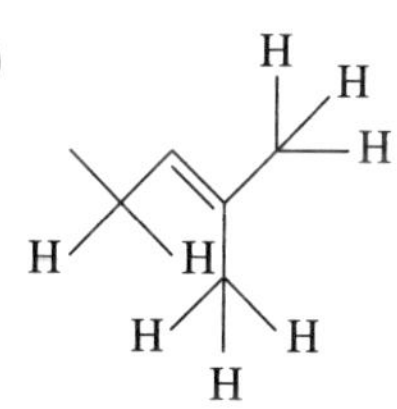

(8 α-Hydrogens)

(iv)

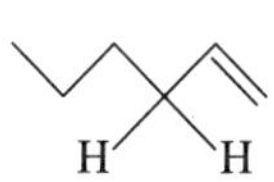

(2 α-Hydrogens)

Ans: (a)

58. (i) $CH_2 = \dot{C}H$ (ii) $CH_3 - \dot{C}H_2$ (3 α-Hydrogens)

(iii) $\begin{matrix} CH_3 \\ CH_3 \end{matrix} \!\!> \dot{C}H$ (6 α-Hydrogens)

(iv) $CH_2 = CH - \dot{C}H_2 \longleftrightarrow \dot{C}H_2 = CH = CH_2$ [Resonance stable]

Ans: (a)

59.

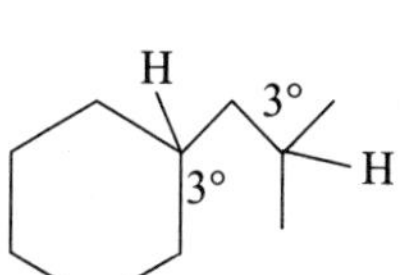

Ans: (b) Two 3° hydrogens are present

Hint: Hydrogen attached to tertiary carbon are called tertiary hydrogens.

61. (ii) $H_2C{=}\ddot{\underset{\cdot\cdot}{O}} \longleftrightarrow H_2\overset{\ominus}{C}{\equiv}\overset{\oplus}{O}$ (Carbon does not form five bonds)

(v)

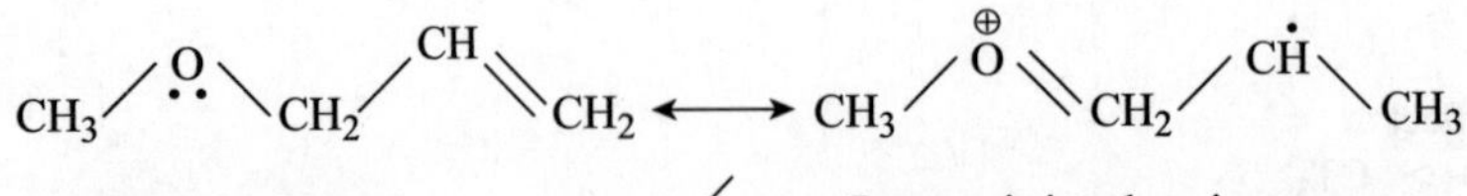

Position of atoms do not change in resonance

Lone pair involves in resonance when they are in conjugation with positive charge or with vacant P or 'd' orbitals.

Ans: (b)

62. (i)

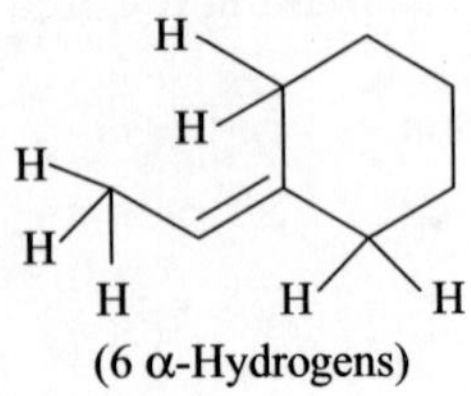

(6 α-Hydrogens)

(ii)

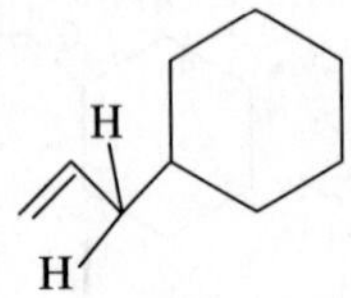

(2 α-Hydrogens)

(iii)

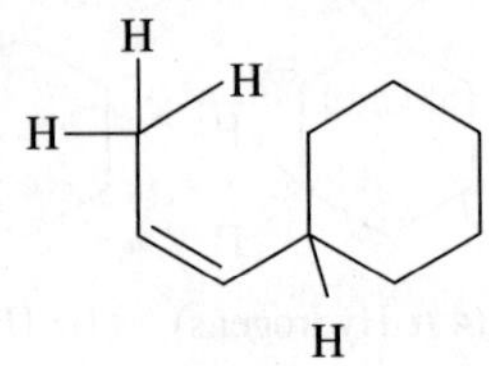

(4 α-Hydrogens)

(iv)

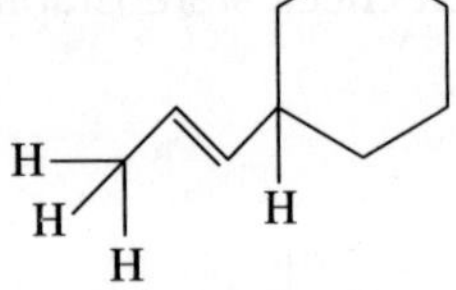

(4 α-Hydrogens)

(v)

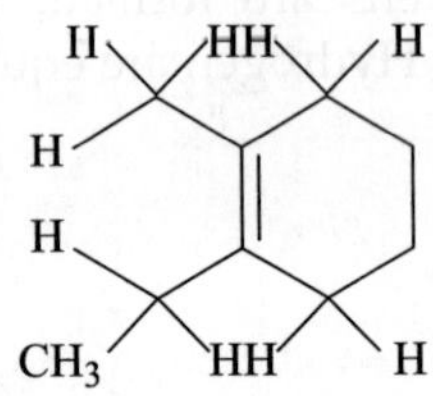

(9 α-Hydrogens)

Ans: (d)

63. (i)

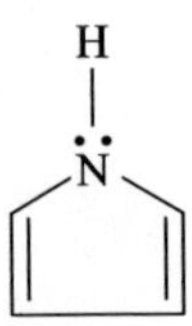

(6πe)

(ii)

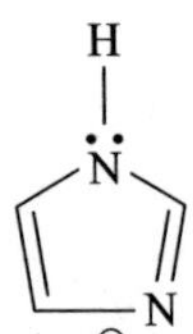

(6πe)

(iii)

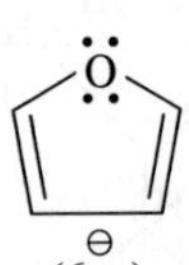

(6πe)

(iv)

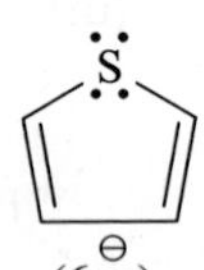

(6πe)

(v)

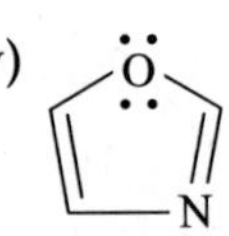

(vi)

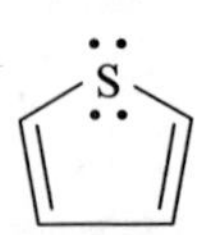

(vii)

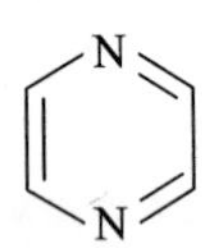

(viii)

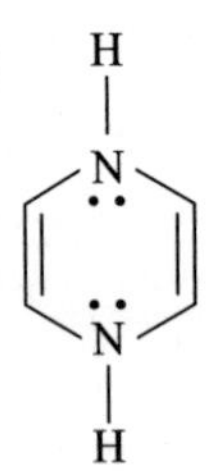

(8πe)

(ix)

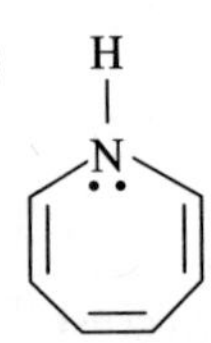

(8πe)

(x)

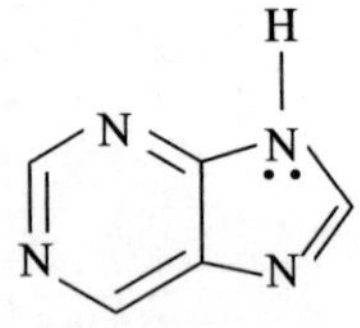

(10πe)

(xi)

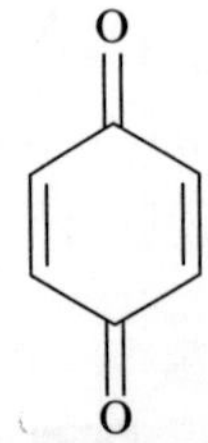

Non-Aromatic

(xii)

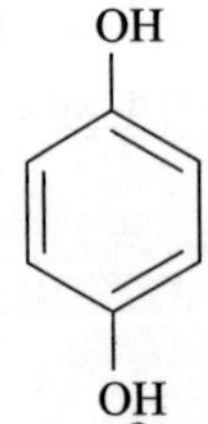

(6πe)

Hint: Lone pairs present on doubly bonded atom are localised hence don't participate in resonance.

Ans: (c)

64. 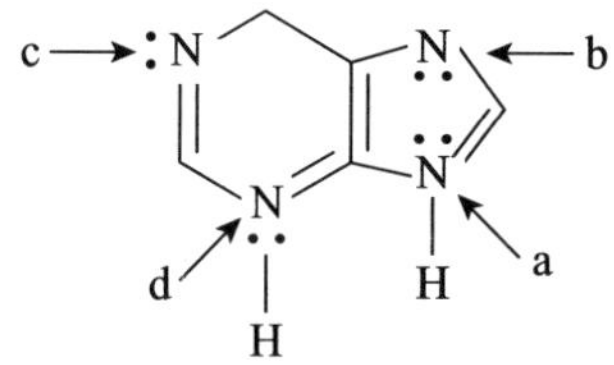

Lone pairs which are involved in resonance will be less available for protonation. Lone pair which is involved inresonance and making the ring aromatic will not be available for protonation.

Ans: (b) Because involved in resonance as well as making the ring aromatic

an indicated atom → [Hydridisation = lp + σ-bond]

65. (i) 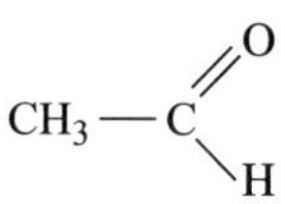 Number of σ-bonds = 3

(ii) 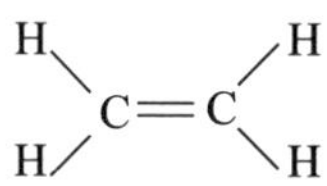 Number of σ-bonds = 3

(iii) 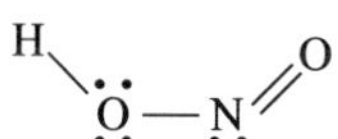 Two σ-bonds + one lone pair

(iv) 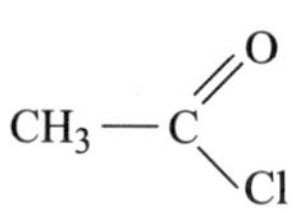 Number of σ-bonds = 3

Ans: (b)

66. (iv) 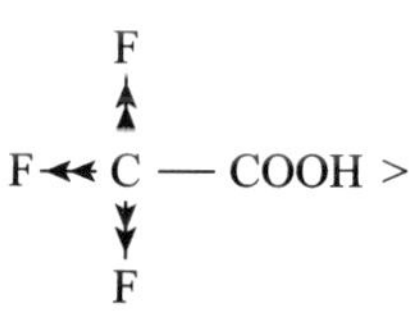 > (iii) 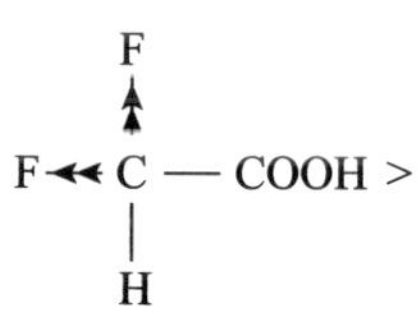 > (ii)  >

(v) CH_3 — COOH > (i) CH_3 — CH_2 — OH

Hint: With the increase in number of atoms showing –I effect acidity increases.

Ans: (a)

67. (a) 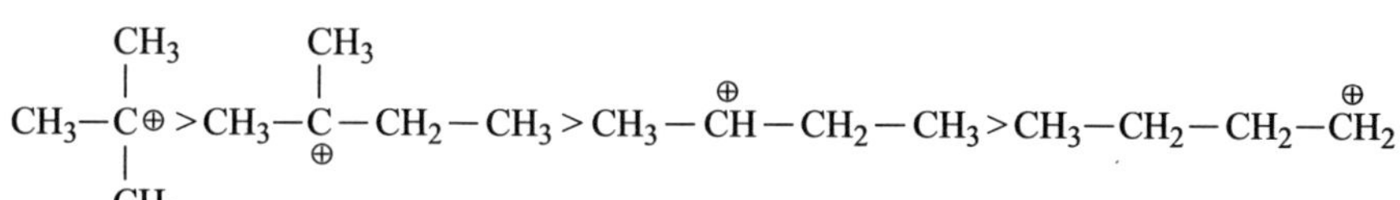

Carbocation having more number of α-Hydrogens will be more stable.

(b) 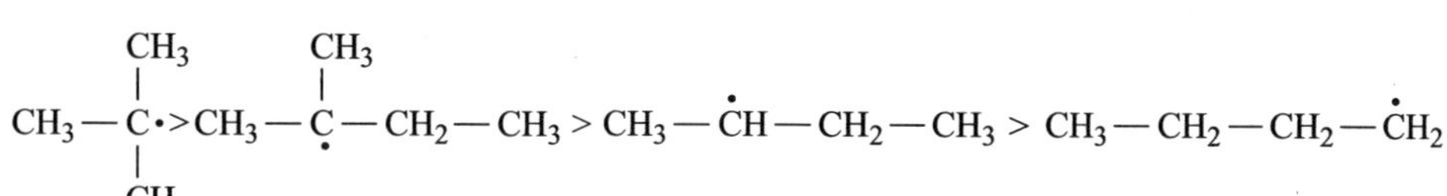

Free radicals having more number of α-Hydrogen are more stable.

(c) 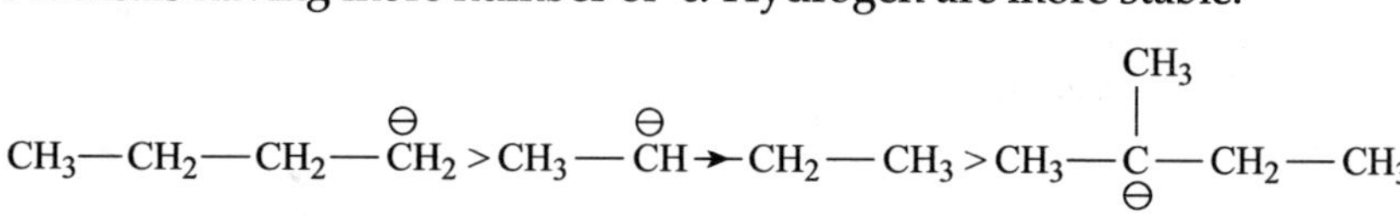

Order of stability of carboanion 1° > 2° > 3°

(d) Assuming single carbene

(i) 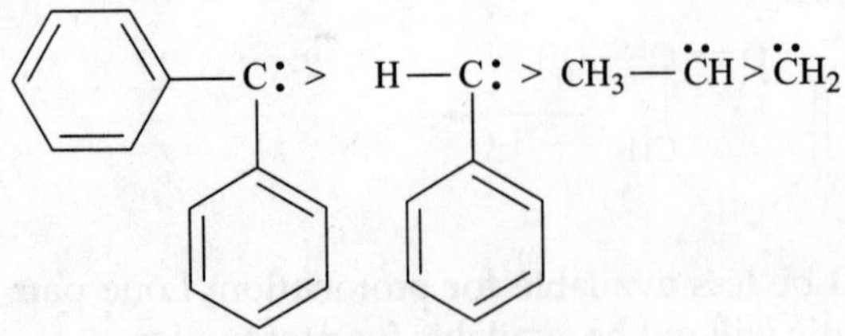

Ans: (c)

68. A weak base has a strong conjugate acid.

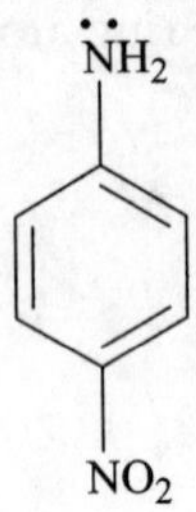

Weakest base in the provided options hence it's conjugate acid will be strong.

Ans: (d)

69. 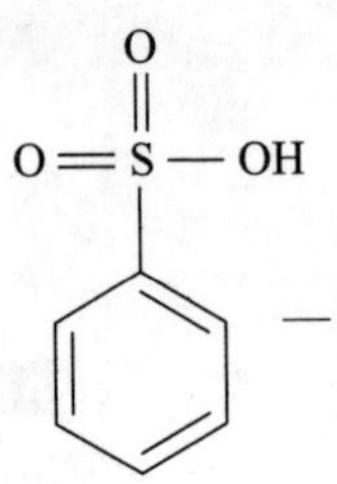 — Benzene Sulphonic acid

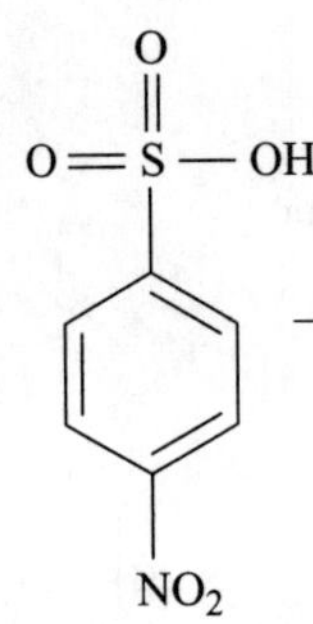 — P – Nitro benzene Sulphonic acid

Ans: (b)

70. Boiling point depends upon H-bonding

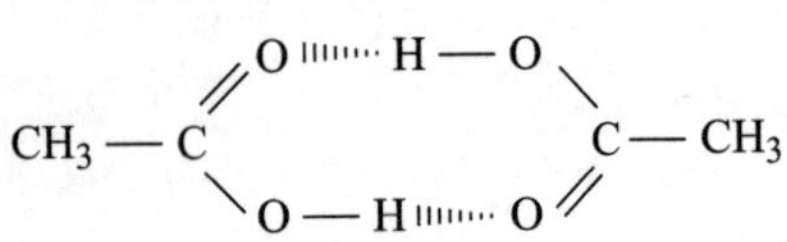

It forms dimmer through intermolecular H-bonding. **Ans:** (d)

71. OH group attached to 2° carbon. **Ans:** (d)

74. Oxygen does not form four bonds, if Nitrogen form four bonds then positive charge will be present on it. **Ans:** (c)

75. (ii) O=C=O (iv) 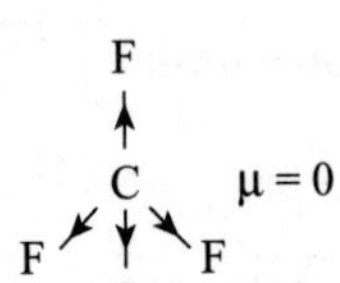 (v) Br — Br (vi) 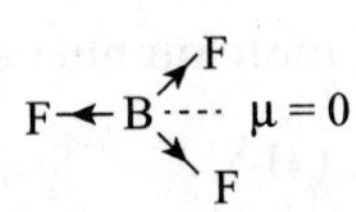

Ans: (a)

76. 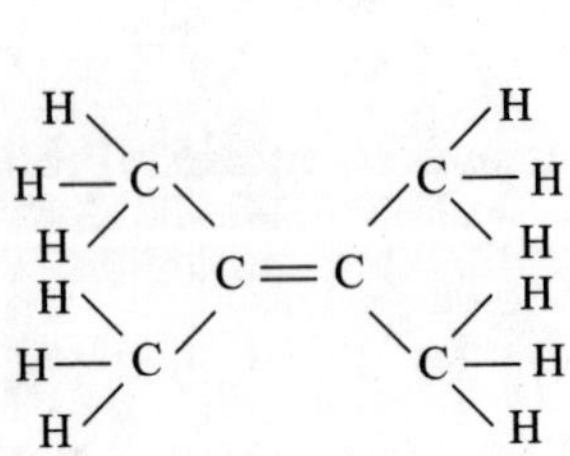

Alkene having more number of α-Hydrogen is more stable

(9 α-Hydrogens)

Ans: (c)

77.

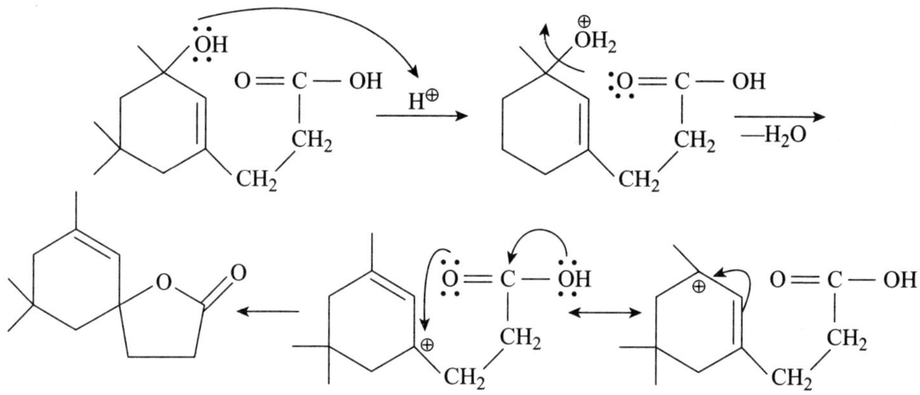

Hint: Attack of $H^{\oplus}$ will take place where stable carbocation is formed. **Ans:** (b)

79.

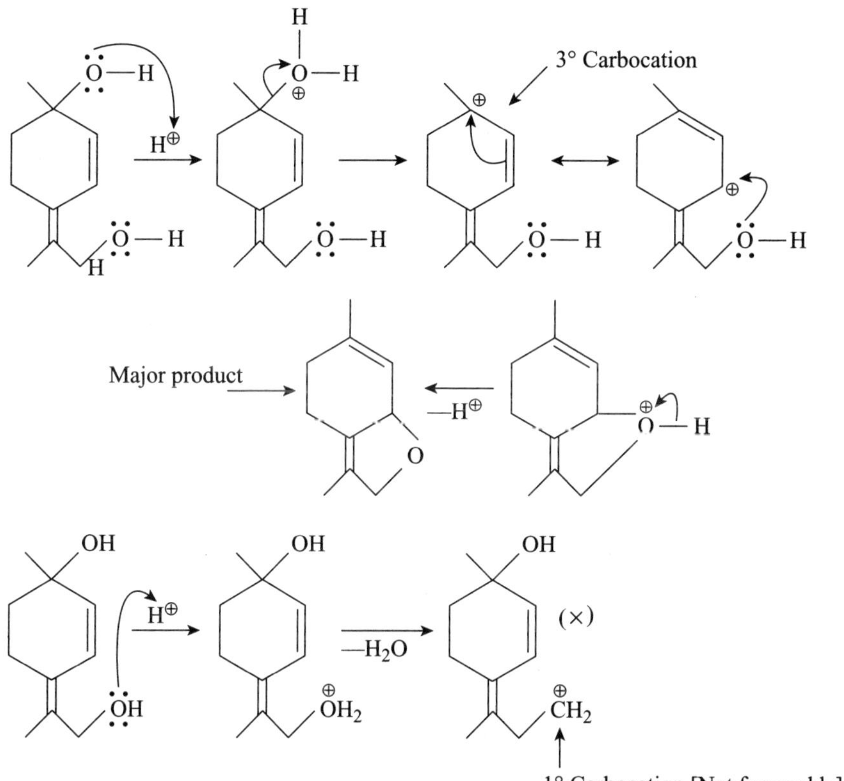

Ans: (c)

82.

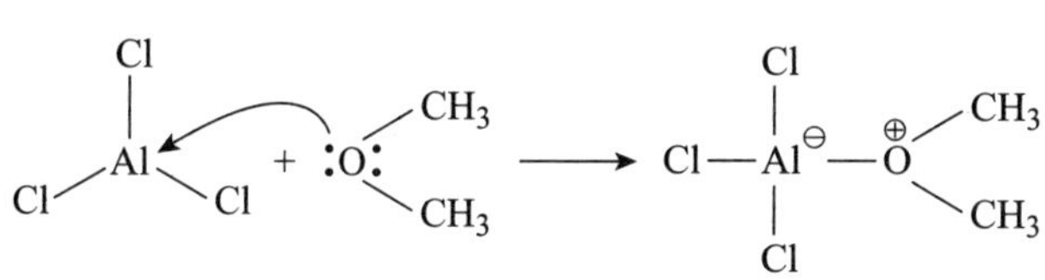

$AlCl_3$ is electron deficient and act as lewis acid. Accepts lone pair from oxygen in its vacant p-orbital.

Ans: (a)

84. (i)

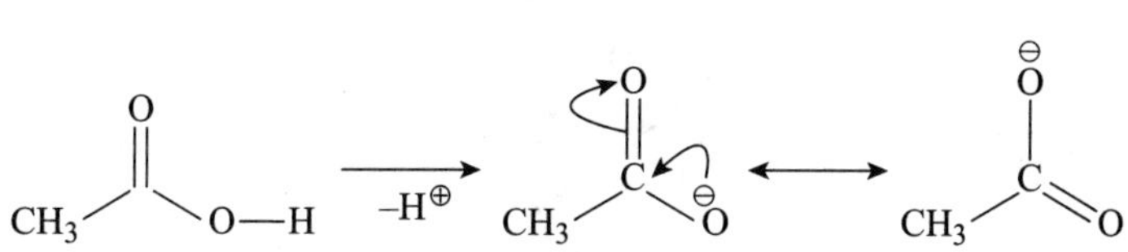

(ii)

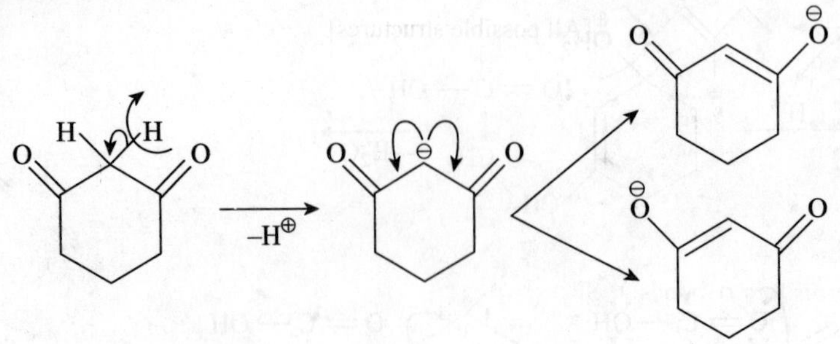

(iii) $CH_3—C\equiv C—H \xrightarrow[-H^{\oplus}]{} CH_3—C\equiv \overset{\ominus}{C}$

Ans: (b)

85. (b) $CH_2{=}\overset{\downarrow}{C}{=}CH—Cl$ (Two σ-bonds ⇒ sp)

(d) $H—C\equiv C—CH_3$ (Two σ-bonds ⇒ sp) [sp]

(e) $CH_3—C\equiv N$ (Two σ-bonds ⇒ sp) [sp]

Ans: (d)

86.

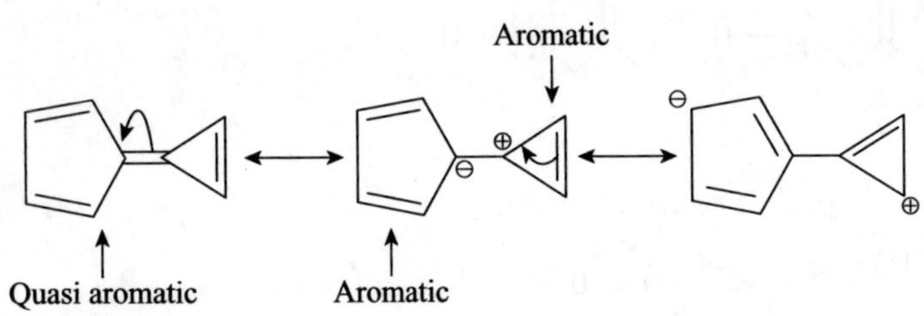

Ans: (b)

87. $CH_3—\ddot{N}(CH_3)—CH_3$ [Three σ-bonds + one lone pair]

Ans: (b)

88. (i)

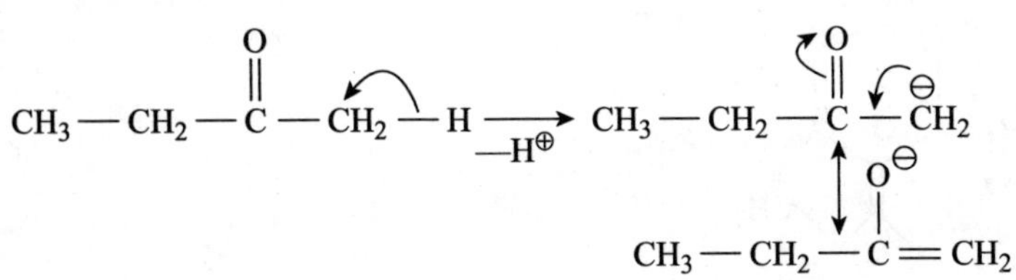

(ii)

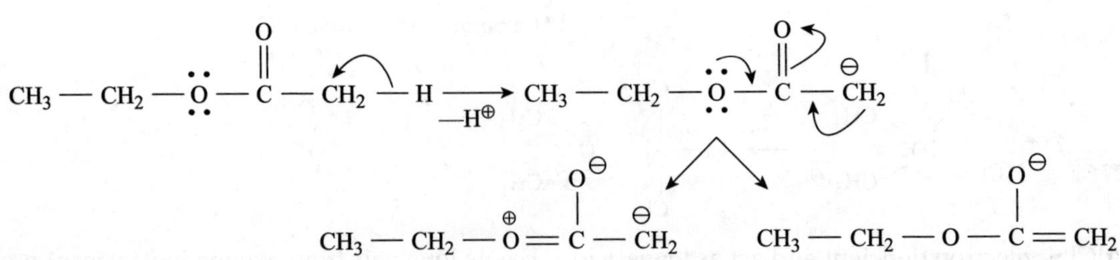

(iii)

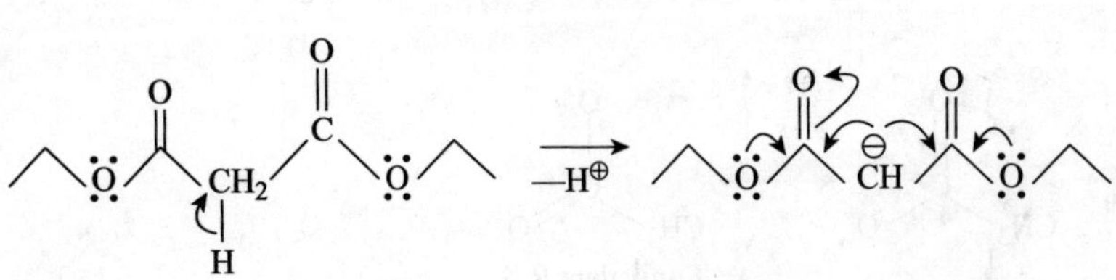

Ans: (d)

89. 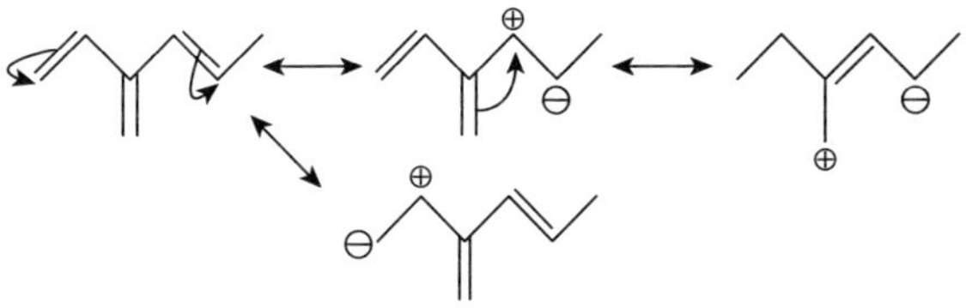 [All possible structures]

Ans: (d)

90. (i) 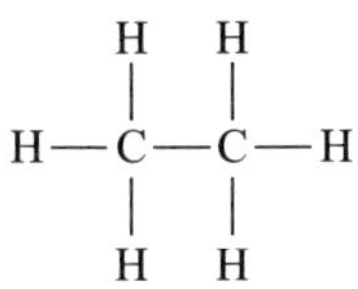 Each carbon forms four σ-bonds ∴ SP^3

(ii) 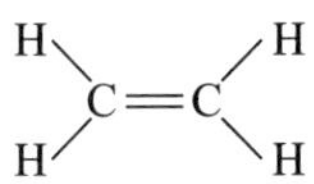 Each carbon forms three σ-bonds ∴ SP^2

(iii) 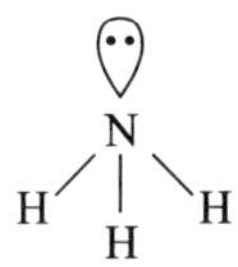 Three σ-bond + one lone pair (SP^3)

(iv) $H-C\equiv C-H$ (Two σ-bonds ⇒ SP)

Ans: (a)

91. 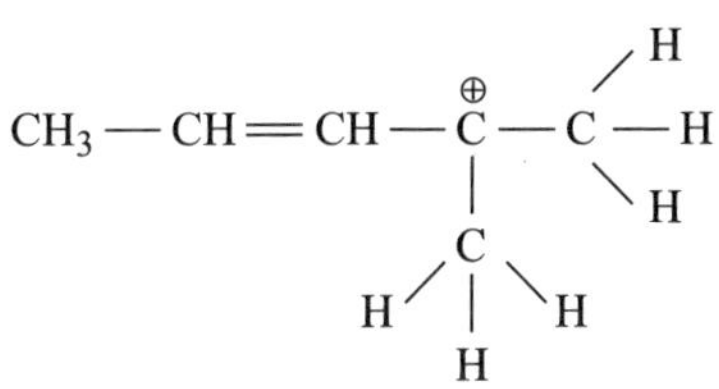

It is in conjugation with double bond as well as it has 6 α-Hydrogens.

Ans: (c)

92. (i) 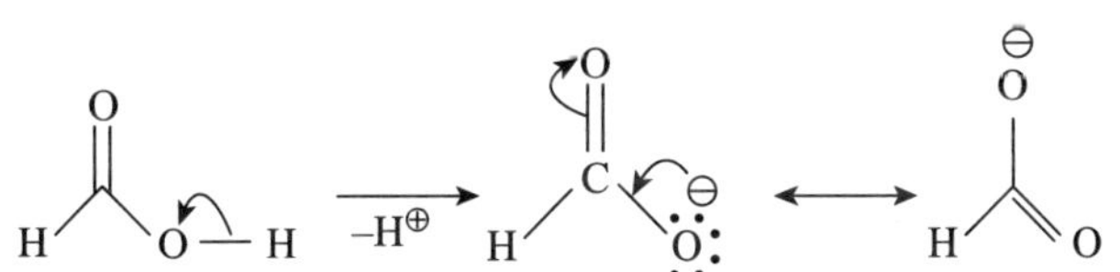

(ii) 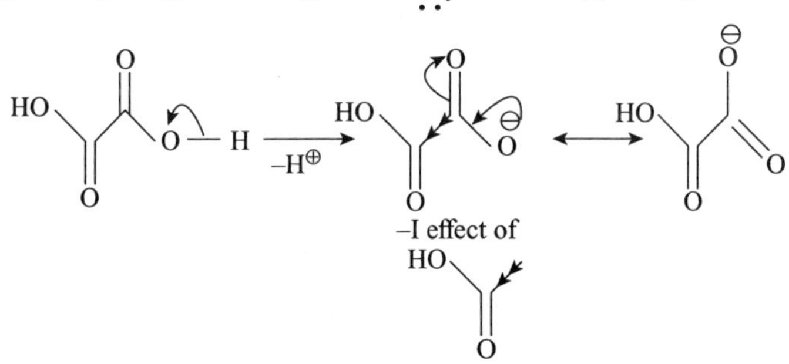

(iii) 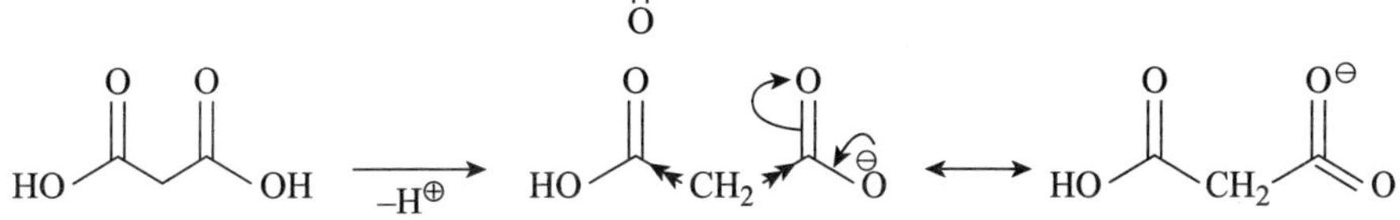

Less –I effect is observed as compared to (ii)

(iv) 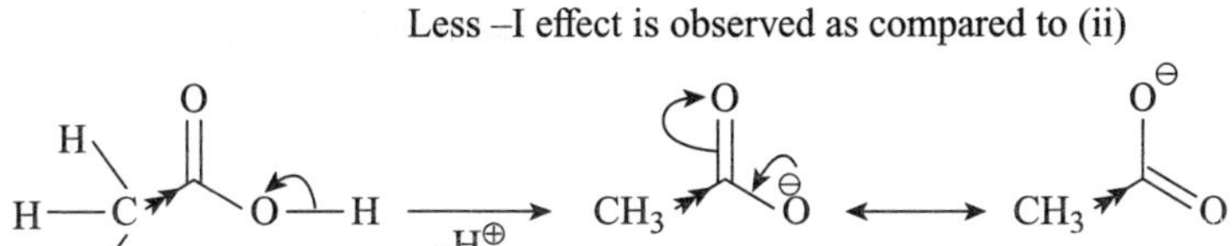

+I effect of $-CH_3$ group

Hence order is ii > iii > i > iv

Ans: (a)

LEVEL 2

1.

$CH_3\overset{\ominus}{O} \longrightarrow CH_3OH$

$\overset{\ominus}{CH_3} \longrightarrow CH_4$

$\overset{\ominus}{NH_2} \longrightarrow NH_3$

$CH_3-\overset{\overset{O}{\|}}{C}-\overset{\ominus}{O} \longrightarrow CH_3-\overset{\overset{O}{\|}}{C}-OH$

Conjugated acids

Hint: A strong base has weak conjugate acid.

∴ Basic order

$$\overset{\ominus}{CH_3} > \overset{\ominus}{NH_2} > CH_3\overset{\ominus}{O} > CH_3-\overset{\overset{O}{\|}}{C}-\overset{\ominus}{O}$$

Acidic order

$$CH_3-\underset{\underset{O}{\|}}{C}-OH > CH_3OH > NH_3 > CH_4$$

Ans: (c)

2.

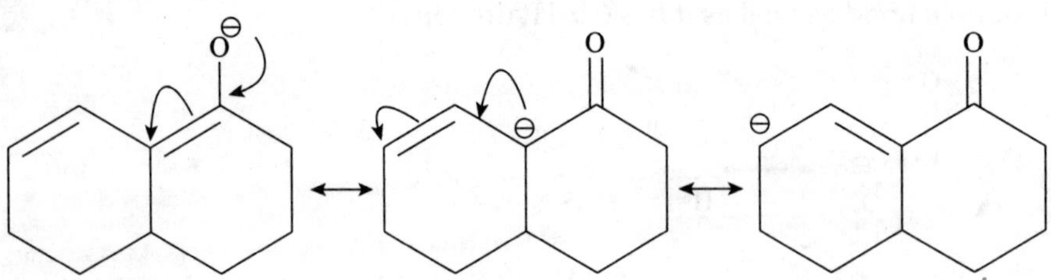

Hint: When negative charge is present next atom of doubly bonded atom, then negative charge will converted into π–bond and π–bond will converted into negative charge. **Ans:** (d)

3.

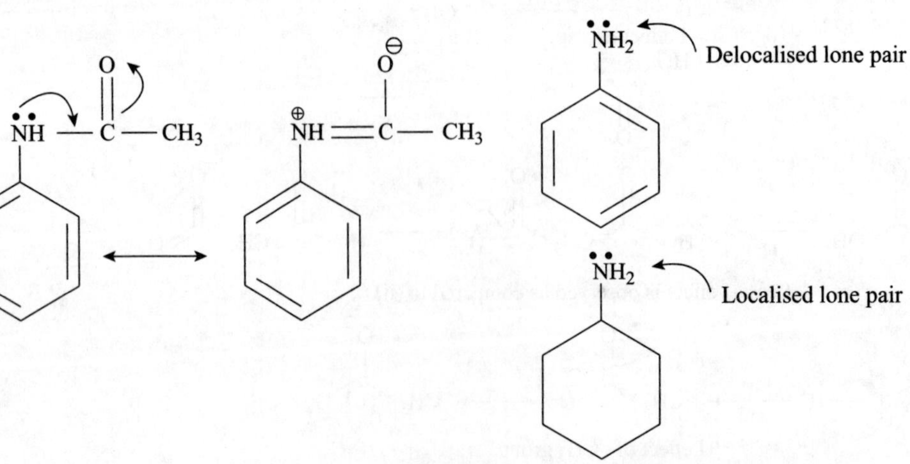

Ans: (b)

5. **Hint:** A strong acid has weak conjugate base acidic order HBr > HF > NH_3 > CH_4

∴ Basic $\overset{\ominus}{C}H_3 > \overset{\ominus}{N}H_2 > \overset{\ominus}{F} > \overset{\ominus}{Br}$ **Ans:** (a)

6. 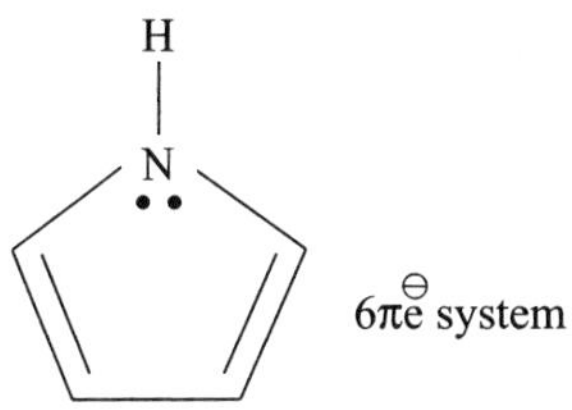

Ans: (b)

8. –C≡N Shows –R effect and gives the acidic nature of the phenol. **Ans:** (c)
9. Count total number of oxygen, carbon, nitrogen and hydrogen in the given formula. **Ans:** (b)

10. 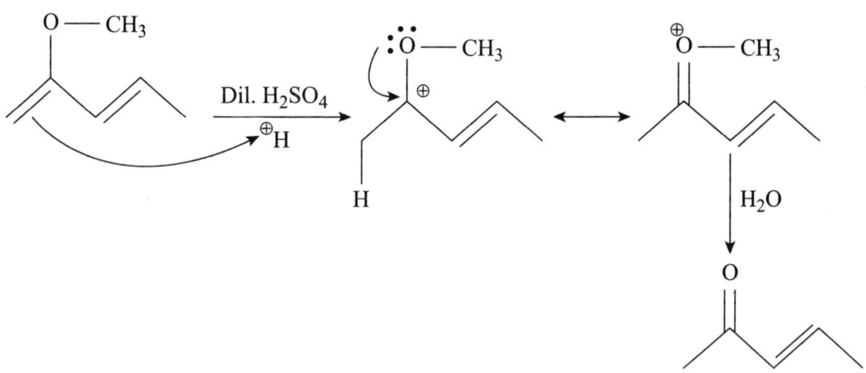

Addition of $H^{\oplus}$ ion takes place in such a way that stable carbocation is formed. **Ans:** (c)

11. 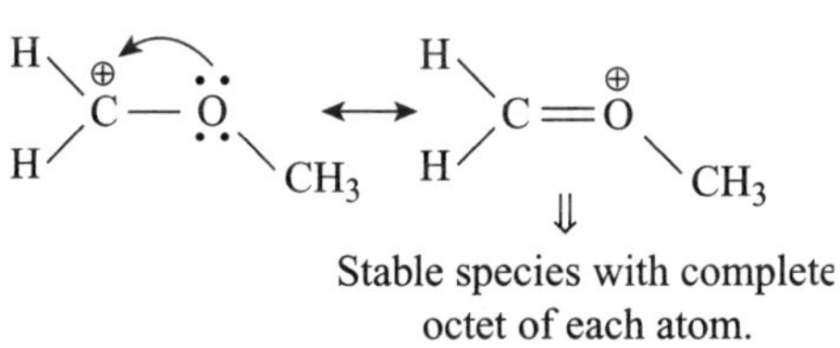

Ans: (b)

12. In a cumulenes when number of π-bonds are even, then terminal $-CH_2$ groups are in perpendicular plane but when odd then present in the same plane.

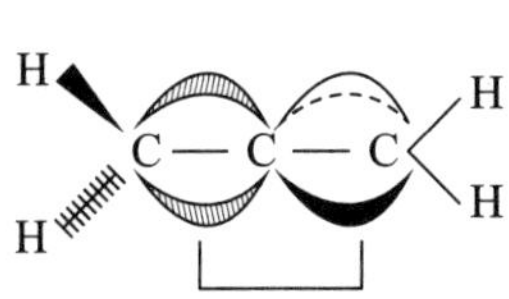

Both the π-bonds are in ⊥ planes. **Ans:** (c)

13. 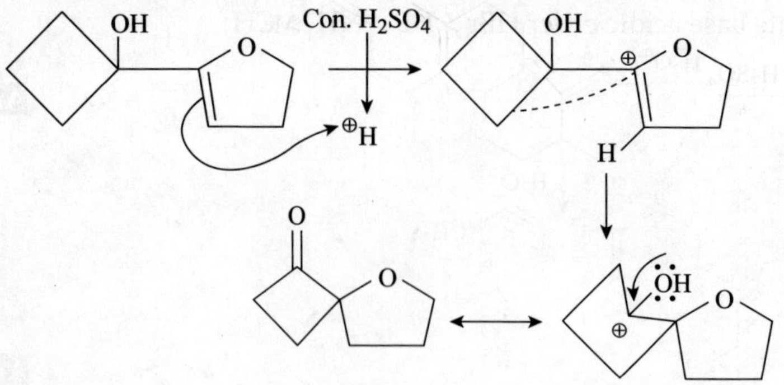

Hint: Addition of H+ takes place where stable carbocation is formed.

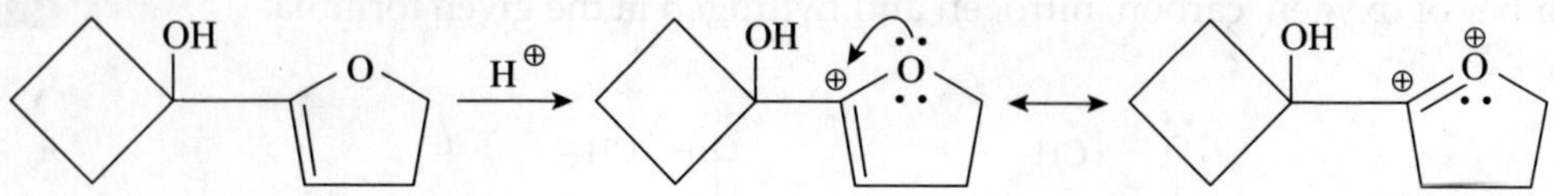

Ans: (a)

15. 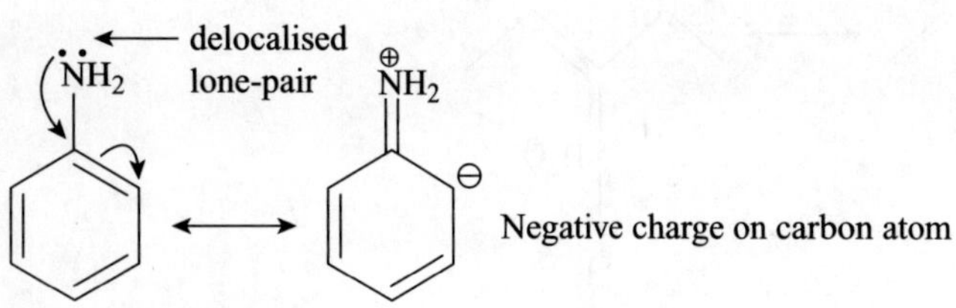

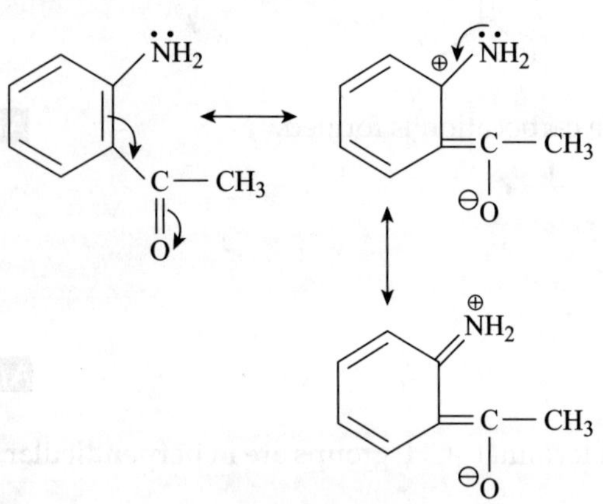

Negative charge on oxygen atom

Hint: $-\underset{\underset{O}{\|}}{C}-CH_3$ group shows –R effect and decreases the basicity.

Ans: (b)

16.

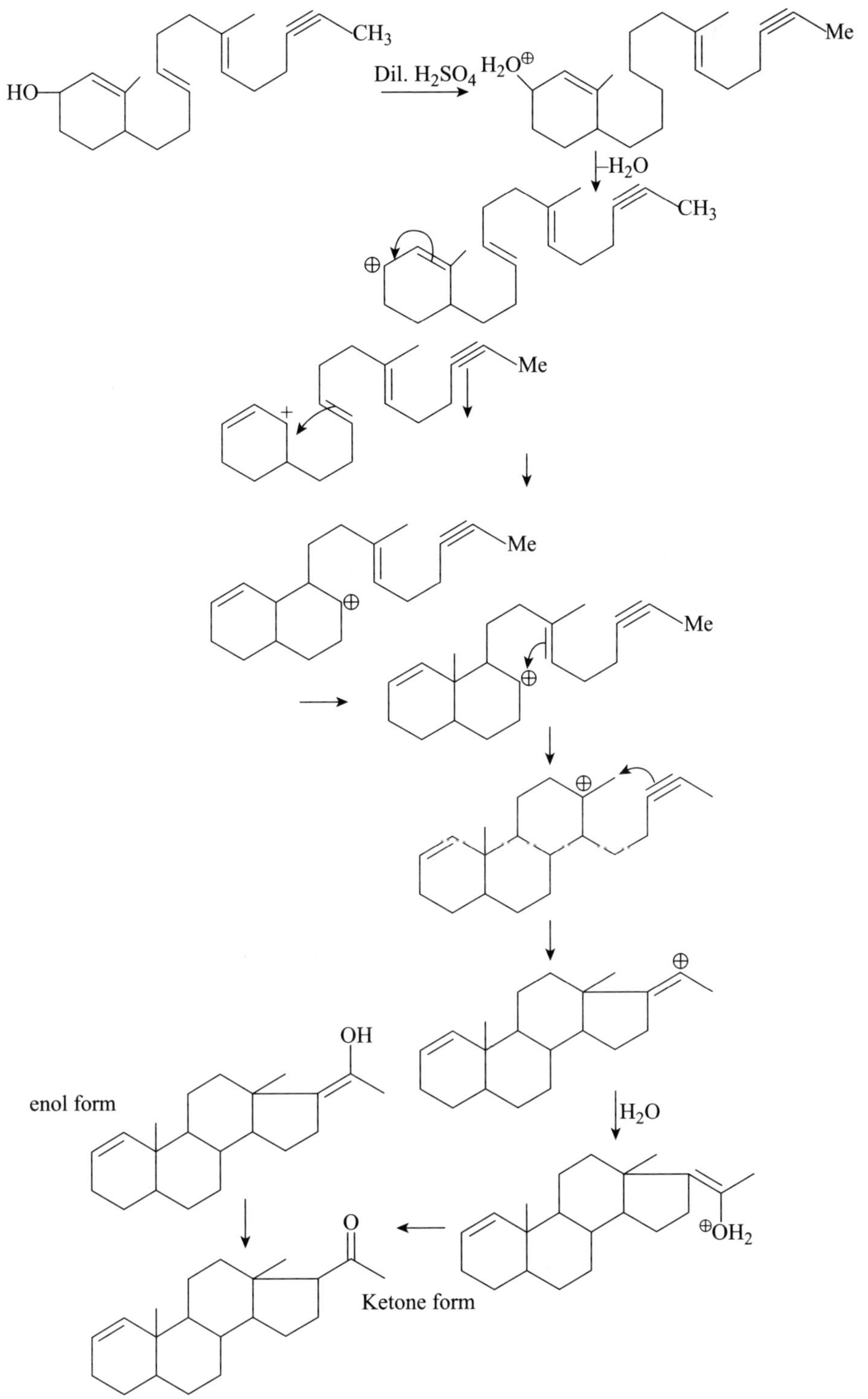
CH3
HO
Dil. H2SO4
H2O⊕
Me
–H2O
CH3
⊕
Me
+
Me
⊕
Me
⊕
⊕
⊕
H2O
OH
enol form
⊕OH2
O
Ketone form

Ans: (d)

17. Check the stability of the anion formed after removal of $H^{\oplus}$ ion.

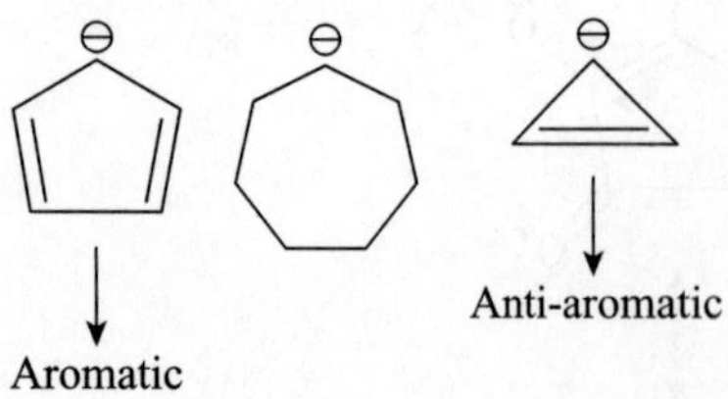

Stability order: Aromatic > Non-aromatic > Anti-aromatic **Ans:** (a)

18.

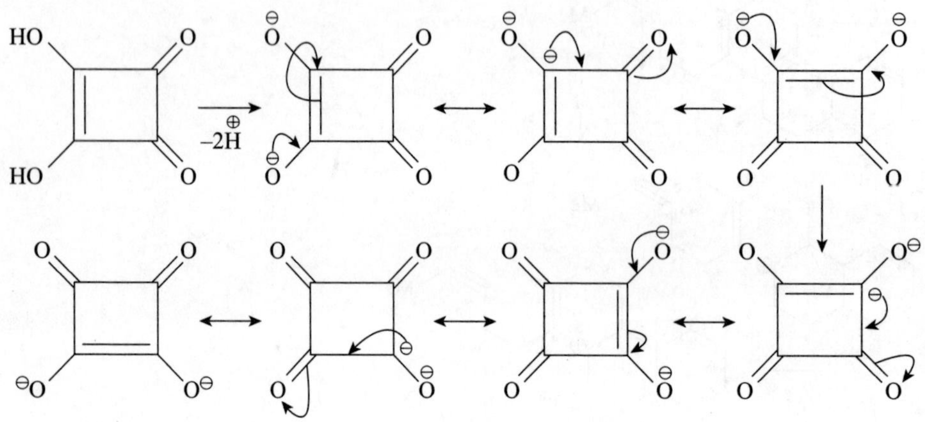

Hence after removal of $H^{\oplus}$ ions compound becomes aromatic, similarly it becomes aromatic after removal of $H^{\oplus}$ ions but negative charge is shared by five oxygen atoms.

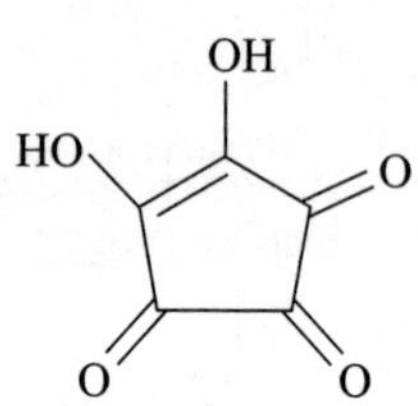

Negative charge is shared by six oxygen atoms after removal of $H^{\oplus}$ ions and it also becomes aromatic.

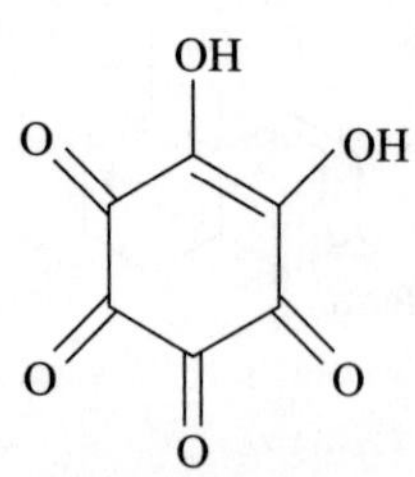

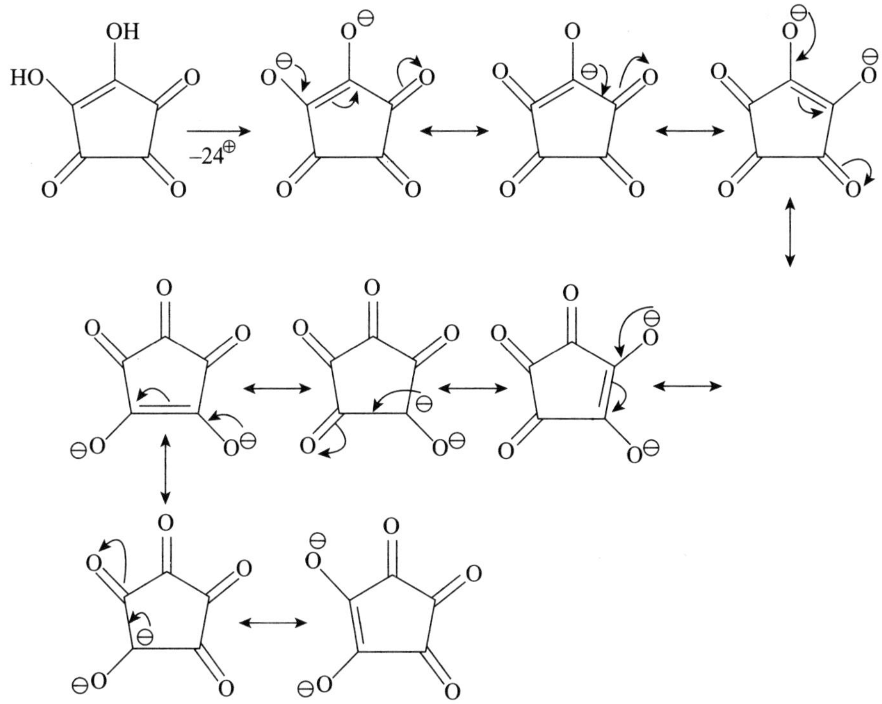

Ans: (b)

19.

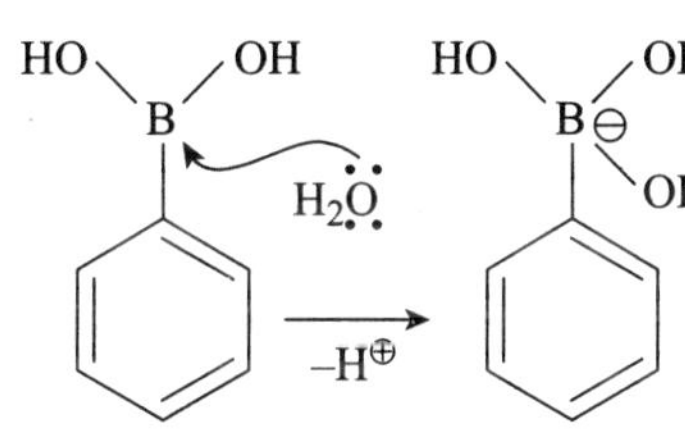

$H^{\oplus}$ ion comes from water

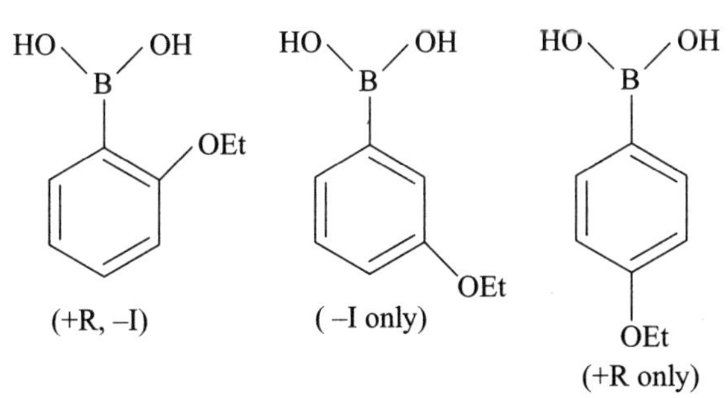

Electron withdrawing group on benzene will increase the electron deficiency on Boron atom through –I or –R and make easy acceptance of OH from water and furnish H+ ion solution.

Ans: (c)

20. 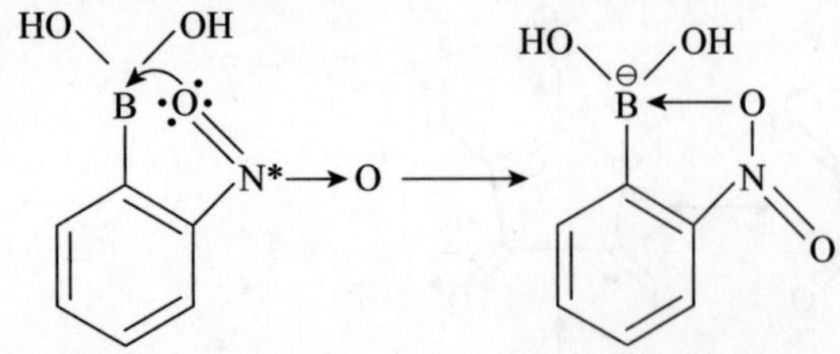

[Lewis acidic strength ↓ Se]

Chelate formation

In this case intramolecular attack of lone pairs of oxygen takes place. **Ans:** (b)

21. 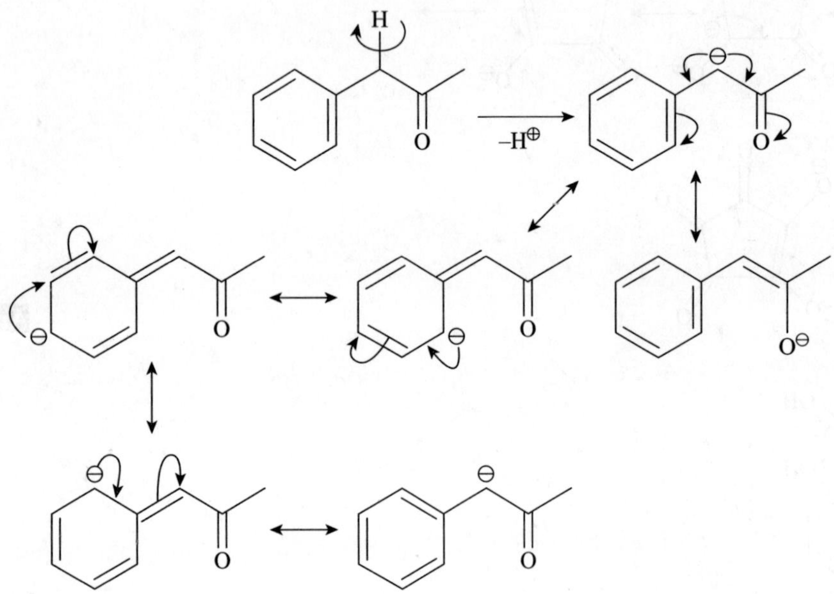

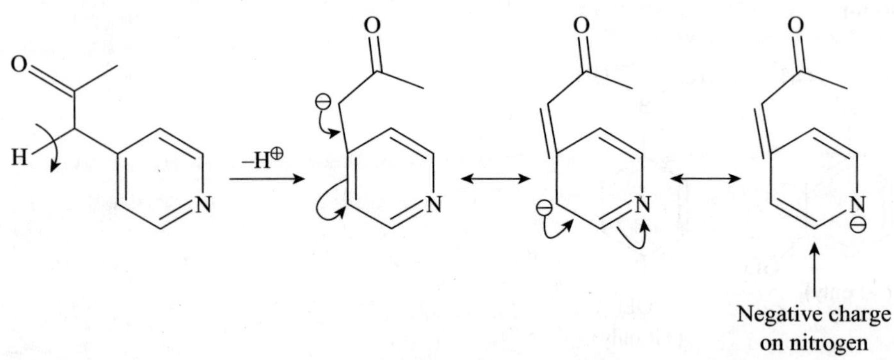

A more electronegative atom can hold negative easily. **Ans:** (b)

22. 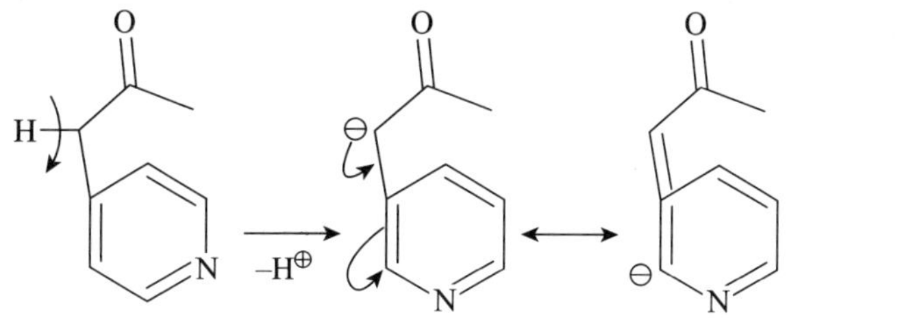

Negative charge is stablised through –I effect of sp hybridised nitrogen

Hint: Acidic strength depends upon the stability of the conjugate base (Anion formed). **Ans:** (a)

23. 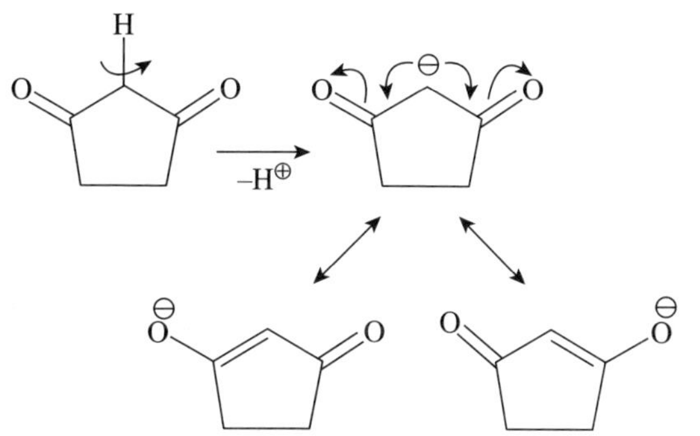

Negative charge is in cross-conjugation with two —C=O bond.

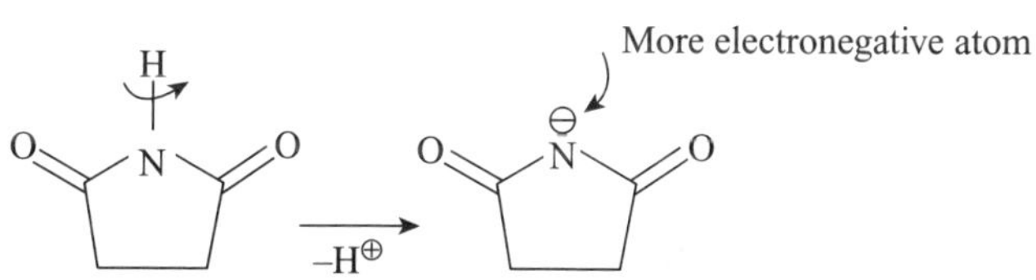

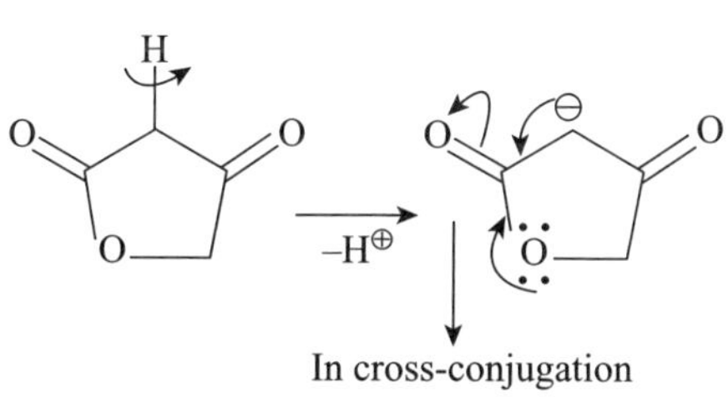

Ans: (b)

24. 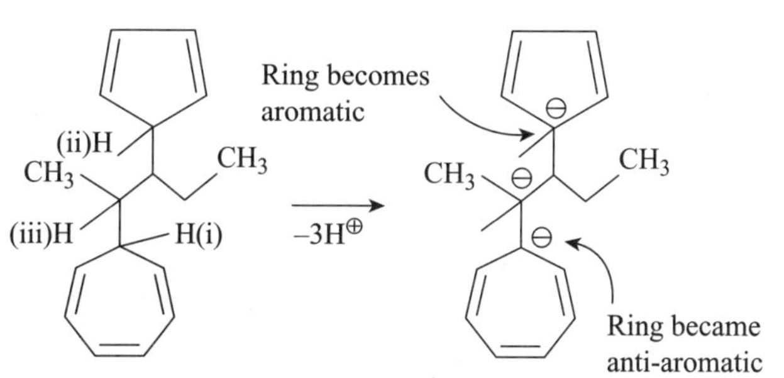

Ans: (b)

25. **Hint:** After removing $H^{\oplus}$ ion check the extent of resonance of negative charge more will be the extent of resonance of negative charge more will be stable.

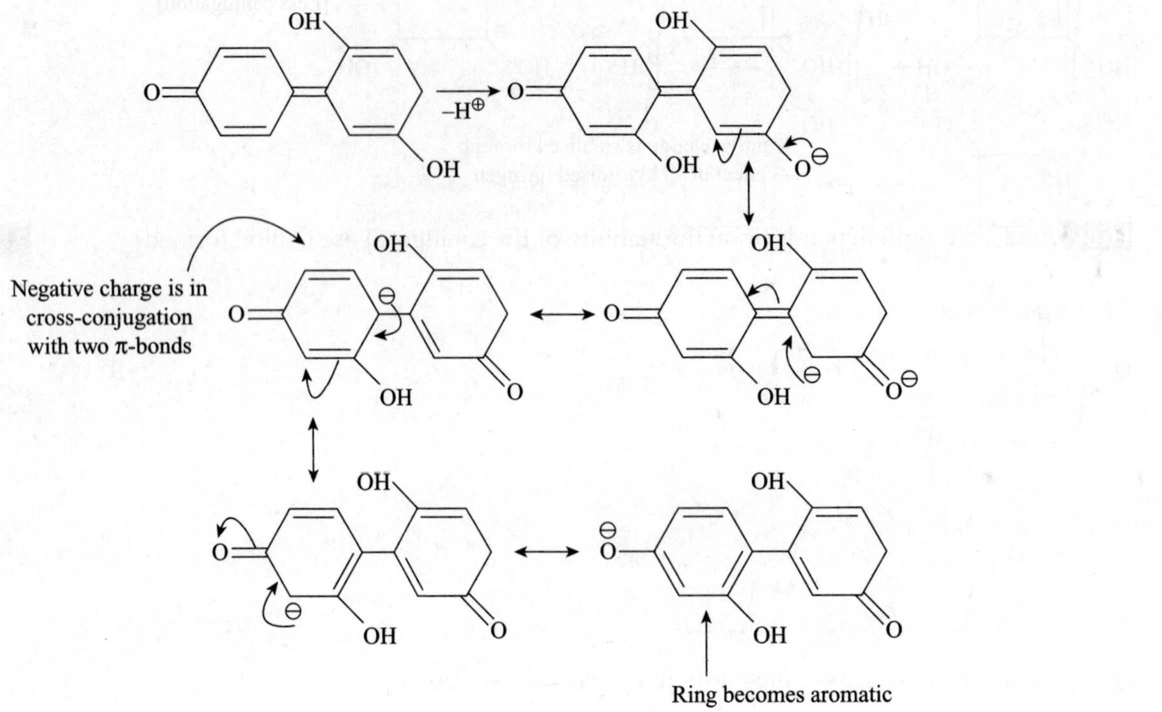

Similarly check for others.

Ans: (d)

27.

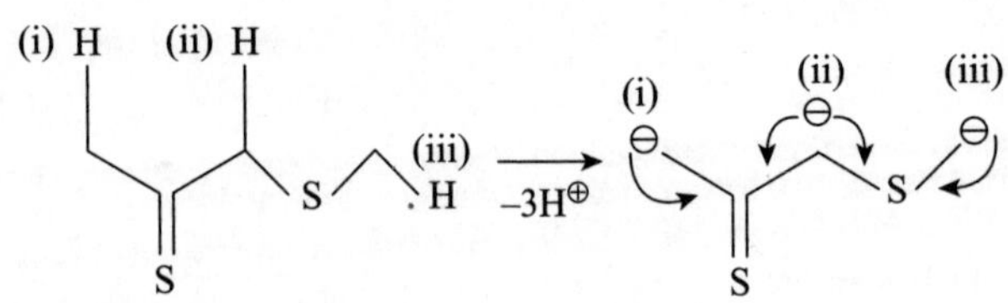

Negative charge (ii) is in cross-conjugation with π-bond and vacant orbital of sulphur.

Hint: Sulphur has vacant –3d orbital and can show –R effect.

Ans: (d)

29.

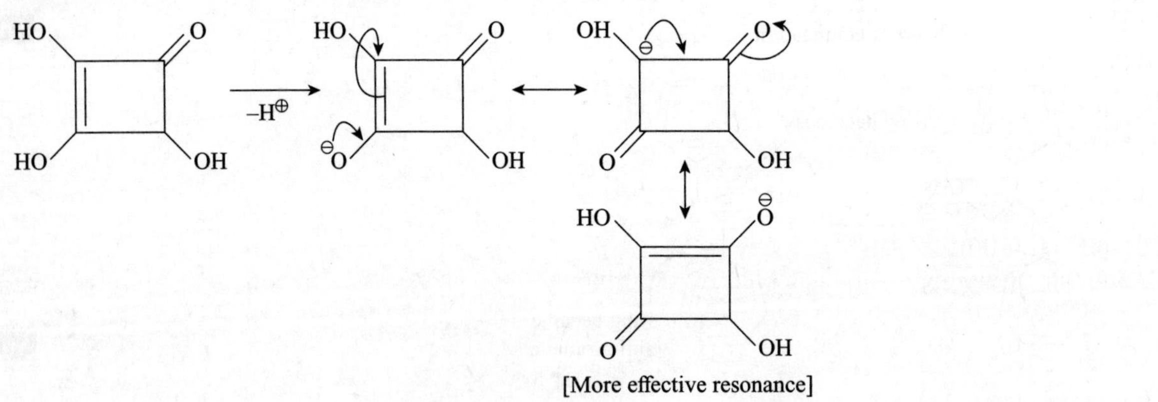

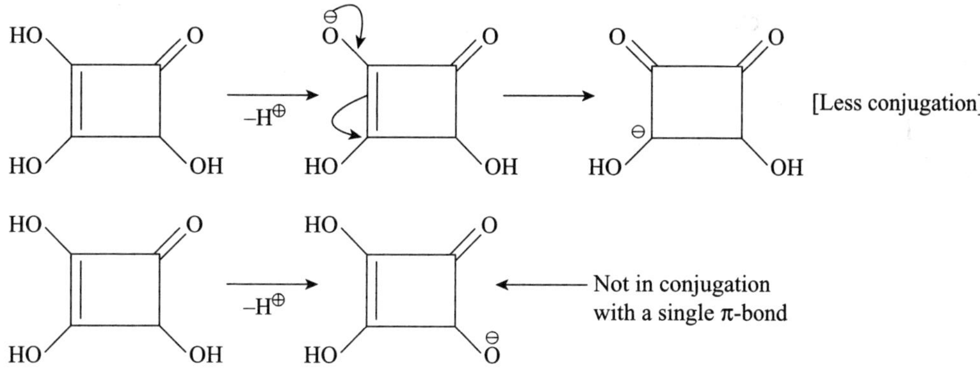

Ans: (a)

30. – Nme$_3$ shows only –I effect

– Phme$_3$ shows –R effect

– CH$_3$ shows +I effect

Ans: (c)

31.

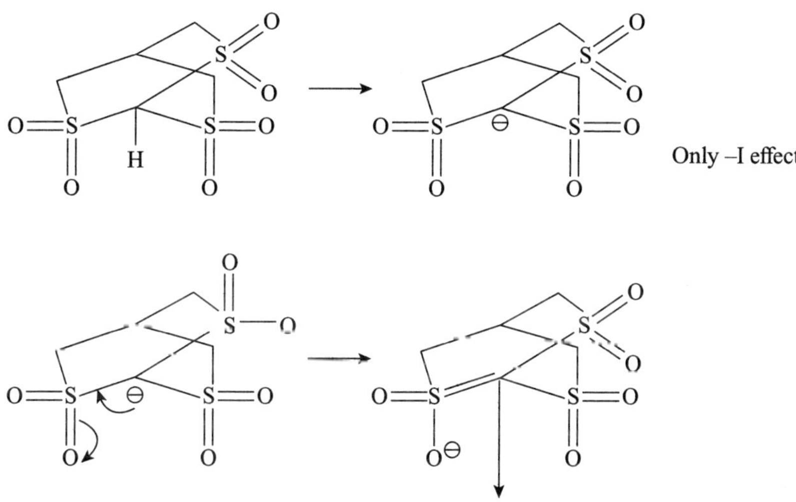

Sulphar has vacant d-orbitals which are more diffused in space hence can form π-bond on bridge position.

Ans: (a)

34.

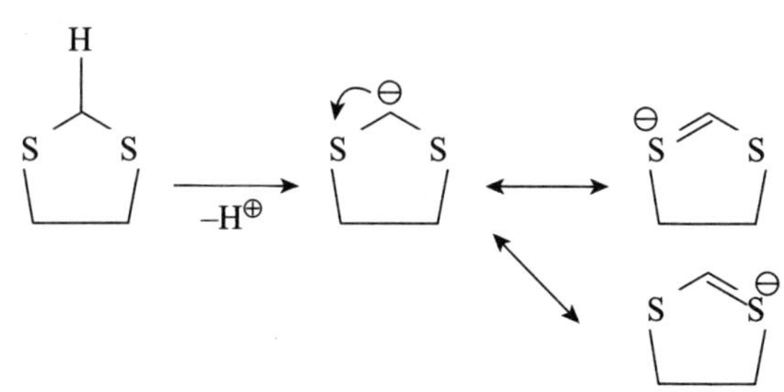

Sulphar shows –R effect due to vacant –3d orbitals.

Ans: (b)

35.

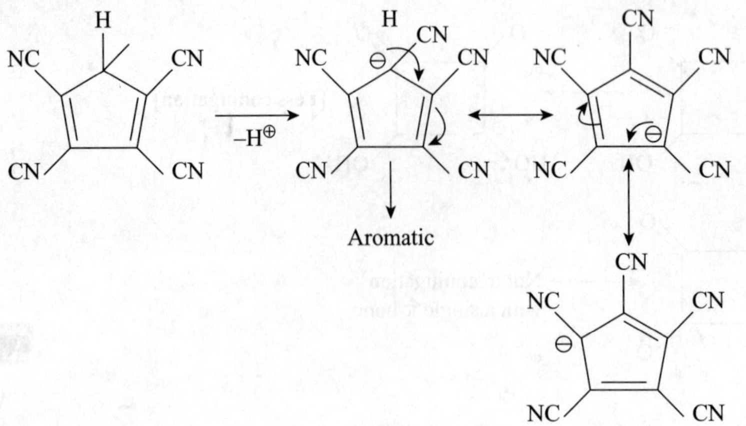

Negative charge is stabilised through –I effect of –CN group at each stage. **Ans:** (d)

36.

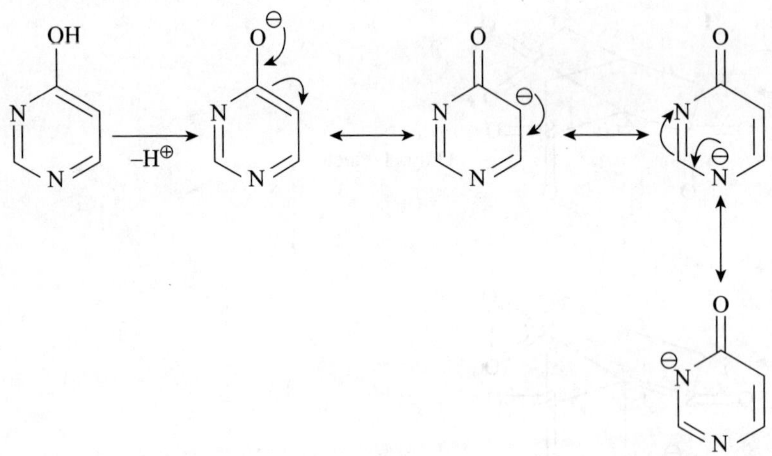

Hint: During resonance negative charge goes on more electronegative atom (E.N of N>C). **Ans:** (a)

37. See Workbook Exercise, Question Number 28.

38.

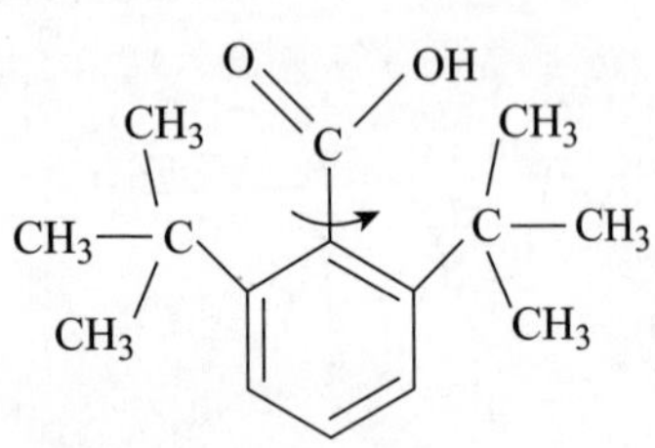

Two bulky groups on ortho position shows more ortho effect (SIR effect). **Ans:** (c)

39. See Workbook Exercise, Question Number 25 and 26.

40.

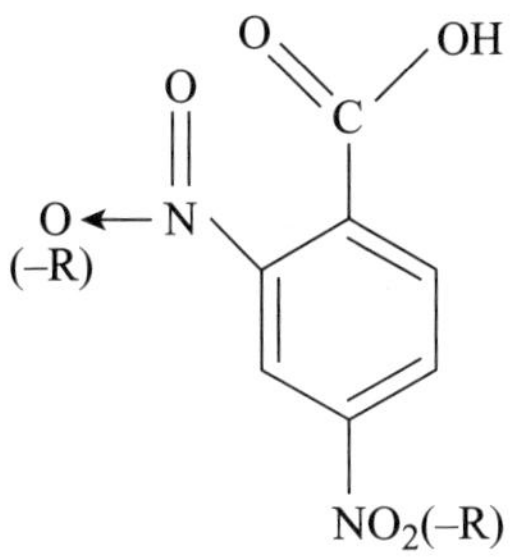

SIR effect is also observed at ortho position (ortho effect). **Ans:** (a)

41.

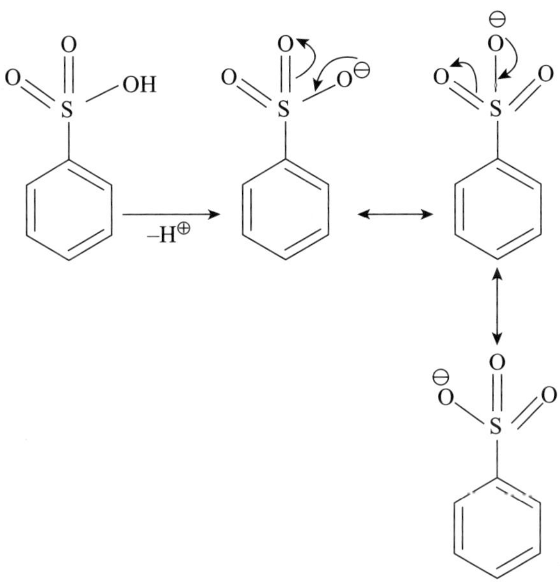

Three equivalent resonating structures.

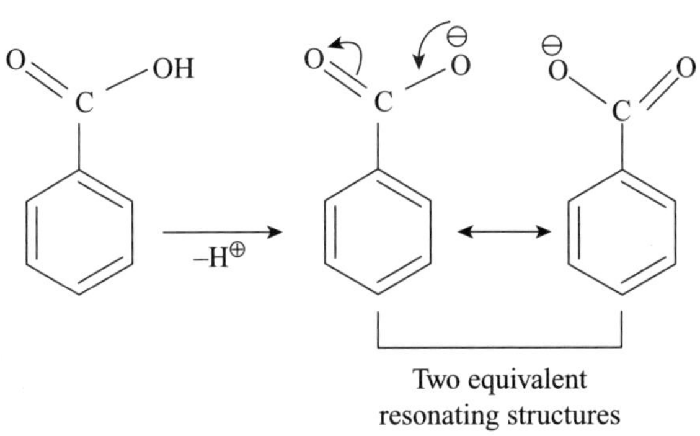

Ans: (d)

42.

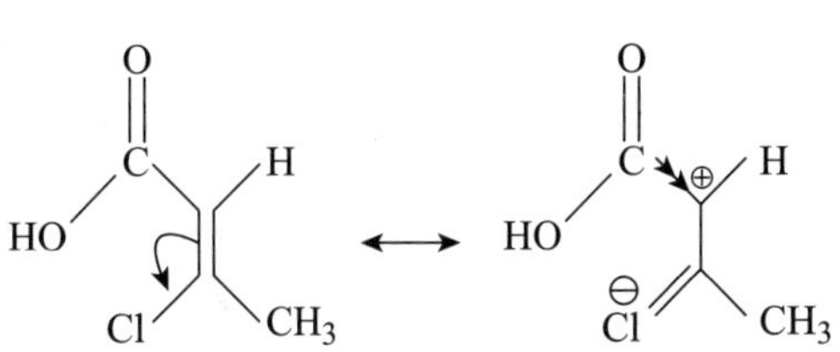

Ans: (b)

43. 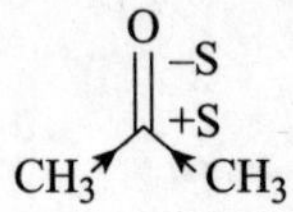

[Note : C=O bond does not show hyperconjugation]

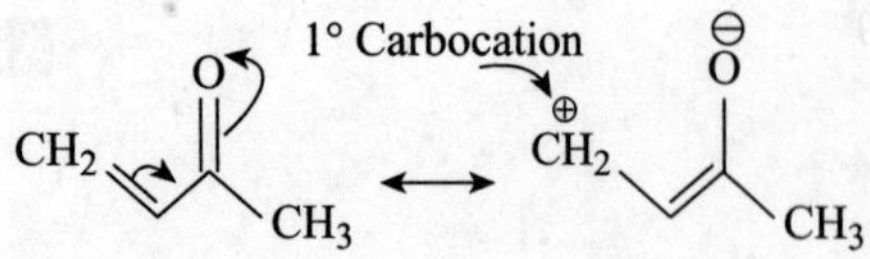

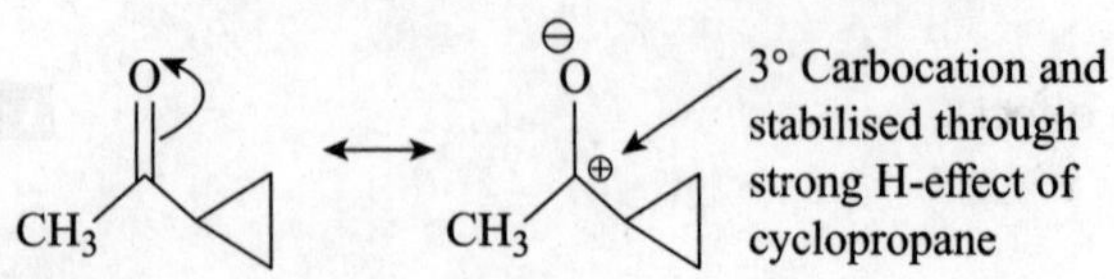

Dipole moment = $q \times l$

Ans: (d)

44. 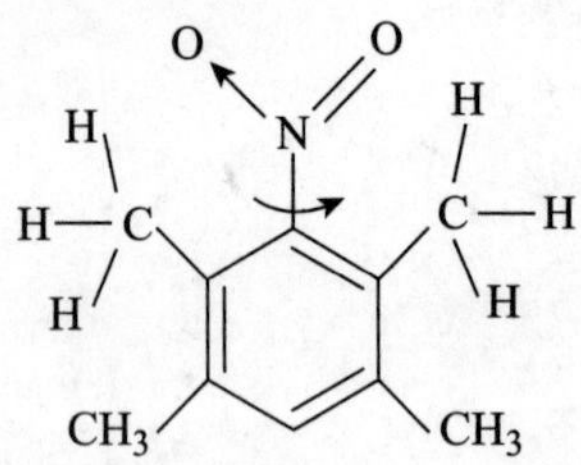

SIR effect due to two bulky groups at ortho position.

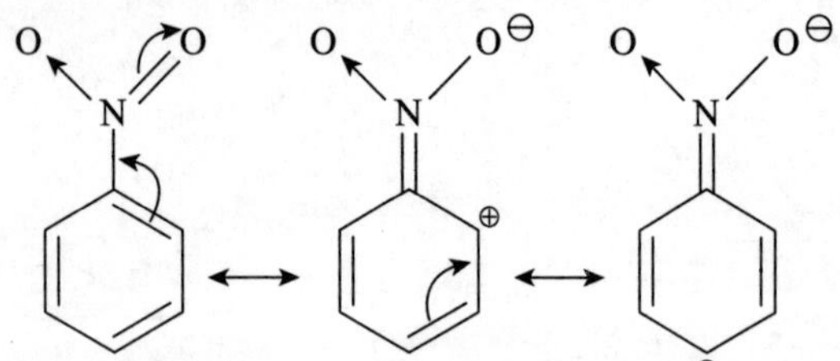

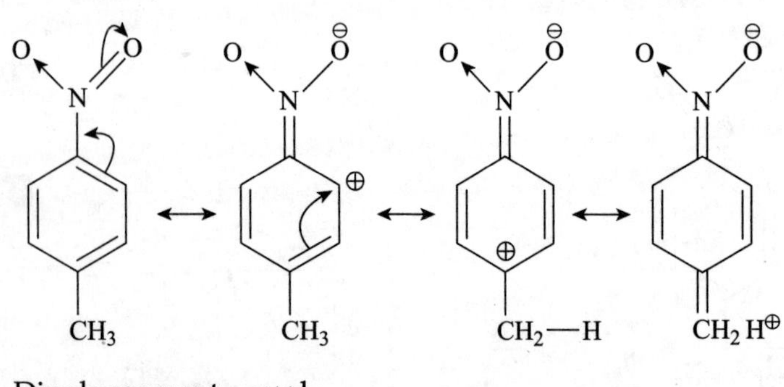

Dipole moment = $q \times l$

Ans: (b)

45. **Hint:** Dipole moment is a vector quantity.

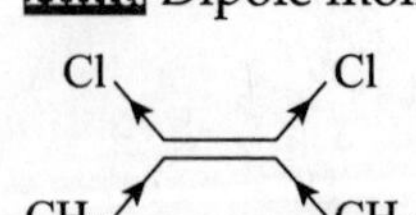

$\mu_1 \neq 0$

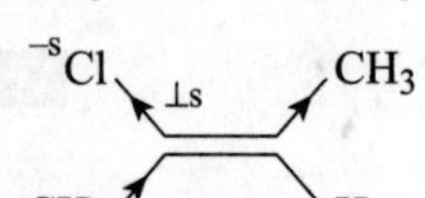

$\mu_2 \neq 0$ $\quad \mu_1 > \mu_2$

Ans: (a)

46. 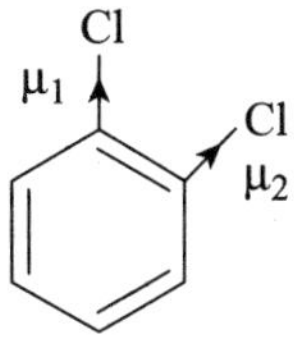

$\mu = \sqrt{\mu_1^2 + \mu_2^2 + 2\mu_1\mu_2 \cos 60°}$
As Q decreases resultant 'μ' increases

Ans: (a)

47. 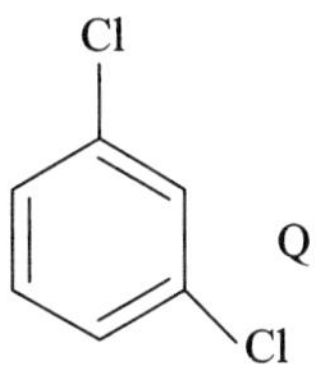

Q = 120° 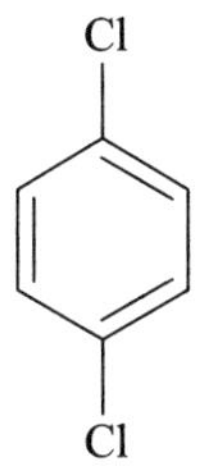

Q = 180°

Ans: (a)

48. See Question Number 17 (Workbook Exercise 2)

50. See Question Number 25 (Workbook Exercise 2)

52. 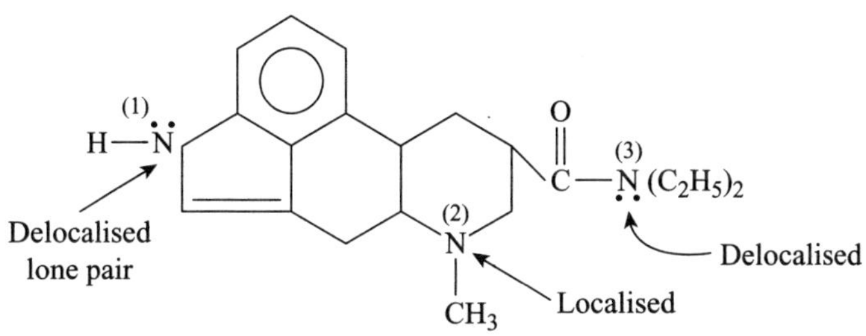

Ans: (b)

54. 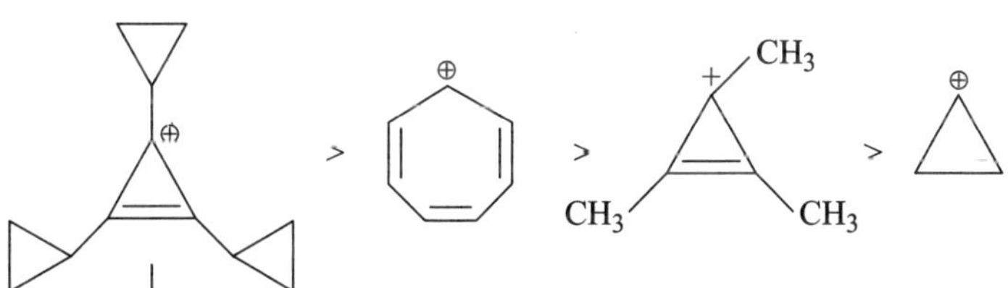

Aromatic as well as it is stablised through dancing resonance of cyclopropane

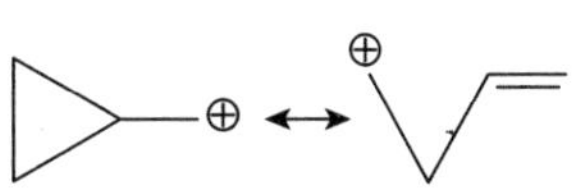

Ans: (a)

55. 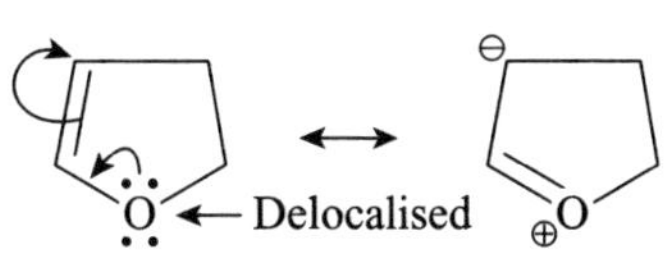

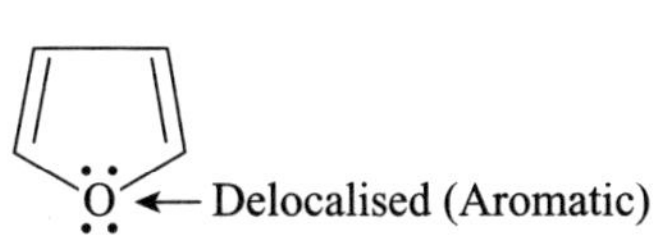

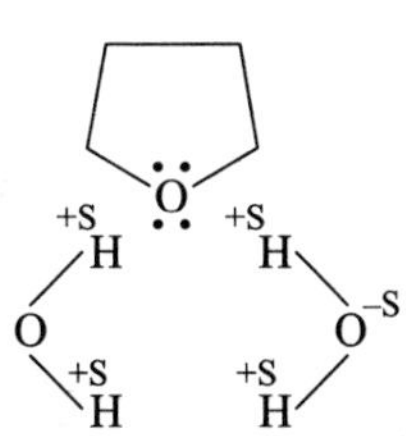

Ans: (a)

56.

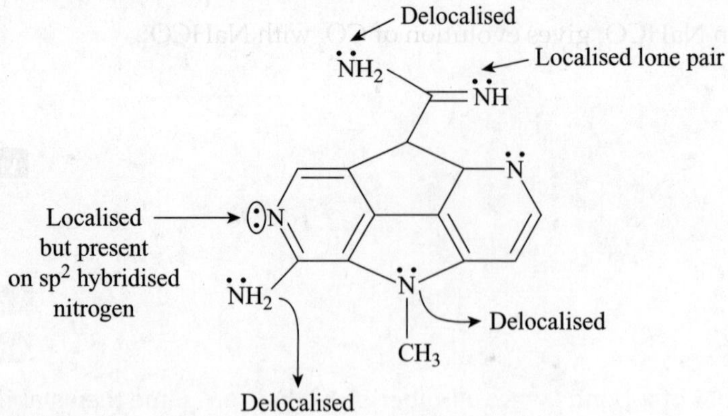

Hint: When the lone pair is delocalised and involved in resonance then check the extent of resonance, if involved in more extended resonance then will be less available for protonation.

Ans: (b)

57. Acidic strength order

$$-\underset{\underset{O}{\|}}{C}-OH > -OH > -\overset{\oplus}{N}H_3$$

Ans: (b)

58. Total charged remains conserved during resonance. **Ans:** (c)

59. If number of carbons are same then heat of combustion depends upon stability of the compound.

$$\text{H.O.C } \alpha \ \frac{1}{\text{stability of compound}}$$

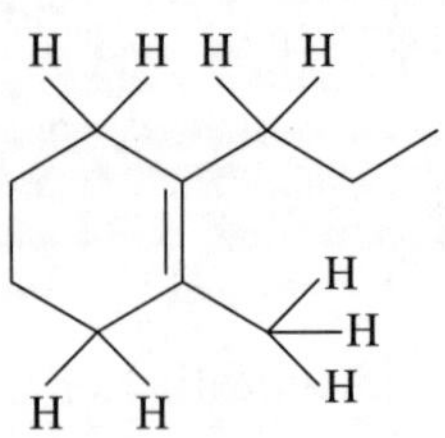

α-Hydrogens

Ans: (a)

60.

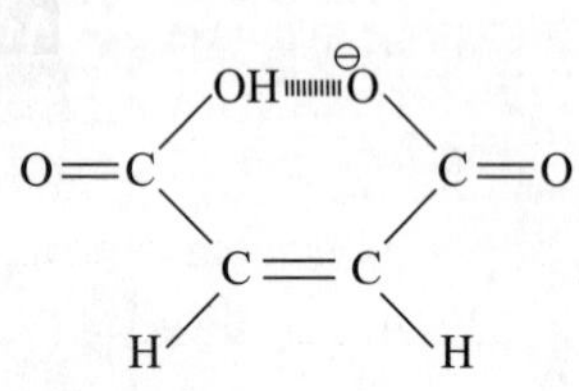

H-bonding after removal of $H^{\oplus}$ ion

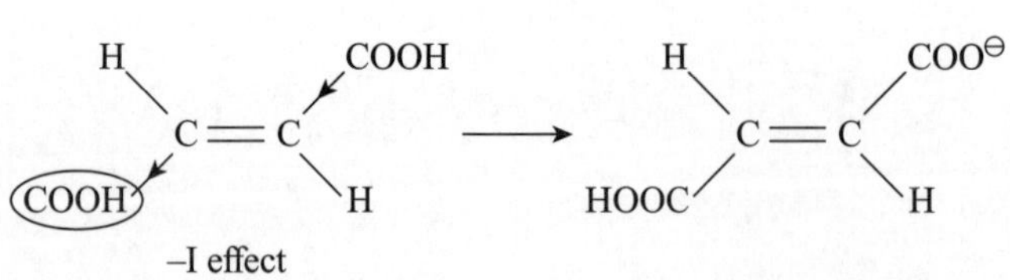

Ans: (d)

61. **Hint:** Acids which are more acidic than $NaHCO_3$ gives evolution of CO_2 with $NaHCO_3$.

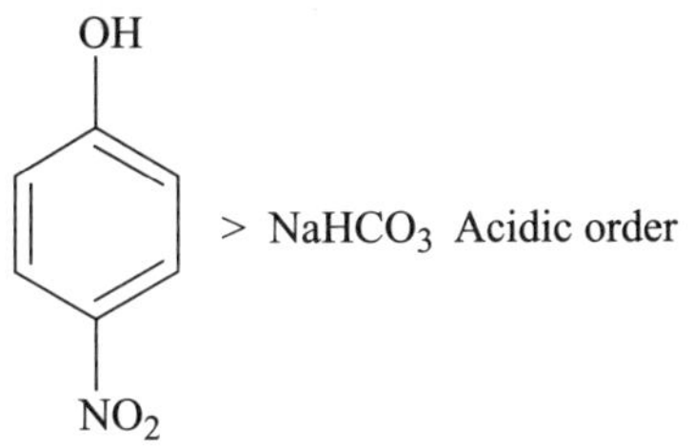

Ans: (d)

63. Resonance energy α extent of resonance. **Ans:** (d)

64. First of all H.O.H depends upon number of π-bonds when number of π-bonds are same then stability of the compound.

$$\text{H.O.C} \propto \frac{1}{\text{stability of compound}}$$

H.O.H does not depend on number of carbon atoms. Alkyne is more stable than alkanes.

Cylindrical cloud of π-electrons

Ans: (d)

65. See solution of Question Number 25.

66.

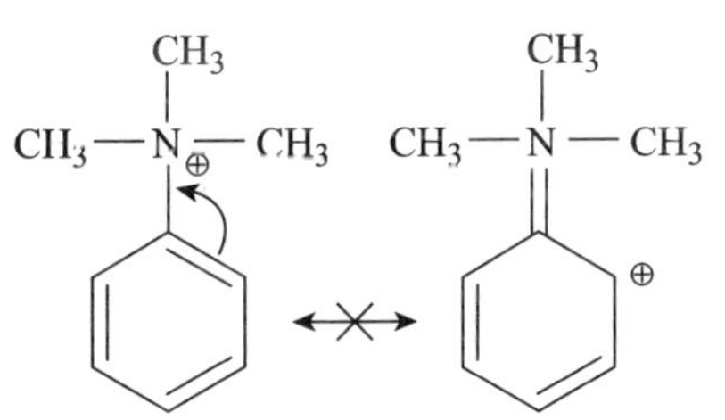

Nitrogen cannot form five bonds

Ans: (a)

67. **Hint:** In carbene carbon form two bonds and one lone pair is present (In case of singlet carbene).

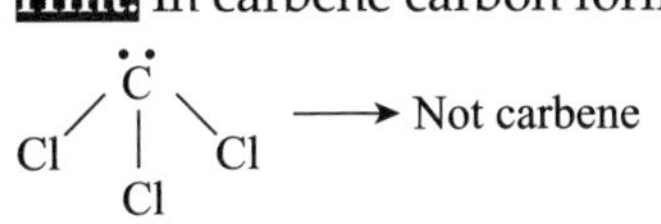

Ans: (d)

68. **Hint:** I_A and II_A group cations do not behave as Lewis acid (electrophile) because they have tendency to loose electrons (Ex. Na^+, $K^{\oplus}$, Mg^{2+} etc.,). $H_3O^{\oplus}$ does not behaves as an electrophile because central atom has complete octet and does not has valent orbitals, similarly $\overset{\oplus}{N}H_4$ is also not an electrophile.

Ans: (b)

69.

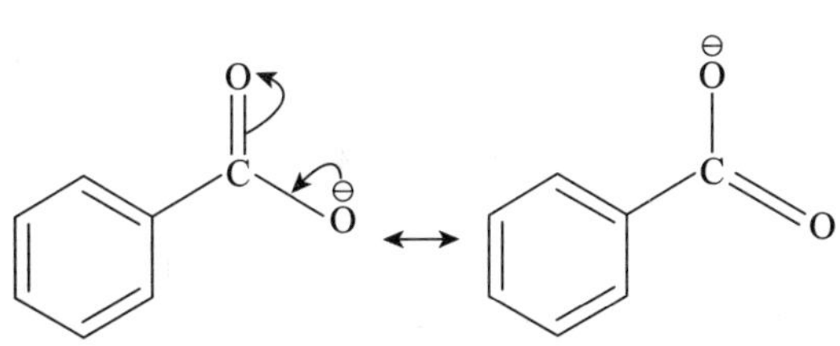

Two equivalent resonating structures (R.S)

Another approach: A strong acid has a weak conjugate base. **Ans:** (c)

70. 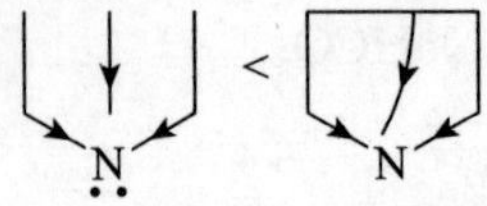

Basicity: No inversion take place

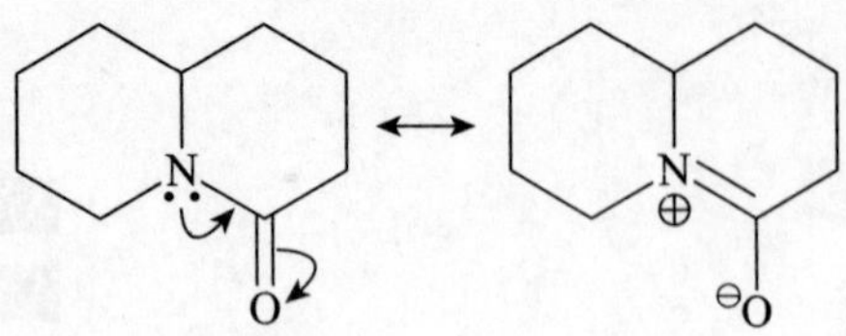

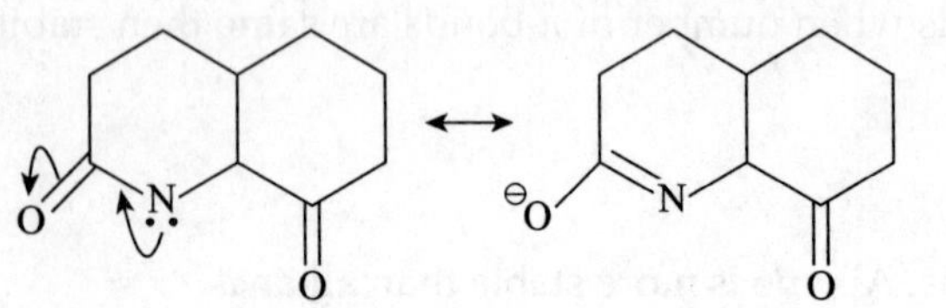

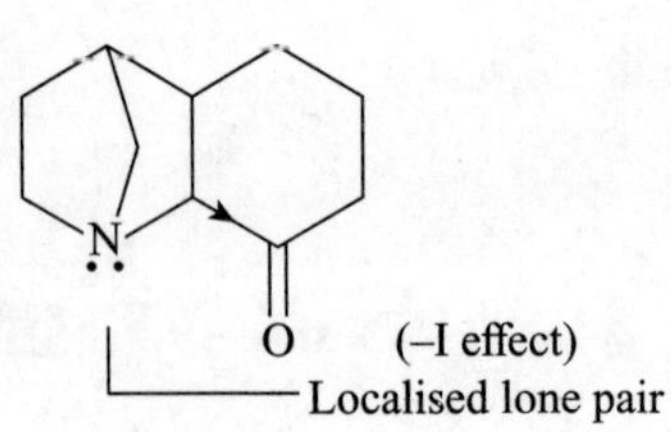

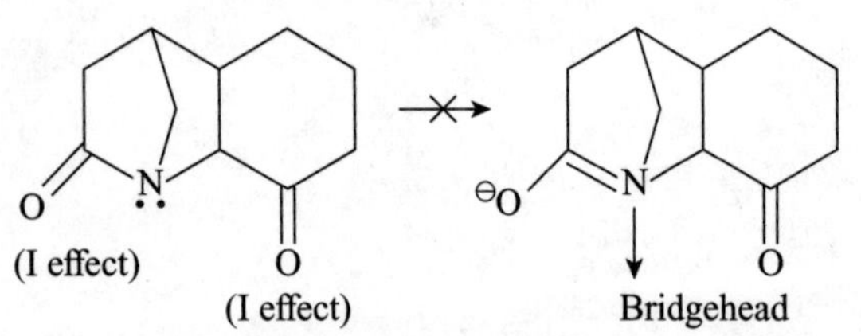

Ans: (c)

71. 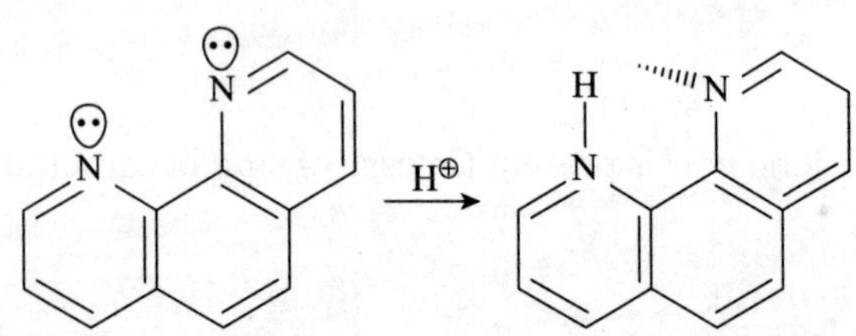

Two lone pairs are close enough,maximum repulsion hence highly willing to get protonated.

After protonation attractive force comes into role (H-bonding).

Hint: Field-effect

Ans: (c)

72. After protonation carbocation is formed, check the stability of carbocation.

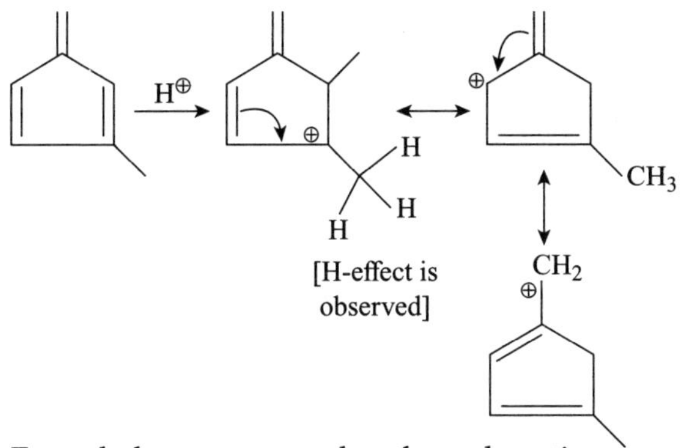

Extended resonance makes the carbocation more stable. **Ans:** (a)

73.

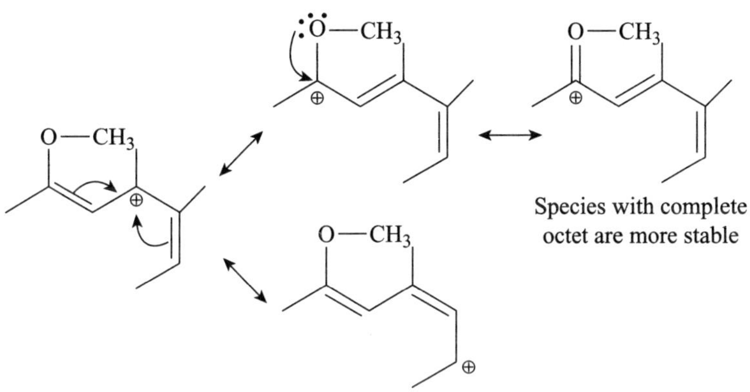

Ans: (d)

74.

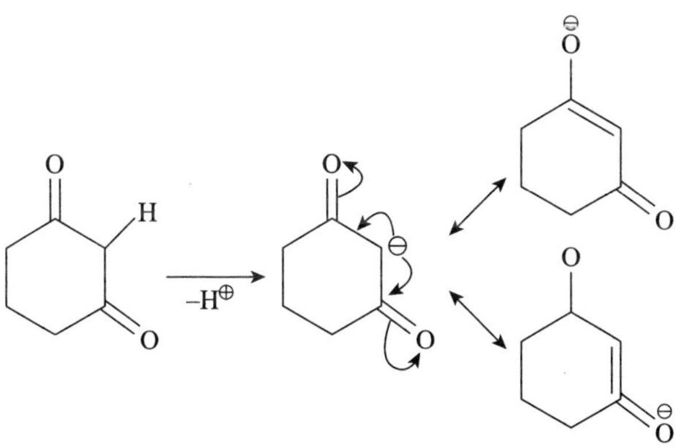

Both the π-bonds are participating 100% in resonance.

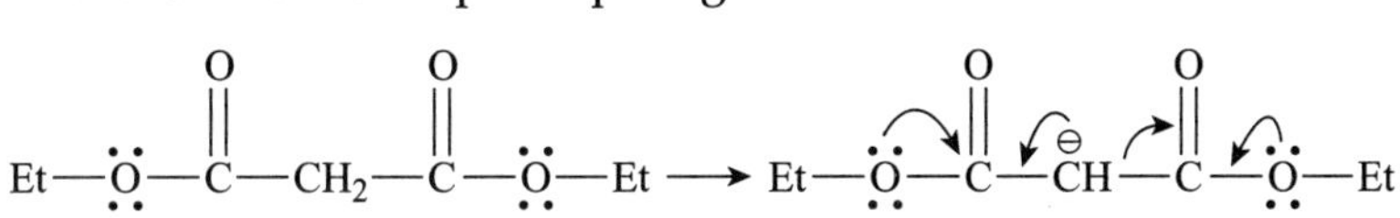

Both the C=O bonds are in cross-conjugation means do not participate with 100% efficiency with negative charge.

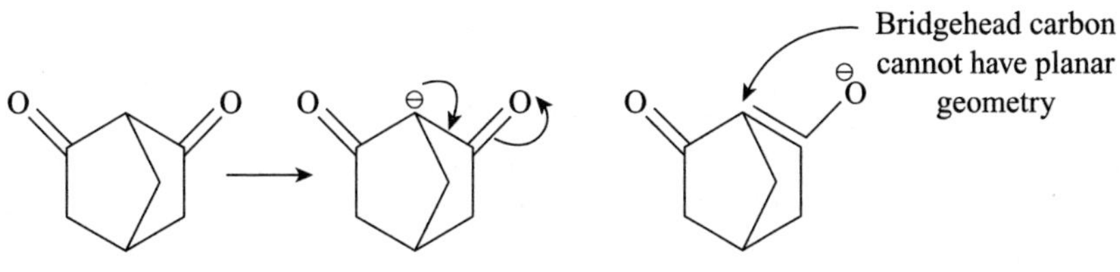

Ans: (a)

79.

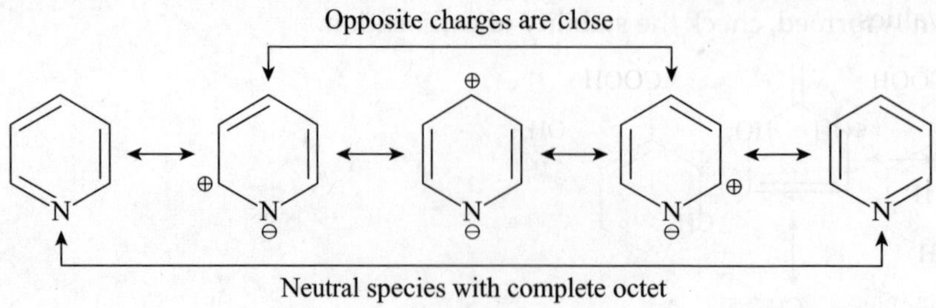

Ans: (a)

80.

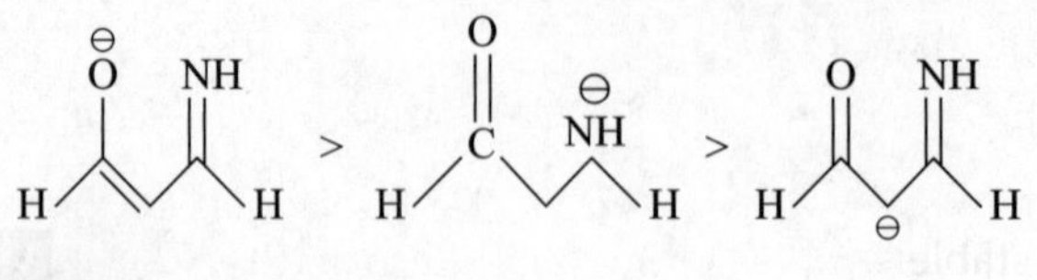

Hint: Negative charge in m ore electronegative is more stable. **Ans:** (b)

81.

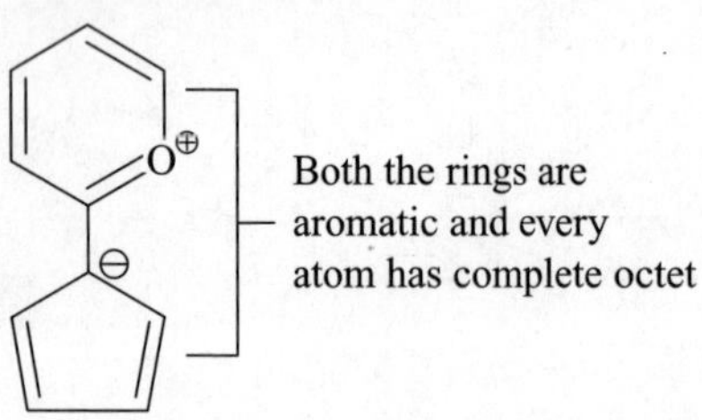

Ans: (b)

83.

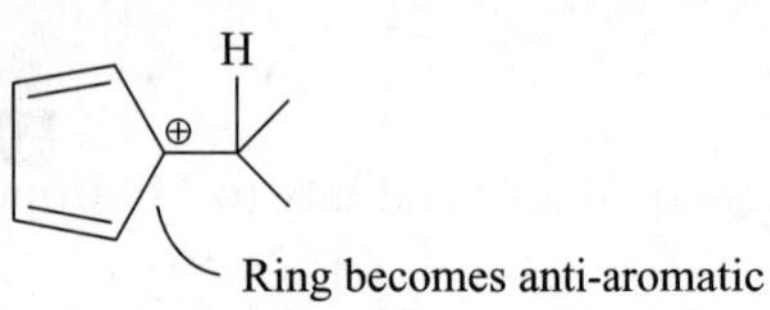

Ans: (a)

84.

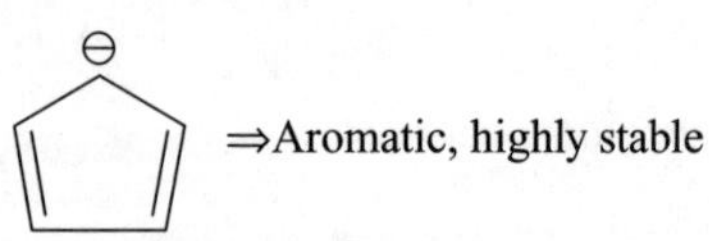

Ans: (c)

85.

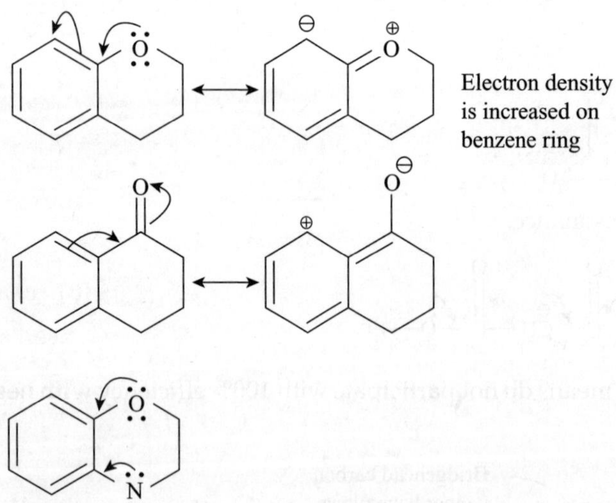

Hint: When lone pair or negative charge is in conjugation with π-bond of benzene ring then electron density is increased on benzene ring, but when polar π-bond positive charge (central atom must have incomplete octet) or vacant orbital then reverse effect is observed. **Ans:** (b)

87. Strong acids have less Pka values.

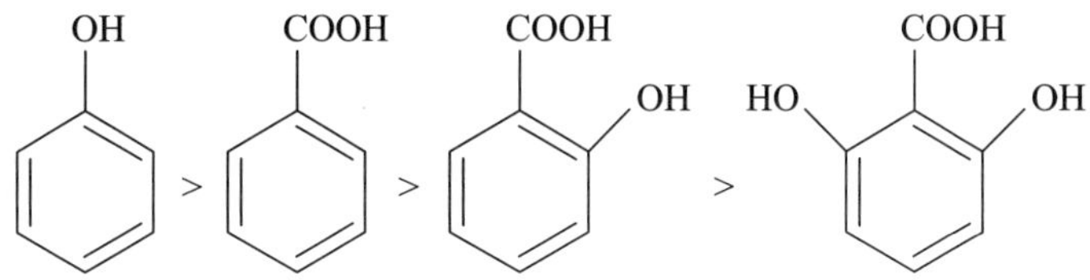

Order of Pka value

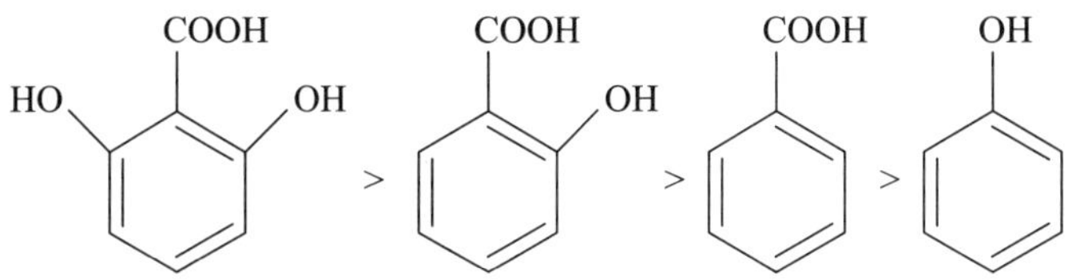

Acidic strength order

$HClO_4 > H_2SO_4 > HNO_3 > HNO_2$

Acidic strength order

Pka value order will be

$HNO_4 > HNO_3 > H_2SO_4 > HClO_4$ **Ans:** (a)

89. 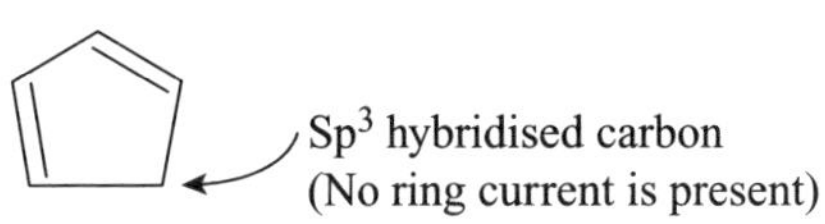

Ans: (d)

90. First of all HOH depends upon number of π-bonds, if π-bonds are equal then stability (α-Hydrogens).

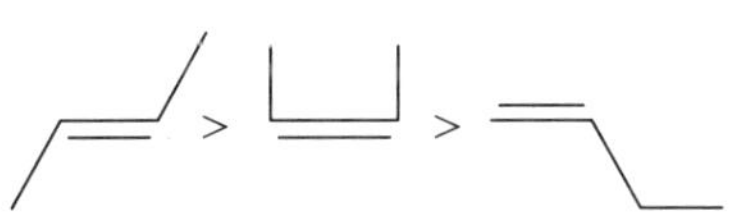

Stability order (when α-Hydrogens are same trans > cis ⇐ stability)

$$HOC \alpha \frac{1}{\text{stability of compound}}$$

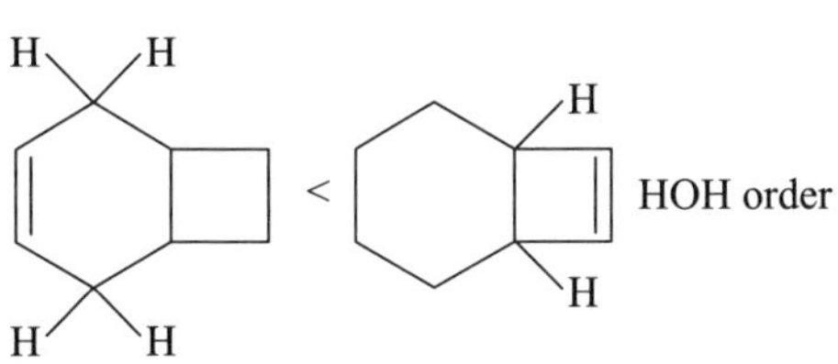

(4α-H) (2α-H)

Ans: (b), (c) and (d)

91. 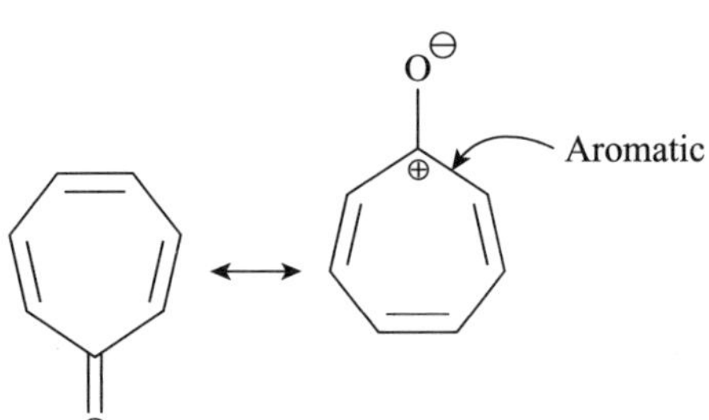

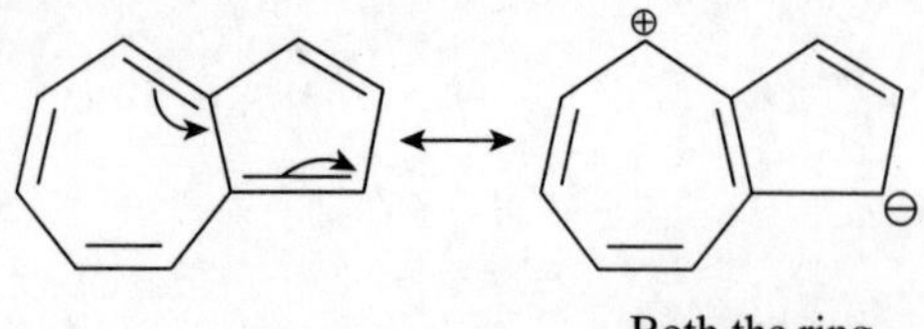

Both the ring
becomes aromatic

Ans: (a) and (b)

92. 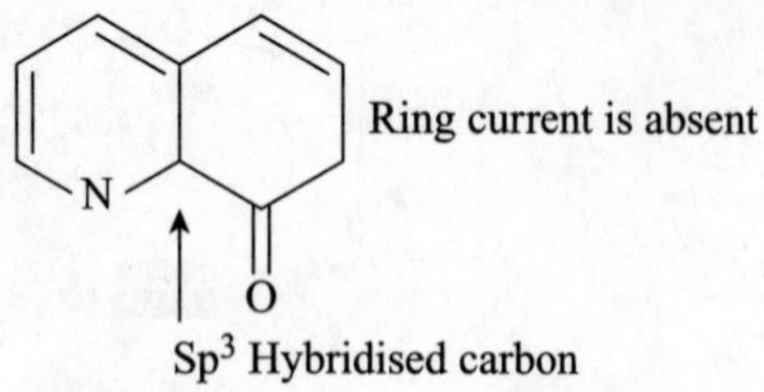

Hint: For aromatic and anti-aromatic compounds ring current must be present. **Ans:** (a), (b) and (c)

93. 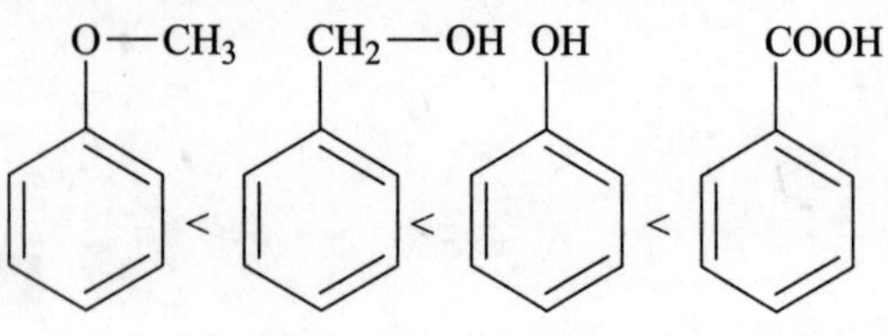

Ether < Alcohol < Phenol < Benzoic acid. (Acidic strength order) **Ans:** (a), (b) and (d)

97. (a) 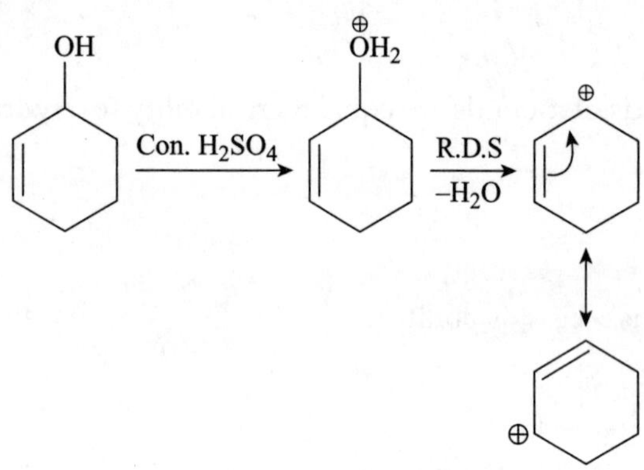

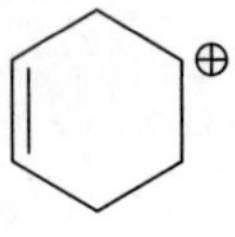

No resonance

(b) 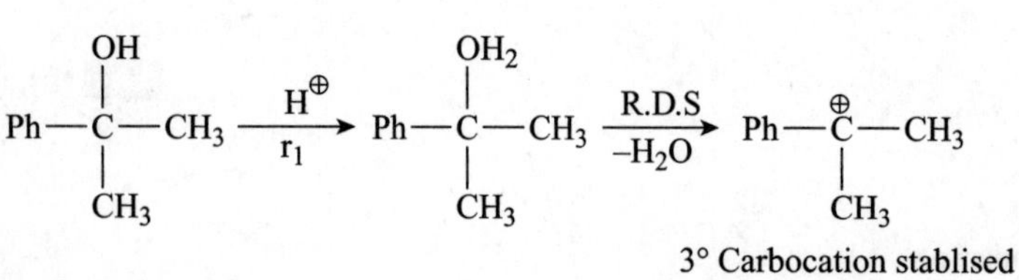

3° Carbocation stablised
by +H and R effect

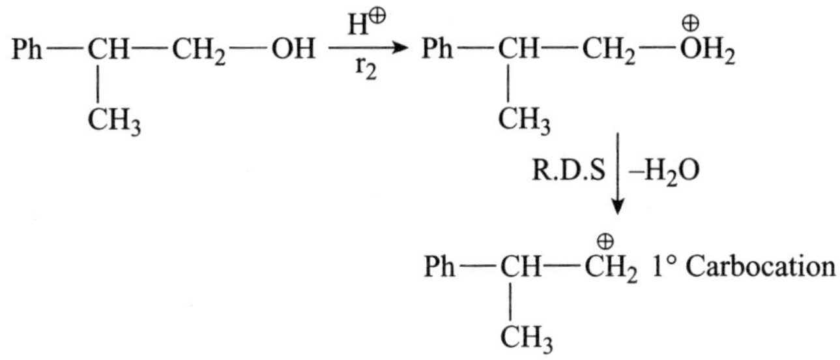

Rate will be very slow because 1° carbocation is merely formed.

∴ $r_1 >>> r_2$ **Ans:** (b) and (c)

103.

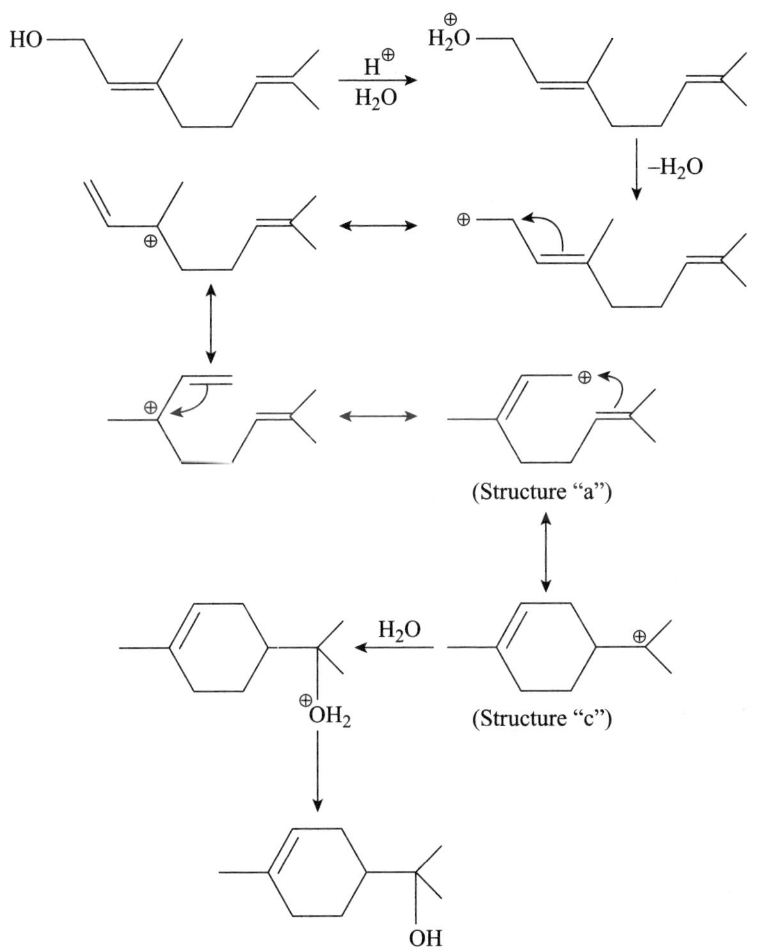

Ans: (a), (c) and (d)

105.

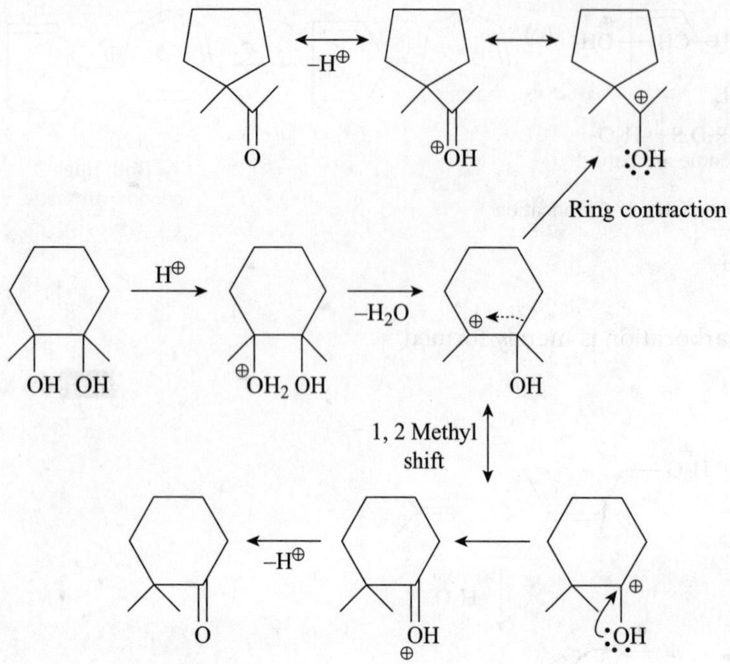

Ans: (b) and (c)

106.

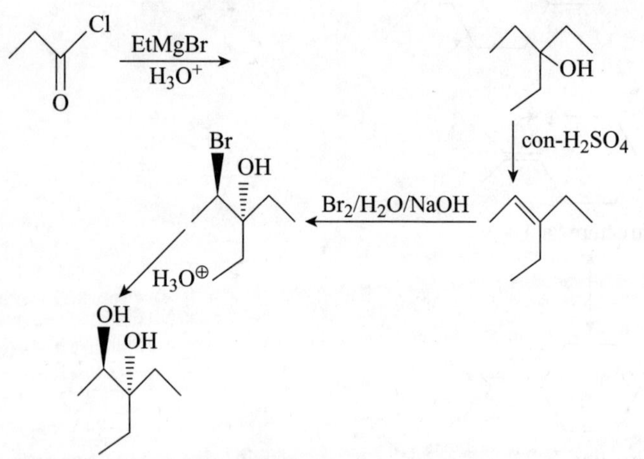

Ans: (a) and (d)

113. **Hint:** Extent of hydrogen bonding increases boiling point increases. **Ans:** (b)

118. (a)

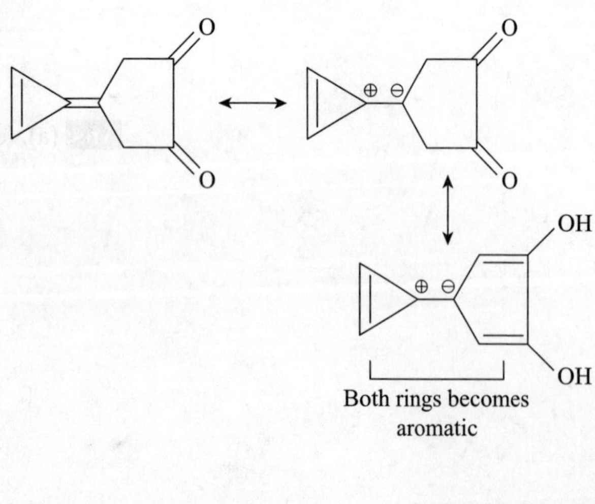

(b) 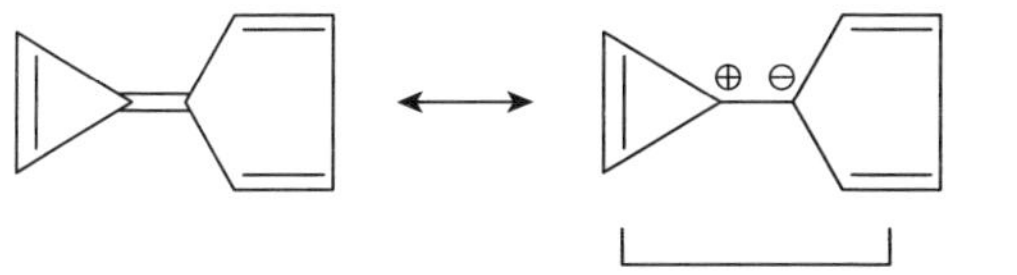

Both rings become aromatic

(c) 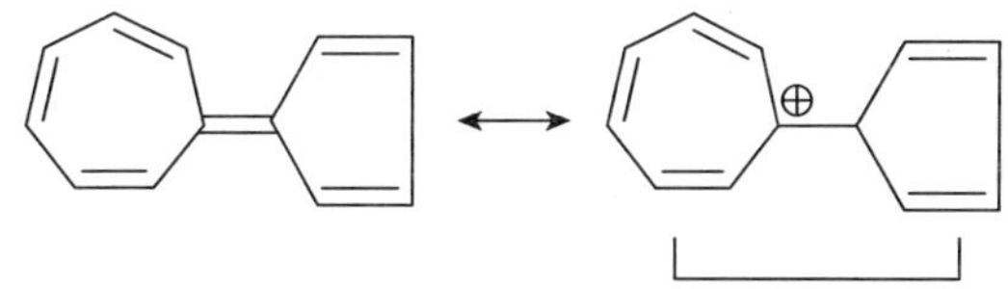

Both rings become aromatic

(d) 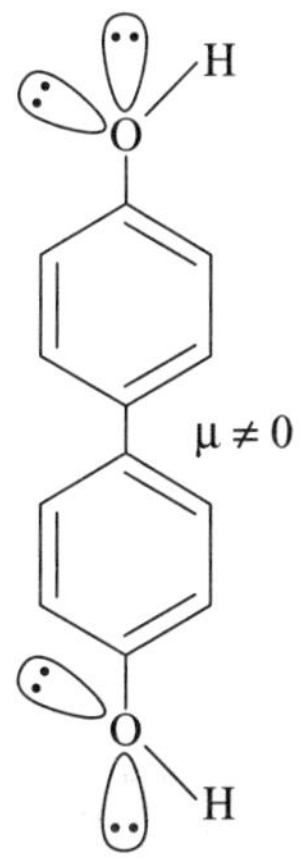

Ans: (a), (b), (c) and (d)

119. (a) 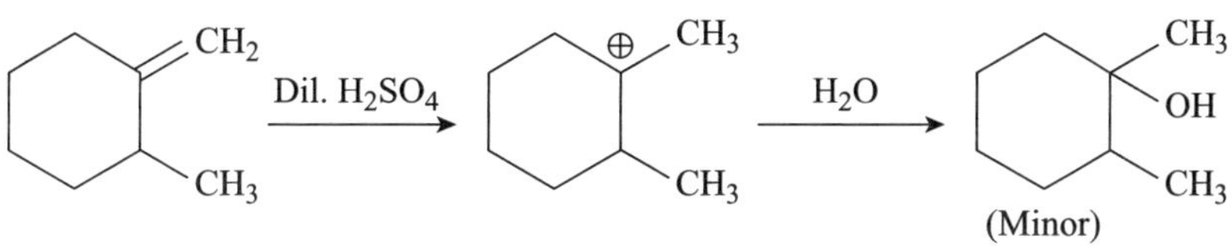

(b) 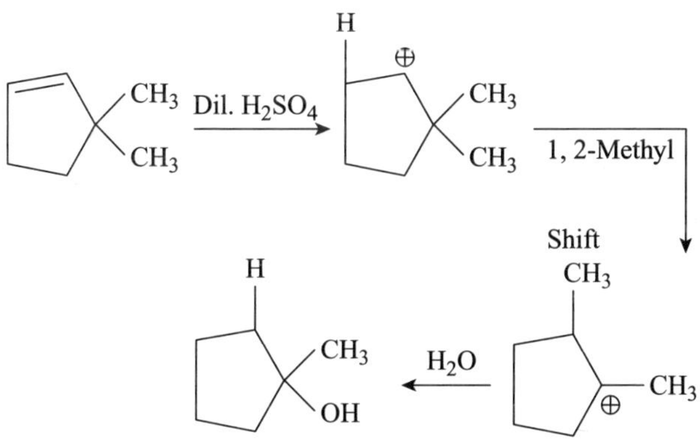

(c) 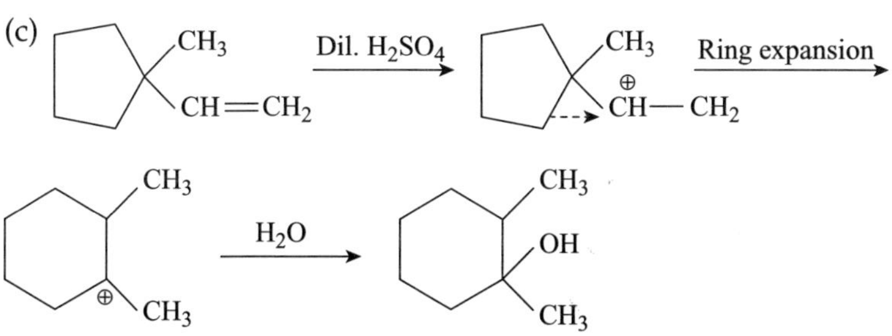

(d) 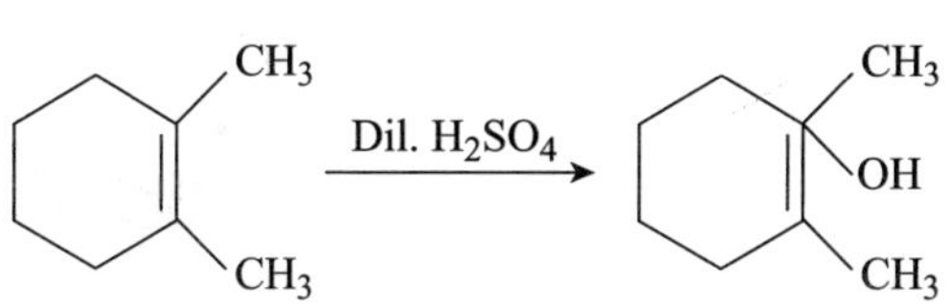

Ans: (a), (c) and (d)

120.

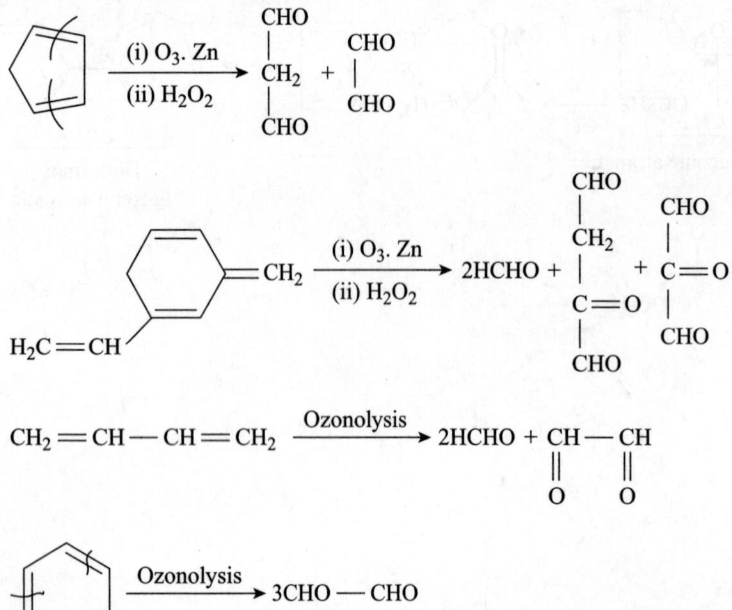

Ans: (a), (c) and (d)

121. In (c) and (d) cyclic transition state is formed. **Ans:** (a) and (b)

122.

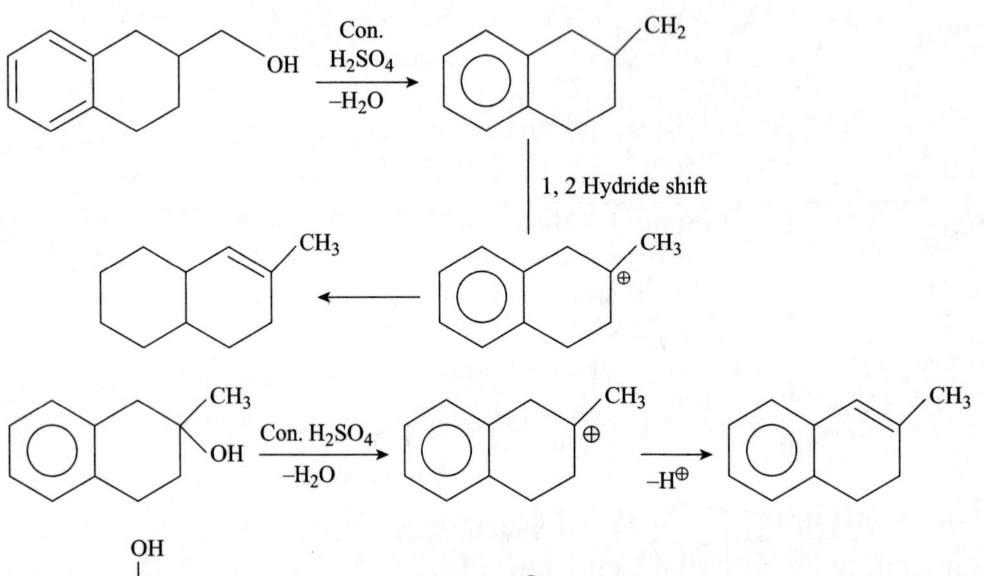

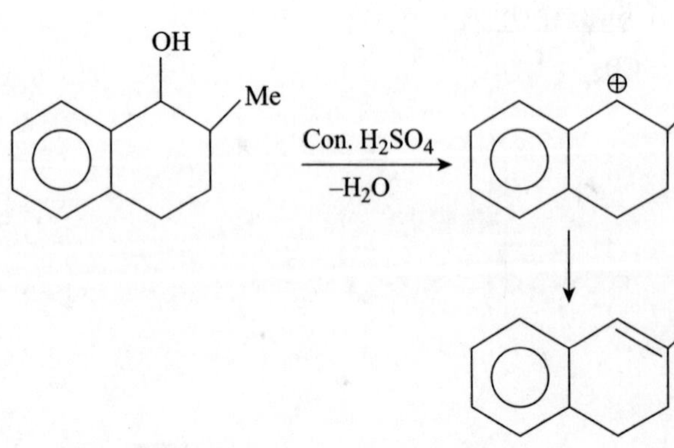

Ans: (a), (b) and (c)

123. (a)

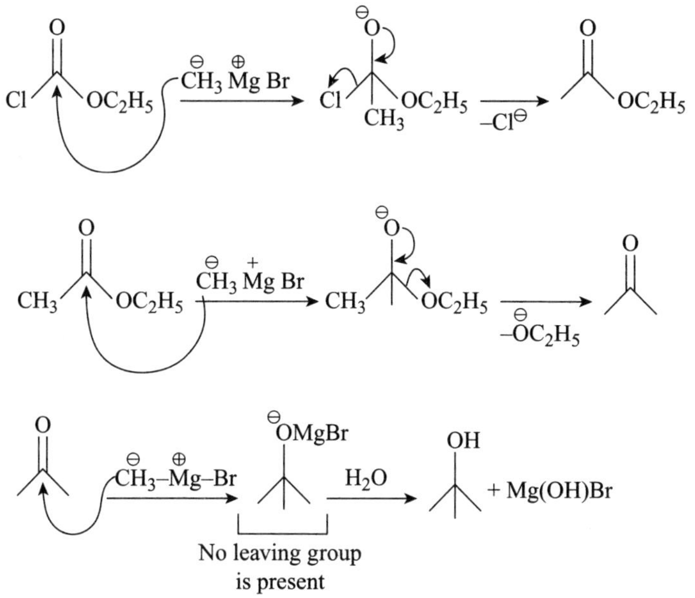

(b)

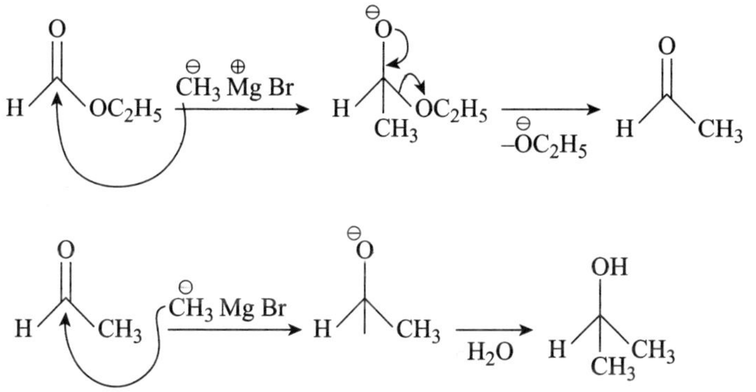

(c)

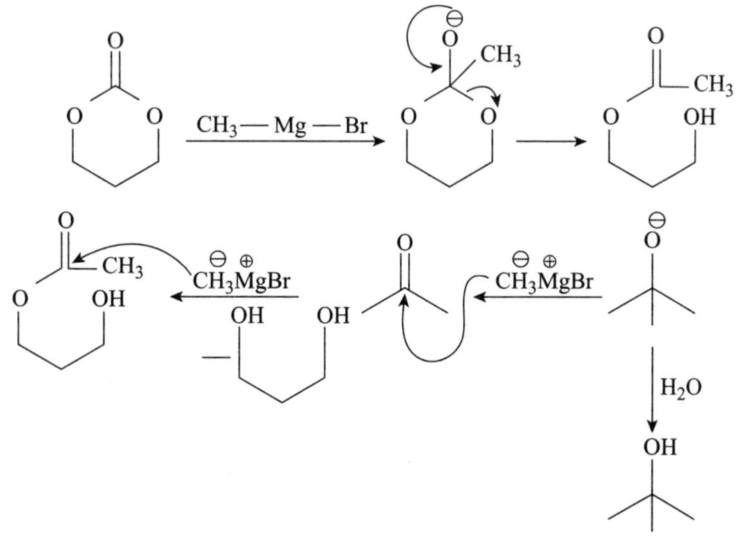

(d)

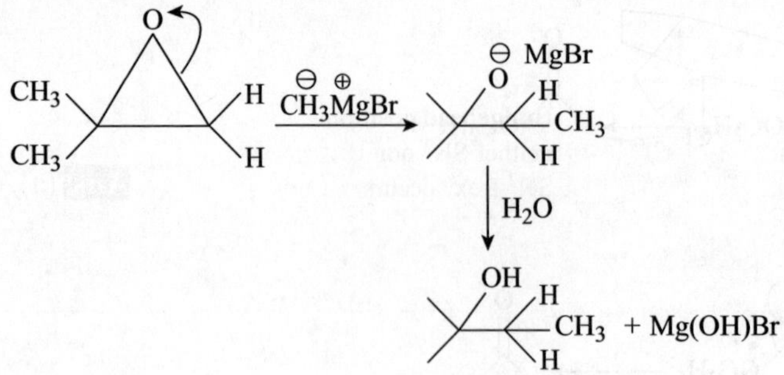

Ans: (a), (c) and (d)

125.

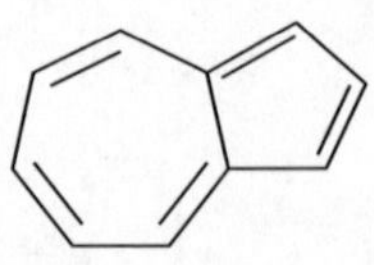

Ans: (a), (b) and (d)

126. (a)

In bicyclo compounds at bridgehead position double bond acquire trans orientation and trans orientation in larger ring is stable

(b)

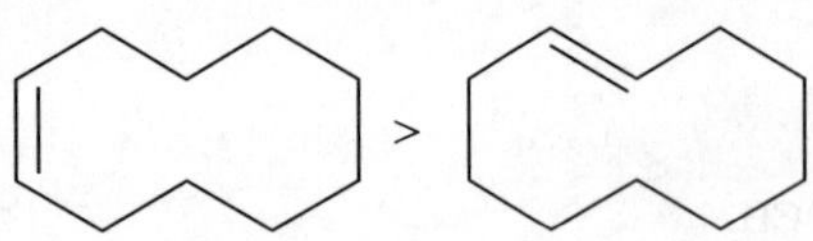

Stability order [up to 11 membered ring cis form is more stable]

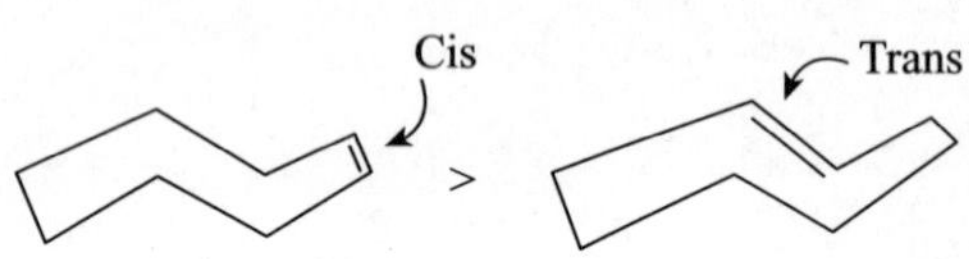

Stability Order

Ans: (b) and (c)

127. (a)

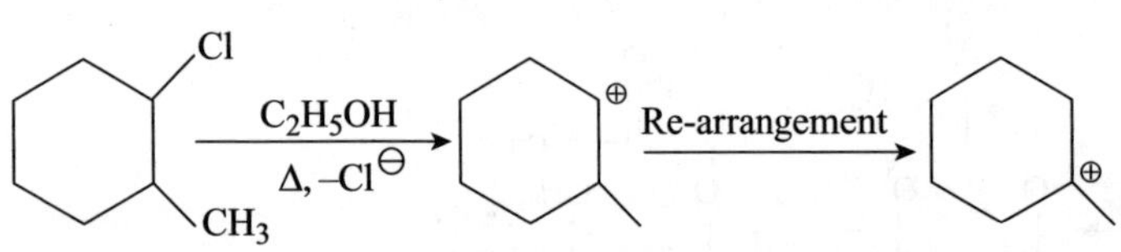

(b)

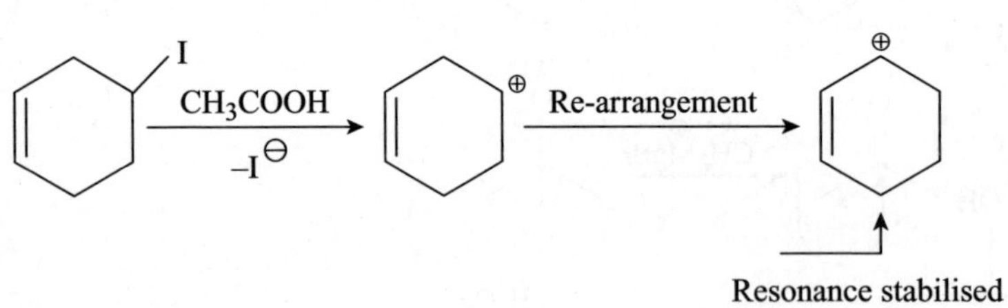

(c)

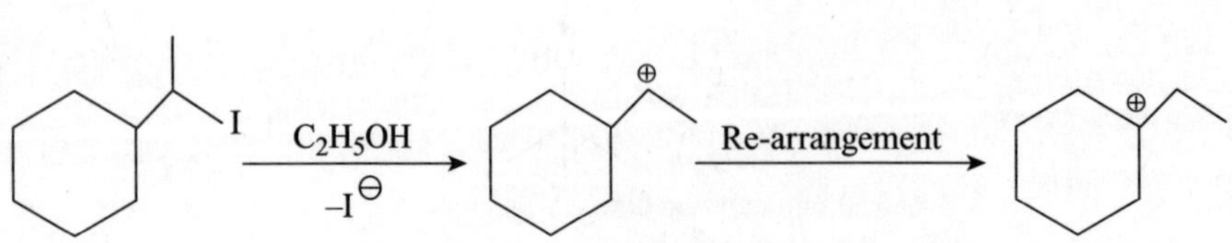

(d) 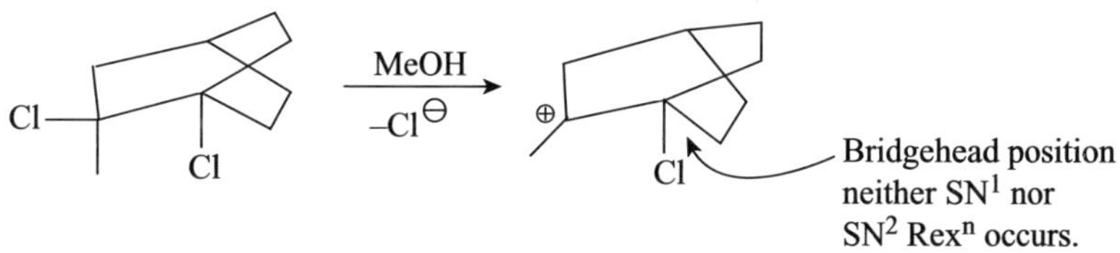

Ans: (a), (b) and (c)

Comprehension Type

Passage 1

129. $HA \rightleftharpoons H^{\oplus} + A^{\ominus}$

$$K_a = \frac{[H^+][A^{\ominus}]}{[HA]}$$

Ka α [H⁺] α acidic strength.

The correct acidic strength order is given below (Ka also follow the same order).

v > ii > i > iv > iii > vi

Ans: (d)

130. Acids which are stronger acid than 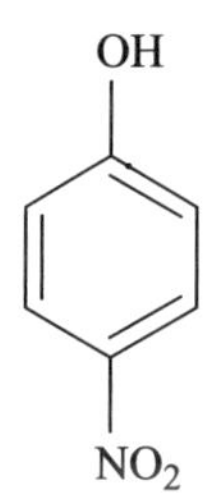 Can give effervescence with $NaHCO_3$

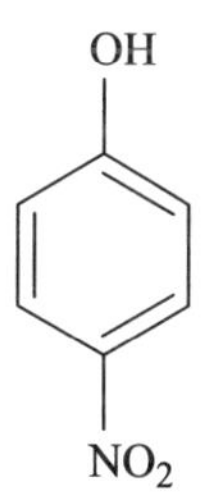 > $NaHCO_3$ Acidic strength order

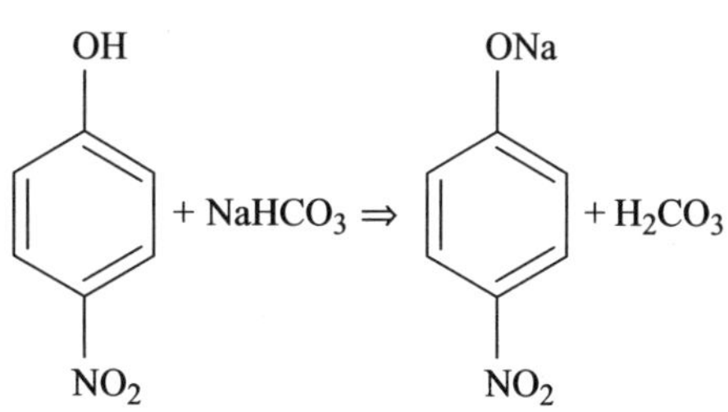

$H_2CO_3 \rightleftharpoons CO_2\uparrow + H_2O$

Ans: (a)

Passage 2

132. **Hint:** Stability of the carbocation α $\dfrac{1}{\text{Electron withdrawing power of the group attached}}$

Ans: (b)

133. Stability of carbonion α electron withdrawing power of the group attached

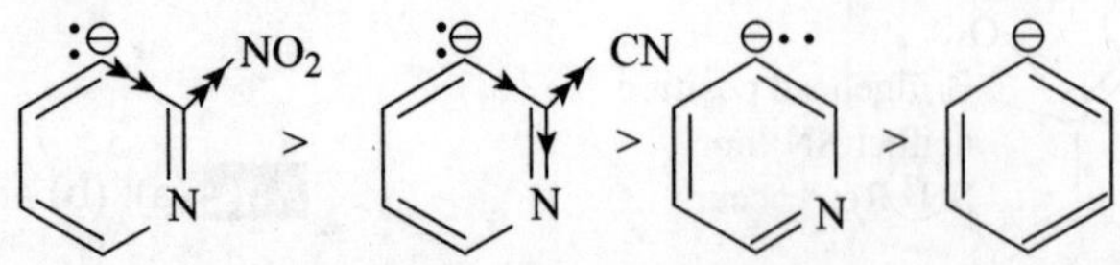

Ans: (d)

134. Acidic strength α the stability of the anion left after removing $H^{\oplus}$ ion

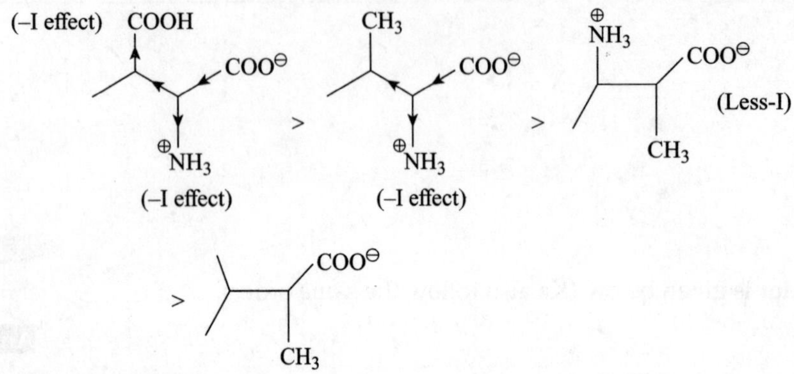

Ans: (c)

Passage 3

135.

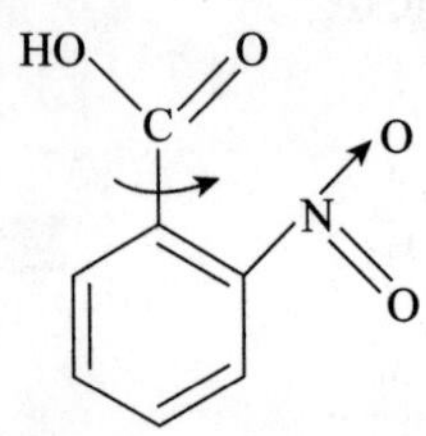

Due to bulky group at ortho position of benzene ring –COOH group goes out of plane [Actually resonance with benzene brakes and only –I effect of benzene is working].

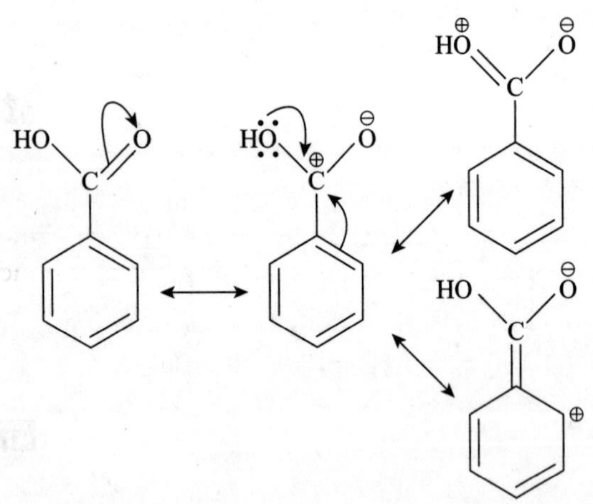

When bulky group is present on ortho position

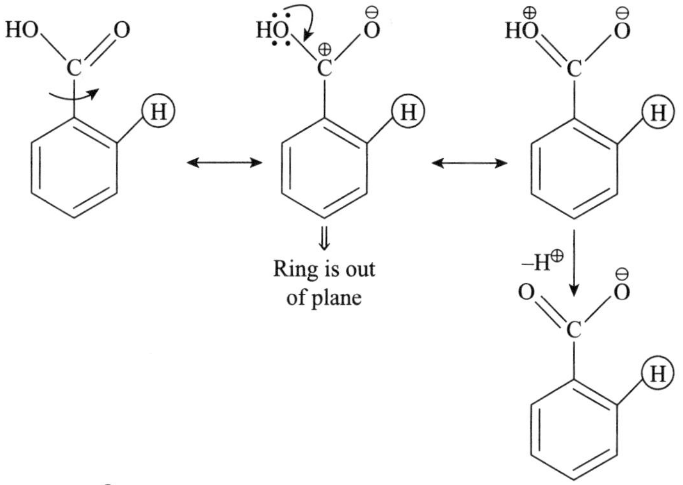

Ans: (a)

136.

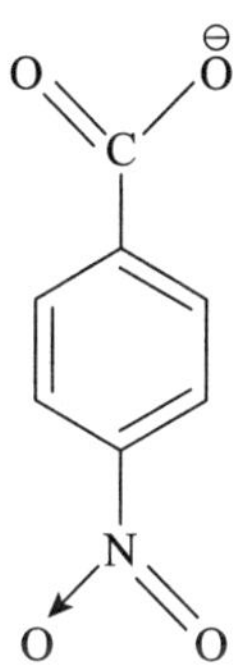

[–M effect]

Ans: (d)

137.

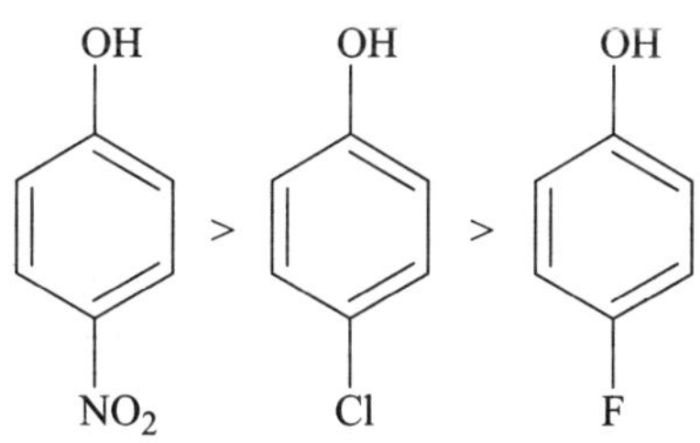

(–M effect) (–M, –I effect) (–I effect)

Ans: (c)

Passage 4

139. Heat of hydrogenation ∝ (directly proportional) number of π-bonds if number of π-bonds are same then depends upon the stability of the alkene (number of bonds of α-Hydrogens) if number of π-bonds are same, then stability of alkene α number of α–H α $\frac{1}{\text{H.O.H.}}$

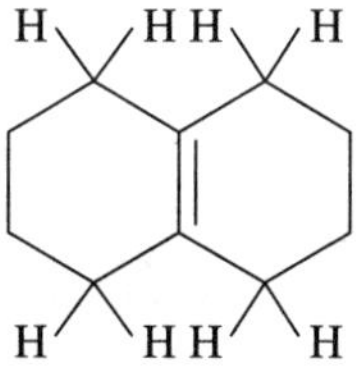

8 α-Hydrogens

Ans: (d)

140.

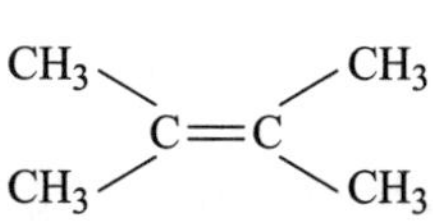

C=C bond length α number of α-Hydrogens

Hint: More will be the α-Hydrogens, more will be hyperconjugation and resulting more C — C single bond character.

Ans: (c)

Passage 5

141. Due to non planarity **Ans:** (c)

142. A compound to be aromatic it must have ring-current and this ring current is formed by the complete cyclic resonance of $(4n + 2)\pi$ electrons.

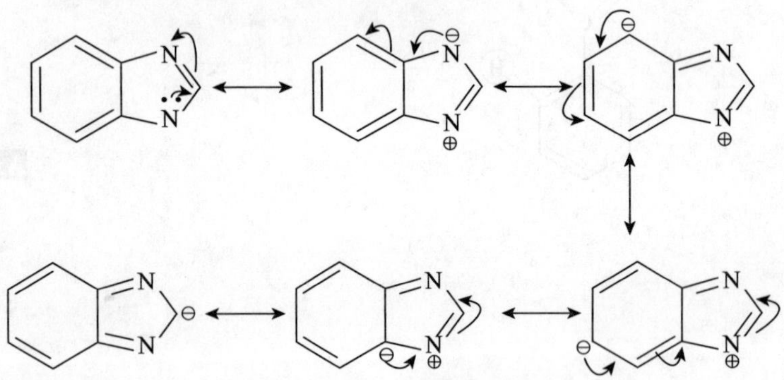

Ans: (b)

143.

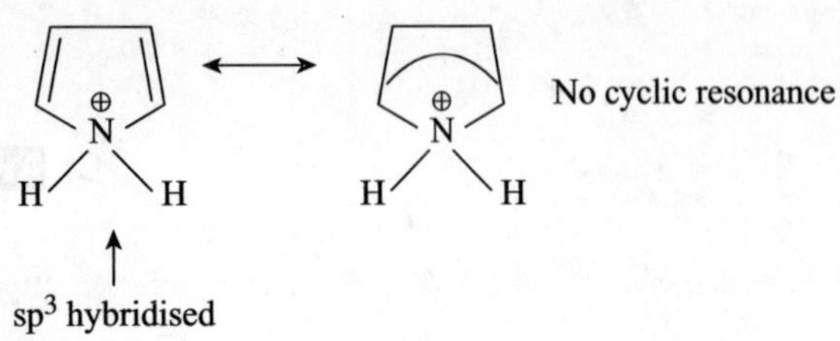

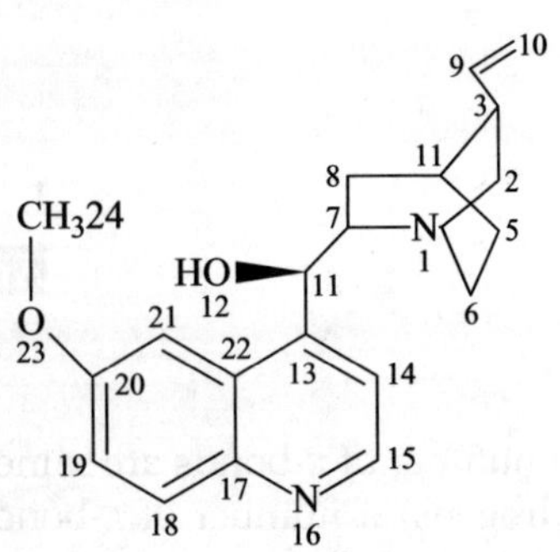

Ans: (b)

144. +1 **Ans:** (a)

145. Hybridisation state = Number of bond pairs and number of line pairs.(α bond only)

N_1 = 1 (lone pair) + 3 bond pair = 4 = sp^3

O_{12} = 2 (lone pair) + 2 σ bonds = 4 = sp^3 **Ans:** (a) and (d)

146. Carbon attached to three carbons **Ans:** (c) and (d)

149. **Hint:** 140 = Number of π-bonds + Number of rings (Generally) **Ans:** (d)

168. Electrophiles CH_3, BF_3, $AlCl_3$, $CaCl_3$, $SbCl_5$, PCl_5, PCl_3 CH_3 — Cl

In CH_3 — Cl, Cl — C bond is polar bond thus carbon have low energy anti-bonding orbitals **Ans:** (8)

169. (a) 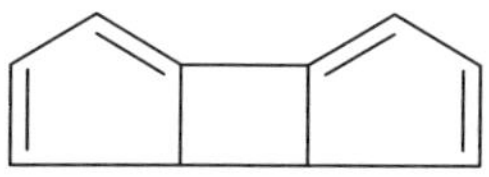 → Non-aromatic

(b) $C_{14}H_{14}$

D.U = 15 one ring + 14 π-bonds

Hint:

(a) 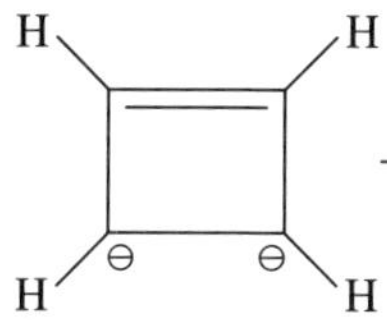 → Aromatic $C_4H_4^{2-}$

(b) 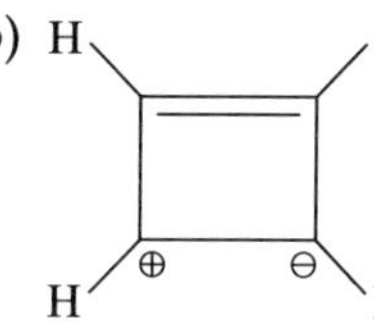 → Anti-aromatic

(c) 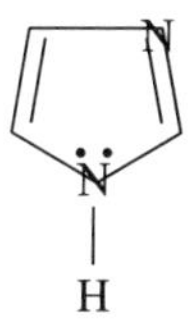 → Aromatic

(g) 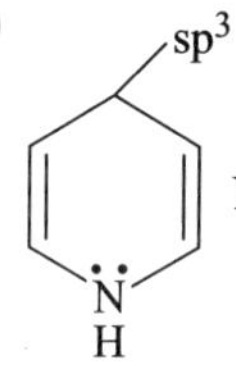 Non-aromatic

(h) 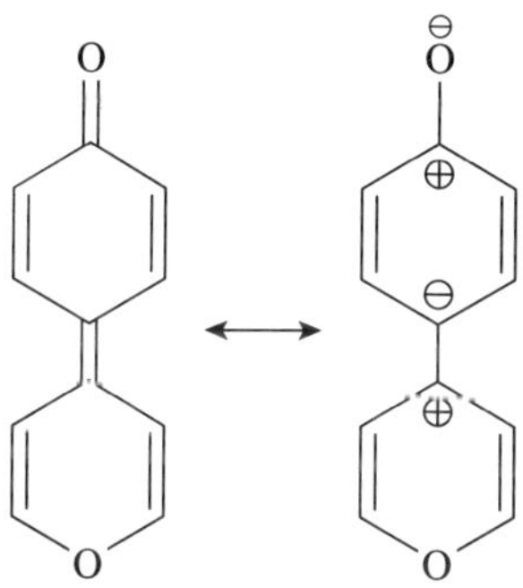 Aromatic

(i) 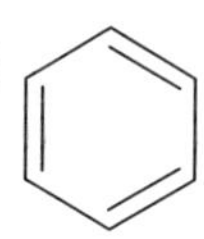 Aromatic

Ans: (5)

170. See Question Number130 (Level 2). **Ans:** (8)

171. (a) 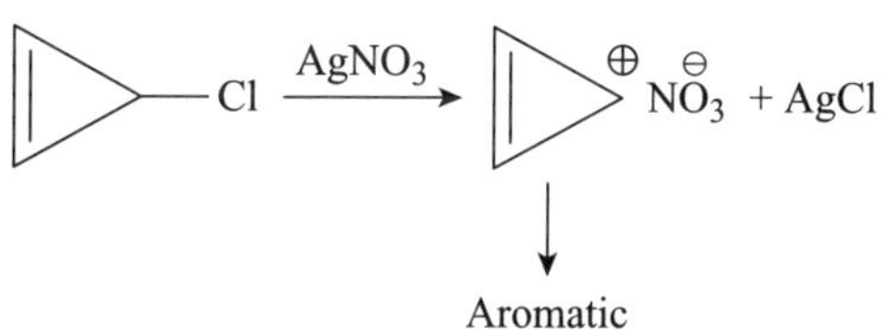

(c) 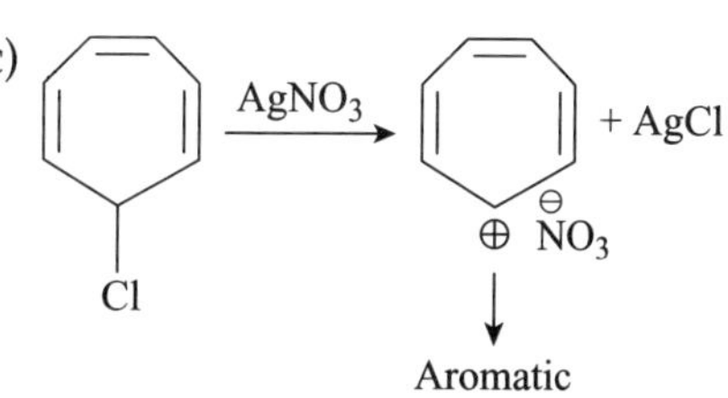

(d) 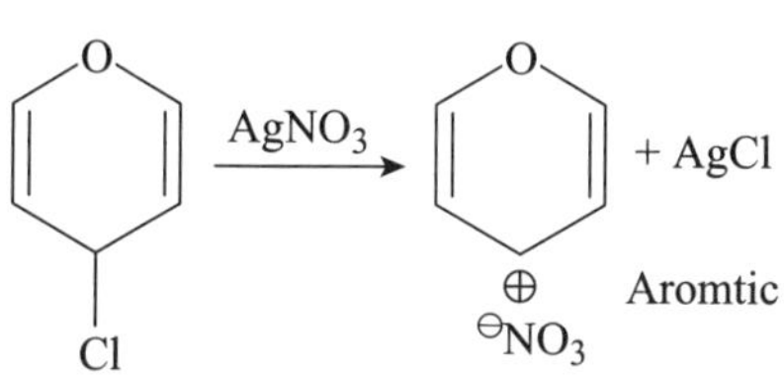

(e) 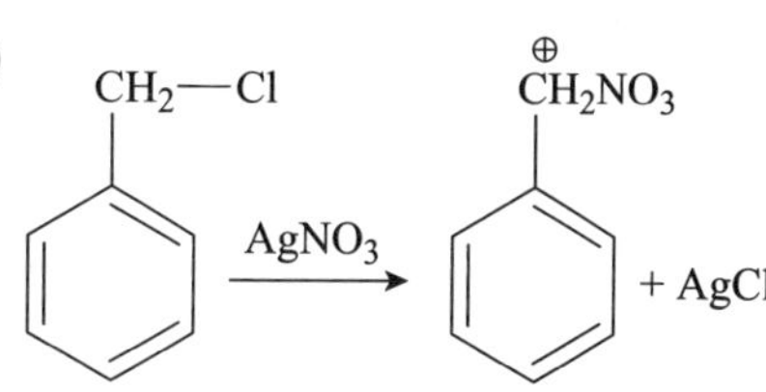

Ans: (4)

172. (i) 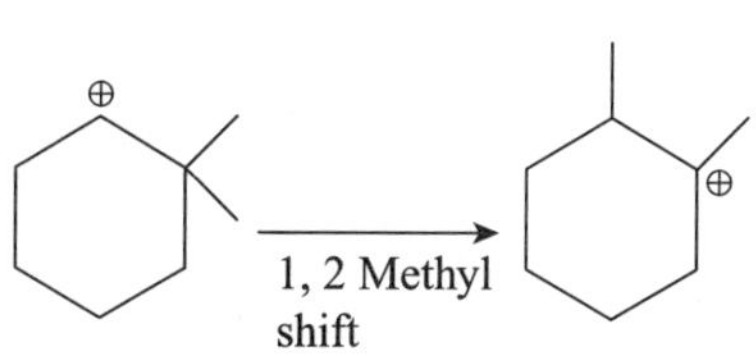

(ii) 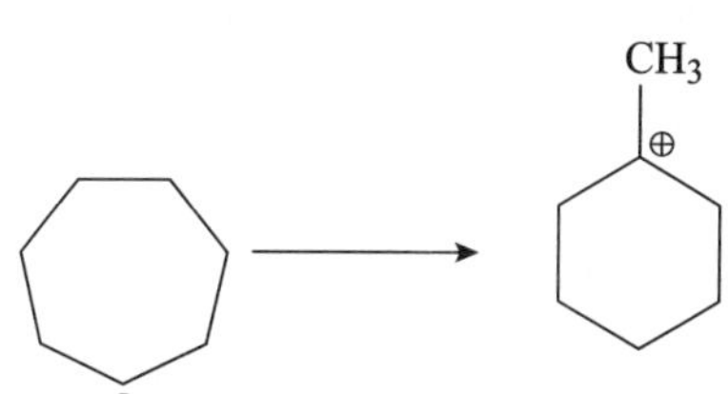

(Ring contraction)

(iii)

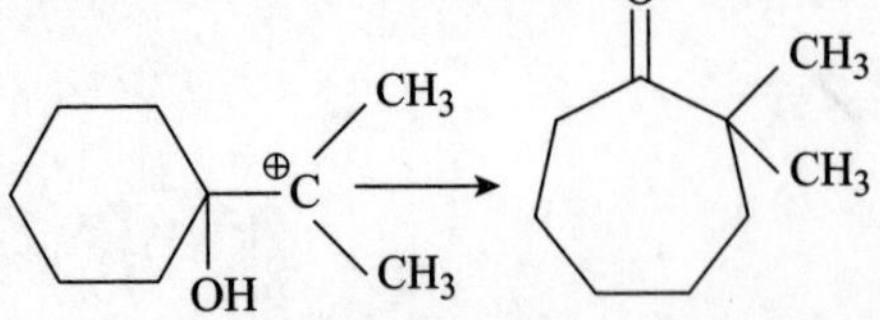

(Driving force is formation of C$=$O bond)

(iv)

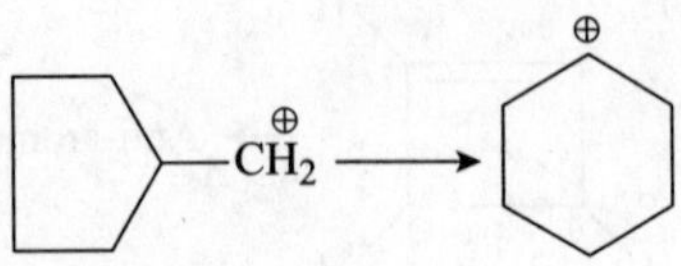

(vi)

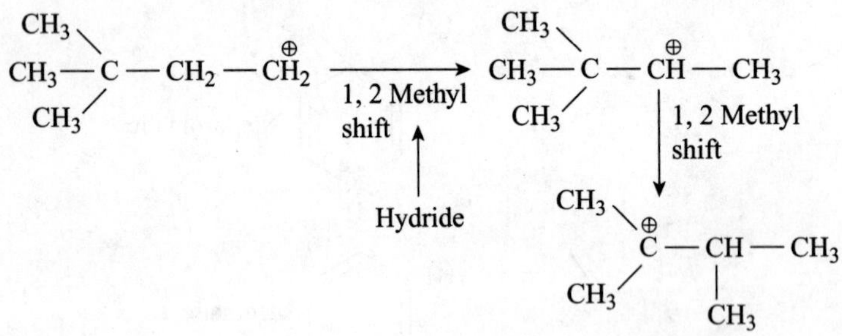

(ix)

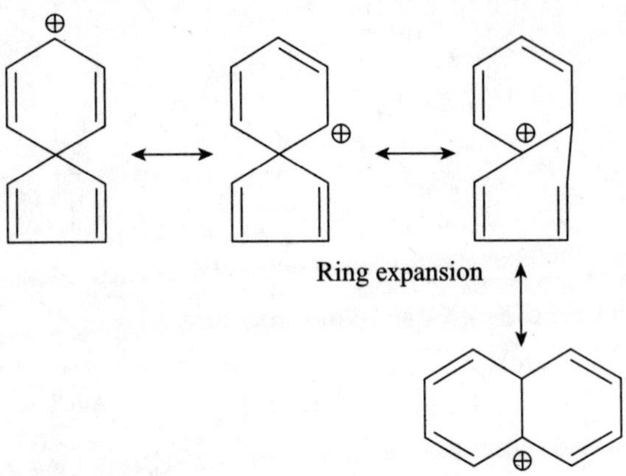

(x)

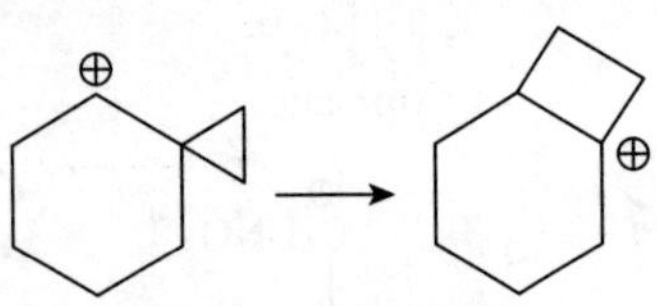

Ring expansion

(xii)

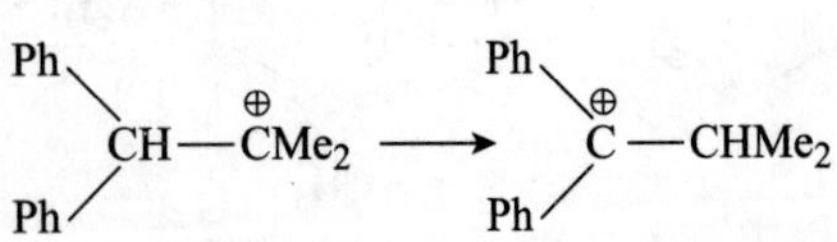

Ans: (8)

173. Compounds which are more acidic than $NaHCO_3$ are soluble in it.

(a)

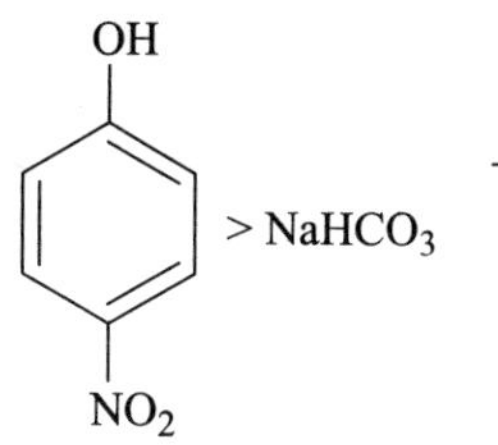

(b)

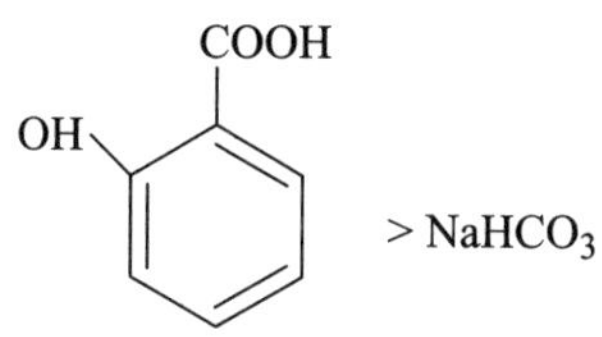

(c)

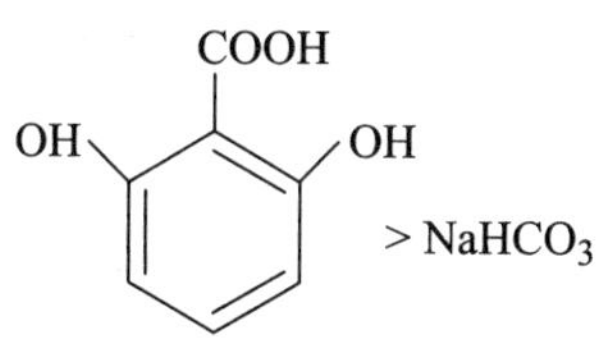

(d)

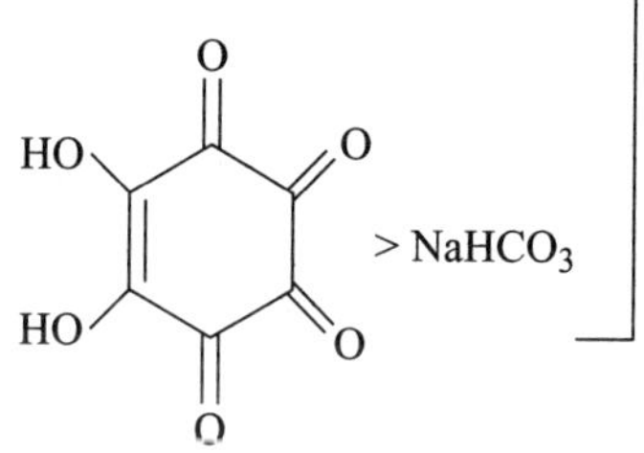

Acidic strength orde [For more details see question numbe 130, level 2]

Ans: (4)

174. Compounds which are more acidic than NaOH are soluble in it.

Hint: NaOH is a base, all acidic compounds are soluble. **Ans:** (7)

175. **Hint:** Compounds which are acidic in nature liberate hydrogen gas on reacting with Na.

Ans: (6)

176.

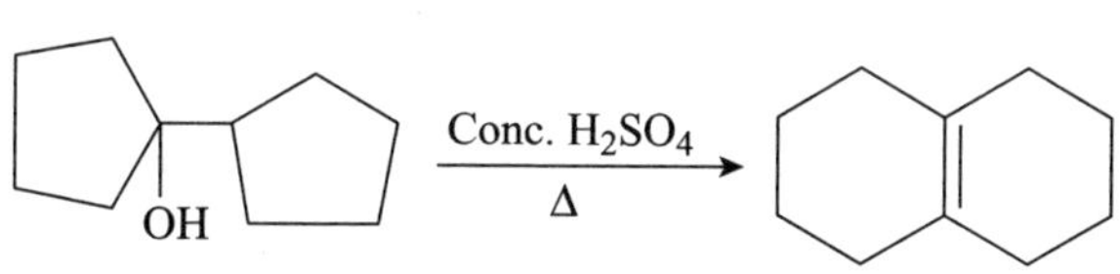

Ans: (5)

WORKBOOK EXERCISE 1

Acidic Strength of Compounds

1. (i) 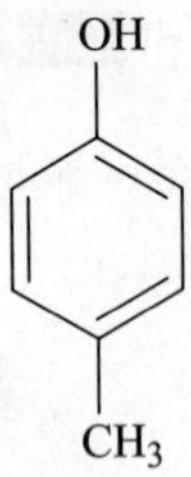 (ii) 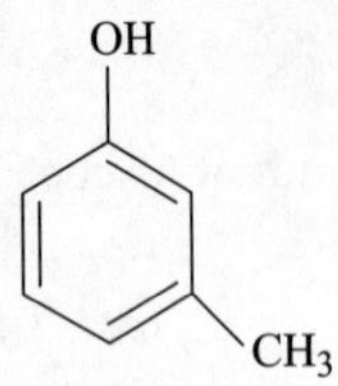 (iii) 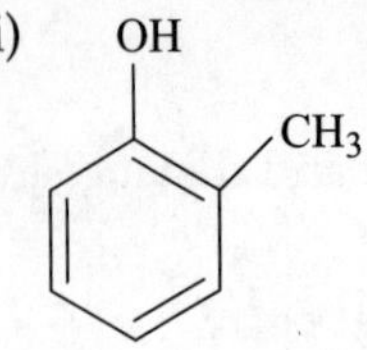

(+H) (+I) (+I, +H)

Ans: ii > i > iii

2. (i) 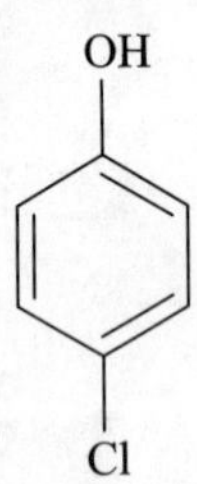 (ii) 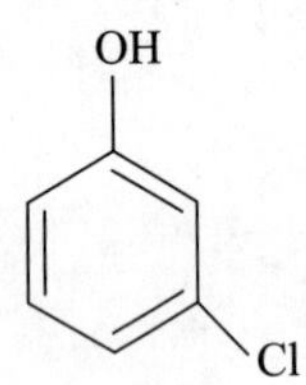 (iii) 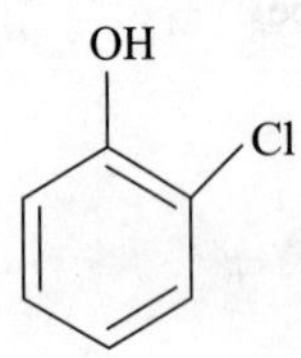

Ans: In case of halogens consider –I effect only iii > ii > i

3. (i) 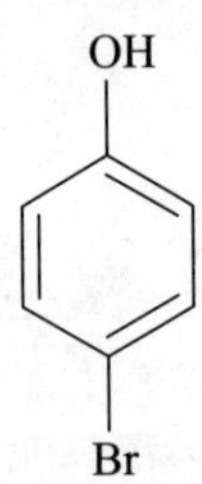 (ii) 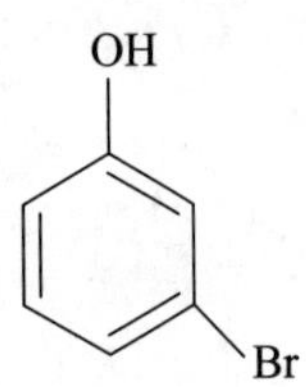 (iii) 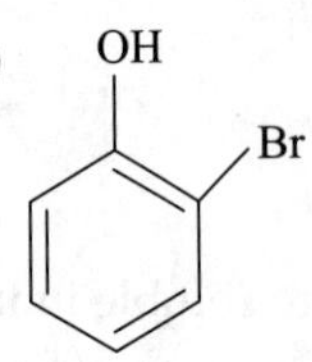

Consider –I effect only

Ans: iii > ii > i

Hint: Distance dependent

4. (i) 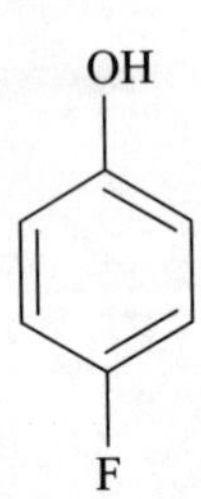 (ii) 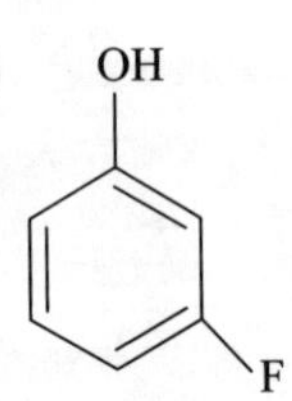 (iii) 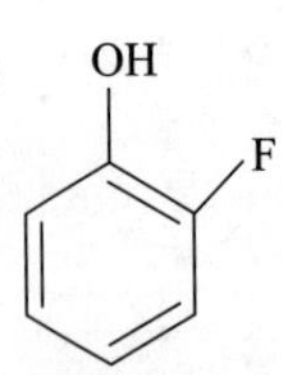

Ans: iii > ii > i

5. (i) 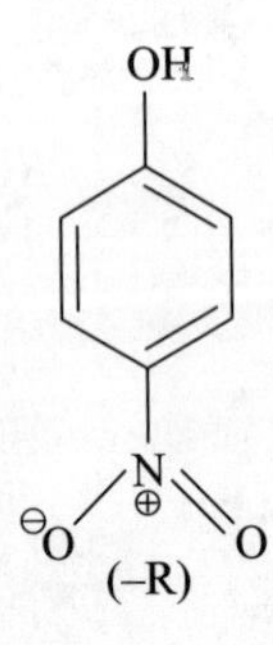 (ii) 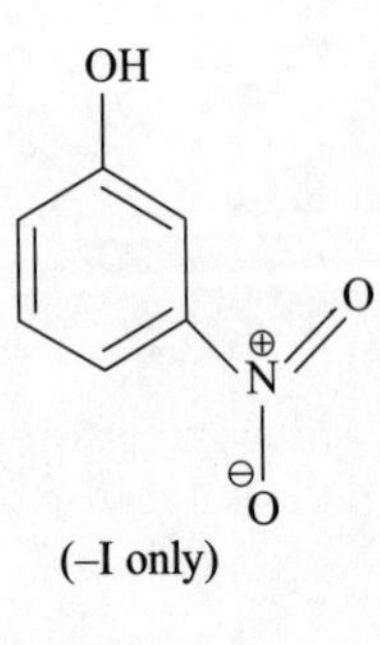 (iii) 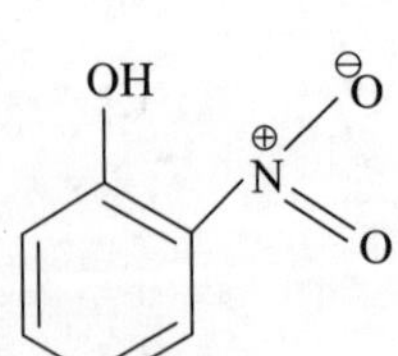

(–R) (–I only) (–R, –I)

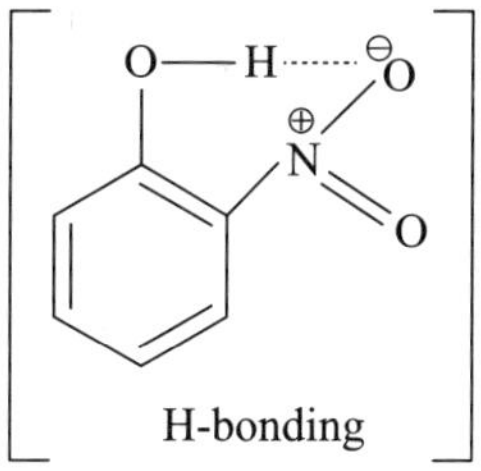

H-bonding

Ans: i > iii > ii

Hint: –R effect always dominates ones –I effect and distance independent (except in case of halogens).

6. (i)

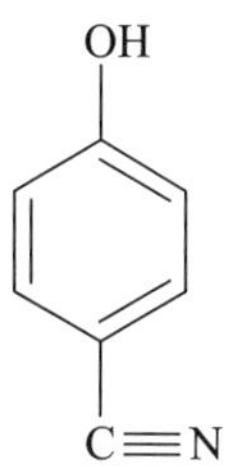

[–R]

(ii)

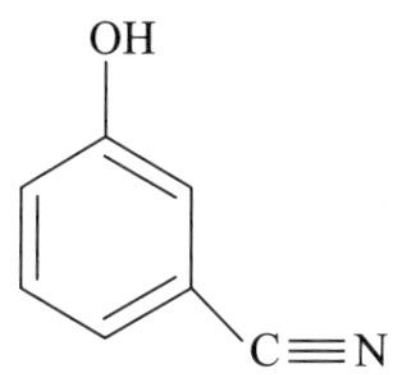

[–I only]

(iii)

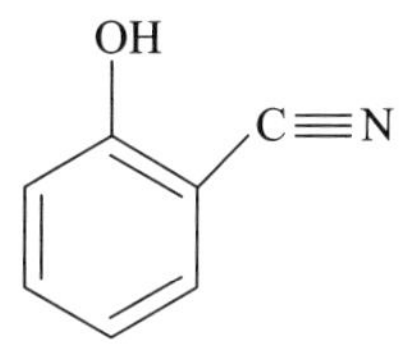

[–I and –R]

Ans: iii > i > ii

7. (i)

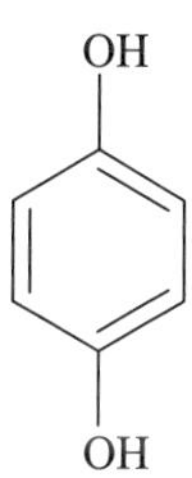

[+R]

(ii)

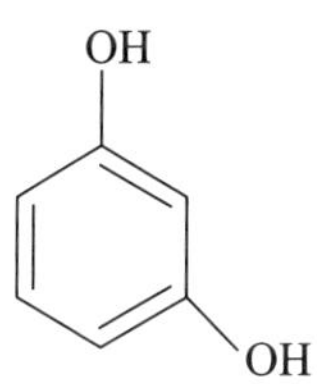

[–I only]

(iii)

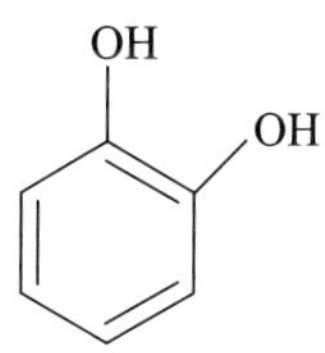

[–I and +R]

(iv)

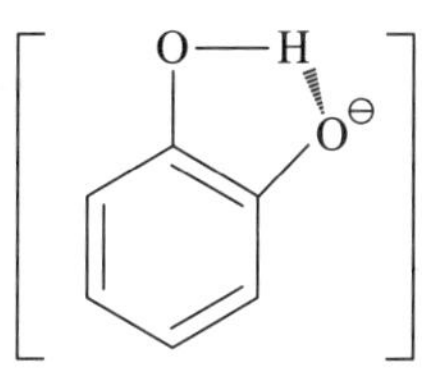

H-bonding effective after releasing one hydrogen

Ans: iii > i > ii

8. (i)

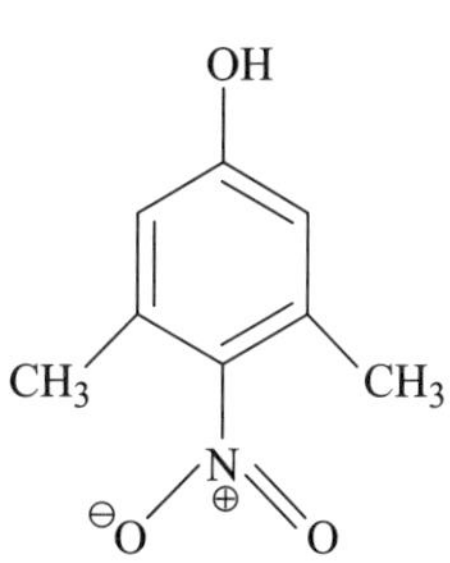

[Only –I effect]

(ii)

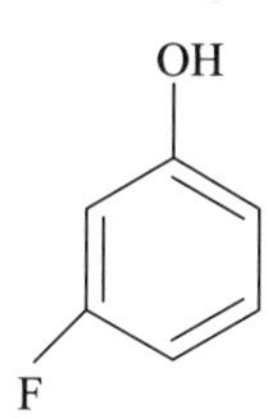

Hint: Due to two bulky groups at ortho position $-NO_2$ (A triangular group) goes out of plane (SIR effect) –C≡ is linear group SIR effect is not applicable.

Ans: iii > i > ii

9. (i) 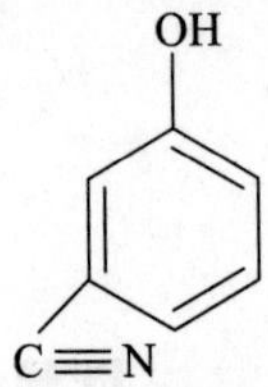 (–I only) (ii) 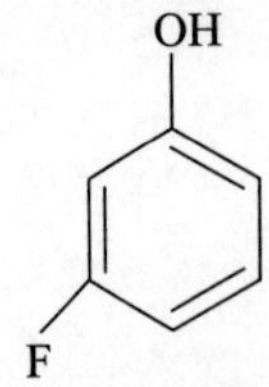 (–I) (iii) 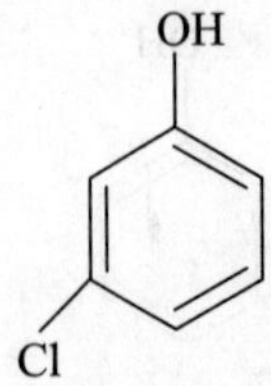 (–I)

Ans: i > ii > iii

12 (i) 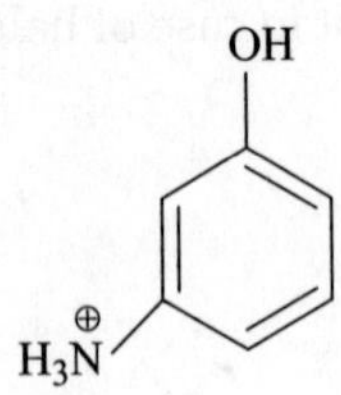 (ii) 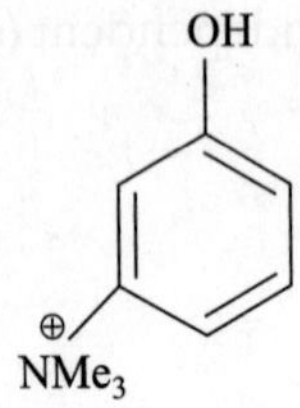 (iii) 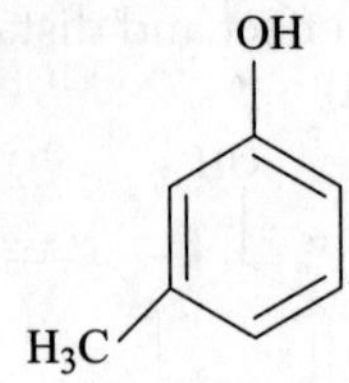 (Only –I effect)

Ans: ii > i > iii

13. (i) 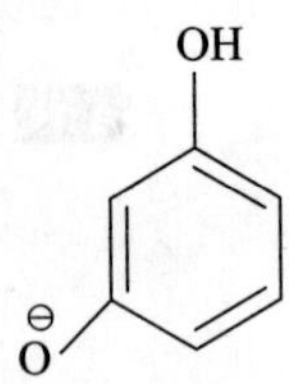 (+I) (ii) 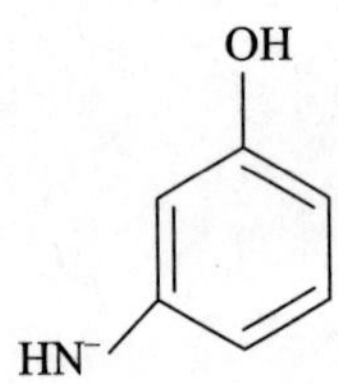 (+I) (iii) 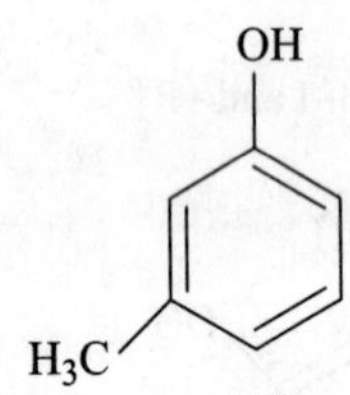 (+I)

Order of +I effect
$-\overset{\ominus}{N}H > \overset{\ominus}{O} > -CH_3$
oxygen is more electronegative so less +I effect

Ans: iii > i > ii

14. (i) 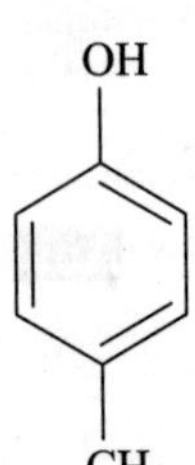
(I)
(+H)

(ii) 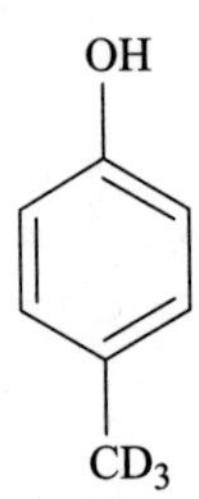
(II)
(+H)

(iii) 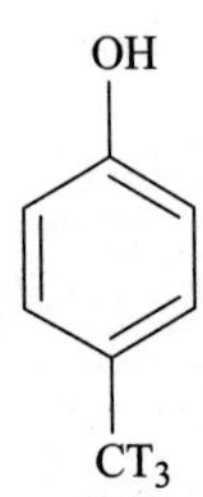
(III)
(+H)

Order of +H effect $-CH_3 > -CD_3 > CT_3$

Ans: iii > ii > i

15. (i) 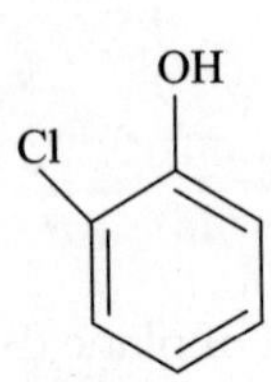 (ii) 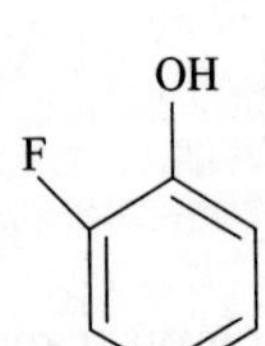 (iii) 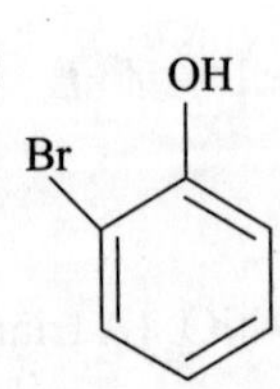

[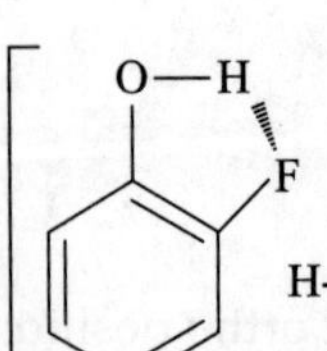 H-bonding]

Ans: i > iii > ii

16. (i) 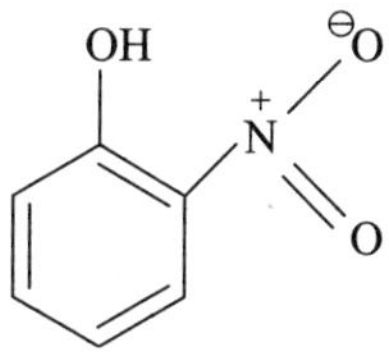 (–R) (ii) 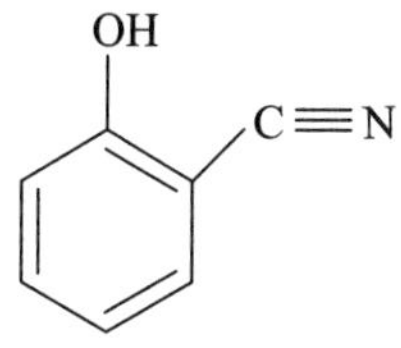 (–R) (iii) 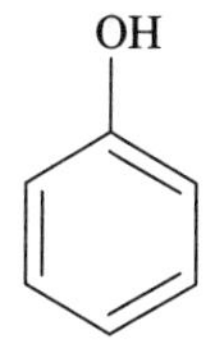

Ans: i > ii > iii

Note: H-bonding concept will come in role when order of –R effect is same.

17. (i) 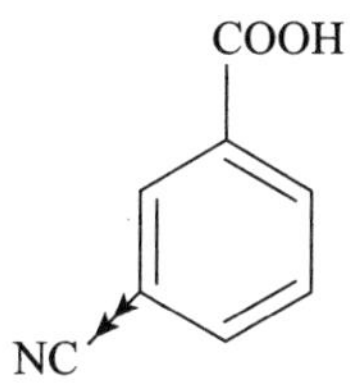 (–I) (ii) 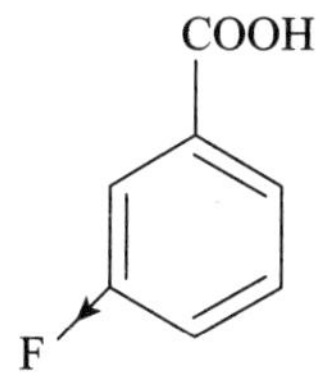 (–I) (iii) 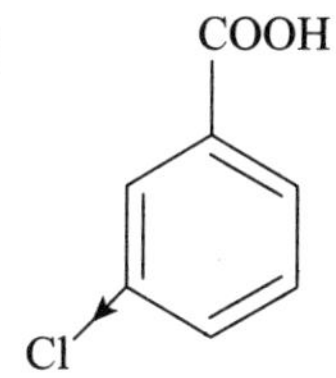 (–I)

Ans: i < ii < iii

18. (i) 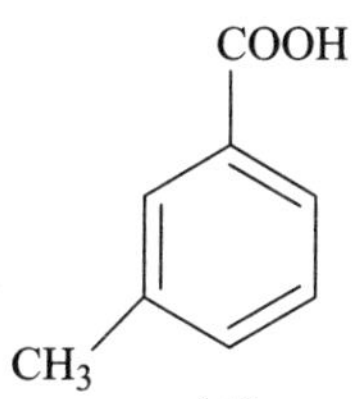 (+I) (ii) 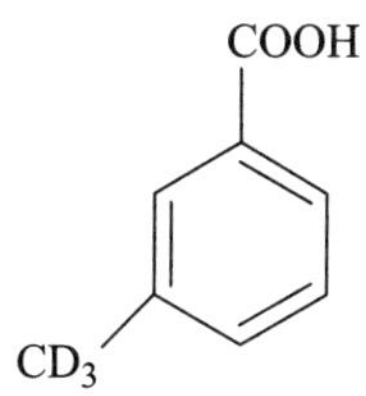 (+I) (iii) 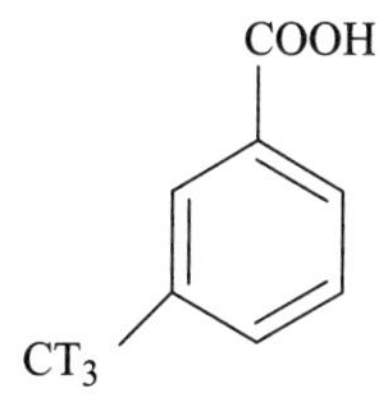 (+I)

Ans: i > ii > iii

Hint: Order of +F effect $-CT_3 > -CD_3 > -CH_3$

19. (i) 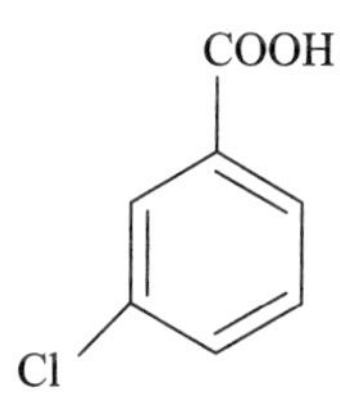 (–I effect) (ii) 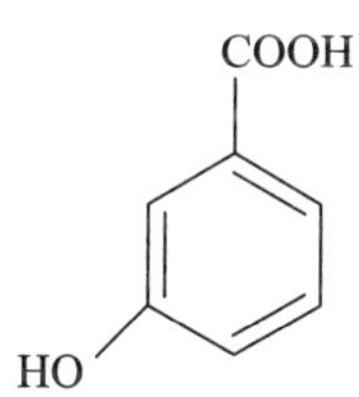 (–I) (iii) 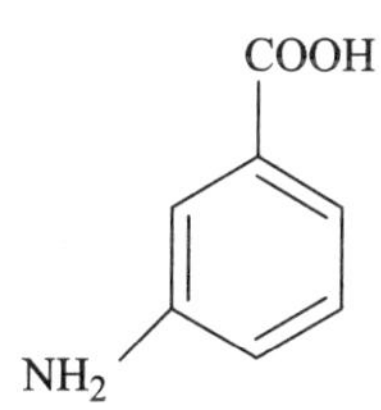 (–I)

Ans: i > ii > iii

Hint: –Cl has more electronegativity than –OH group [it is –OH group not oxygen atom alone].

20. (i) 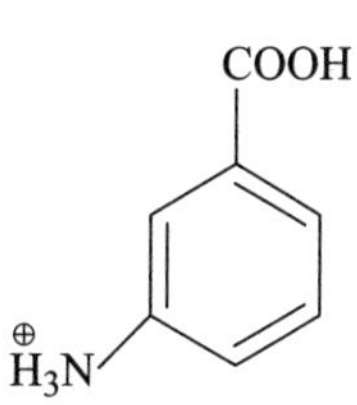 (ii) 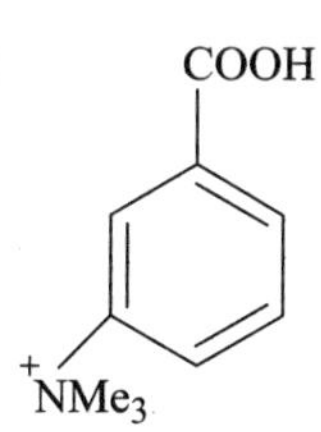 (iii) 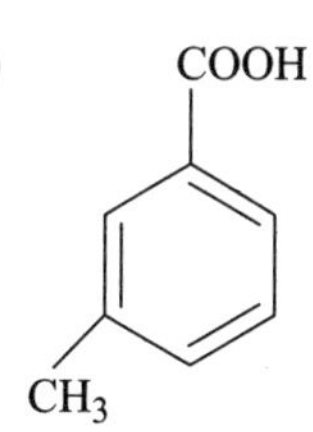

Ans: i > ii > iii

Hint:

21. (i) 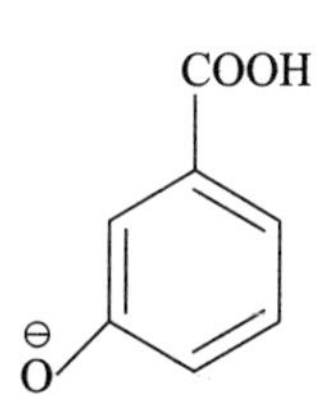 (ii) 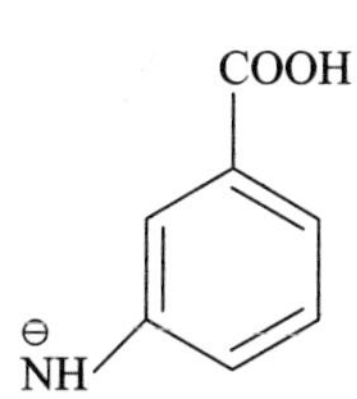 (iii) 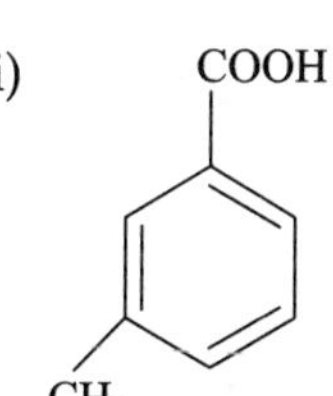

Ans: iii > i > ii

Hint: Orer of +I effect –NH > –O > $-CH_3$

22. (i)

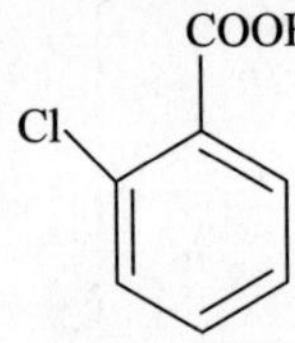

(–I effect)

(ii)

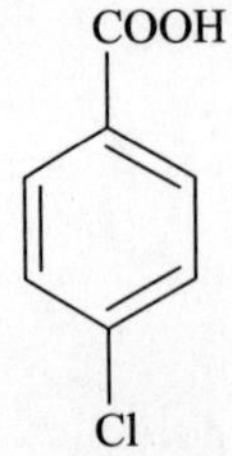

(–I effect)

(iii)

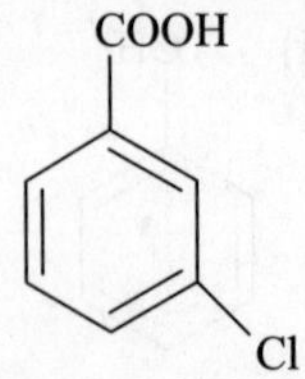

(–I effect)

Ans: i > iii > ii

Hint: –I effect distance dependent

23. (i)

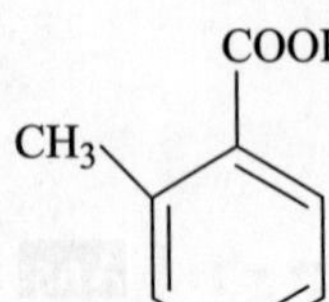

(+H and +I)

(ii)

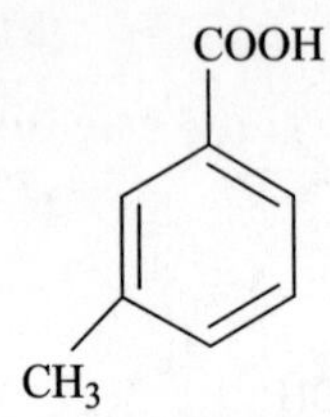

(+I only)

(iii)

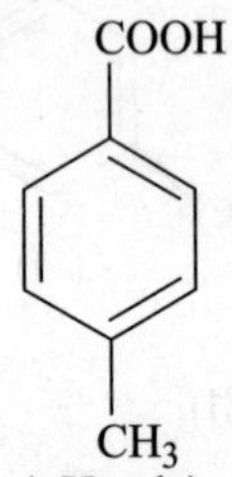

(+H only)

Hint:

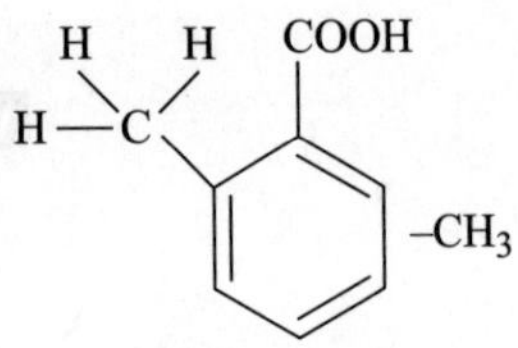

Group present at ortho position. Hence ortho effect is applicable

Ans: i > ii > iii

24. (i)

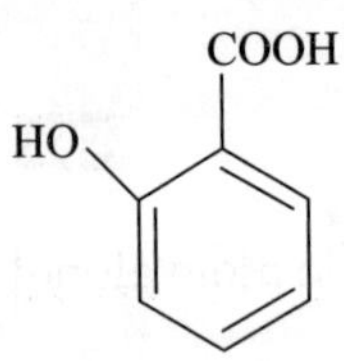

(ii)

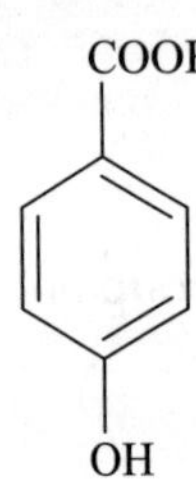

(iii)

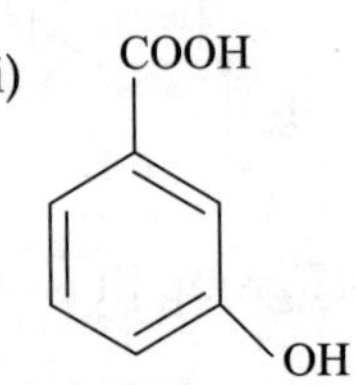

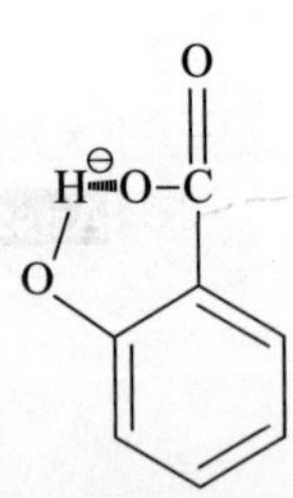

Effective H-bonding after releasing $H^{\oplus}$ from –COOH

Ans: i > iii > ii

25. (i)

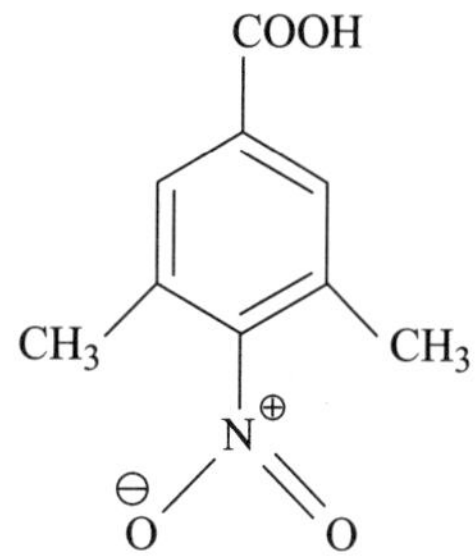

(ii)

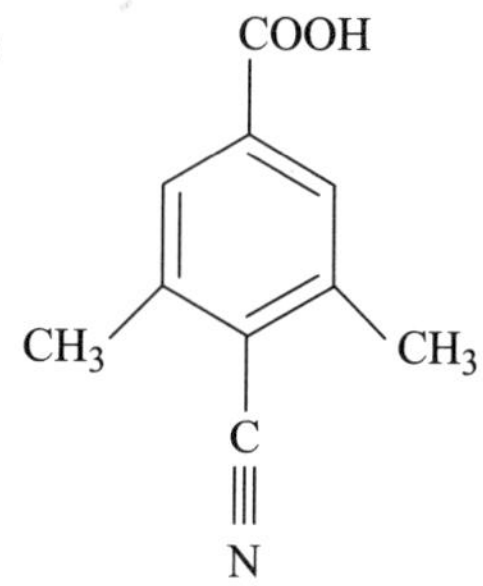

Hint: Same as Question Number 8.

Ans: ii > i

26. (i)

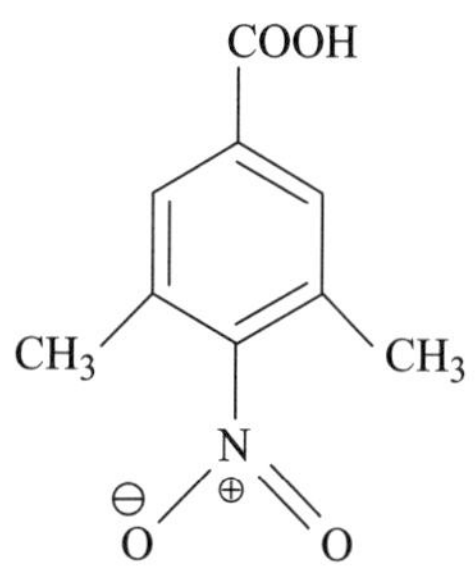

(ii)

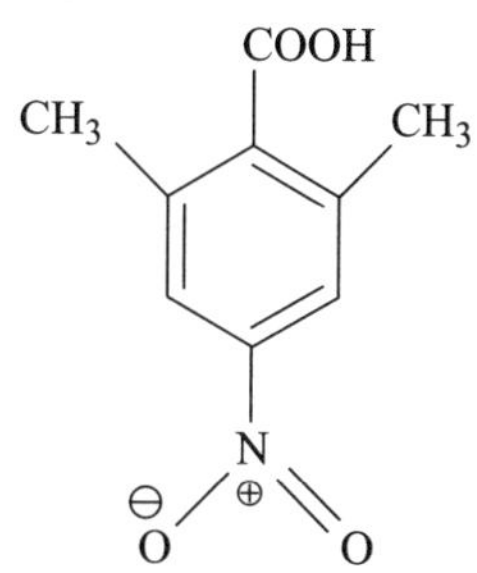

In (i) –NO2 group goes out of plane but in (ii) –COOH group goes out of plane. In both the cases only –I effect of –NO_2 group is acting but in (ii) there is no +R effect of benzene ring whereas in (i) it is present.

Ans: ii > i

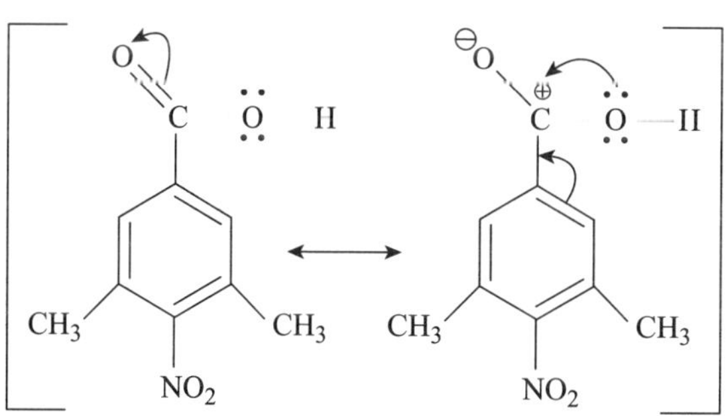

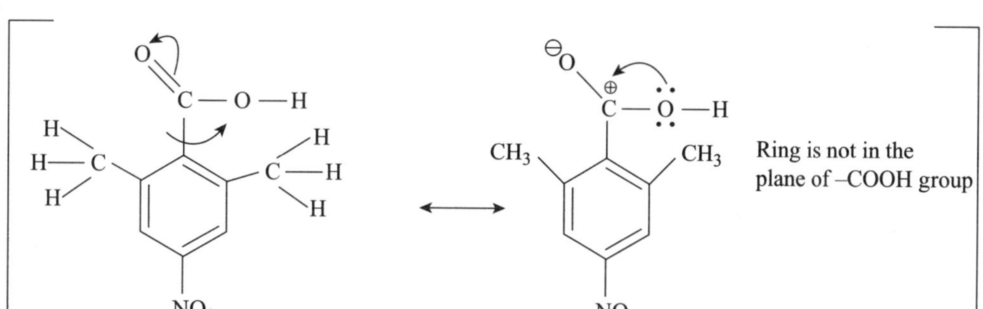

27. (i)

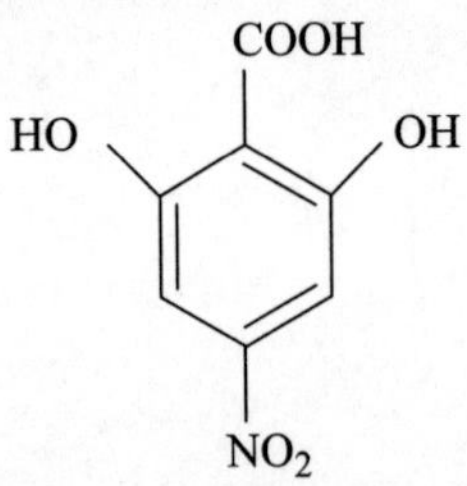

[+R of two –OH and –R of –NO2]

(ii)

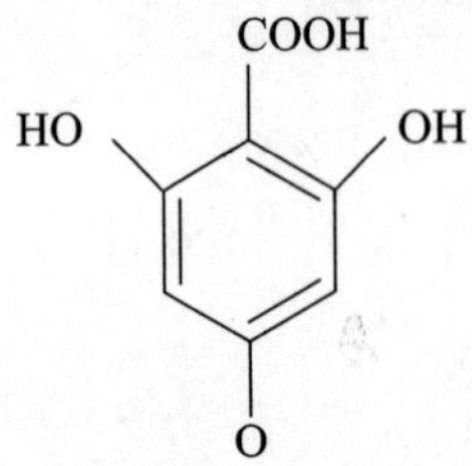

[Only +R of two –OH]

Ans: i > ii

28. (i)

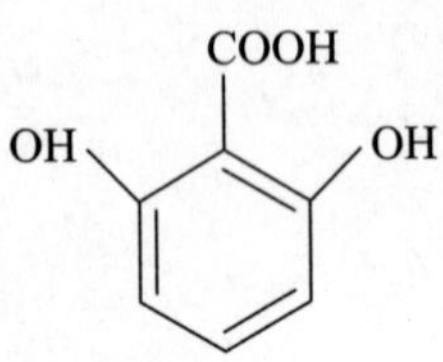

(ii)

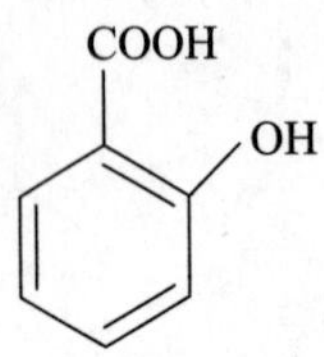

Hint:

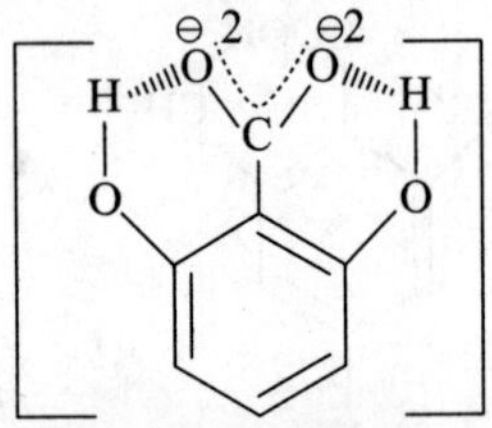

More stablisation through H-bonding

Ans: i > ii

29. (i)

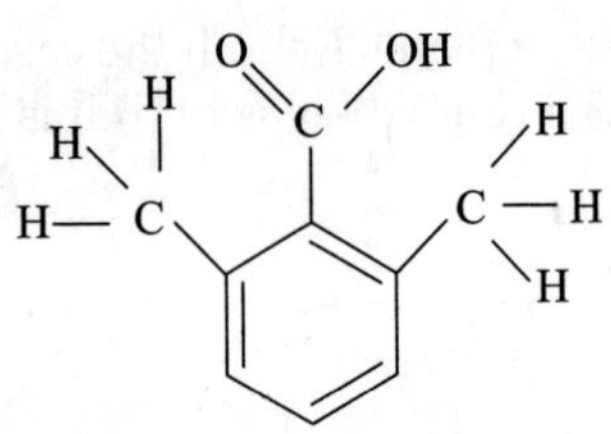

(ii)

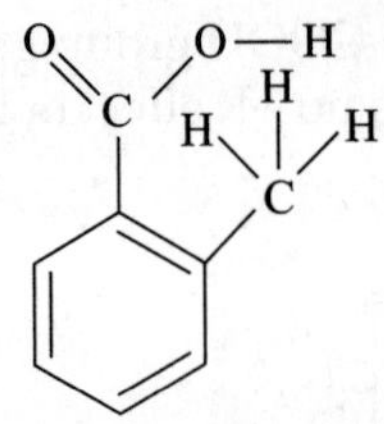

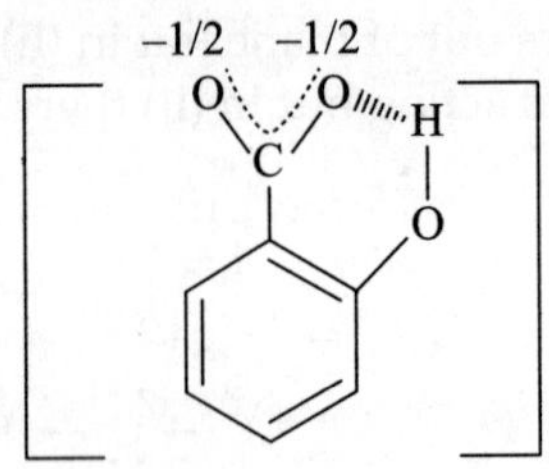

Hint: Two $-CH_3$ group at ortho position will repel –COOH group more effective from the plane of benzene ring.

Ans: i > ii

30.

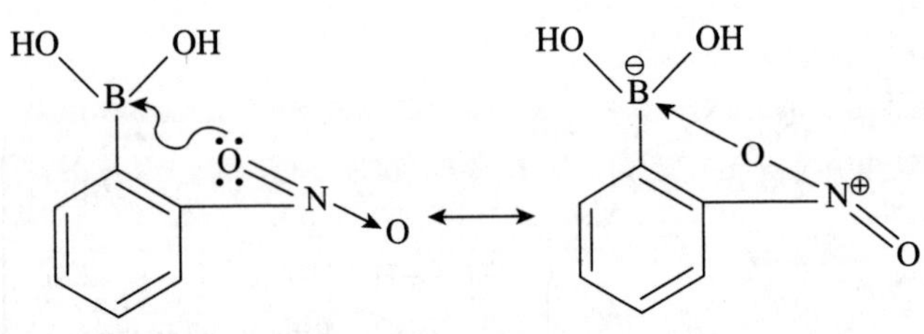

31. (i)

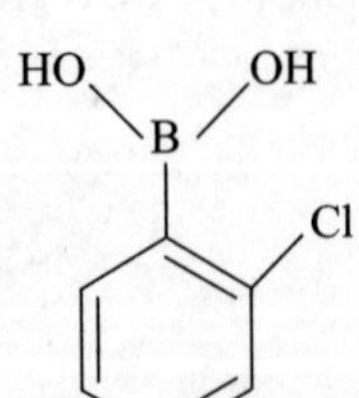

(ii)

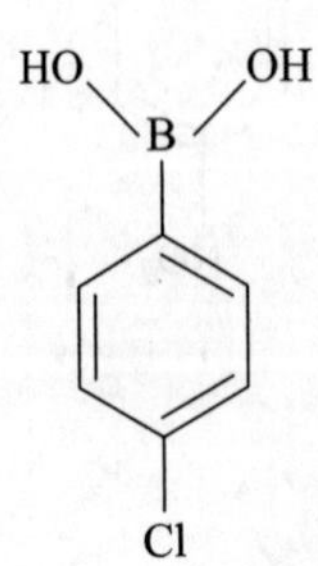

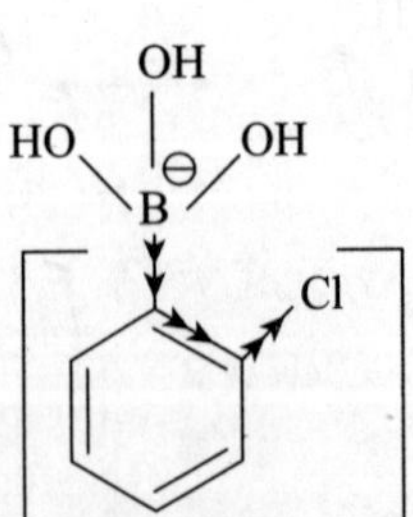

–I effect stabilises conjugate base

Ans: i > ii

32. (i)

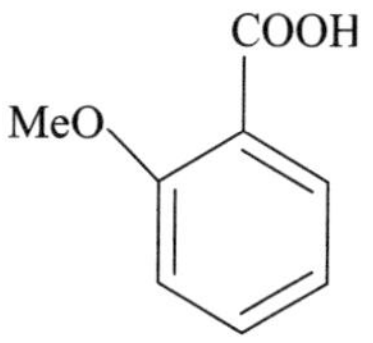

(+R and –I)

(ii)

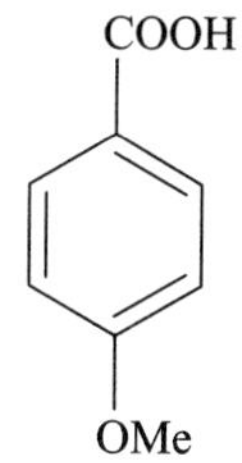

(Only +R effect)

[–I effect is distance dependent]

Ans: i > ii

33. (i)

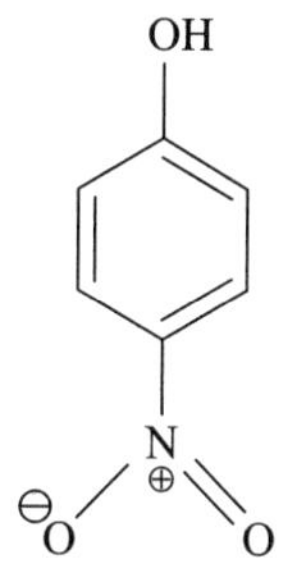

[–R of one $-NO_2$]

(ii)

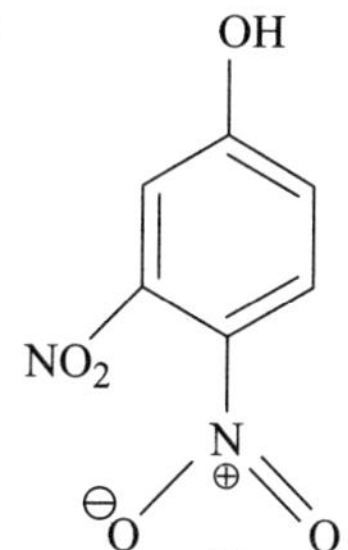

II

[–R of one $-NO_2$ and –I of one $-NO_2$]

(iii)

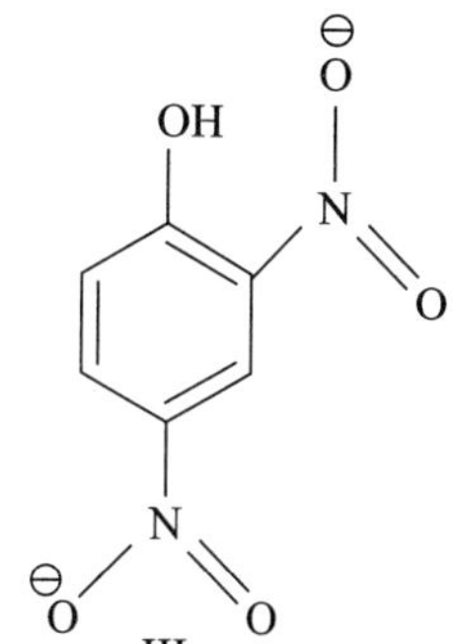

III

[–R of two $-NO_2$ groups]

Ans: iii > ii > i

Note: H-bonding concept will come in role only when –R effect is same.

34. (i)

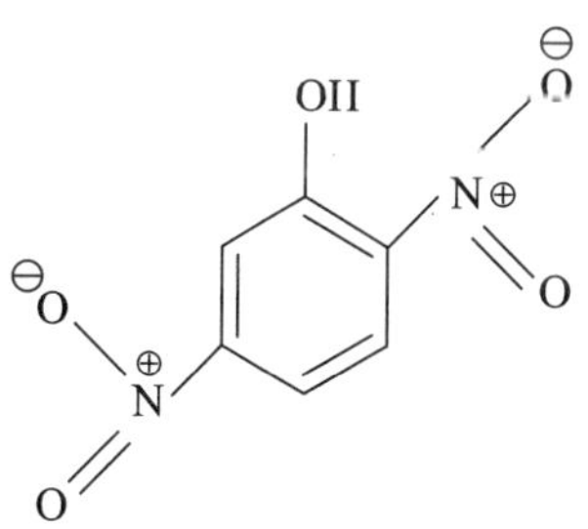

[–R of one $-NO_2$ group –I of one $-NO_2$ group]

(ii)

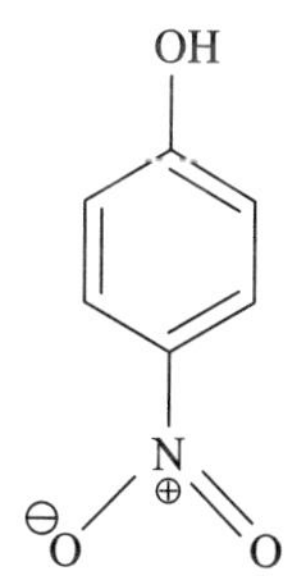

[–R of one $-NO_2$ group]

(iii)

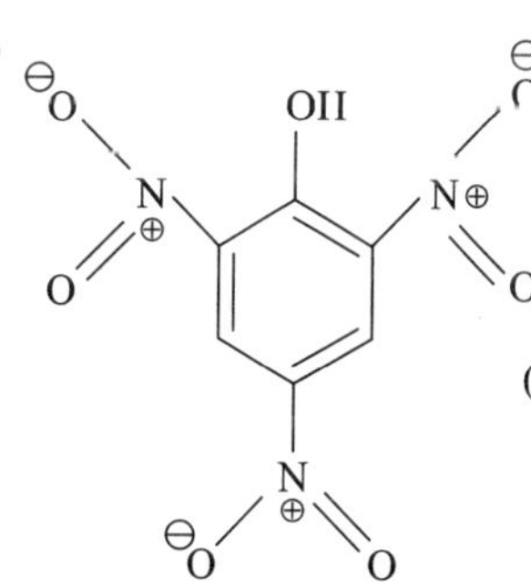

(Picric acid)

[–R of 3 $-NO_2$ group]

Ans: iii > i > ii

35. (i)

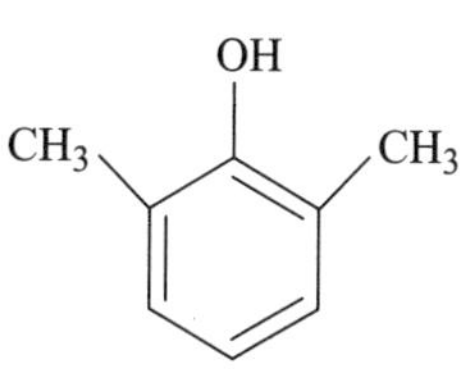

[+H effect of two $-CH_3$]

(ii)

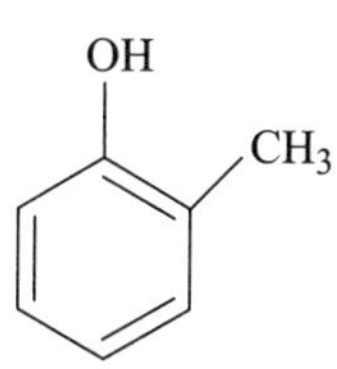

[+H effect of one $-CH_3$]

(iii)

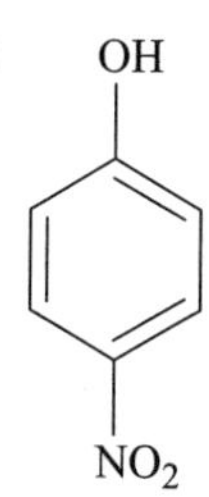

[–R effect of $-NO_2$]

Ans: i > ii > iii

Note: –OH group is a linear group so no –SIR [SIR- steric hindrance in resonance].

36. (i) 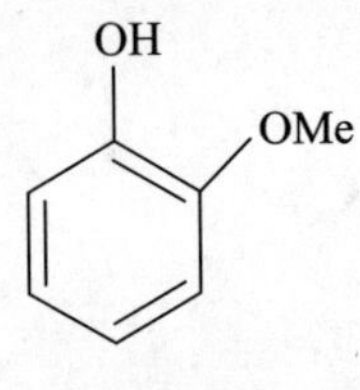 (+R and –I)

(ii)  (+R)

(iii) 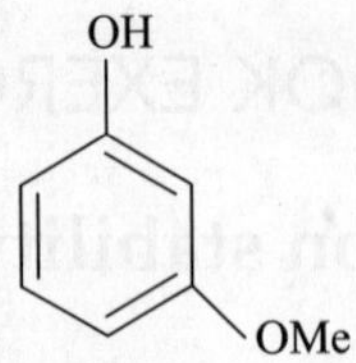 (–I only)

Ans: iii > i > ii

37. (i) 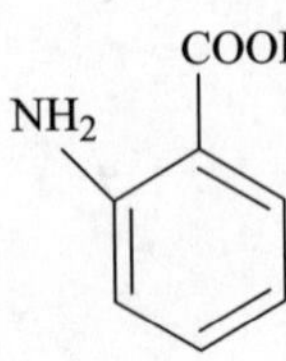 1

(ii) 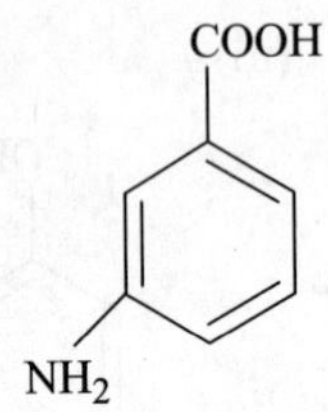 2

(iii) 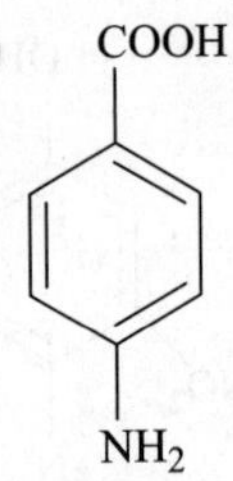 3

Ans: ii > iii > i

In (i) ortho is present still it is less acidic because as soon as $H^{\oplus}$ is released from —COOH, it attack on $-NH_2$ group near by to it and form.

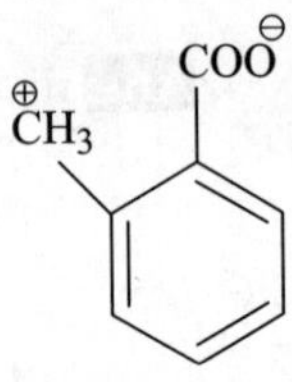

but it does not happen in case of meta and para substituted.

38. (i) 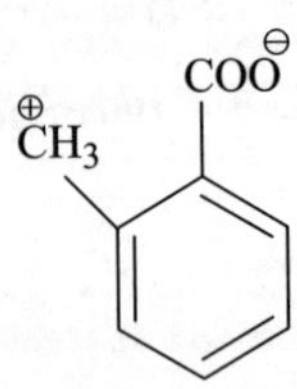

(ii) 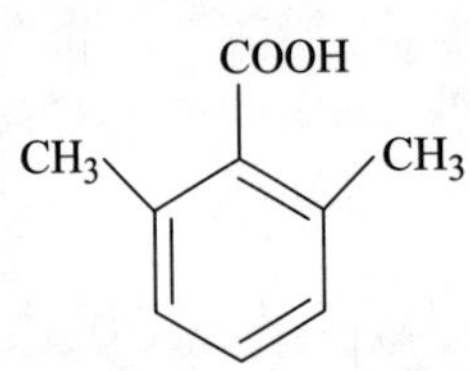

(iii) 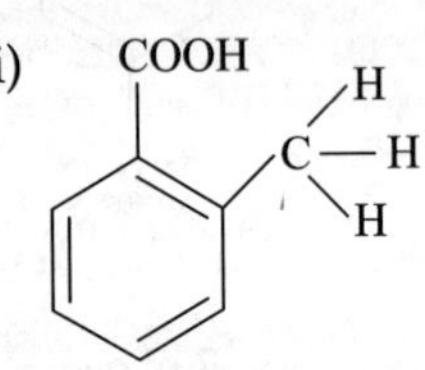

Hint: 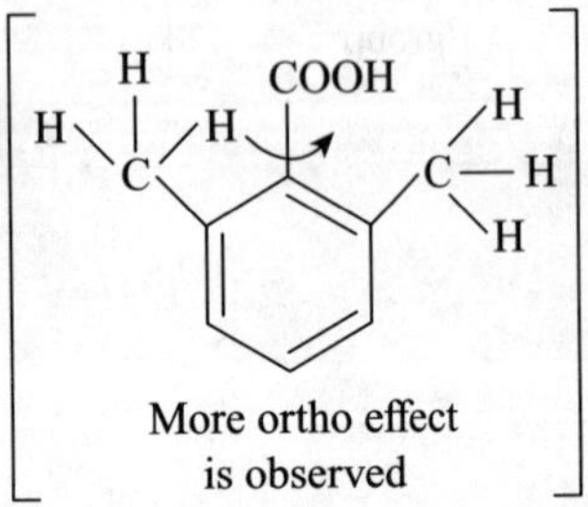

More ortho effect is observed

Ans: ii > iii > i

WORKBOOK EXERCISE 2

Carbocation stability order

1.

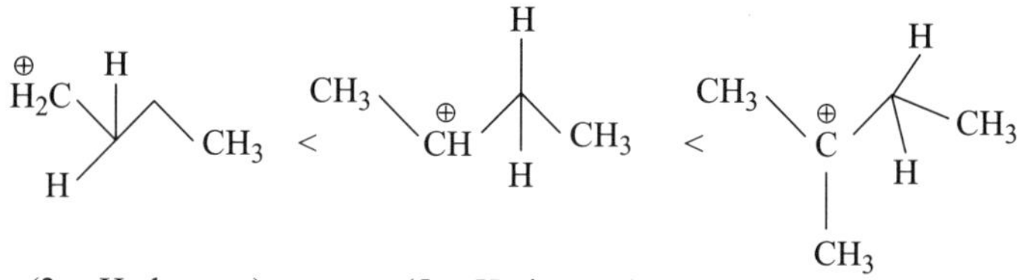

(2 α-Hydrogens) (5 α-Hydrogens) (8 α-Hydrogens)

Hint: Stability of carbocation α number of α-Hydrogens.

2. Calculate the number of α-Hydrogens.

3. (i)

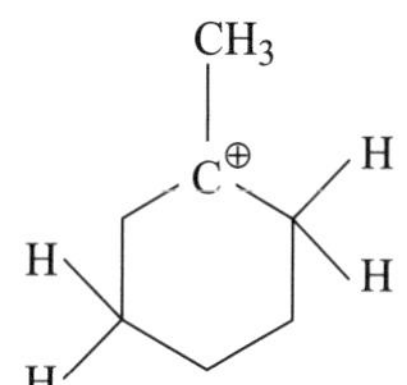

1

(2 α-Hydrogens)

(ii)

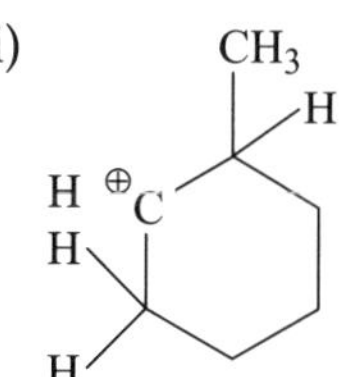

2

(3 α-Hydrogens)

(iii)

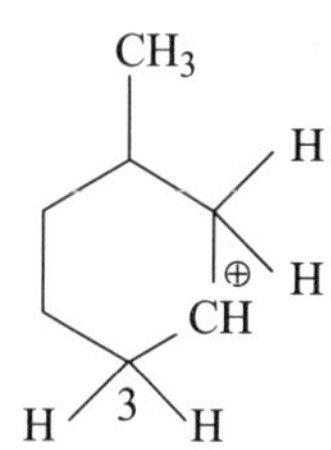

(4 α-Hydrogens)

Order of stability i > iii > ii

4. Find out the number of α-Hydrogens.

6. (i)

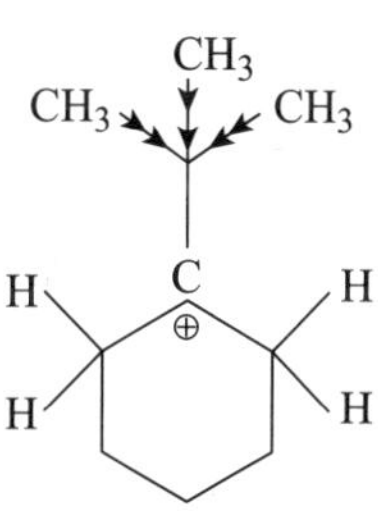

(ii)

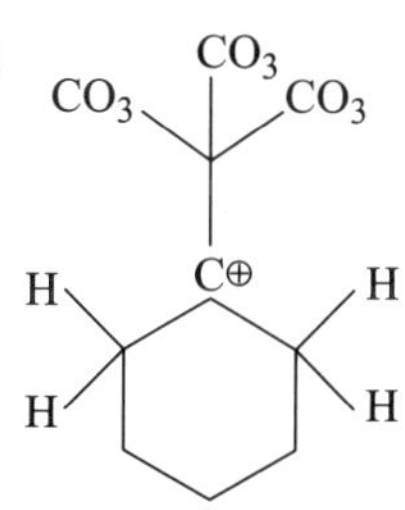

2

(iii)

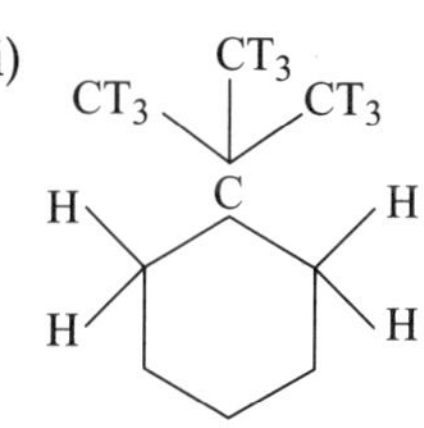

3

Each structures have equal number of α–Hydrogens now stability is decided on the basis of I effect

$-CT_3 > -CD_3 > -CH_3$ **Ans:** iii > ii > i

7. (i)

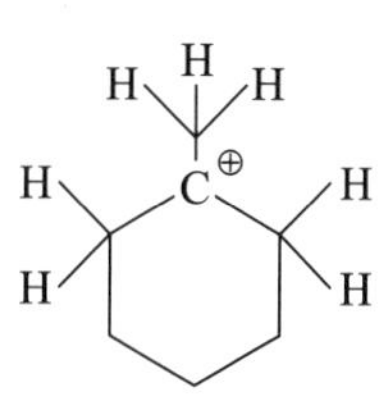

1

(ii)

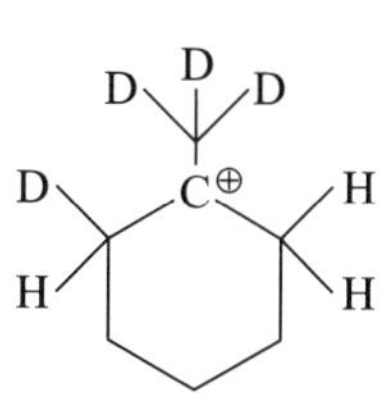

2

(iii)

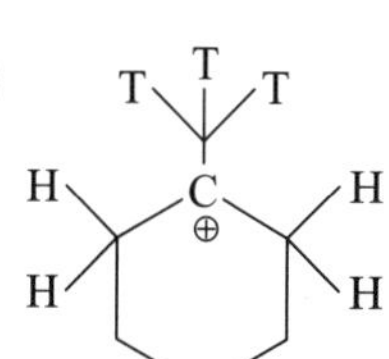

3

Each structure has 4 α-Hydrogens but different number of –D and –T [Order of hyperconjugation effect].

Note: It is reverse of +I effect order.

8. (i)

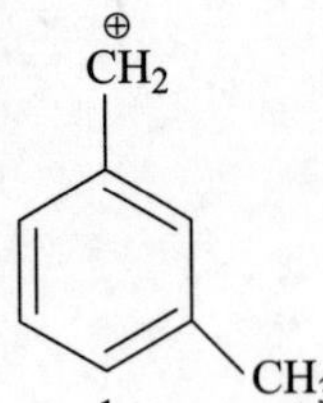

1

(ii)

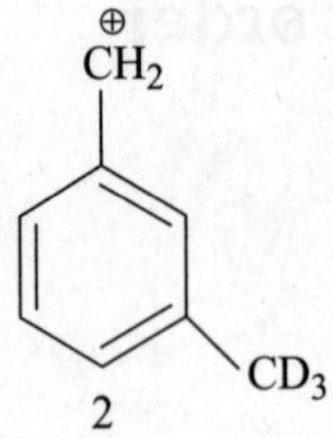

2

(iii)

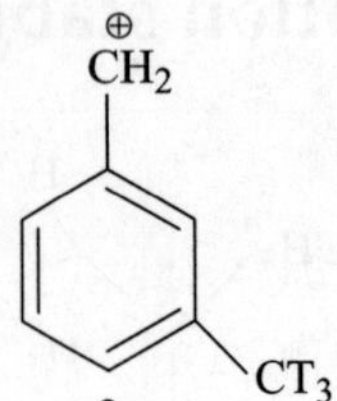

3

At meta position only I effect is applicable (operate) order of +I effect

$-CT_3 > -CD_3 > CH_3$ **Ans:** iii > ii > i

9. (i)

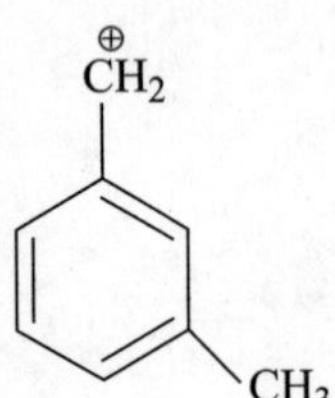

1

(ii)

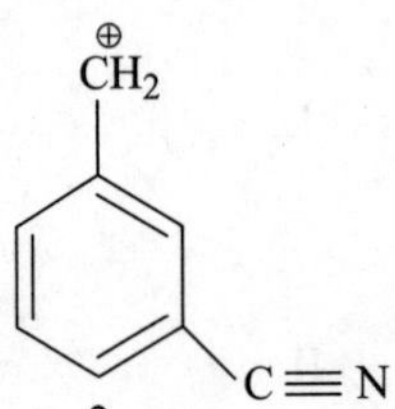

2

(iii)

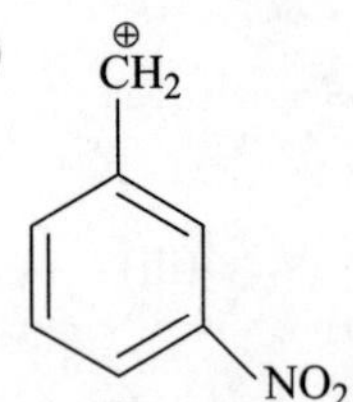

3

–I effect showing groups decreases the stability of carbocation. **Ans:** i > ii > iii

10 (i)

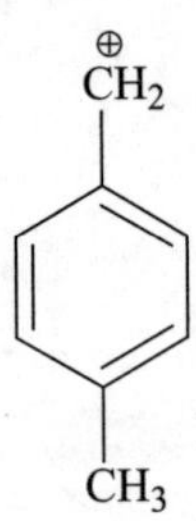

1

(ii)

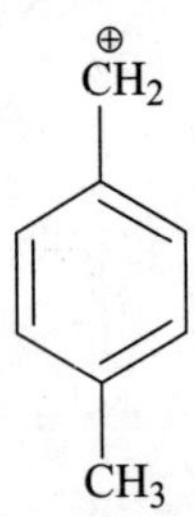

2

(iii)

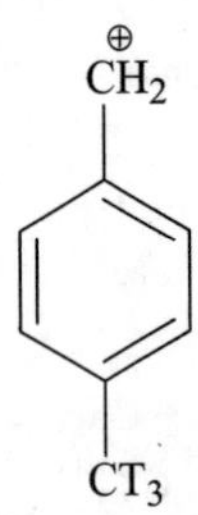

3

At para position +H effect is operating order of +H effect $-CH_3 > -CD_3 > -CT_3$ **Ans:** i > ii > iii

12. (i)

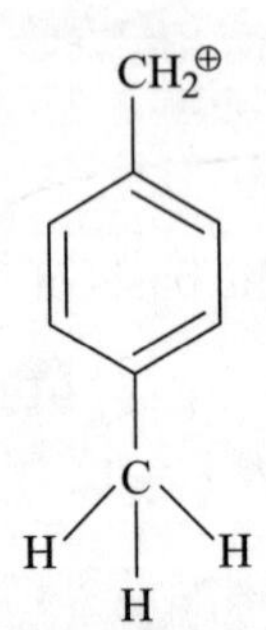

(3 α-Hydrogens)

(ii)

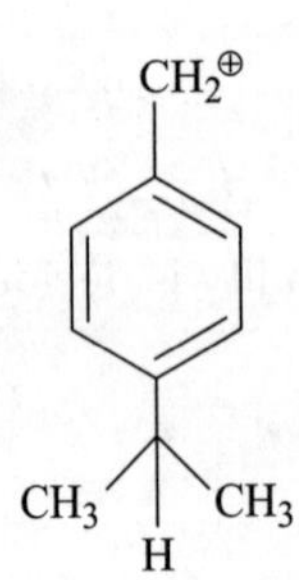

(1 α-Hydrogens)

(iii)

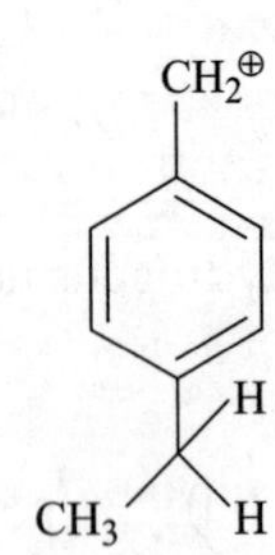

(2 α-Hydrogens)

Stability order i > iii > ii

13. Find out the number of α-Hydrogens and decide the stability.

14.

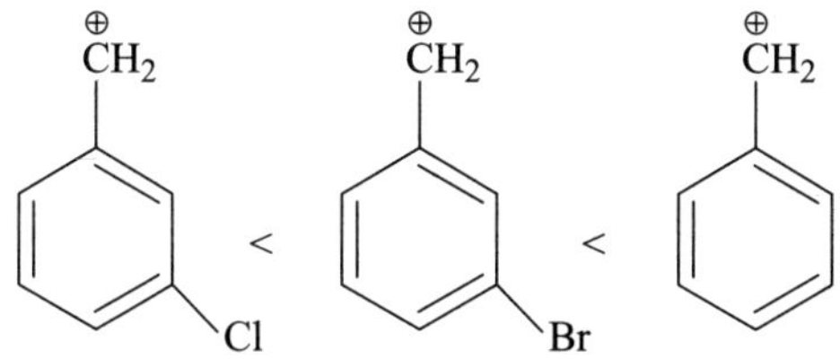

1 (–I effect)   2 (–I effect)   3

15.

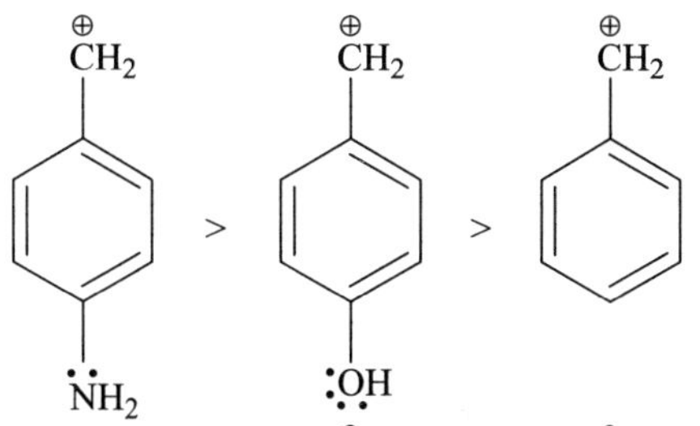

1   2   3

Order of +R effect $-NH_2 > -OH$

17.

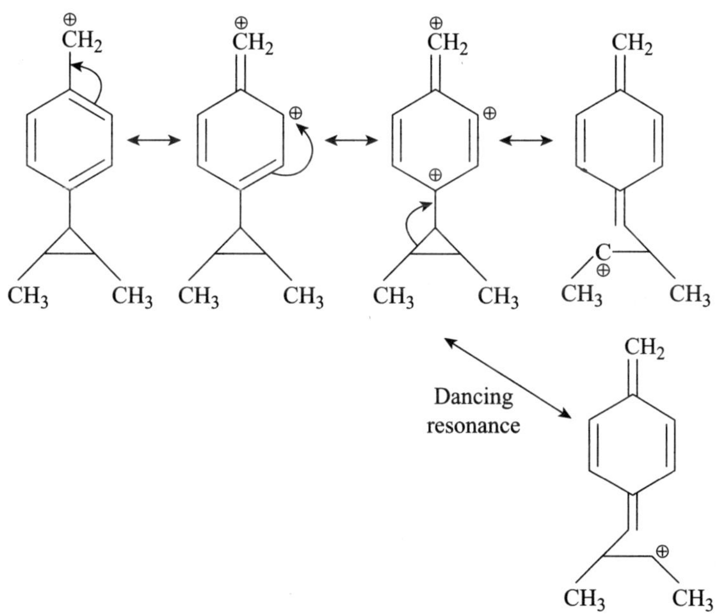

19.

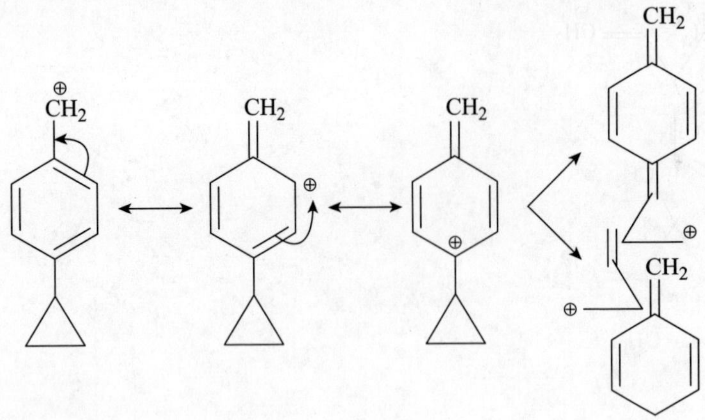

Hint: Check the stability of carbocation formed during the dancing resonance.

20. (i)

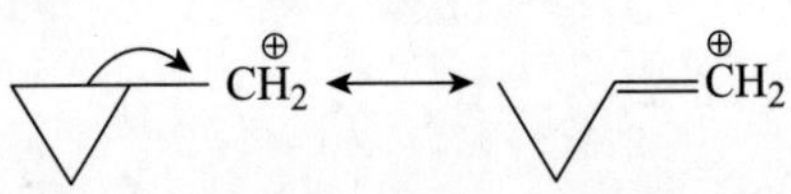

(ii)

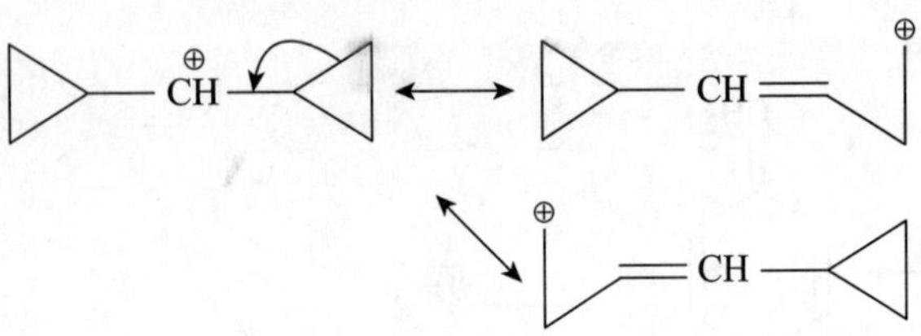

(iii)

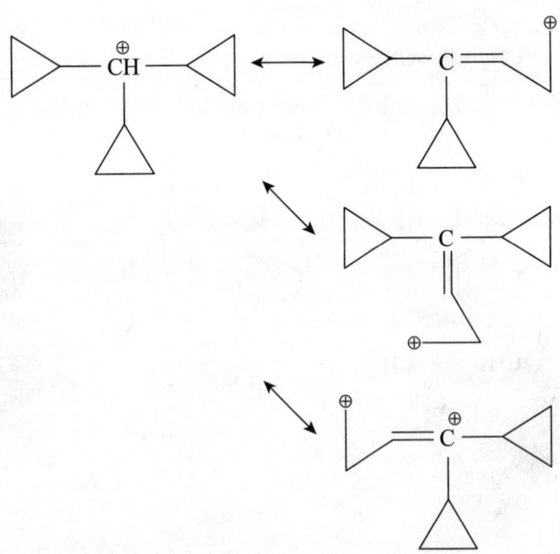

Hence stability order iii > ii > i

21. (i)

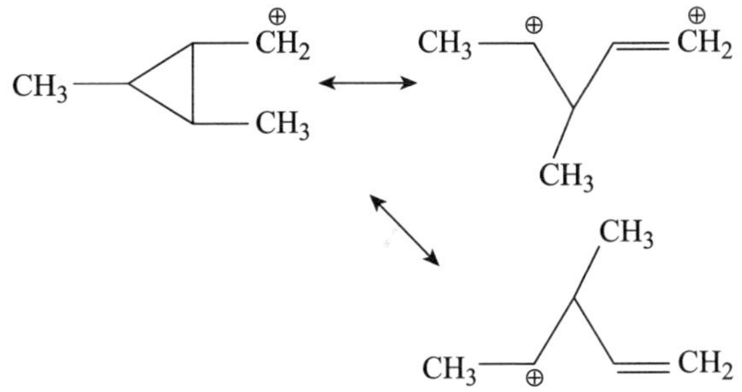

(ii)

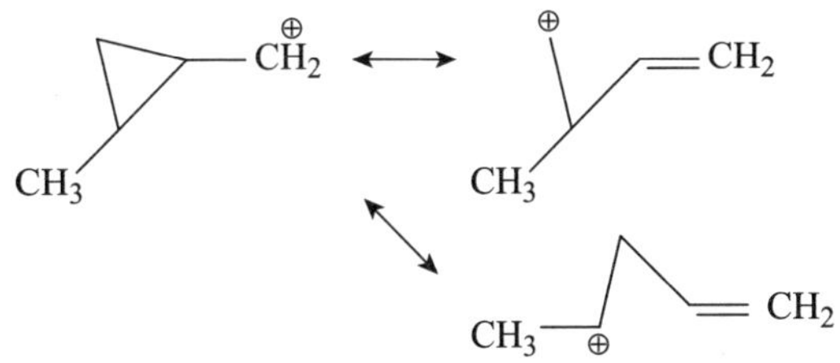

(iii)

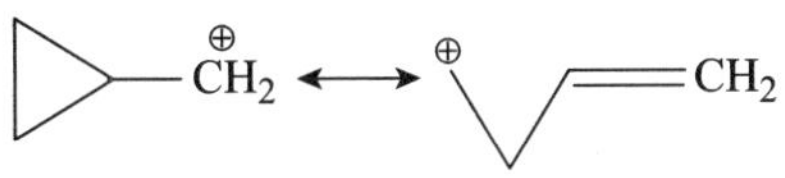

22.

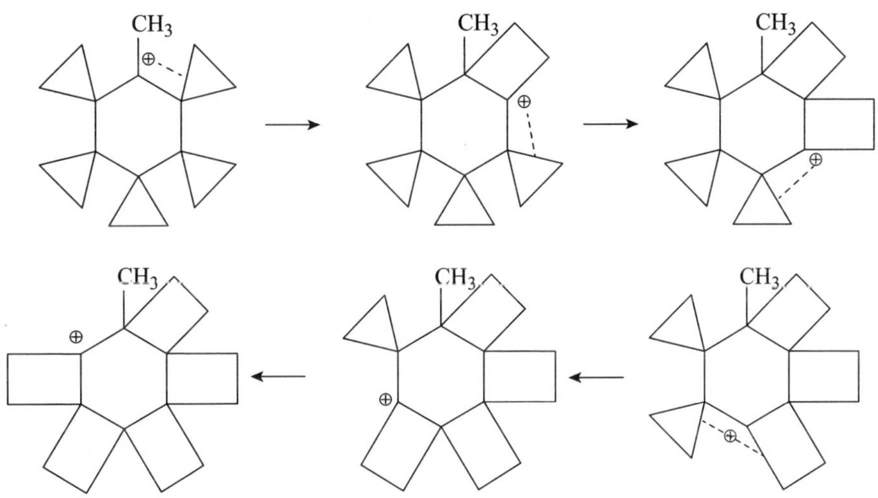

Hence (ii) is more stable than (i).

24. (i)

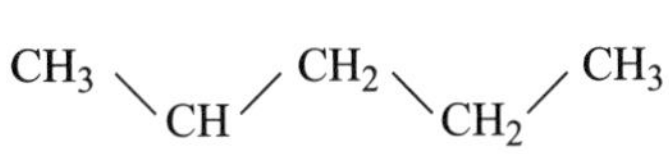

(ii)

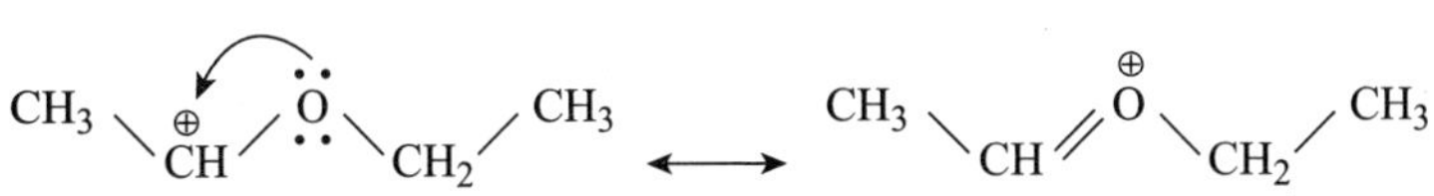

(3 α-Hydrogens)

Species with complete octate are more stable

(iii)

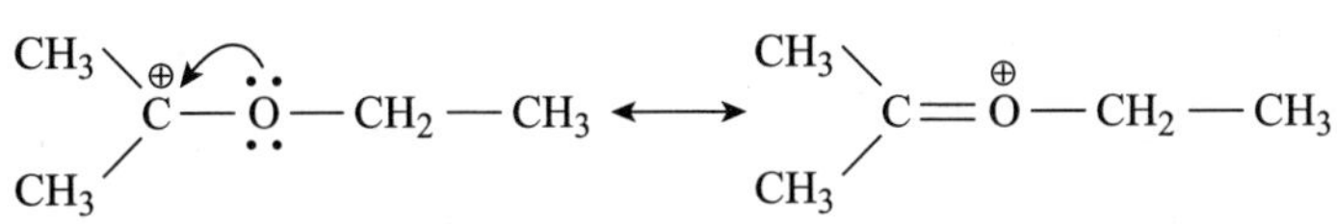

(6 α-Hydrogens)

Stability order iii > ii > i

26. (i) 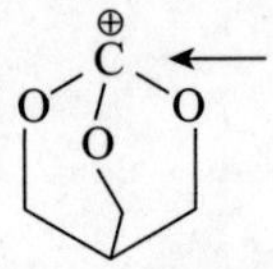 On bridgehead (highly unstable)

(ii) 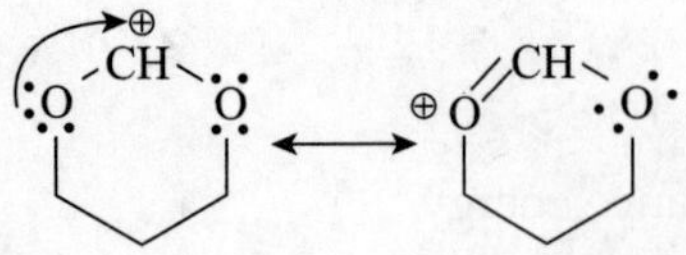 [Resonance through oxygen]

(iii) 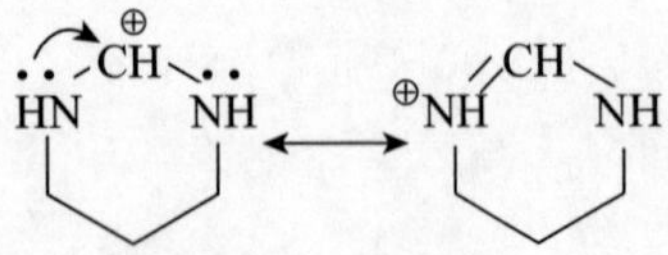 [Lone pair of nitrogen is involved in resonance (less electronegative than oxygen)]

WORKBOOK EXERCISE 5

1. 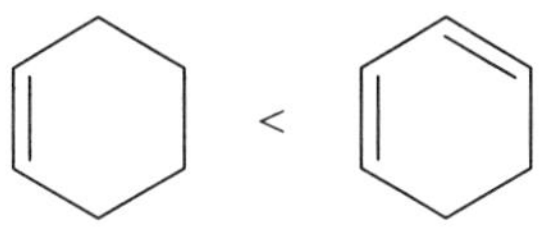

Hint: If π-bond is involved in resonance, it will have more resonance energy.

2. 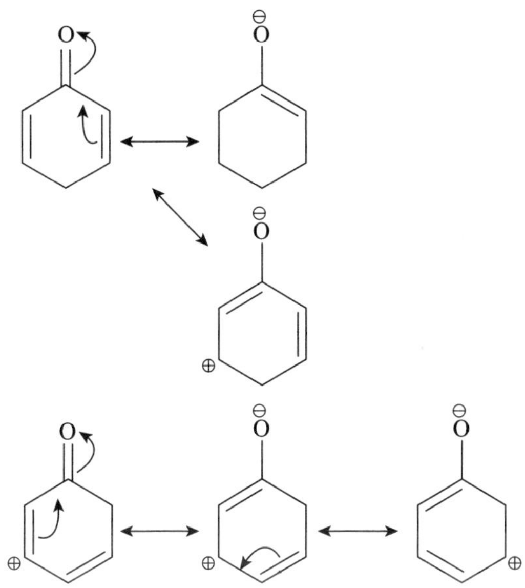

Hint: Extended resonance provides more stability than cross-resonance.

5. 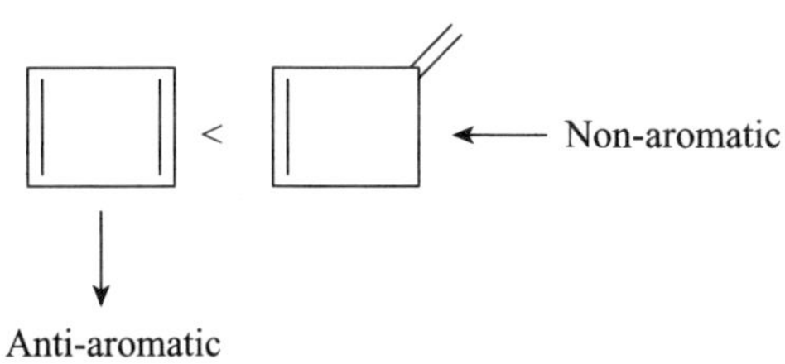

Hint: Resonance energy order ⇒ Aromatic > Non-aromatic > Anti-aromatic.

6. 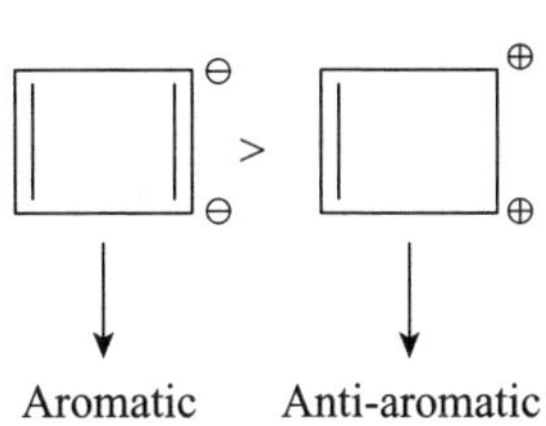

8. 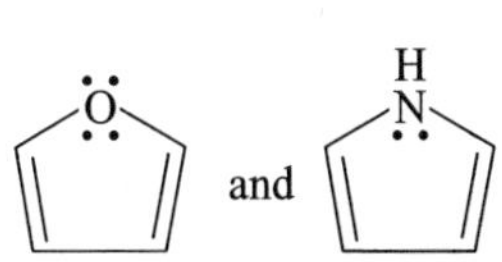

both are aromatic but oxygen is more electronegative than nitrogen hence when lone pair of nitrogen is involved in cyclic resonance provides more stability.

9. 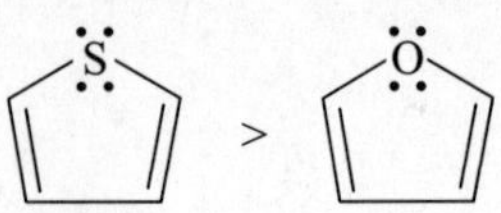

Hint: 'S' have l.p. and '3d' orbital so give more Resonance.

12. 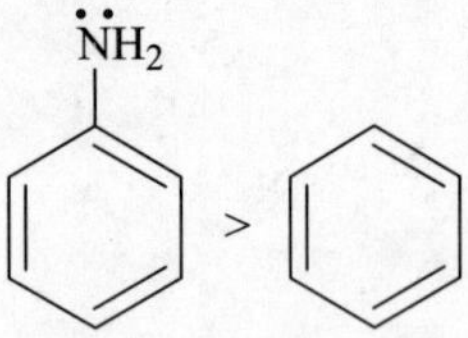

Resonance energy depends upon extent of conjugation.

13. 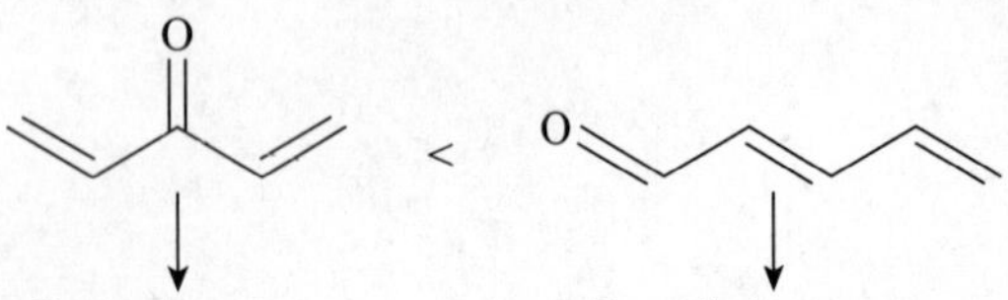

Cross-resonance Extended-resonance

Hint: Extended resonance provides more stability.

17. 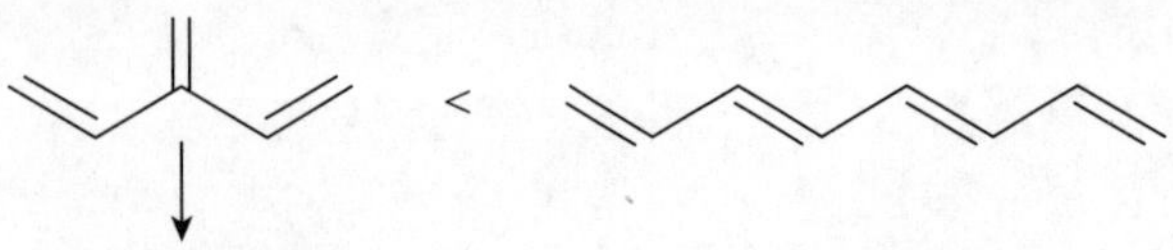

3-π bonds are in cross-conjugation

4-π bonds are in extended-conjugation

19. 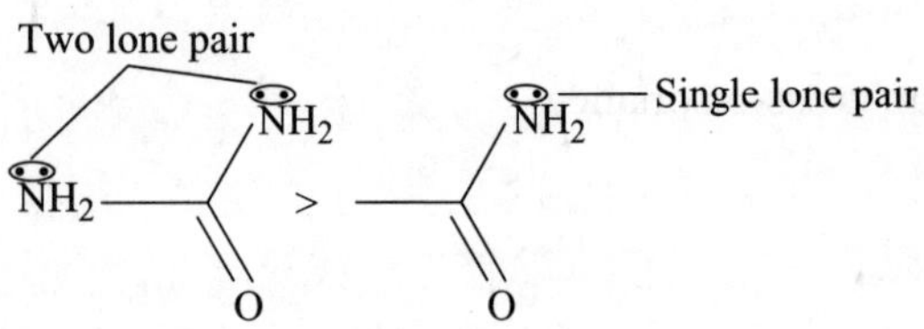

20. 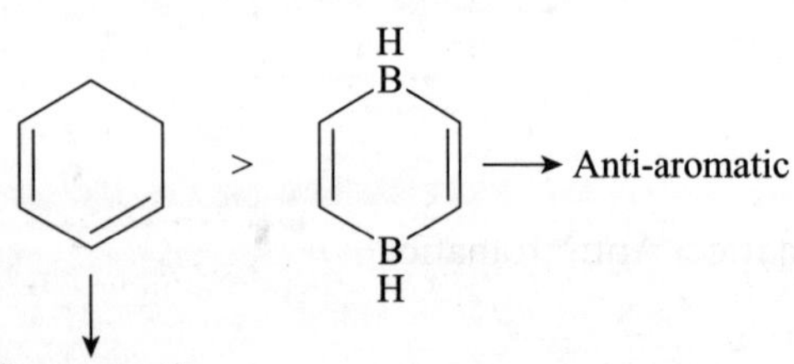

Non-aromatic

21. See explanation of Question Number 8.

22. 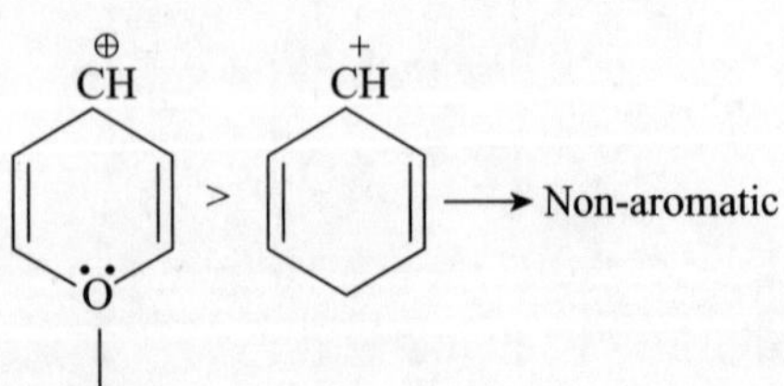

Aromatic

Hint: Resonance energy order: Aromatic > Non-aromatic.

23.

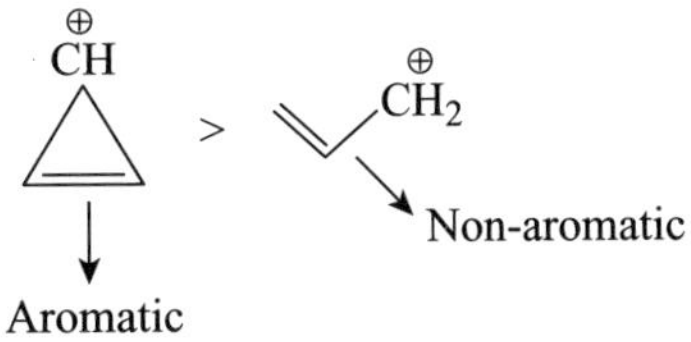

24.

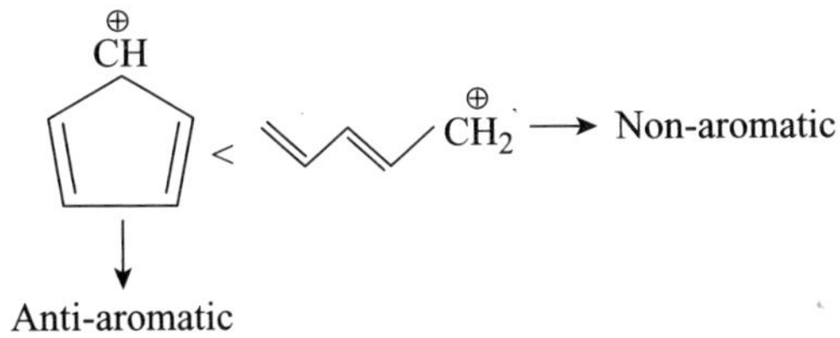

25.

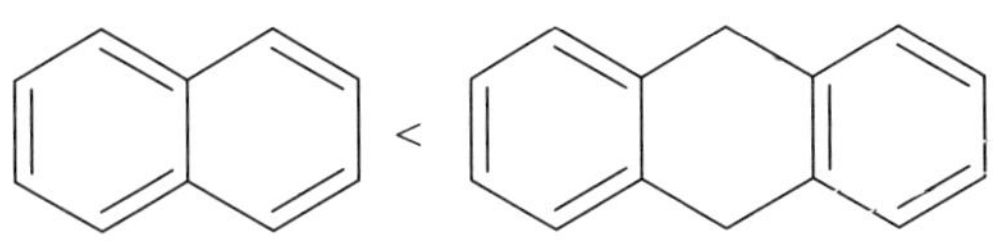

27.

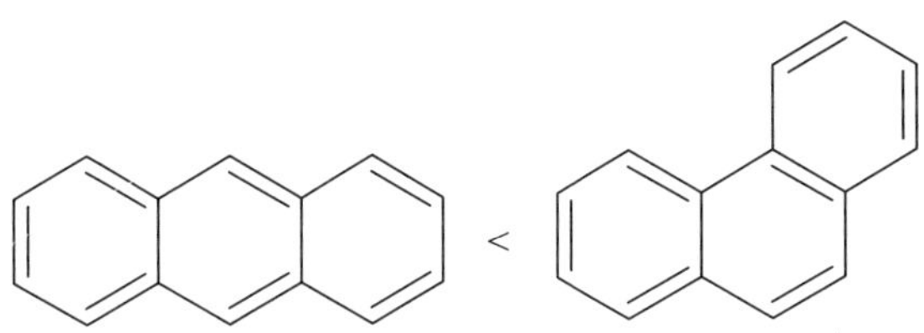

28.

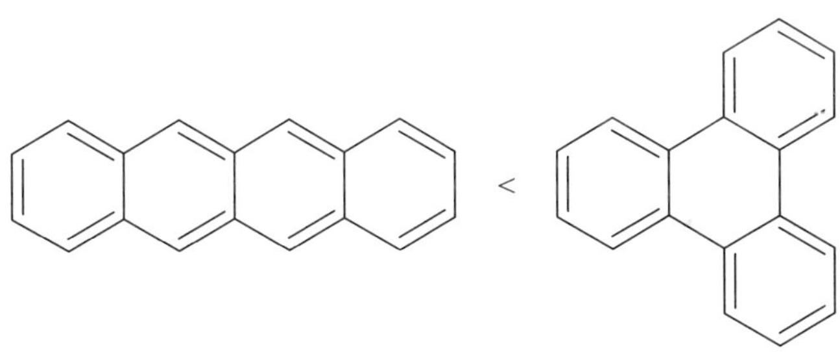

33.

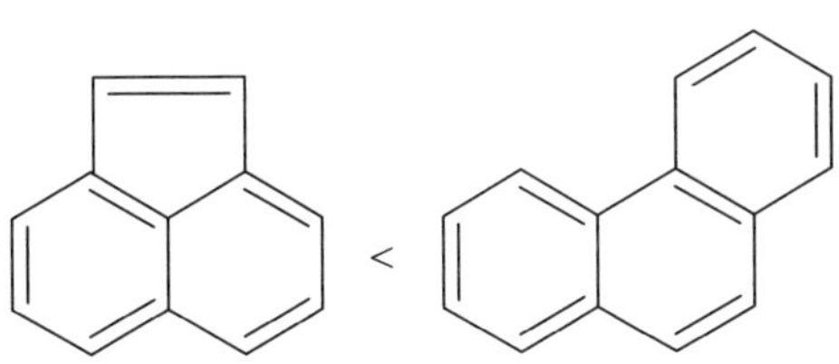

CHAPTER 2

Isomerism

LEVEL 1

1.

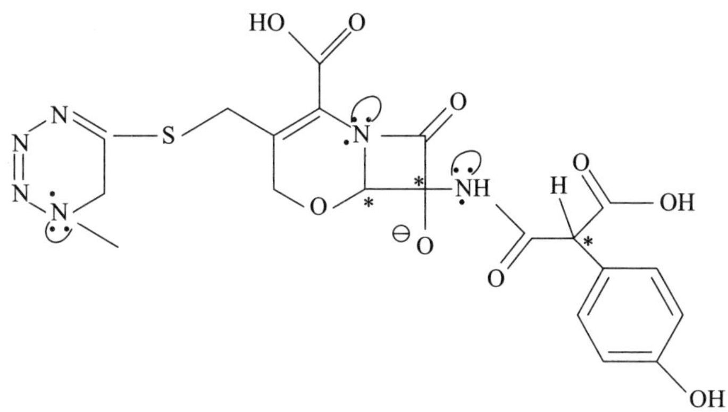

Indicated LP of nitrogen are delocalised, So N is not chiral centre here, as SP^2 is hybridised. **Ans:** (b)

2.

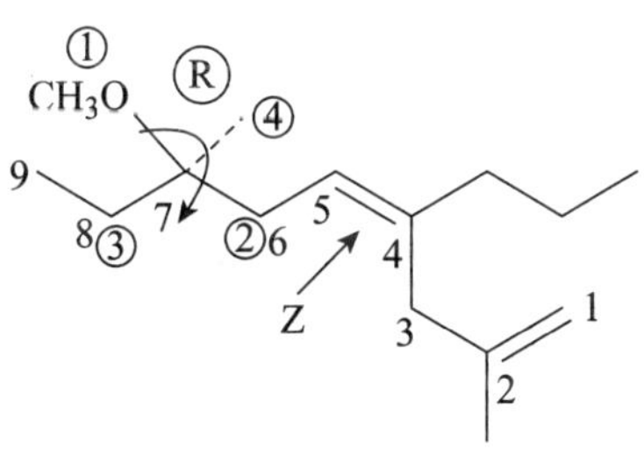

Ans: (a)

3. When methyl is at position, it feels 1, 3 – diaxial repulsion with H making compound unstable. **Ans:** (d)

4.

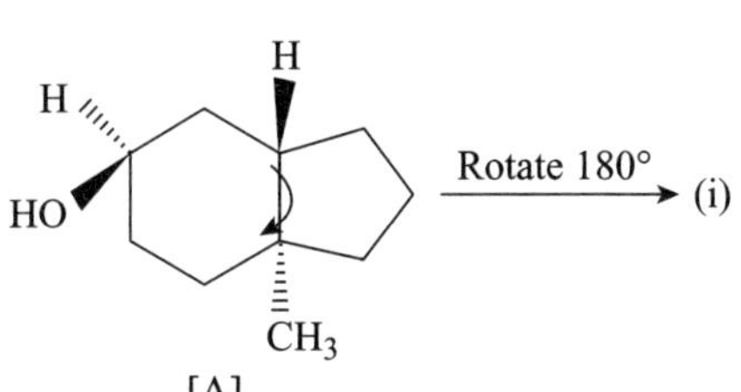

A and (i) are identical

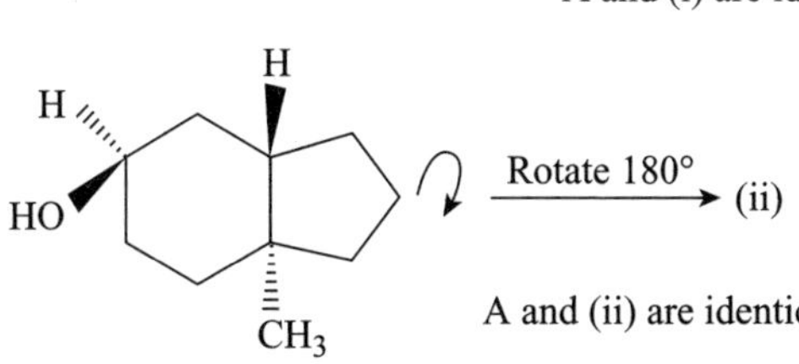

A and (ii) are identical

(iii) Is mirror image of A, so (A) and (iii) are enantiomers.

See (i) and (v) there is change of configuration at only 1 chiral centre, therefore (i) and (iv) are diastereomers. Since (i) is identical with (A); (A) and (iv) are diastereomers. **Ans:** (d)

5. DU = 0

Primary Amine :

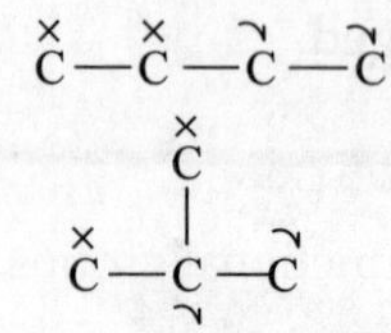

: 2

: 2

④

Total = [×-Not Possible]
[⌐-Possible]

Secondary Amine :

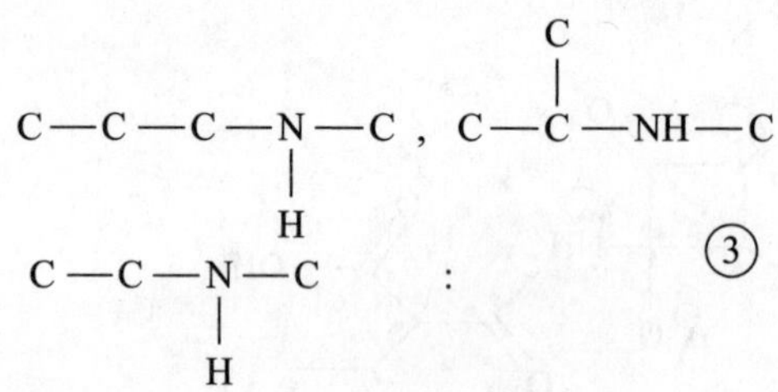

: ③

Tertiary Amine :

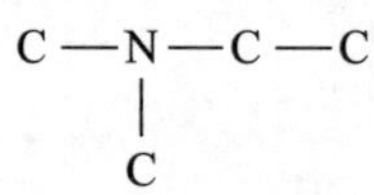

: ①

Total : [8]

Ans: (b)

7.

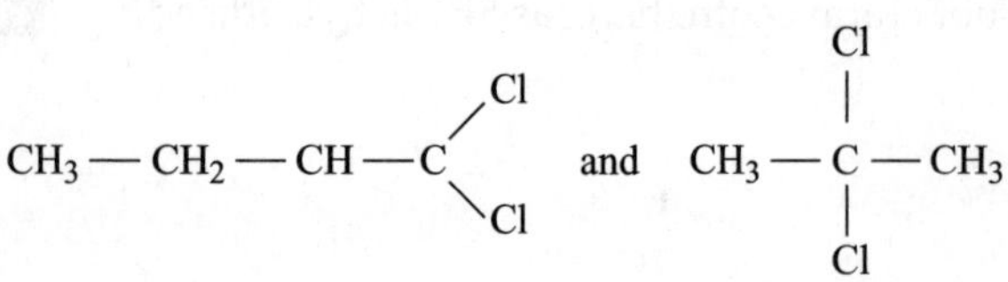

Ans: (b)

9.

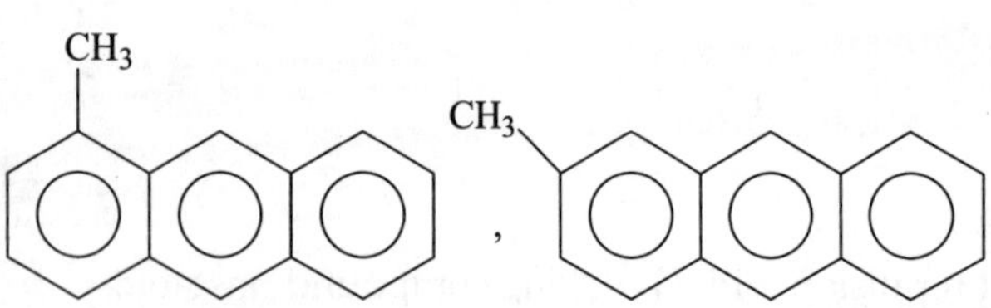

and

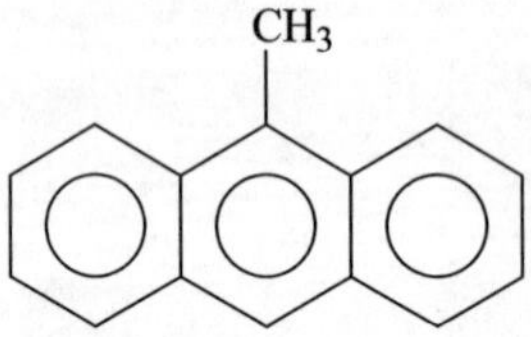

Ans: (b)

10. $(CH_3)_3N$ ⟶ 3° Amine
$CH_3CH_2CH_2NH_2$ ⟶ 1° Amine
Functional isomer

Ans: (c)

11. Molecular formula should be same to show isomerism. **Ans:** (d)

12.

Primary :

Secondary :

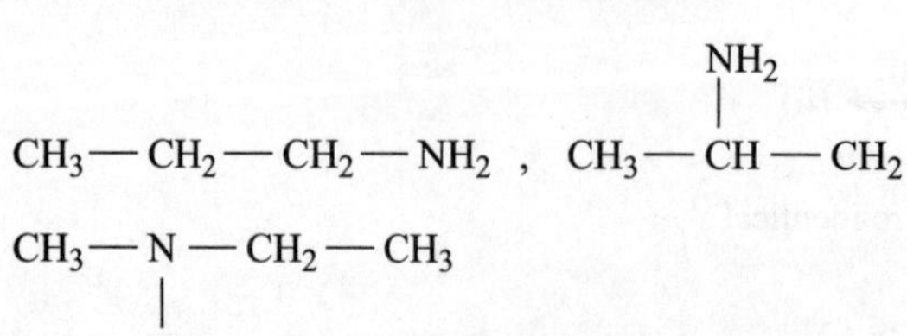

Tertiary : $CH_3—\underset{\underset{CH_3}{|}}{N}—CH_3$

Ans: (c)

14. Different structure as position of bromine is changed. Ans: (a)

16. $—\overset{\overset{O}{\|}}{C}—OH$ and $—O—\overset{\overset{O}{\|}}{C}—H$ are different functional groups.

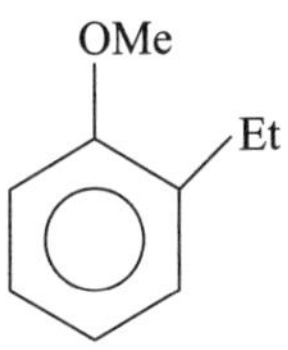 and 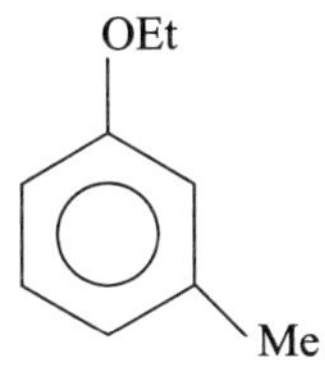 have different arrangement of carbons along each side of polyvalent functional groups.

3° Amine and 2° amine are treated as different functional groups. Ans: (c)

18. (ii) and (iv): Arrangement of carbon changed on one side of polyvalent functional group. Functional isomerism is given priority in case of other type of isomerism occurs. In compound having same molecular formula, if N of carbon vary in main chain or side chain then compounds are chain isomers of each other.

(i) and (iv): Different functional group, so functional isomers. Ans: (d)

20. a, b, c each having DU = 2, while (d) has DU = 3 molecular formula of a, b and c is same while is not of (d).

Ans: (d)

22. 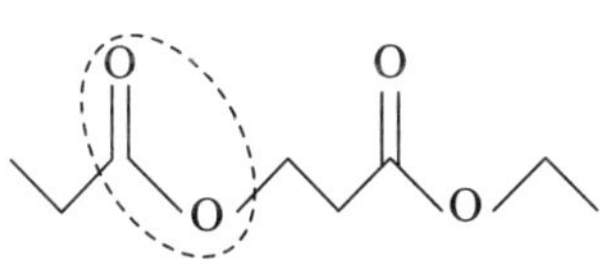

(i)

and 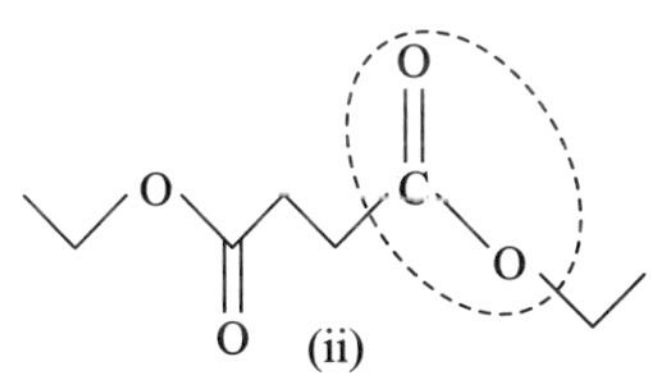

(ii)

different arrangement of carbon at each side of $—\overset{\overset{O}{\|}}{C}—O—$ Ans: (a, c)

23. (a) $—\overset{\overset{O}{\|}}{C}—$ is polyvalent functional group and 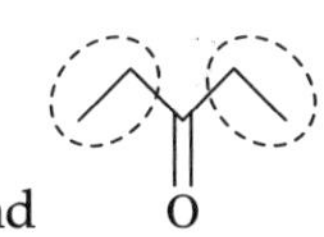 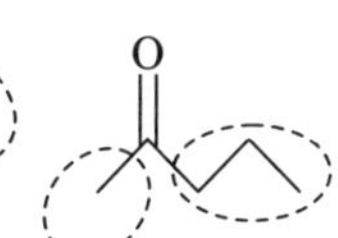

have different arrangement of carbon on each side of $—\overset{\overset{O}{\|}}{C}—$ Ans: (a, b)

24.

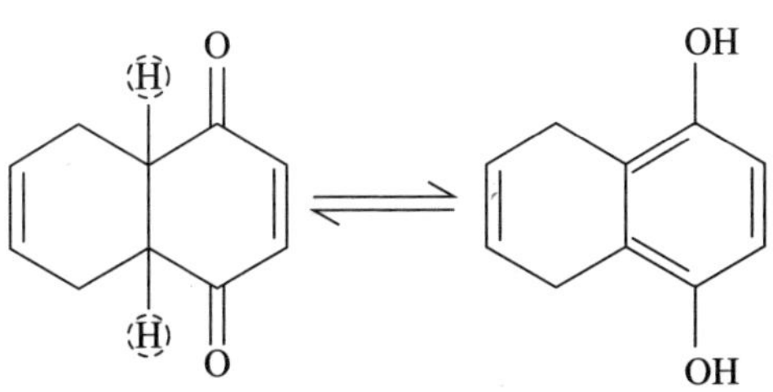

Keto Enol

Ans: (d)

25. (a)

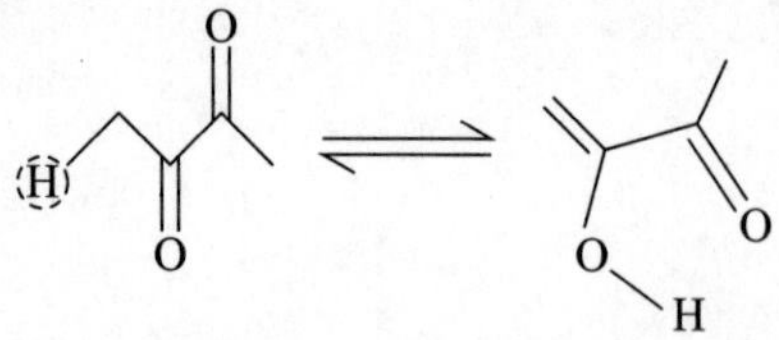

(b)

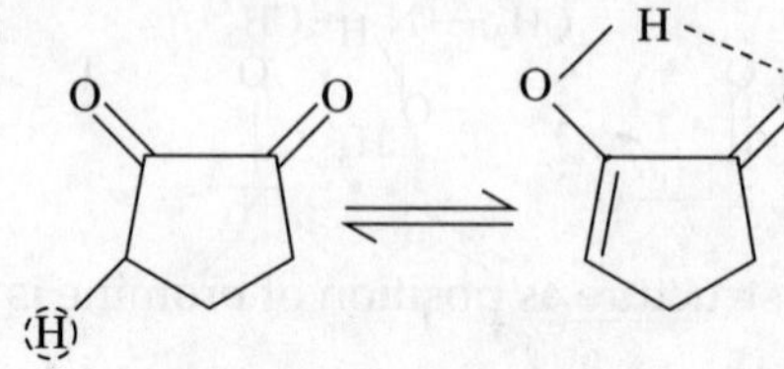

Ans: (d)

26.

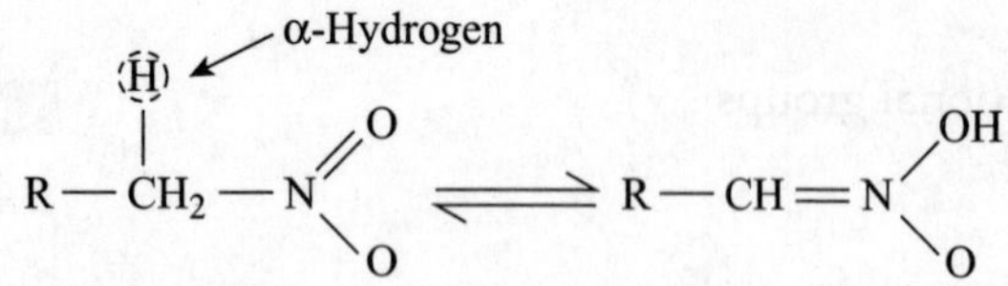

To show tautomerism, on carbon next to electron. Withdrawing group atleast acidic hydrogen should be there.

Ans: (d)

28. (d)

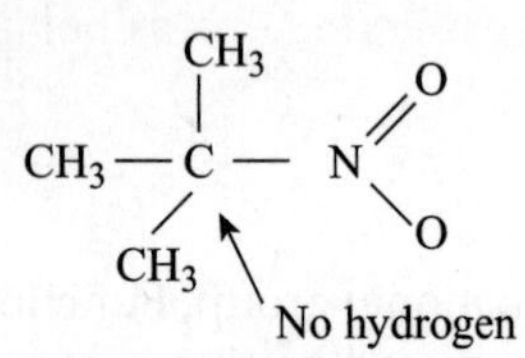

Ans: (d)

30. (a)

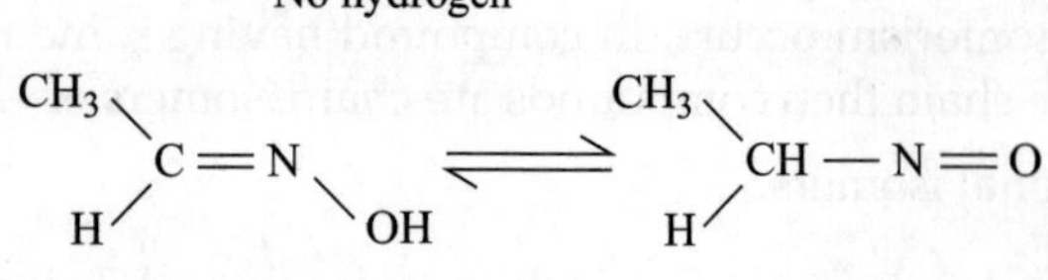

(b)

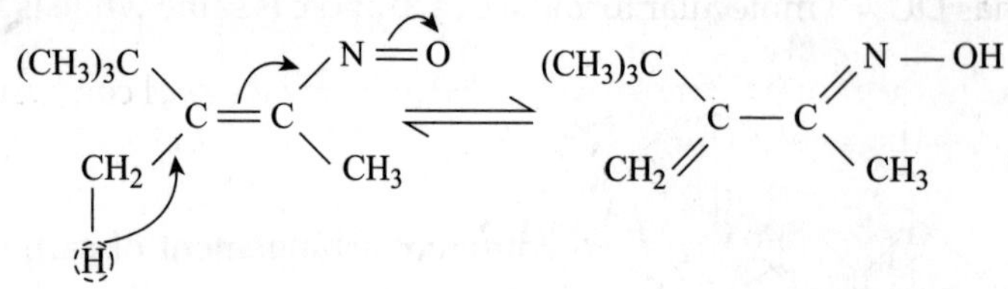

(d)

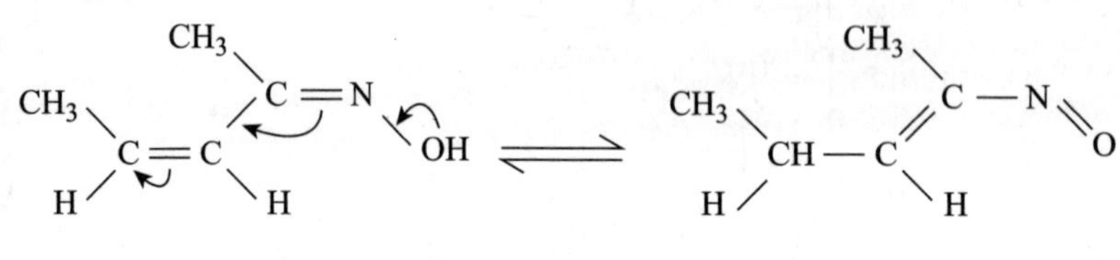

Ans: (c)

32. (a)

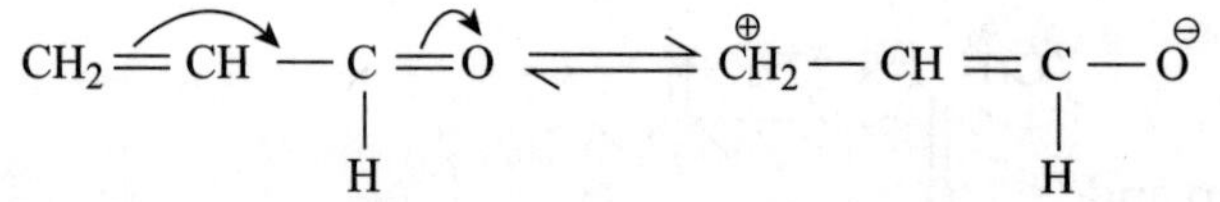

Only flow of e- took place. They are resonating structures of each other.

Ans: (a)

33.

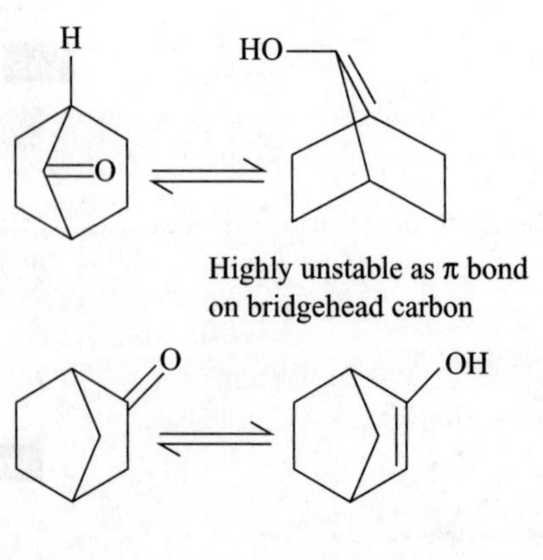

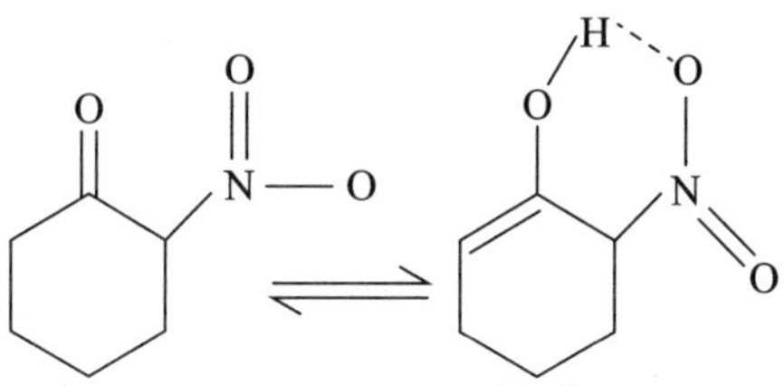

Enol
Stabilised by H-bonding

Ans: (b)

34. 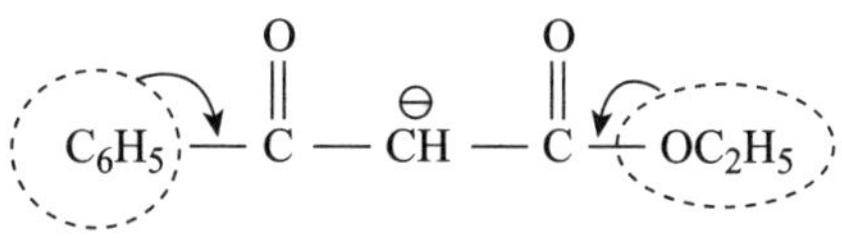 here due to extra conjugation of $-C_6H_5$, conjugation of negative charge with $-\underset{\underset{O}{||}}{C}-$ decreases leading to less stability of negative charge that decreases enol content.

Ans: (d)

35. (i)

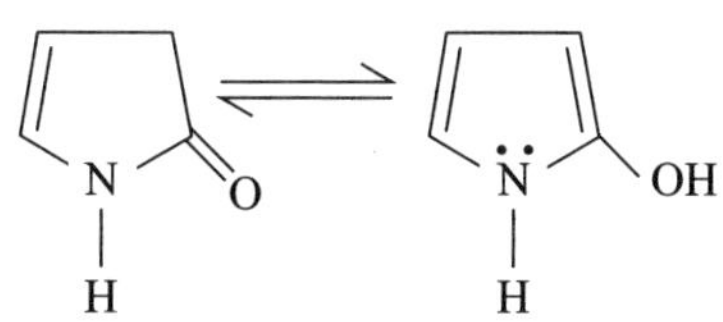

Aromatic
(got stabilised)

Ans: (b)

36. Negative charge stabilised by electron withdrawing nature of $-NO_2$; that increases enol content. While in (n) negative destabilised by electron donating nature of $-OCH_3$.

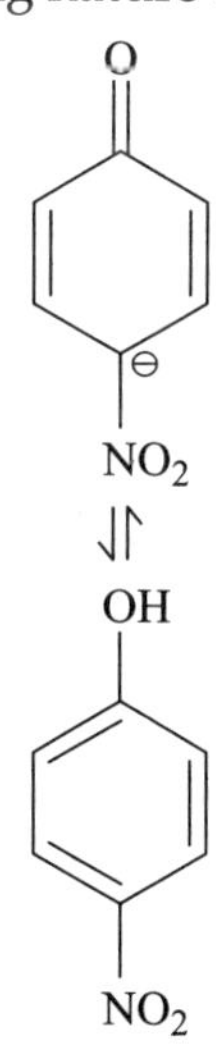

Ans: (a)

40. In ring more tendency of rotation of atom around single bond leads to development of strain in ring.

Ans: (d)

42. First condition for showing cis–trans across π-bond is to show geometrical isomerism.
Second condition: Atleast one group should be similar at each terminal of C — C.

Ans: (a)

45. Number of product should be only 2.

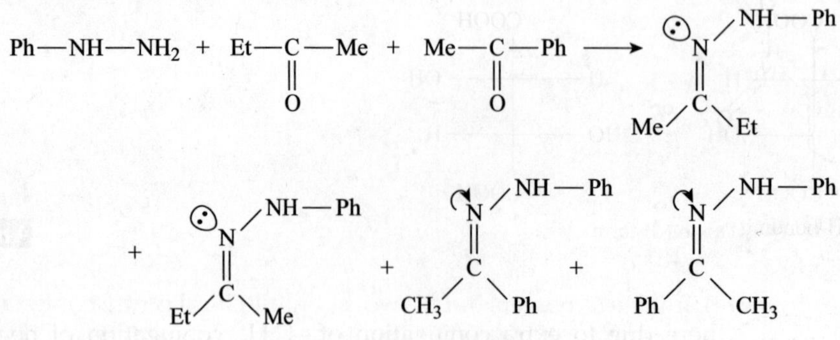

[4 products]

(b) 4 products :

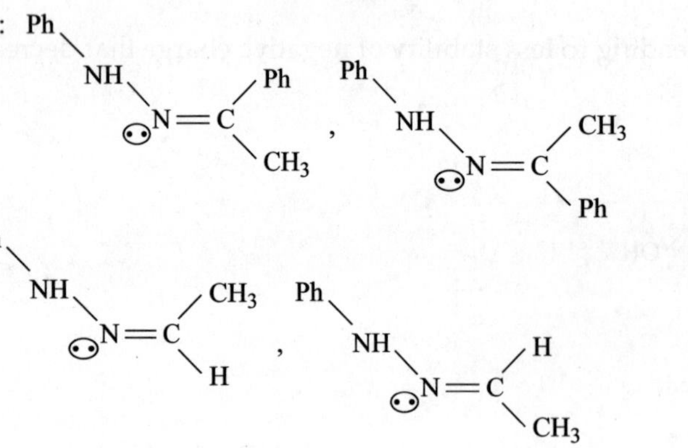

(c) 3 products :

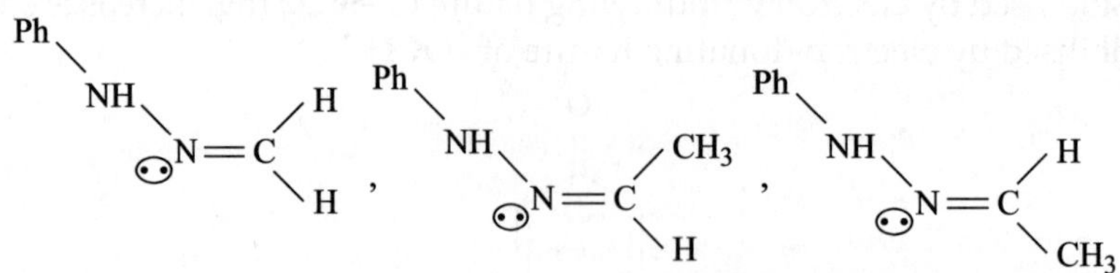

(d) 2 products :

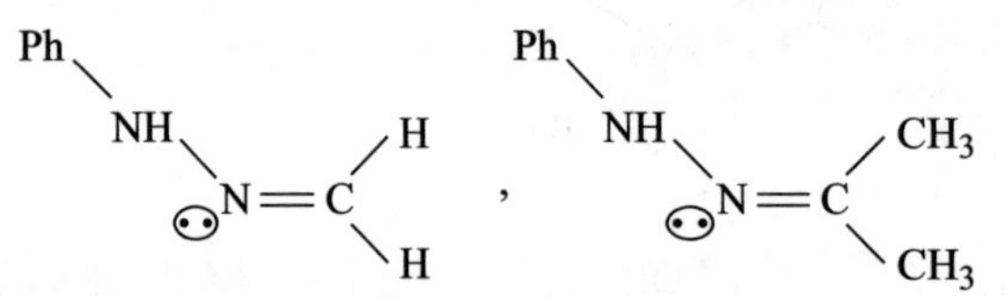

Ans: (d)

48.

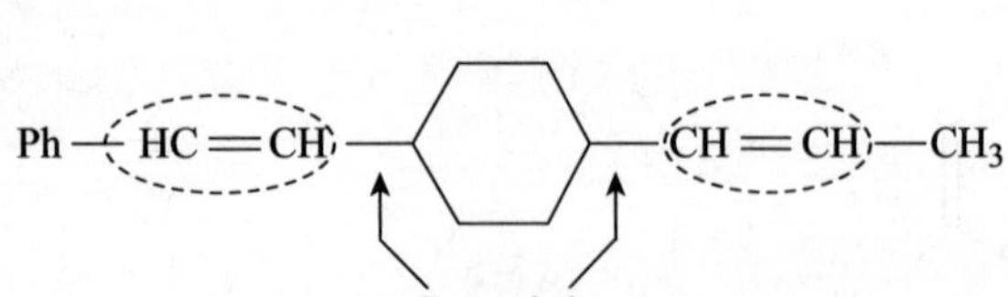

: n = 3

Ans: (d)

49.

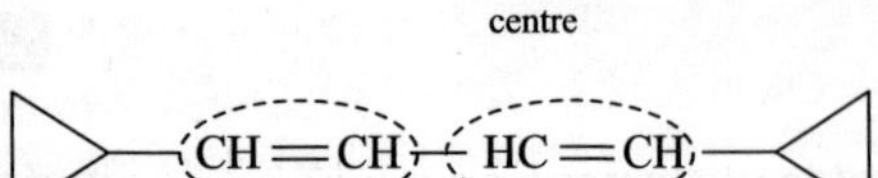

: n = 2 and symmetrical

$$GI : 2^{n-1} + 2^{n/2-1} = 3$$

Ans: (c)

50. n = 2 and unsymmetrical; GI = 2^n = 4. **Ans:** (a)

51.

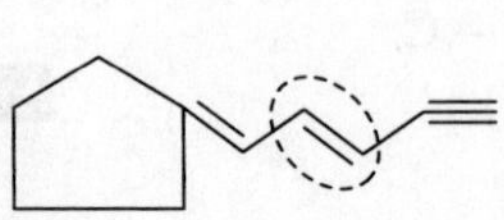

Ans: (a)

52. Both cis-2-butene and trans-2-butene gives butane on hydrogenation. **Ans:** (d)

56.

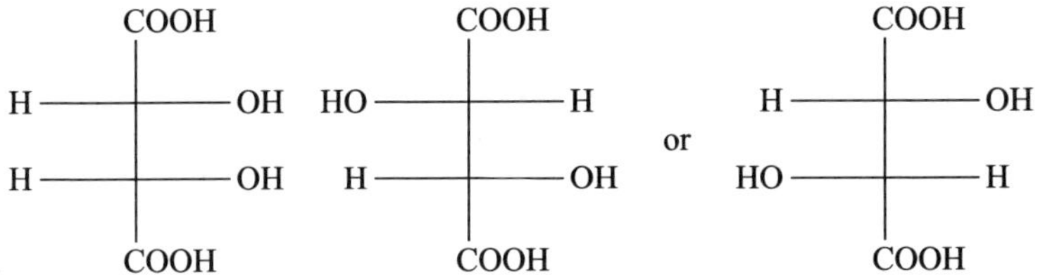

Meso tartaric acid [A] Dexto (d) form [B]

Therefore A and B (any one structure) are diastereomers out of two, at only 1 chiral centre configuration changed. **Ans:** (c)

58. $CH_3-CH_2-\overset{*}{C}H(Br)-CH_3$ that is

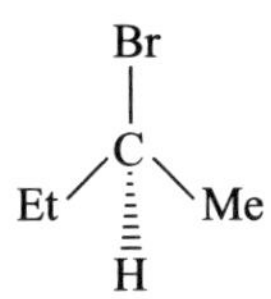

and

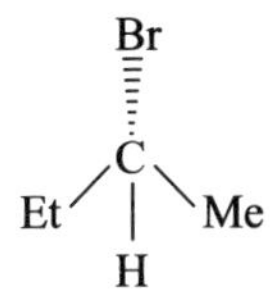

Ans: (b)

60. n = 2 and symmetrical compound

Optical isomers = 2^{n-1} (Optically active) $2^{n/2-1}$ (MeSO)

Ans: (b)

61.

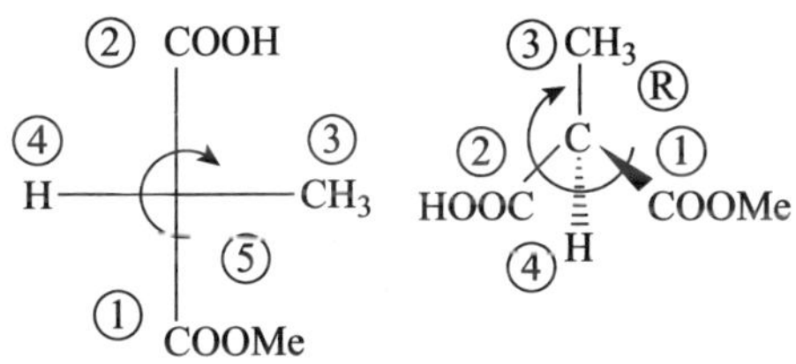

Ans: (c)

62.

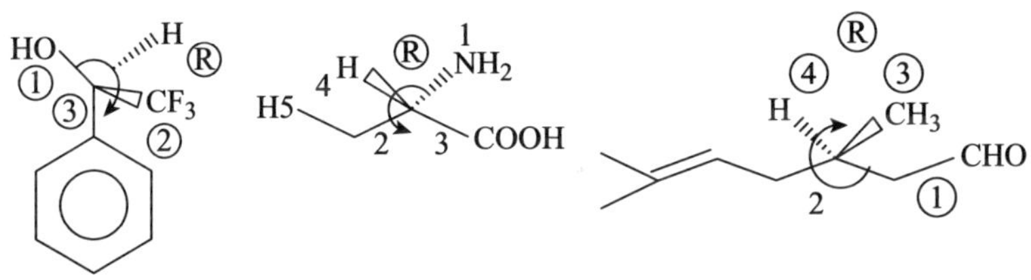

Ans: (a)

63.

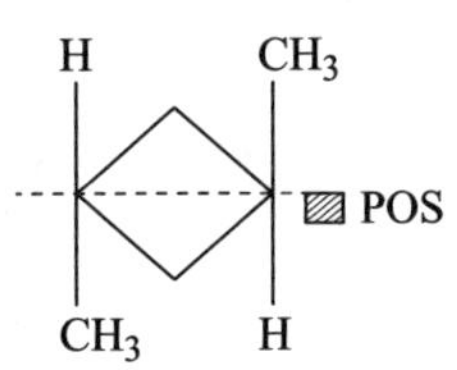

POS y
COS y

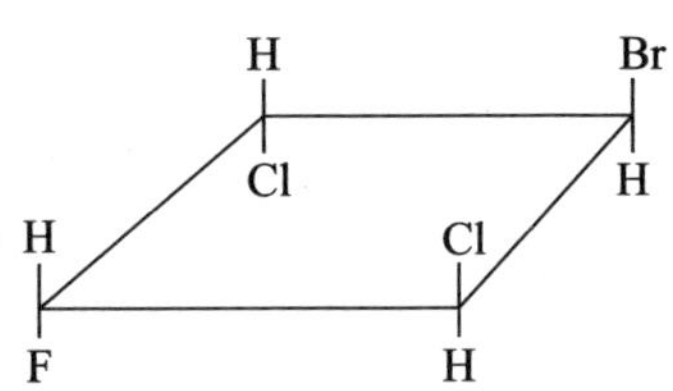

POS x
COS x

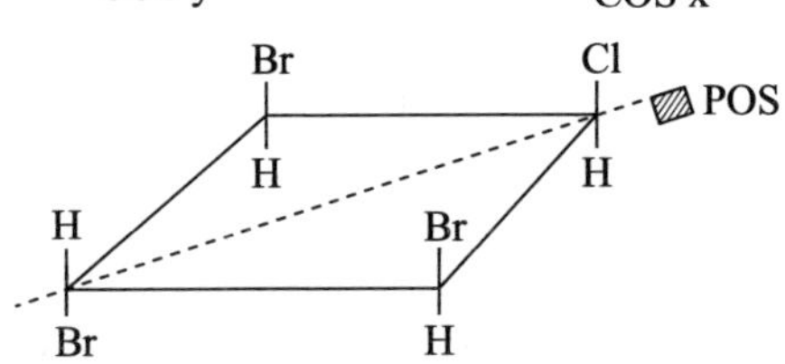

COS x

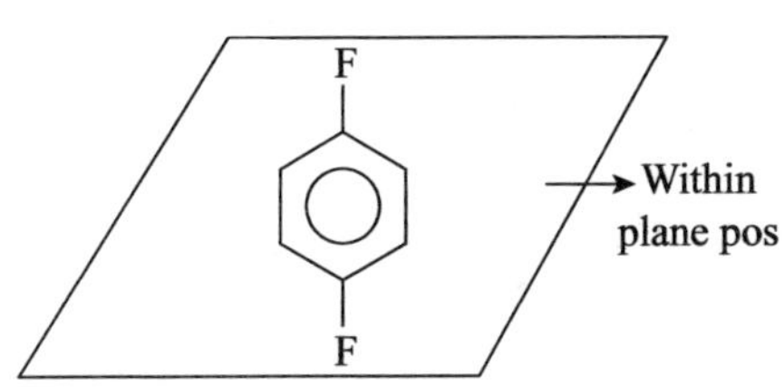

COS y

Ans: (c)

66. 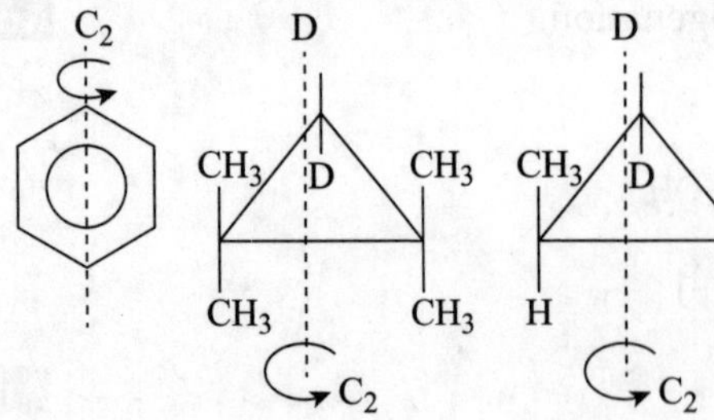

Ans: (c)

67. n = 3 (symmetric odd system with pseudo centre)

Number of stereoisomers = $2^{n-1} = 2^2 = 4$ **Ans:** (b)

71. $CH_3-CH=CH-\overset{*}{C}H(Cl)-CH=CH-Et$

Unsymmetrical system n = 3

Number of stereoisomers = $2^n = 8$ **Ans:** (d)

73. Optical purity = $\frac{6-4}{10}\times 100 = 20\%$ = Enantiomeric excess.

Specific rotation for 100% optical purity = 13.5 unit

Specific rotation for 1% optical purity = $\frac{13.5}{100}$

Specific rotation for 20% optical purity = $\frac{13.5}{100}\times 20 = +2.7$ unit.

Ans: (c)

74. Enantiomeric pairs have same physical properties (i.e., BP)

Therefore, They can not be separated by fractional distillation. **Ans:** (d)

75. H-bonding in gauche form, makes it more stable than anti. **Ans:** (c)

77. (i) $CH_3-CH-(CH_3)_3$ (ii) $CH_3-CH_2-CH(CH_3)-CH_3$

These are chain isomers (constitutional isomers). **Ans:** (c)

79. Torsional strain comes into picture in eclipsed form of n-butane. **Ans:** (a)

81. 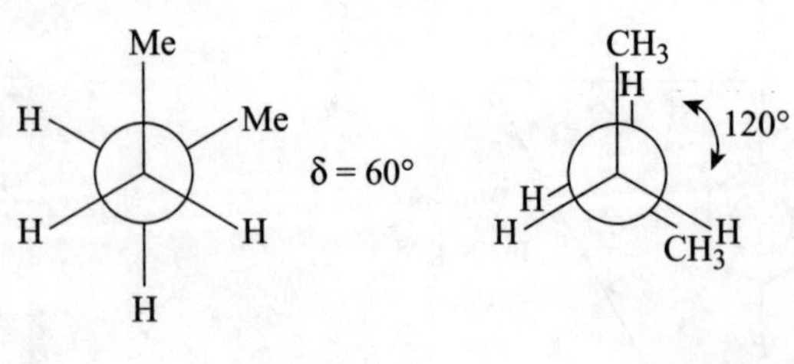

Ans: (b)

82. 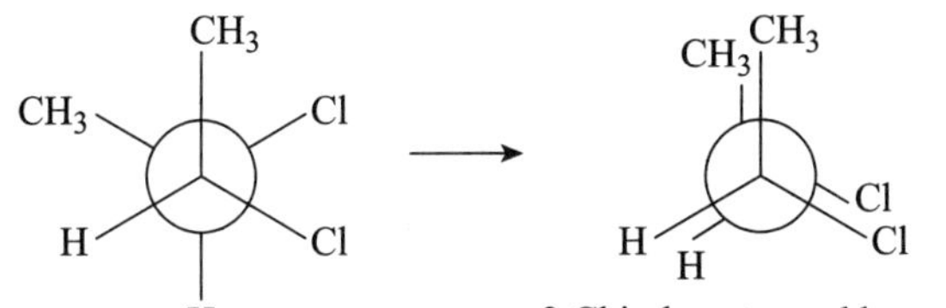

2 Chiral centre and has pos
∴ Meso compound (Achiral)

Ans: (a)

83. 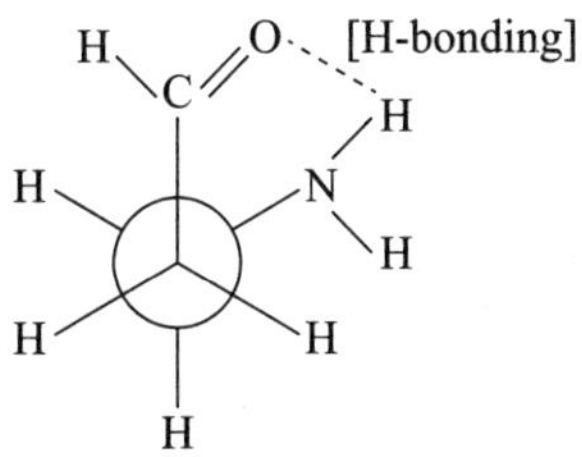

Ans: (a)

85. 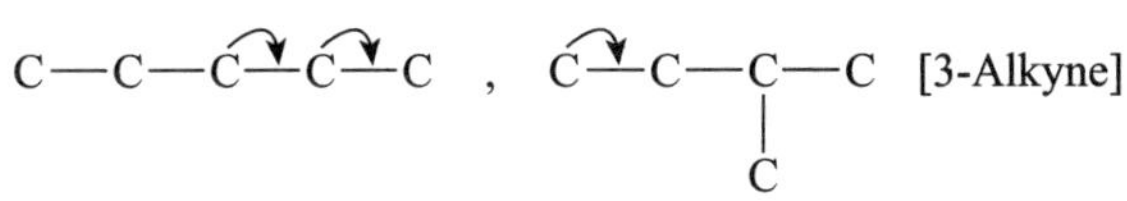

Ans: (b)

86. 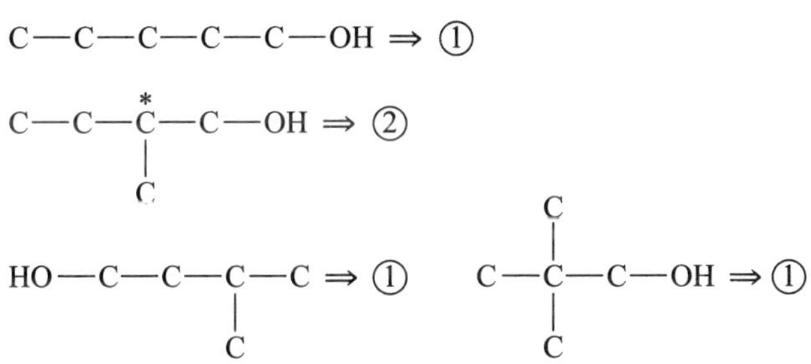

Ans: (d)

87. 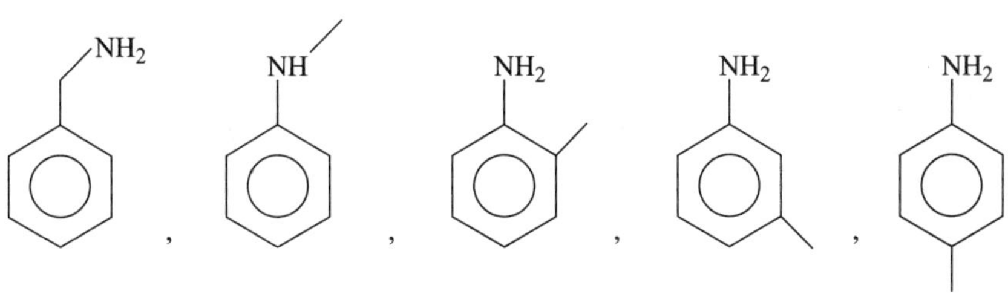

Ans: (c)

91. 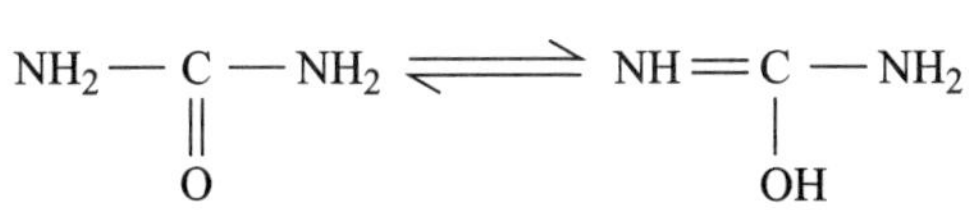

Ans: (c)

92. $CH_3—CH_2—CH_2—CH_3$ and $CH_3—CH(CH_3)—CH_3$ are chain isomers.

Minimum 4C required to show structural isomerism in alkane.

Ans: (d)

93. $\boxed{6+0}$: C — C — C — C — C — C

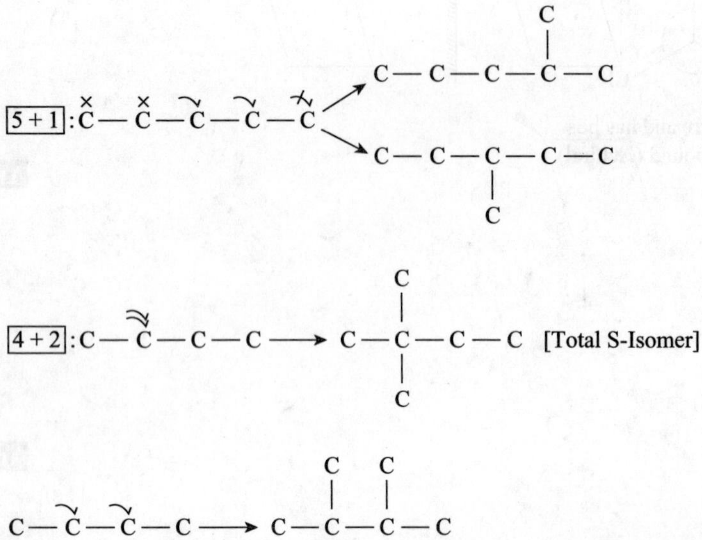

Ans: (b)

95. 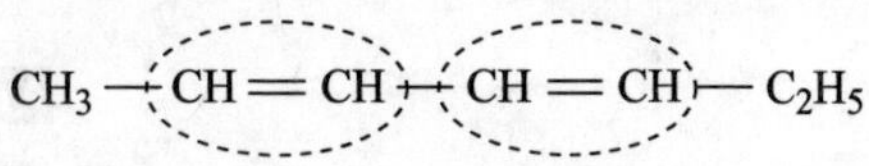

Unsymmetrical compound with n = 2
Number of GI = 2^2 = 4

Ans: (a)

97. 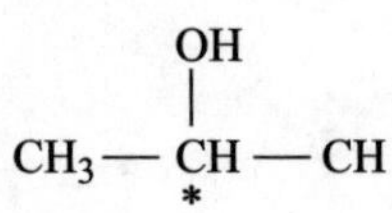

Number of stereoisomers = 2^1 = 2

Ans: (b)

98. 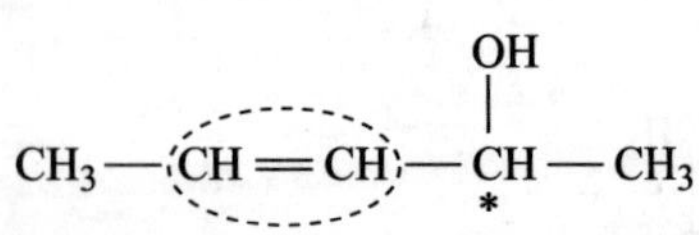

Ans: (a)

99. 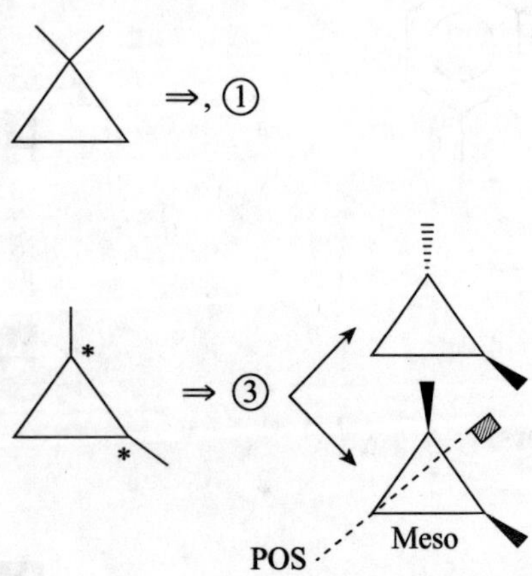

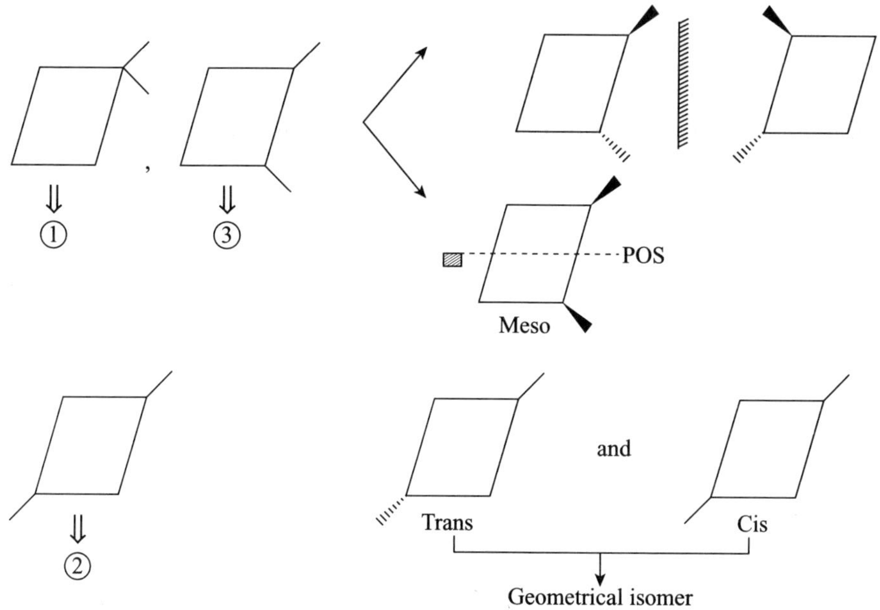

Ans: (d)

101.

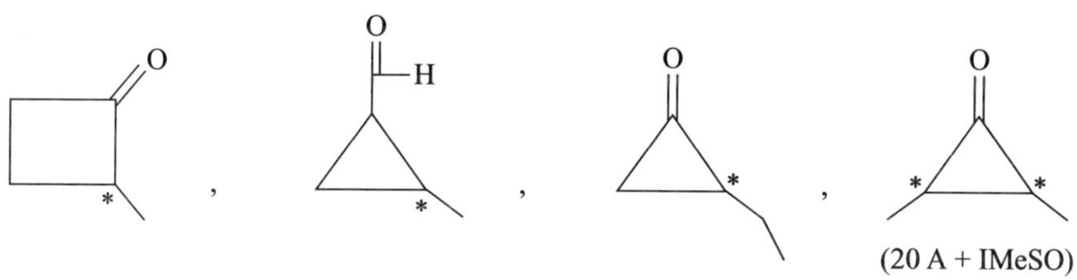

Ans: (a)

102.

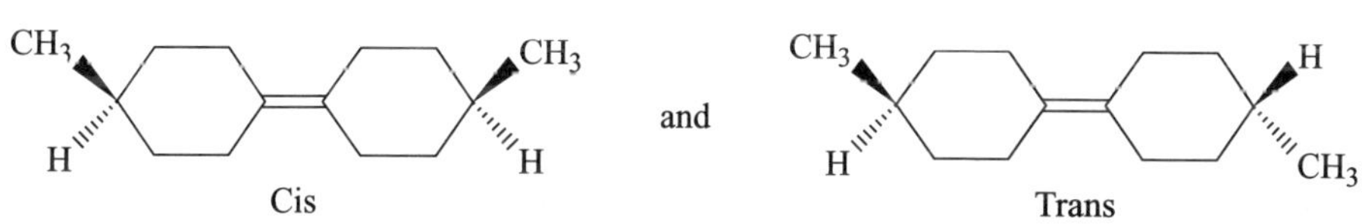

Ans: (a)

103.

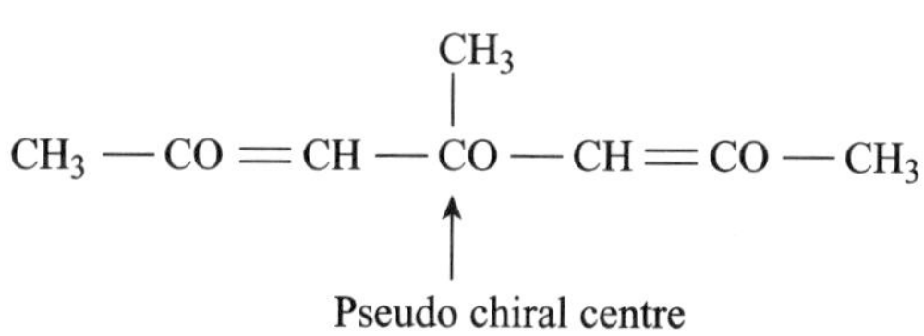

Symmetrical system with odd number of stereocentre including pseudo chiral centre. Number of stereoisomers = $2^{n-1} = 2^{3-1} = 4$.

Ans: (a)

110.

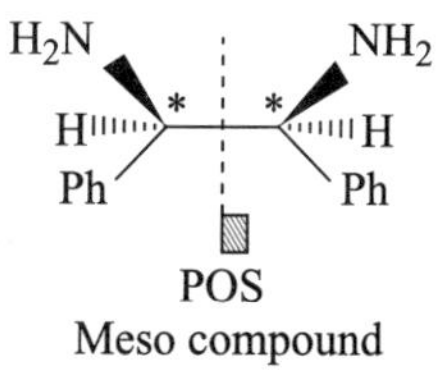

Ans: (c)

112.

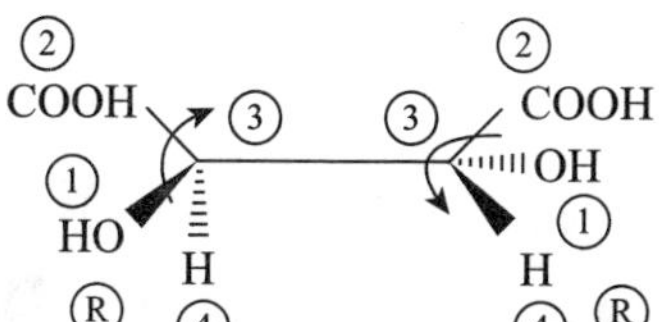

Ans: (a)

LEVEL 2

1. In chair form 1, 3, 5 axial repulsion exist between –C(me)3, –H and –H; making it less stable.

Ans: (b)

2. 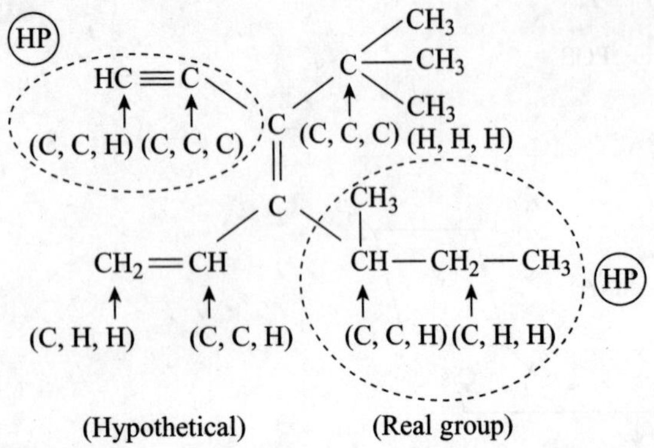

[HP-High priority]

Ans: (a)

3. + H effect: $-CH_3 > -CD_3$
+ I effect: $-CH_3 > -CD_3$

Ans: (b)

4. (i) → COS

(iv) → 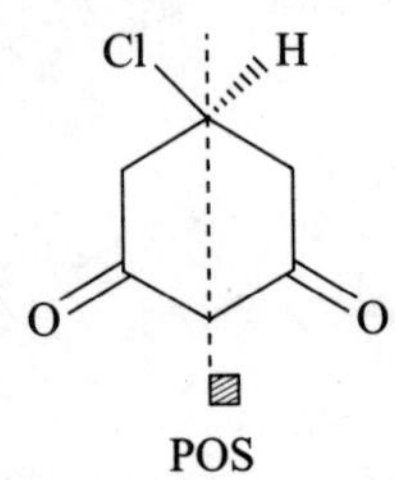

Ans: (b)

5. 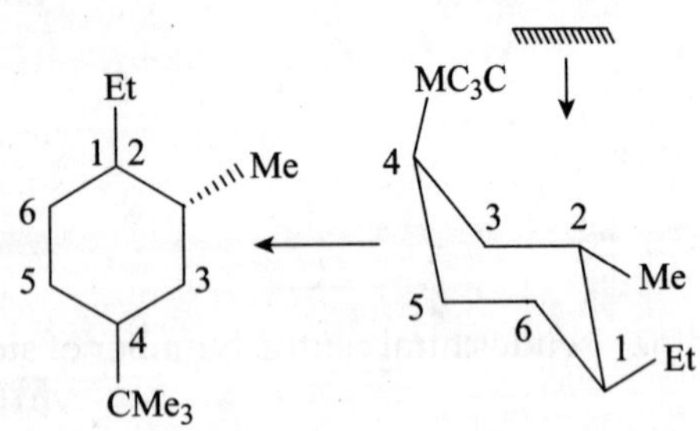

– Et and Me are trans

– Et and CMe_3 are Cis

Ans: (b)

Ans: (c)

6. For Meso, atleast two chiral centre should be there.

7. (a) → Diastereomers
(b) → Functional
(c) → Metamers
(d) → Metamers

Ans: (b)

8. (b) 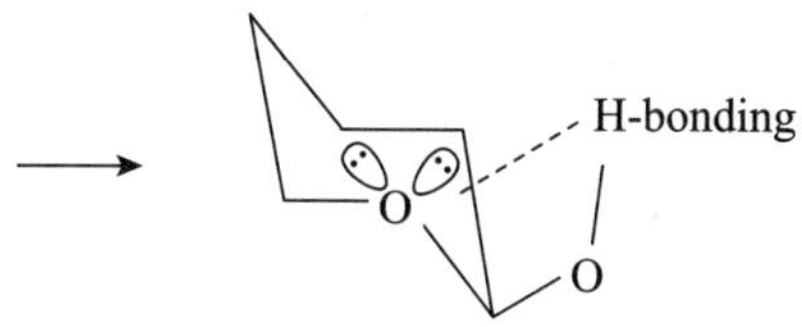

(d) 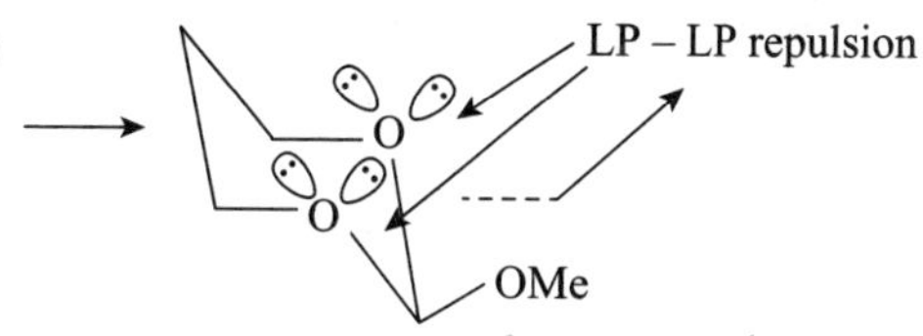

Ans: (b)

10. 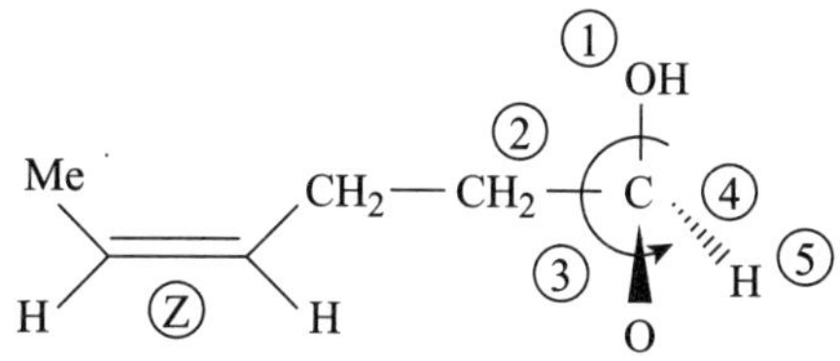

Ans: (d)

11. 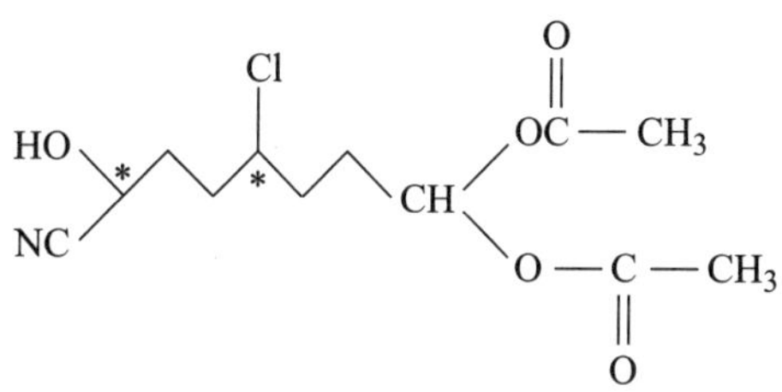

Ans: (b)

12. 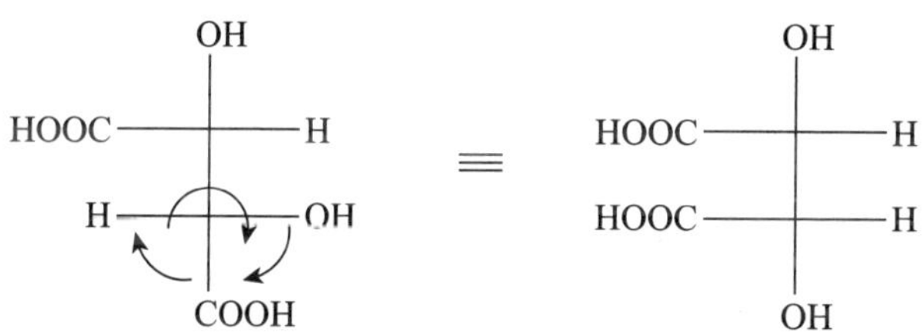

Meso compound

Ans: (d)

13. 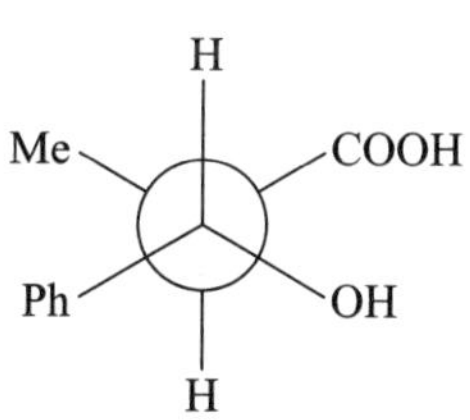

Staggered

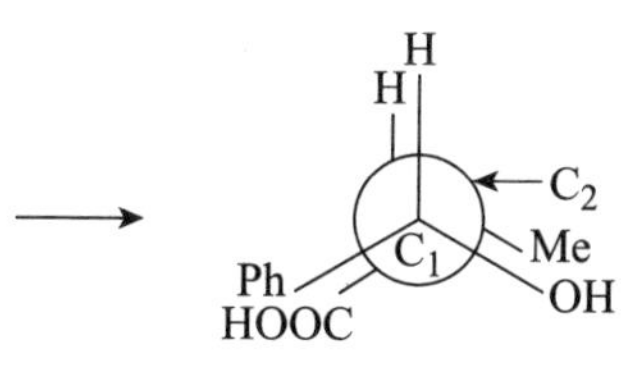

Eclipsed form

≡

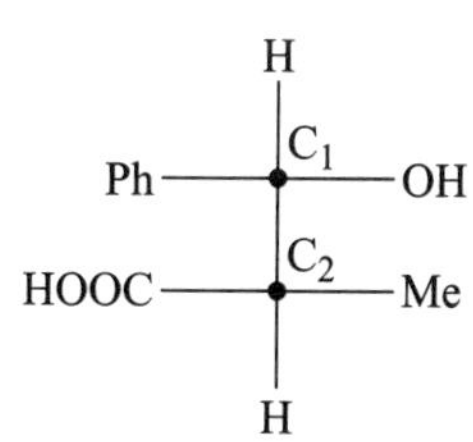

Fischer form is eclipsed form. Therefore make eclipsed form in Newman projection first.

Ans: (c)

14. (i) 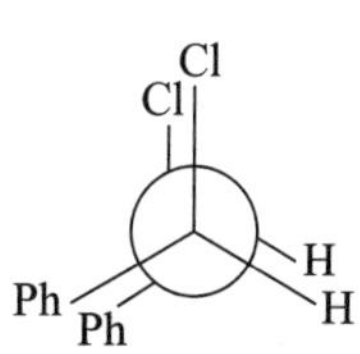

has POS (2 chiral centres are present) so meso compound

Ans: (d)

15. Optically active is used for resolution of racemic mixture.

(a) 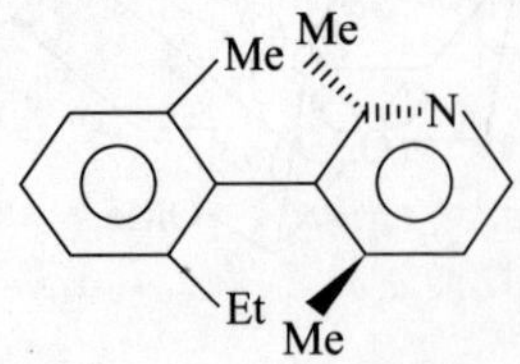

No POS
No COS
therefore optically active

Ans: (a)

16. Optical purity : $\frac{|7\quad 5|}{12} \times 100 = 16.66\%$ [l form]

Specific rotation of 100% pure compound = 13.5°

Specific rotation of 1% pure compound $= \frac{13.5}{100}$

Specific rotation of 16.66% pure compound

$= \frac{13.5}{100} \times 16.66 = 2.25°$ levo rotation

Ans: (a)

17. Heat of combustion α Number of carbon α instability (if number of c, same are). **Ans:** (b)

18.

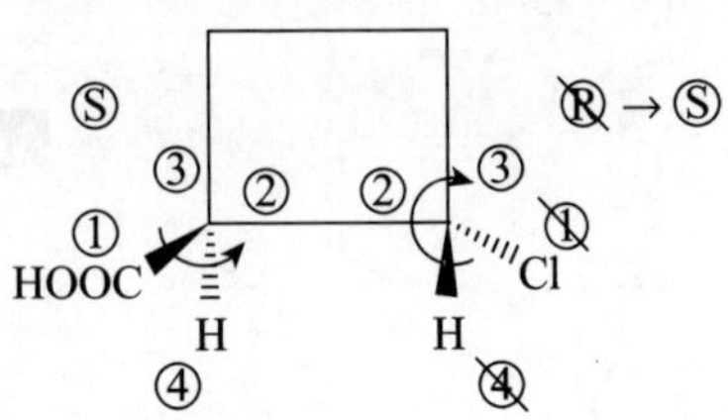

Ans: (c)

Ans: (c)

19. Position isomers

21.

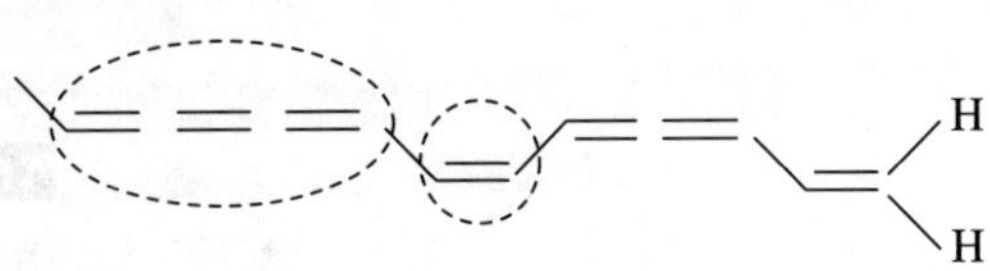

n = 2

Number of GI = $2^2 = 4$

Commulene with even number of π-bond does not show GI.

Ans: (b)

23.

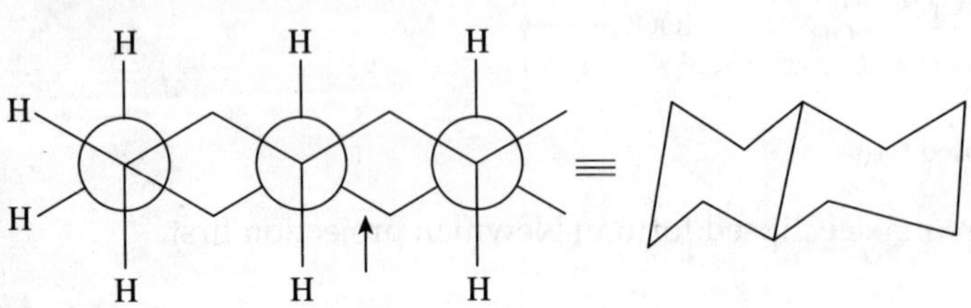

Trans decalin

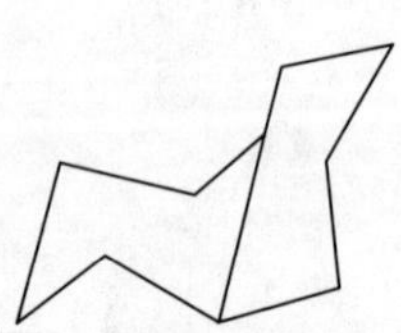

Cis decalin

Ans: (b)

24. (a) 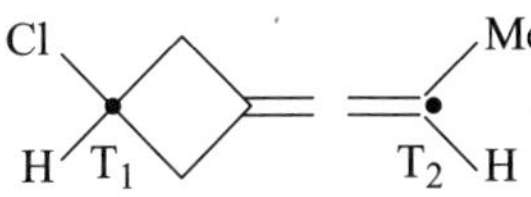 and 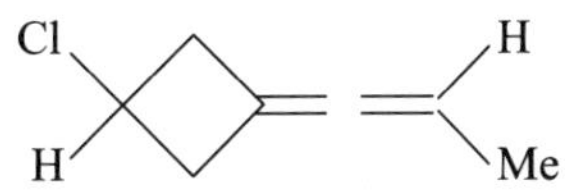

Cis Trans

(b) 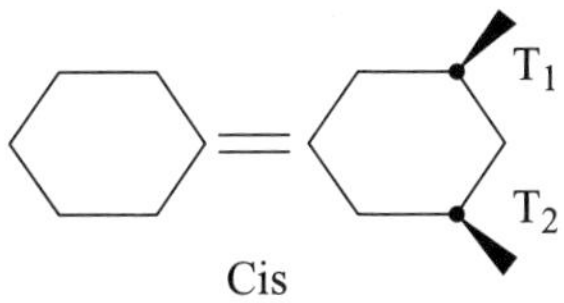 and 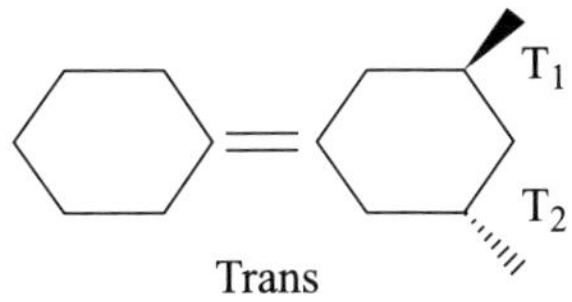

Cis Trans

(c) 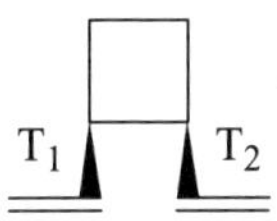 and 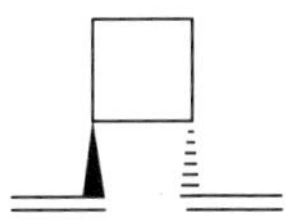

Cis Trans

T1, T2 : Terminals across which GI is possible.

Ans: (a, b and d)

25. Chair: 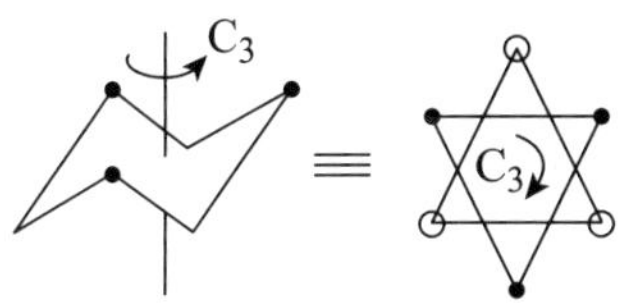

Ans: (c)

27. (iii) 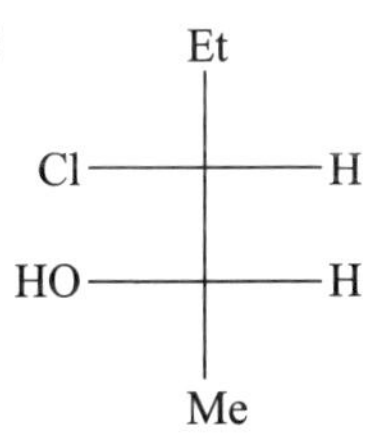 (iv) 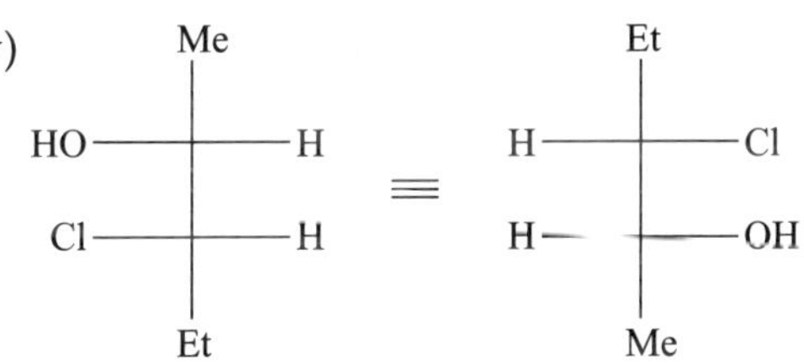

Ans: (b)

28. $C_4H_{10}O : C_nH_{2n+2}O$

Alcohols: 

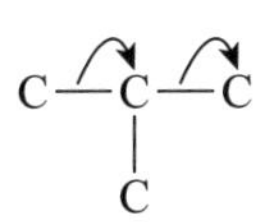

Total: (4) Isomeric alcohol

Ether: 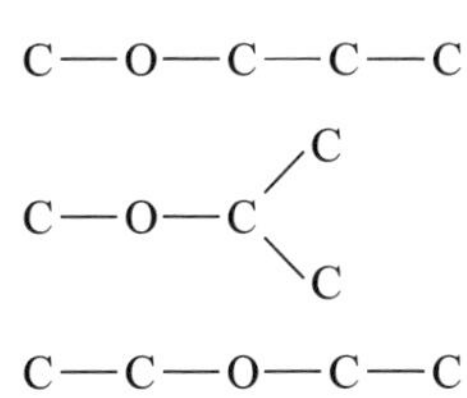

Total: (3) Ether

Ans: (b)

29. (a) 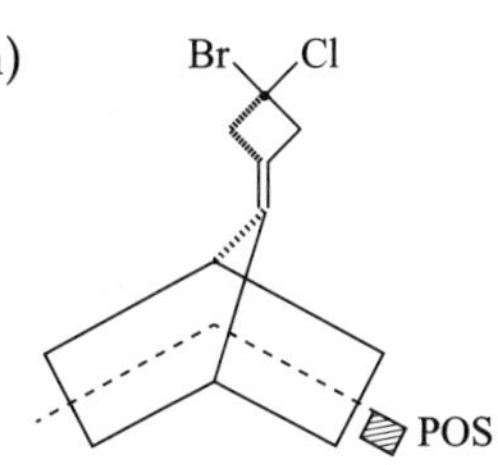

(b)

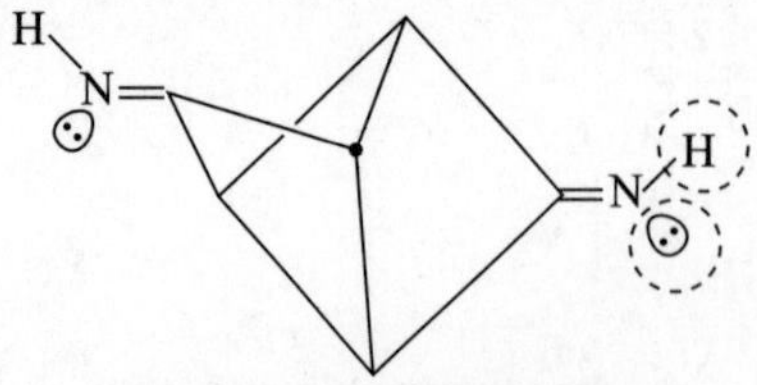

POS not possible as H and LP are in different directions

Ans: (b and c)

30. (a)

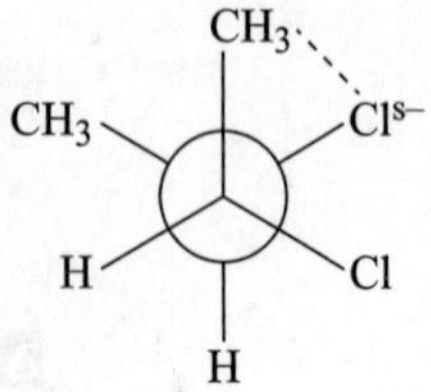

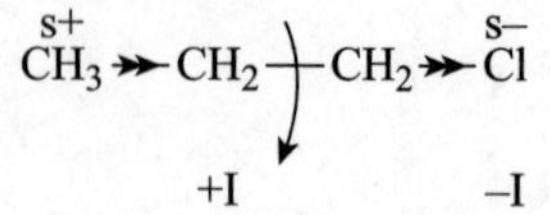

(b)

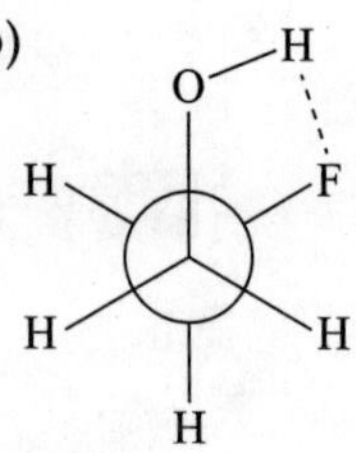

(d)

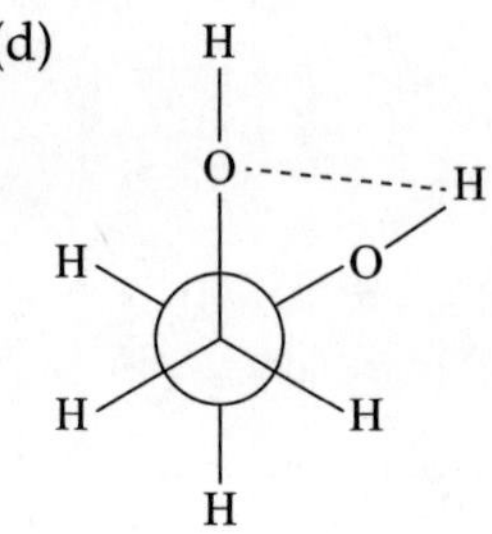

Ans: (a, b and d)

31. (a)

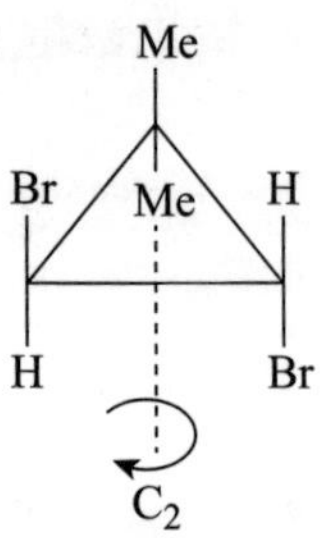

POS x

(b)

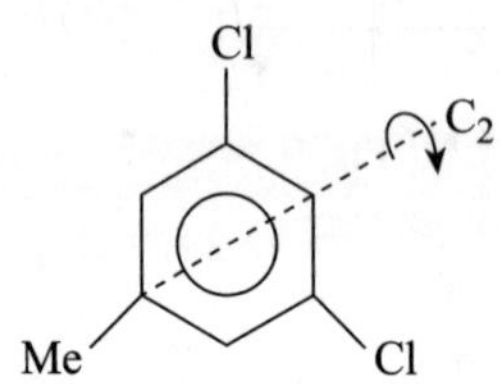

Within plane POS is present

(d)

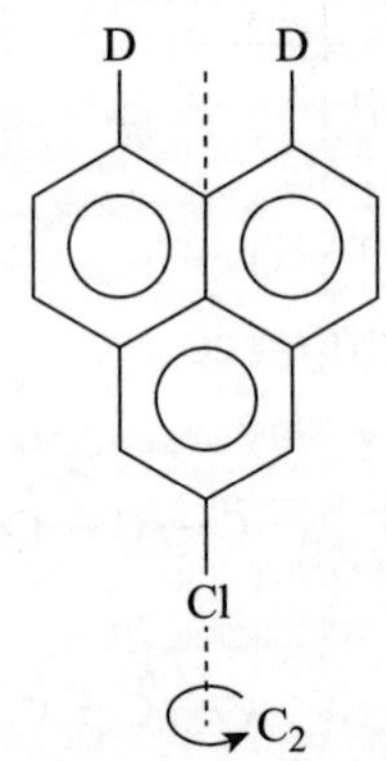

Within plane POS is present

Ans: (b and d)

32. (a) → C_3 **(about axis perpendicular to plane)**

(b) → C_3 (c) → C_3 (d) → C_3

Ans: (a, b, c and d)

33. 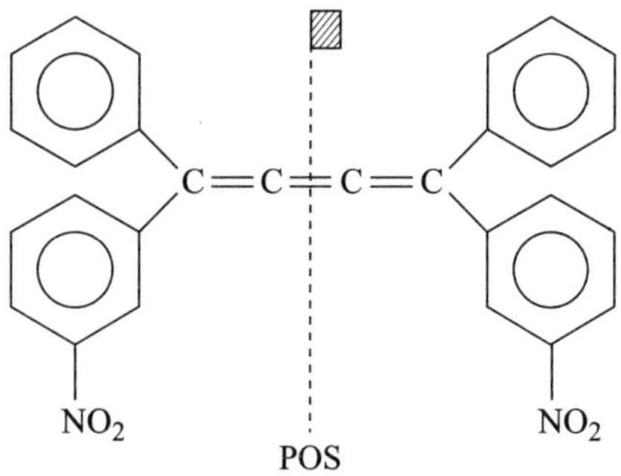

Achiral molecule

Ans: (a, b and c)

34. (a), (c), (d): C_3 (about axis perpendicular to plane)

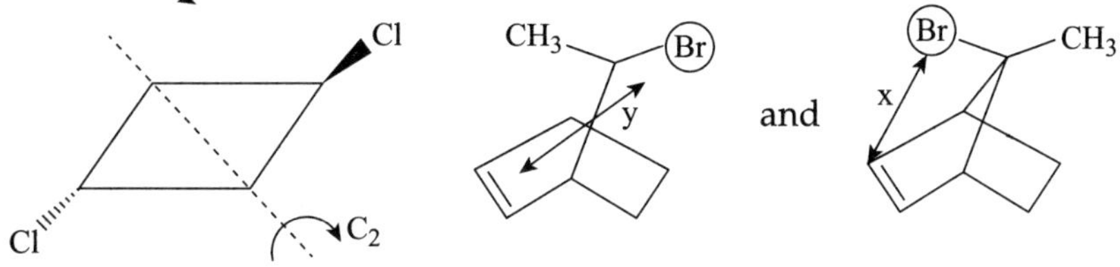

Distance: y ≠ x

Ans: (a, c and d)

35. (a) GI(diastereomers): as they are stereoisomers which are not mirror image of each other.

(b) → (d) Mirror image.

(c) Out of 4 chiral centres, only 2 changed their configuration.

Ans: (b, c and d)

36. (a) No POS No COS. Therefore, optically active.

(d) Due to repulsion between LP of chlorine molecule becomes non-planar, leads to absence of POS. Therefore, molecule becomes chiral.

Ans: (a, c and d)

37. (a)

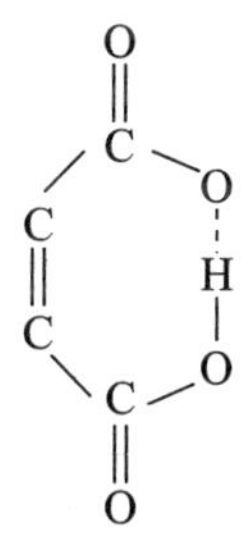

Cis but-2-ene dioic acid

Stabilised by H-bonding, therefore stronger acid when compared to trans but-2-ene dioic acid.

(c)

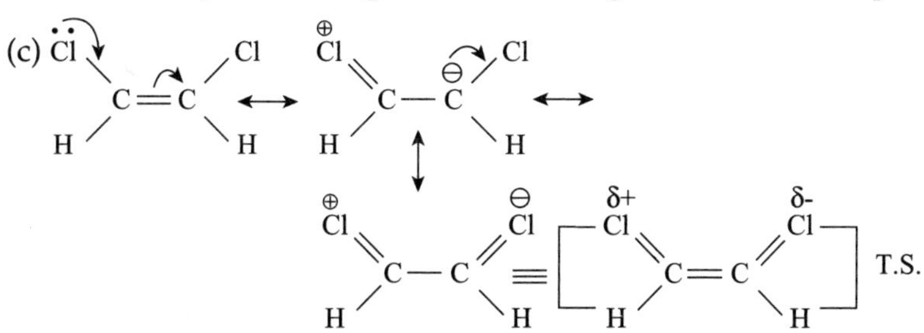

In cis form attraction between both chlorine (as S+ on Cl, and S– on other Cl) leads to stability.

Ans: (a, b and d)

38. (a), (d) No POS and No COS

(b)

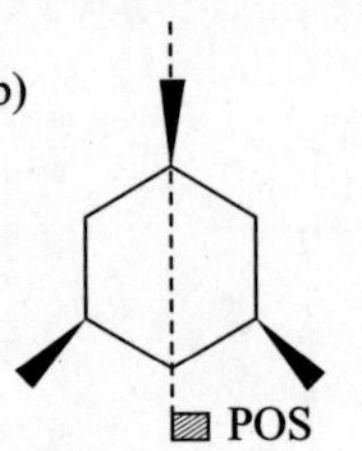

(c)

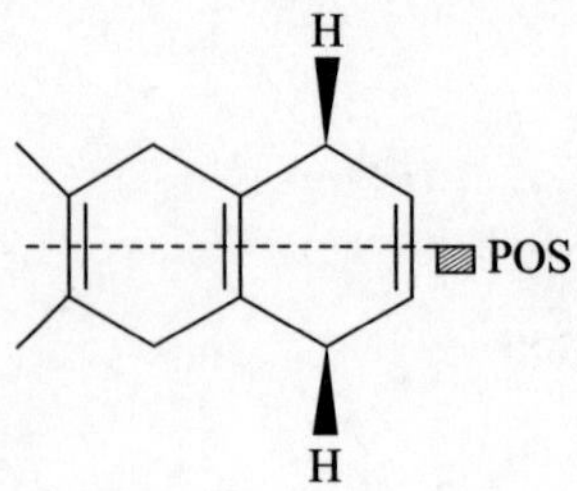

Ans: (a and d)

39. 1, 11 : Functional isomer. **Ans:** (b and c)

40. (a) H-bonding in Gauche form leads to stability.

(b)

Stable :

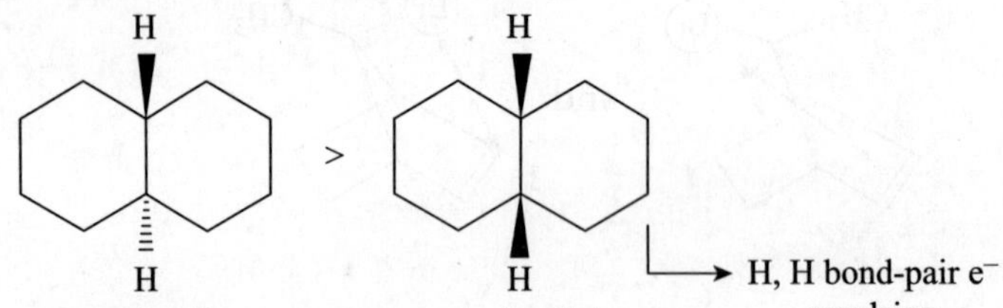

trans decalin cis decalin

Ans: (b and d)

41. (b) Ester and carboxylic acid are functional isomers.

(c) 3°–Amine and 2°–Amine are functional isomer of each. **Ans:** (a and d)

42. (a) Aldehyde and carbonyl halide (halo carbonyl) are FGI.

(b) Metamers

(c) $H-\overset{\overset{O}{\|}}{C}-O-Me$ (Ester) and $CH_3-\overset{\overset{O}{\|}}{C}-O-H$ (Carboxylic acid) are FGI

(d) $CH_3-\underset{\underset{O}{\|}}{C}-CH_3$ (Ketone) and 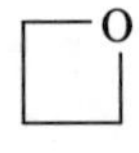 (Cyclic ether) are FGI.

Ans: (a, c and d)

44.

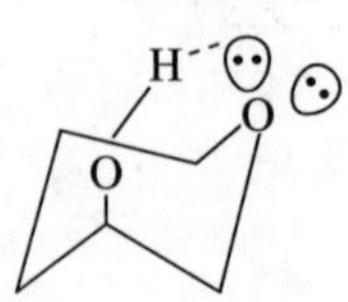

stabilised by H-bonding

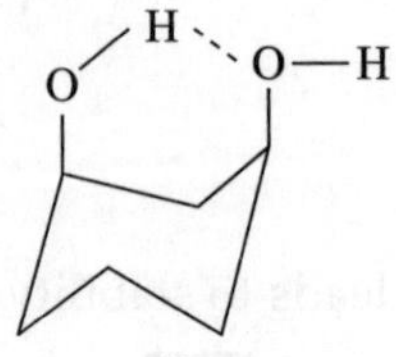

stabilised by H-bonding

Ans: (a and b)

45. (a)

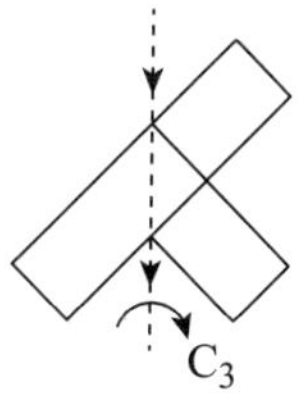

(b)

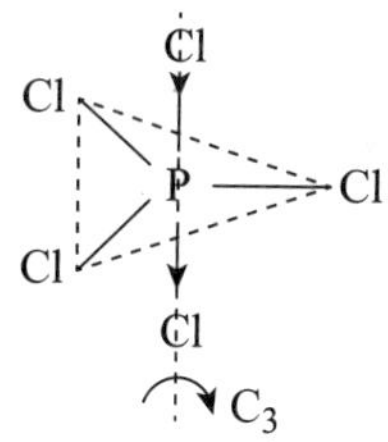

(c)

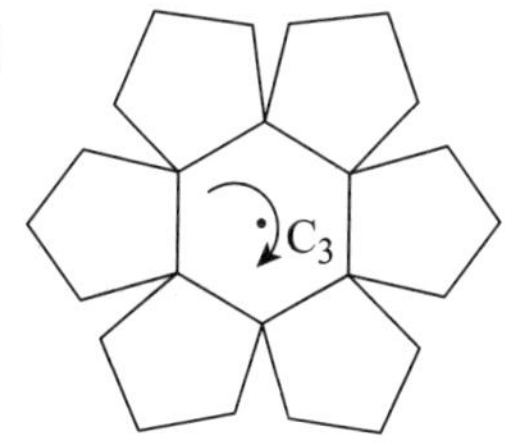

(d)

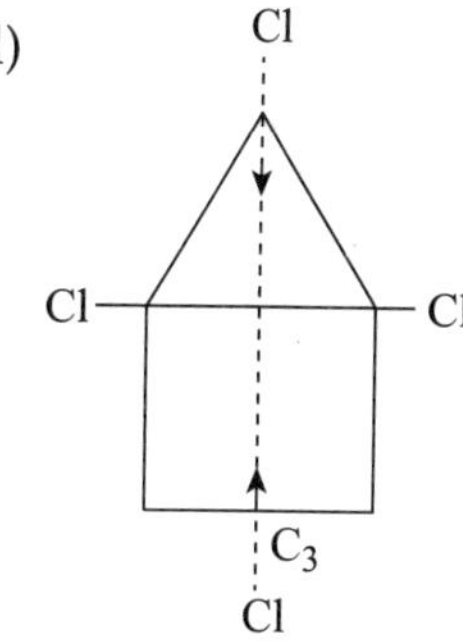

Ans: (a, b, c and d)

46.

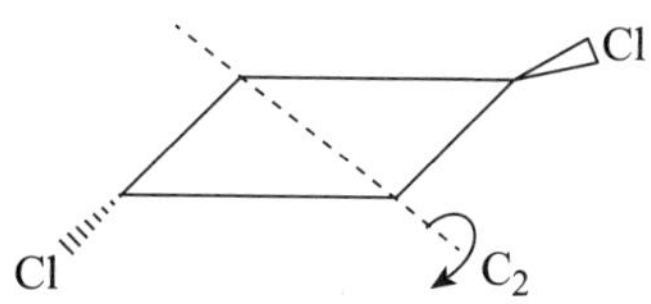

POS x
COS y
AOS y

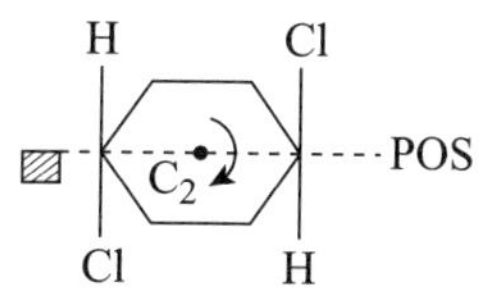

POS y
COS y
AOS y

[y = yes
x = no]

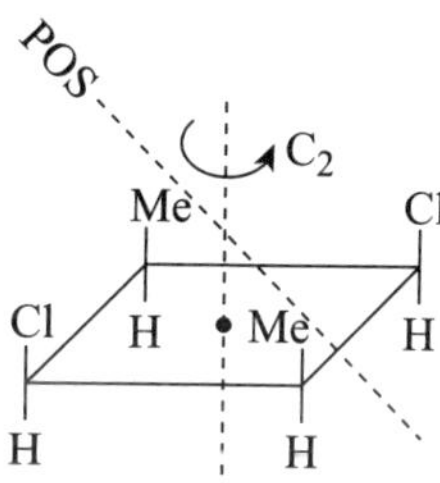

POS y
COS x
AOS y

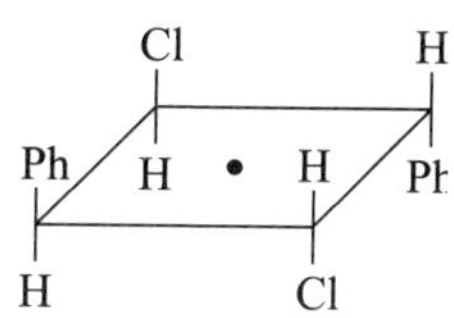

POS x
COS y
AOS x

Ans: (a and b)

47. (a), (b) Geometrical isomers (diastereomers)

(c) Stereoisomers which are not mirror image of each other are called as diastereomers.

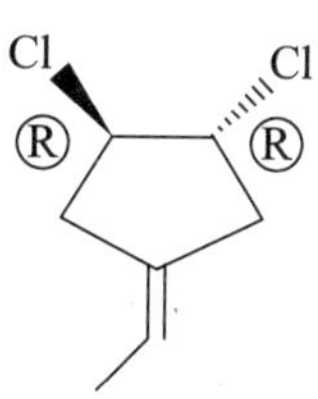

and

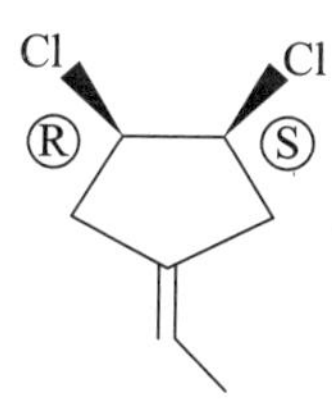

are diastereomers

Ans: (a, b, c and d)

48. (a) Planar molecule, has within plane POS.

(b)

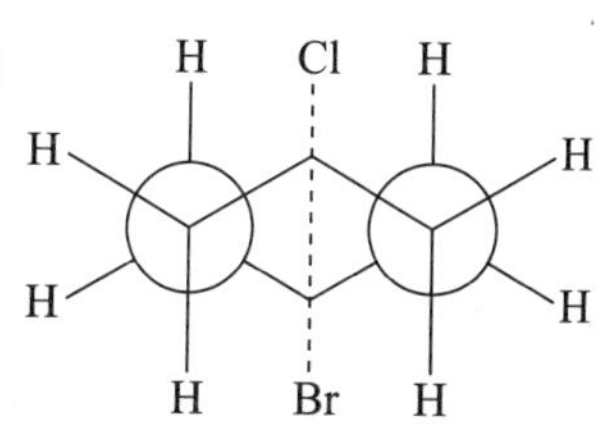

POS

Ans: (c and d)

50. (c) Functional isomers

Ans: (a, b and d)

51. (a)

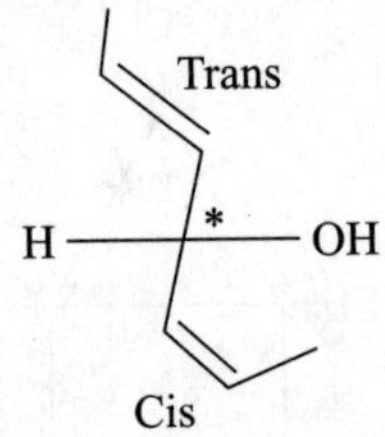

(b)

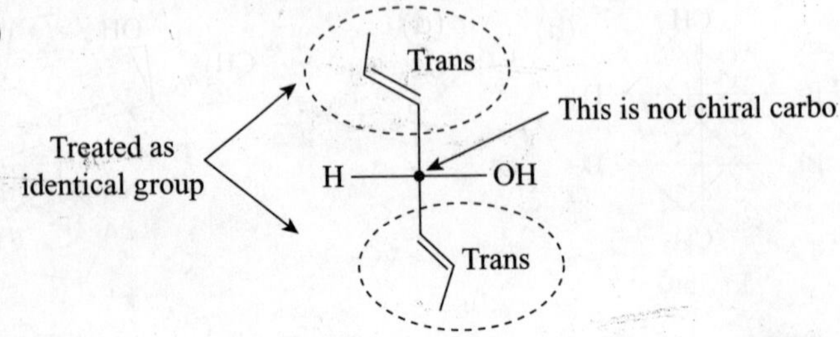

Ans: (a, c and d)

52. (a)

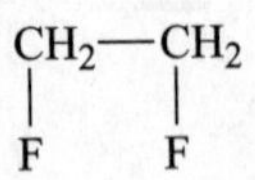

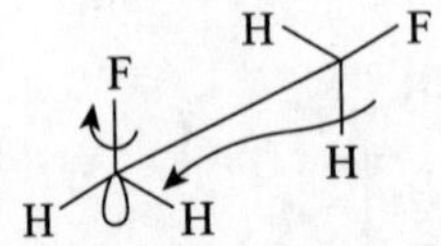

Delocalisation of bond-pair e⁻ of C–H bond in parallel ABMO of C–F is possible; leads to stability in gauche form.

(b)

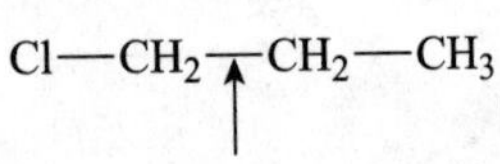

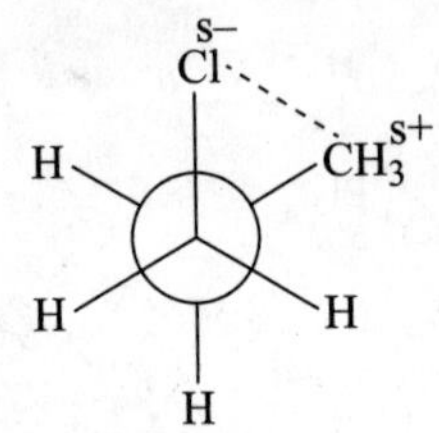

Due to +1 nature of –CH_3, s+ created on it while –I nature of –Cl induce s– on it.

Ans: (a, b, c and d)

53. (a)

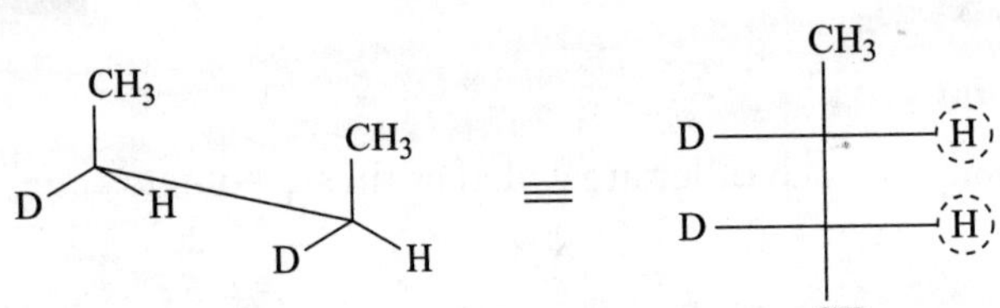

Erythro

(b)

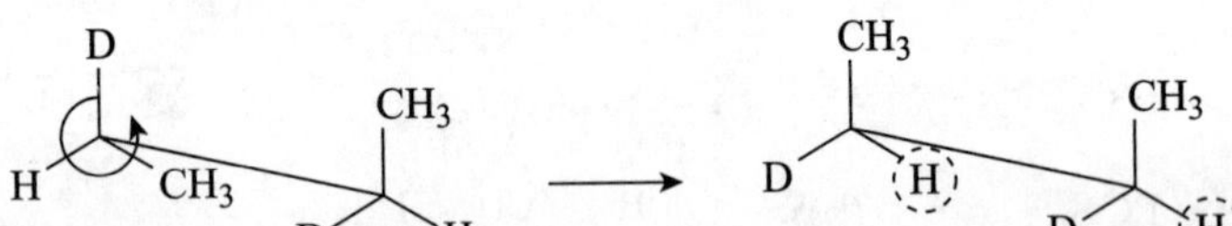

Erythro

(d)

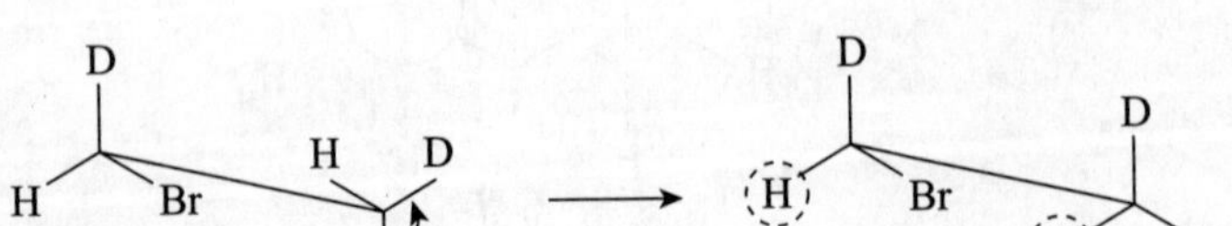

Erythro

Ans: (a, b, c and d)

54. (c)

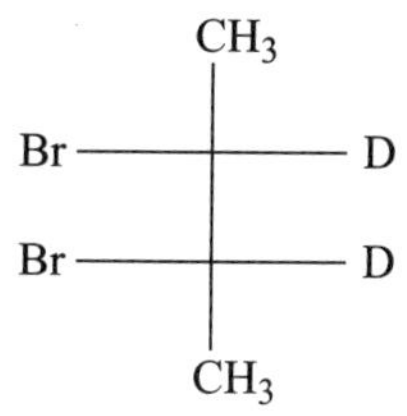

Erythro

(d)

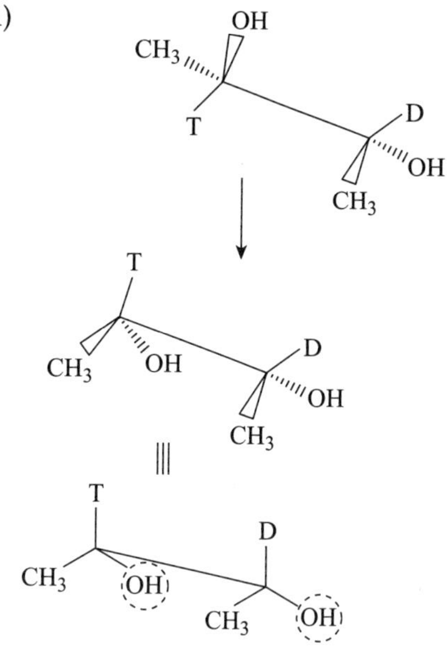

Ans: (a, b, c and d)

56. (a)

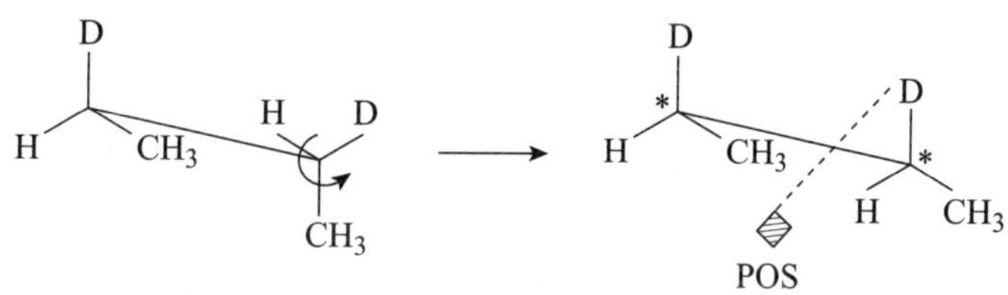

(b), (c): Also have same POS.

Ans: (a, b and c)

57. (a) has only 1 chiral centre. For meso compound atleast two chiral centres should be present along with POS or COS.

Ans: (b and d)

58. (a)

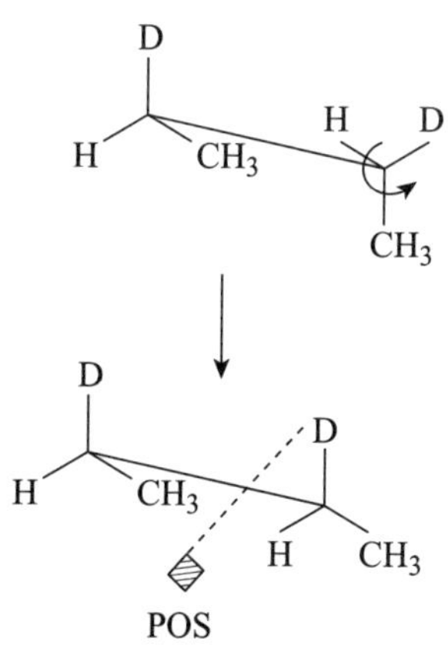

(b)

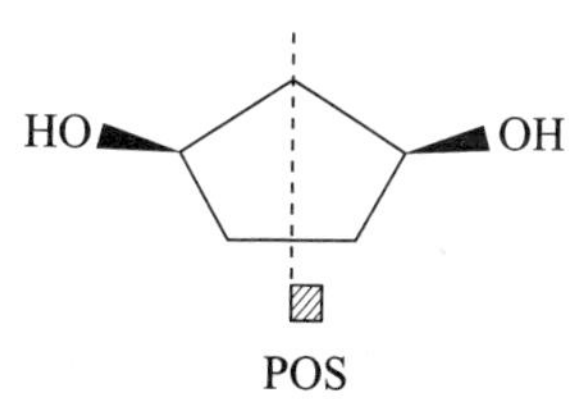

(d)

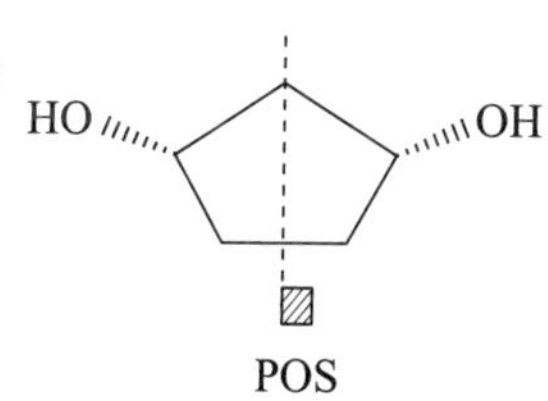

Ans: (a, b and d)

59. (a)

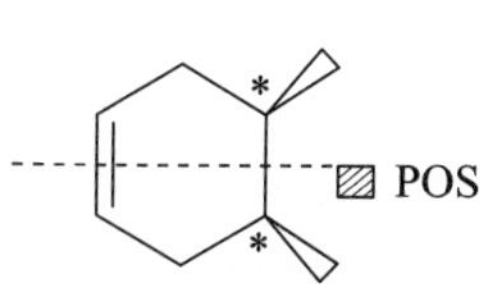

(c)

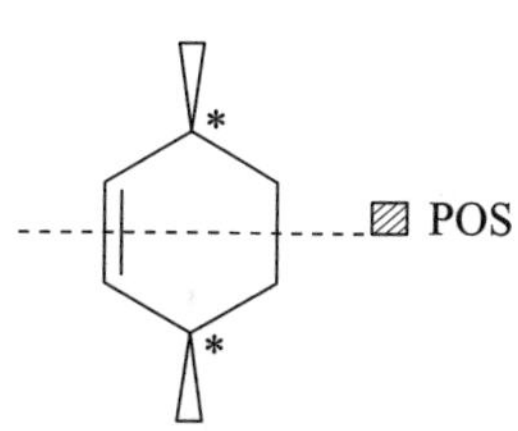

(d) No chiral centre

Ans: (a, b and c)

60.

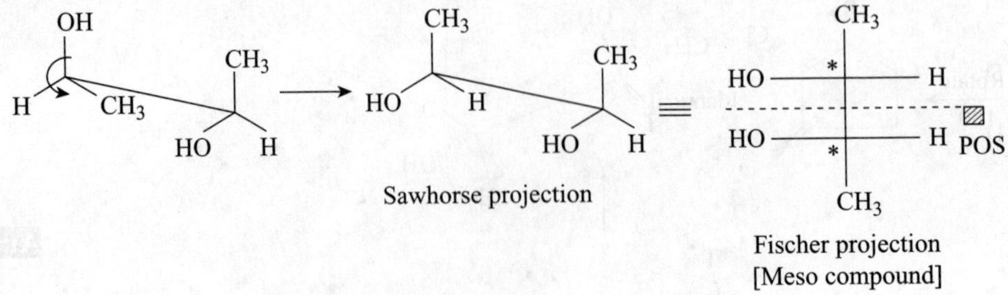

Ans: (a, b and c)

61.

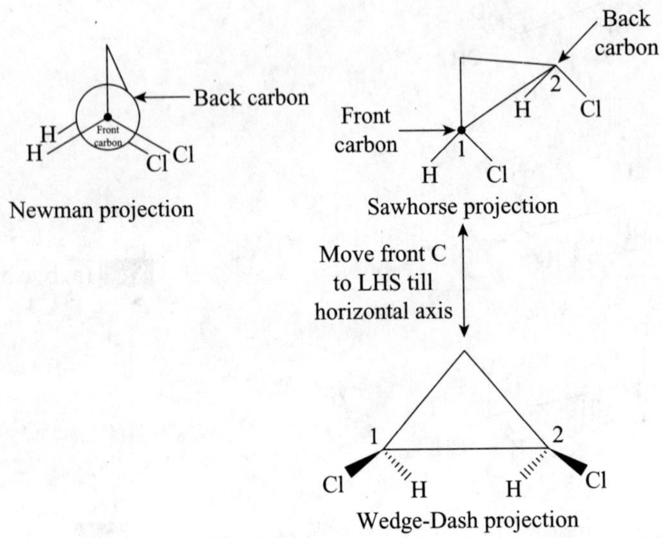

Ans: (c and d)

62. (a) 1, 4 trans (1, 4 : Indicate position of CN)

(b) 1, 3 cis

(c) 1, 4 cis

(d) 1, 1

Ans: (a and c)

63. (a) 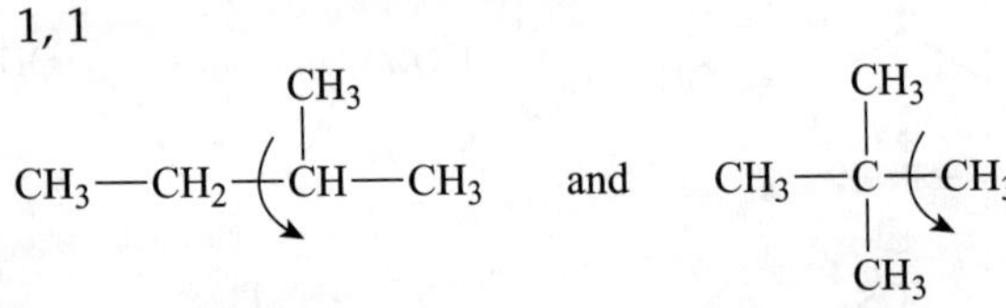

: Chain ISO

(b) 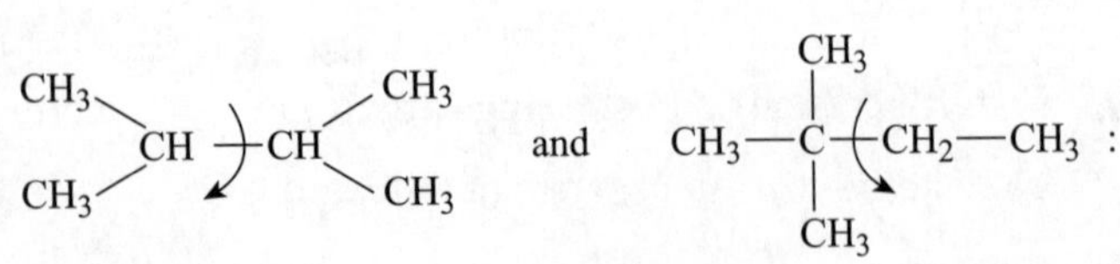

: Position ISO

(c) 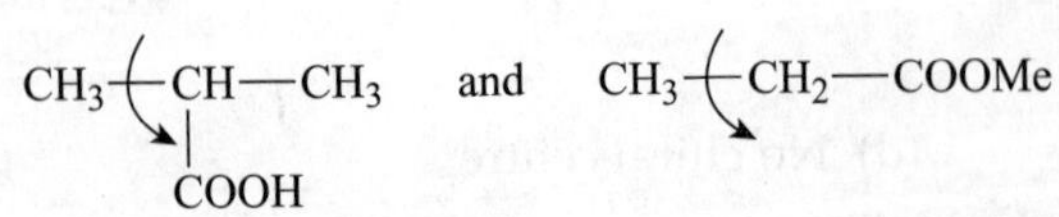

: Functional ISO

(d) Diastereomers

Ans: (a)

64. (b) 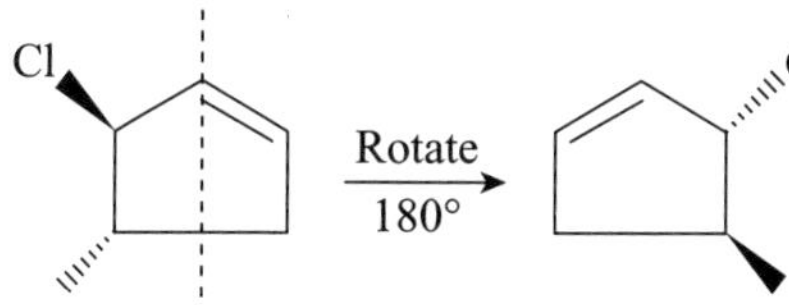

Identical

(c) Position isomer

(d) Enantiomers

Ans: (a)

65. (a) 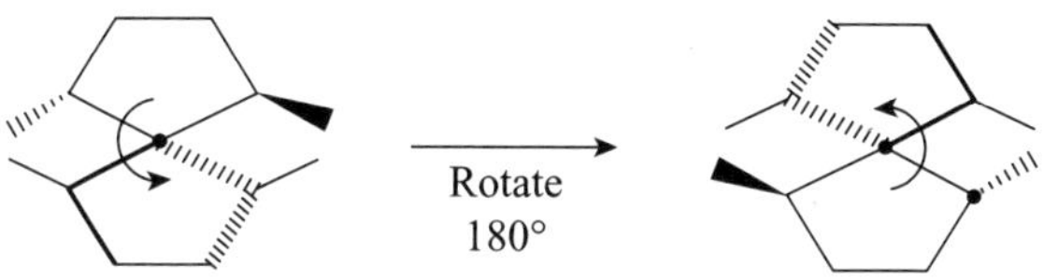

: Identical

(b) Identical:  rotate 180°

Axis
perpendicular to plane

Ans: (c and d)

66. (b) Enantiomers: Mirror image about top mirror

(c) Enantiomers

(d) Identical: 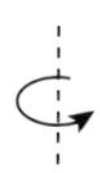 rotate 180°

Vertical axis
(in plane)

Ans: (a)

67. (a) Diastereomers: Stereoisomers that are not mirror image of each other.

(b) Identical:  rotate 180°

(c) Identical: Both are meso form of same compound.

(d) Enantiomers

Ans: (a)

68. (a) Geometrical isomers

(b) Identical

(c) Identical  rotate 180°

(d) Enantiomers: Mirror image.

Ans: (d)

69. (a) Optically active as POS and COS are absent because all cyclic rings are not in same plane.

(b) Helical structure like on screw is optically active due to absence of POS and COS.

(c), (d): Within plane POS present

Ans: (a and b)

71. (a) Enantiomers: Mirror image about top mirror

(b) Geometrical isomer

(c) Diastereomers

(d) Enantiomers: Mirror image

Ans: (a and d)

74.

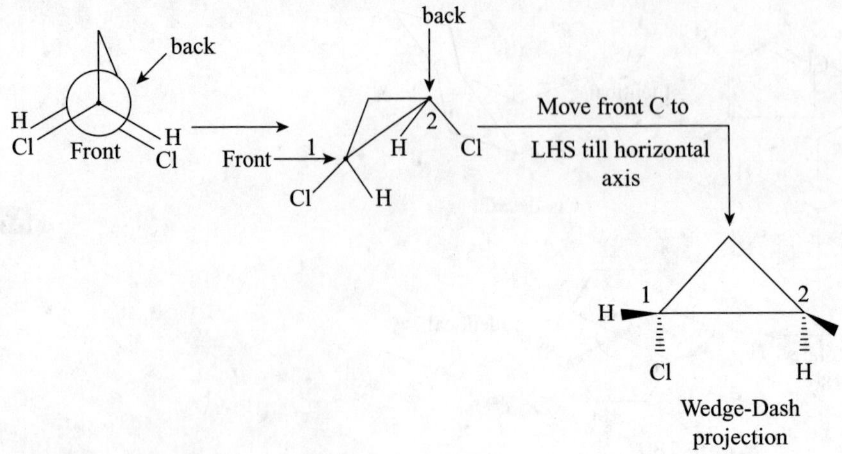

(d) Refer Question Number 61.

Ans: (b and c)

75. (c)

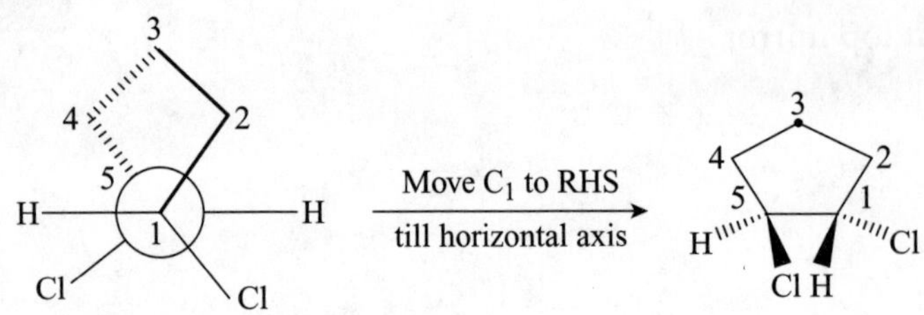

(d)

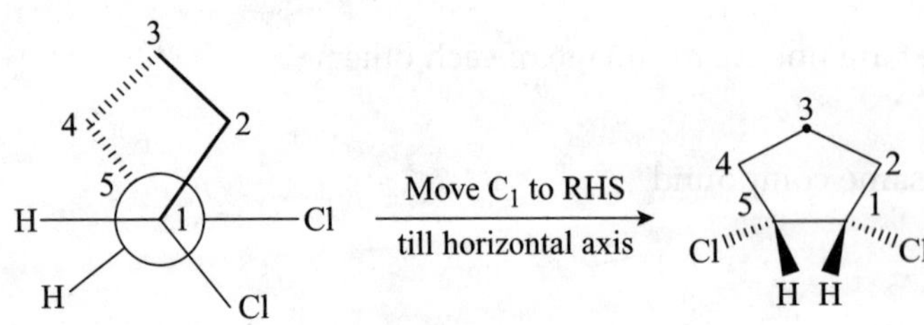

Ans: (b and c)

76. (b)

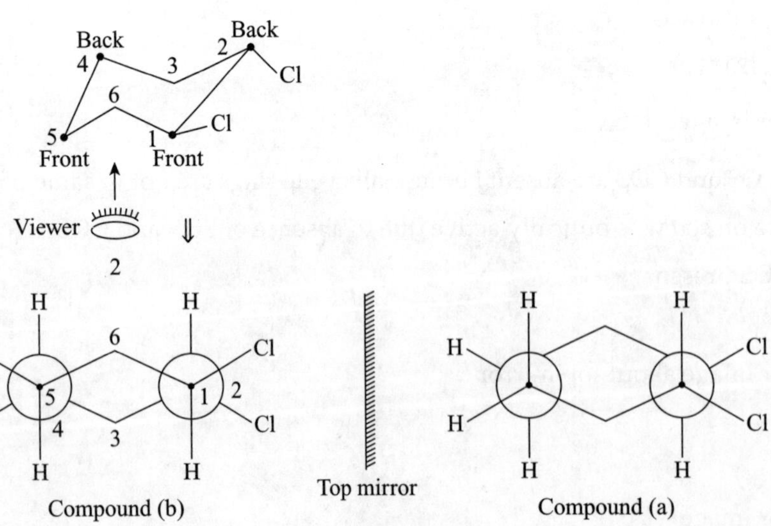

Ans: (a and b)

77.

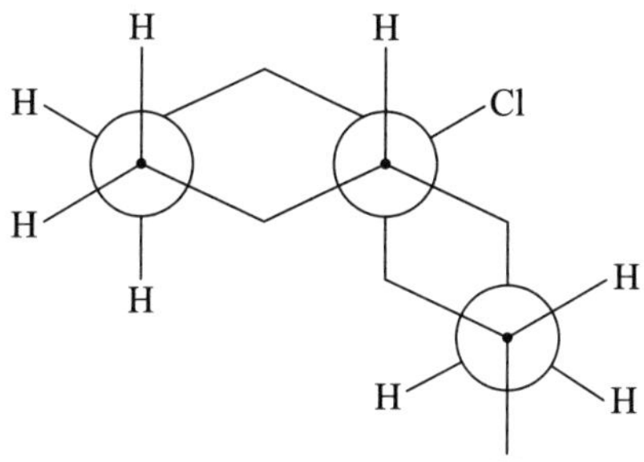

Cis-decalin

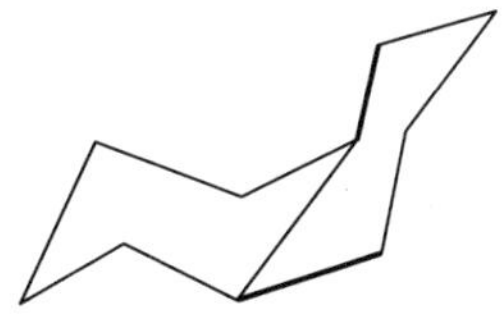

Cis-decalin

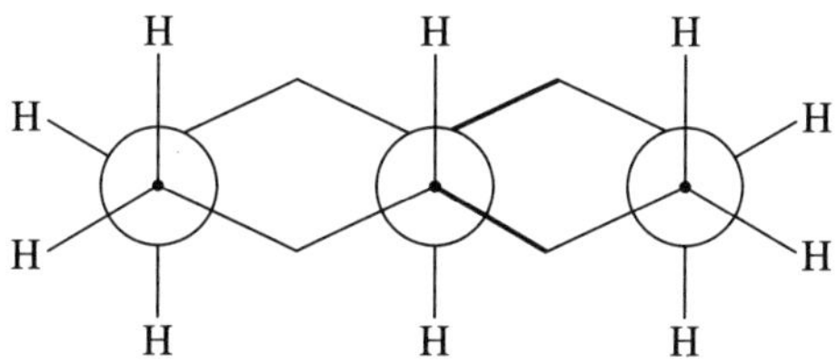

Trans-decalin

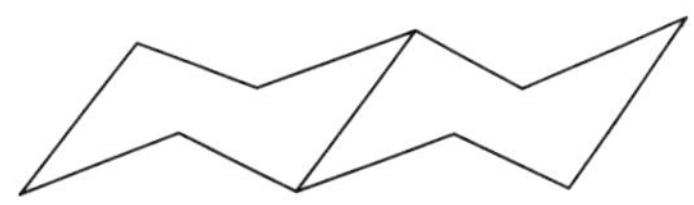

Trans-decalin

Ans: (b and c)

78. Zn dust provides anti-elimination of halogens to give alkene.

(a)

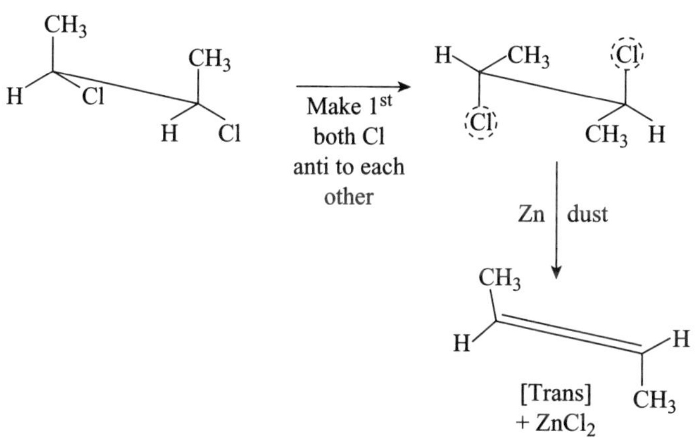

(b)

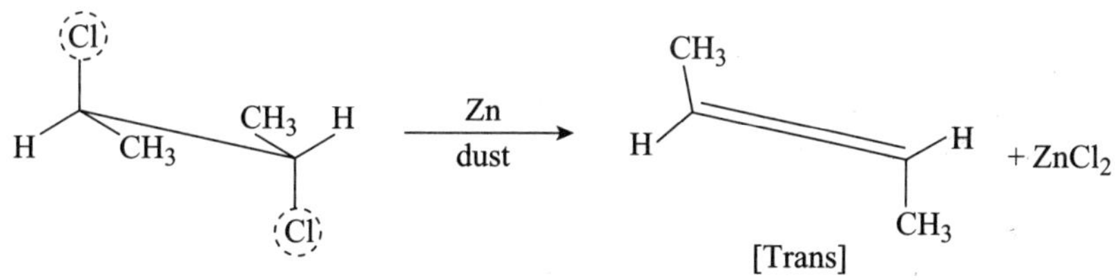

First make Cl anti to each other

(c)

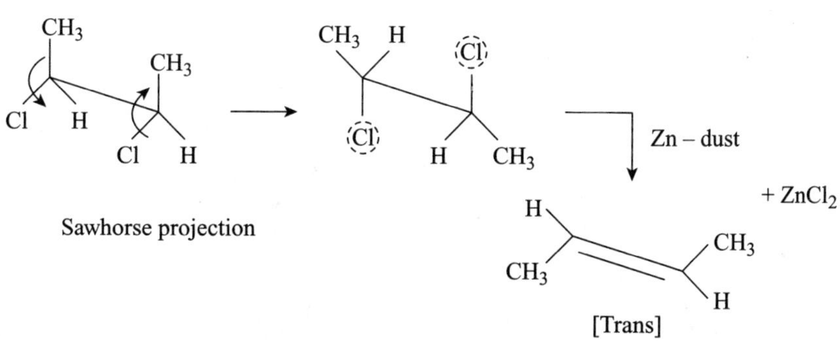

Sawhorse projection

(d)

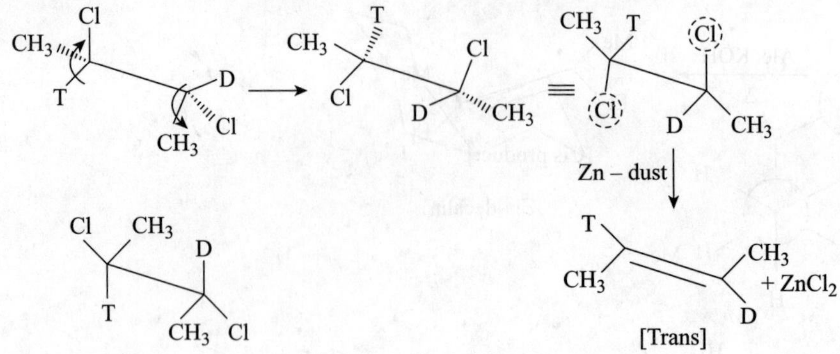

Ans: (a, b, c and d)

79. (a)

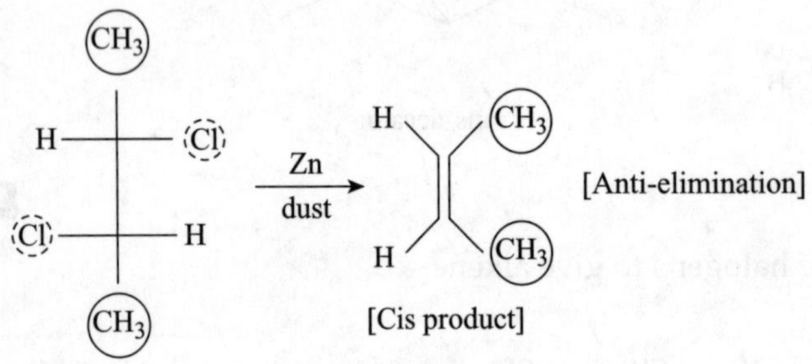

In Fischer angle between both (top and bottom) $-CH_3$ is 0°

Ans: (a, b and c)

80. (b)

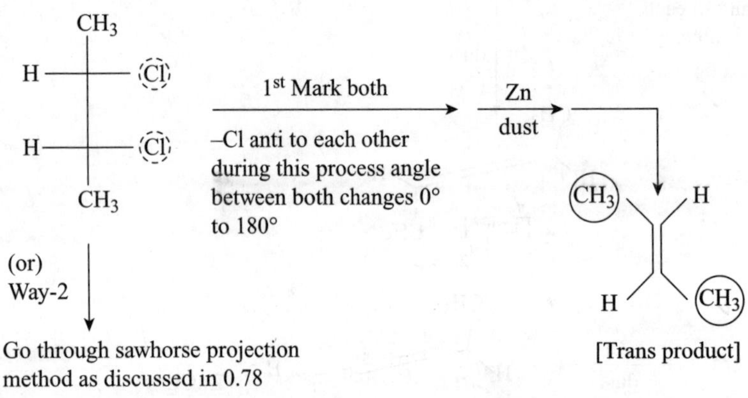

Ans: (a, b and d)

84. Alcoholic KOH, Δ gives anti-elimination of HCl via E2 mechanism.

(a)

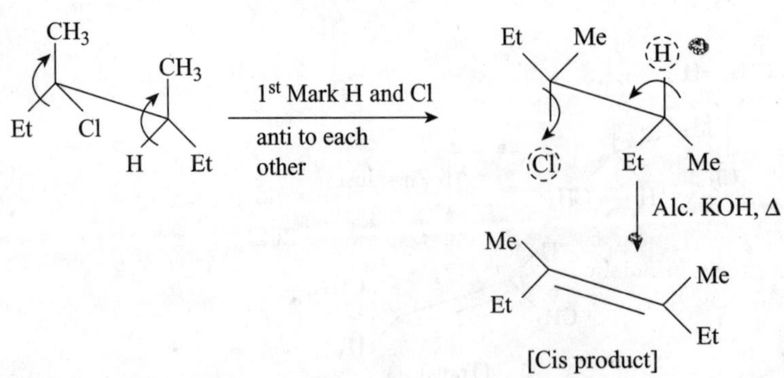

(b)

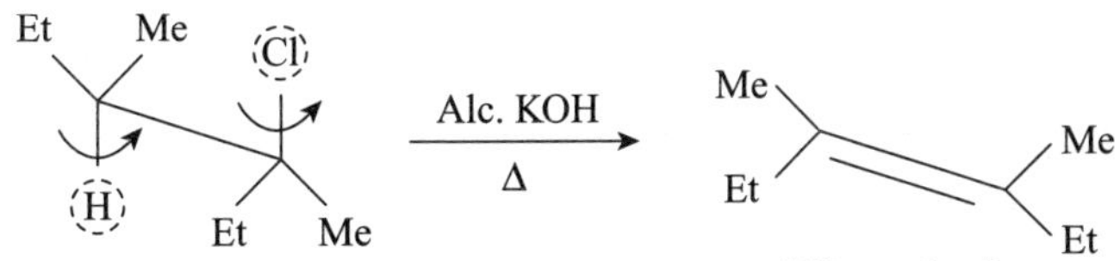

[Cis product]

(c)

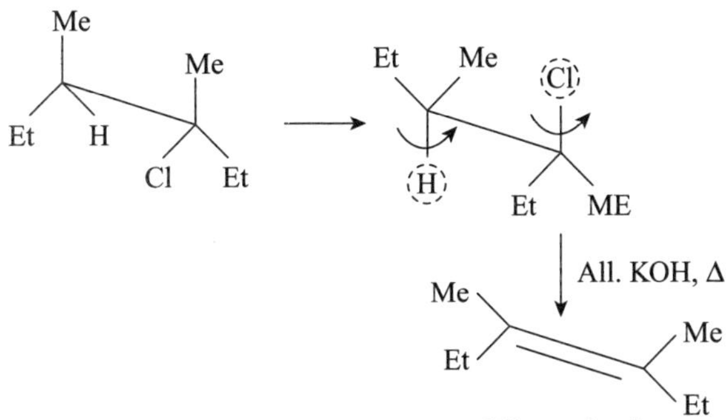

[Cis product]

Ans: (d)

Comprehension Type

Passage 1

89. (b)

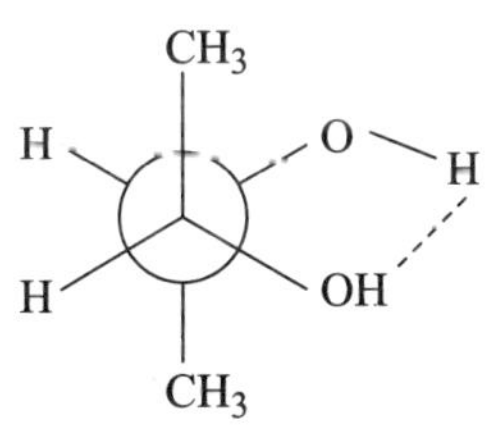

Gauche stabilised by H-bonding

Ans: (b)

Passage 2

92.

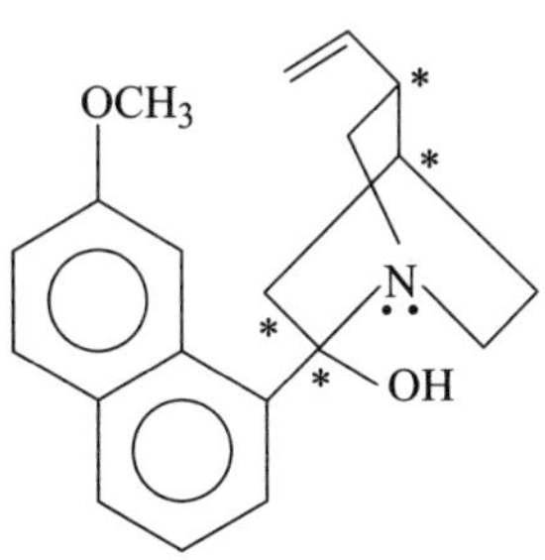

Chiral carbon: 4

Chiral Centre: 5 (including nitrogen)

Ans: (b)

93. Chiral compound **Ans:** (d)

94. (a)

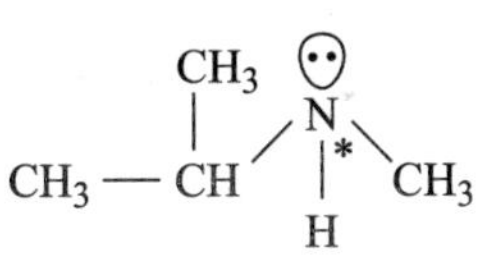

Exist in (±) from that's racemic mixture form due to umbrella inversion (flipping).

Ans: (c)

96. (c) Conformers **Ans:** (a and b)

97. 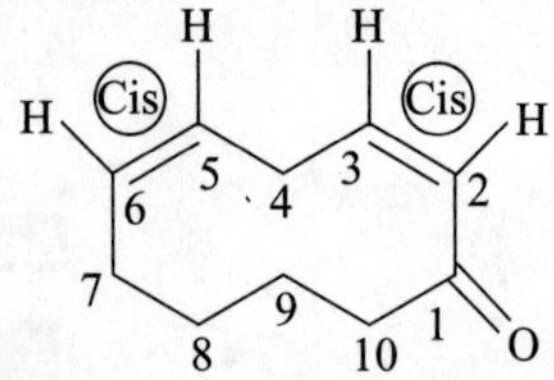 and 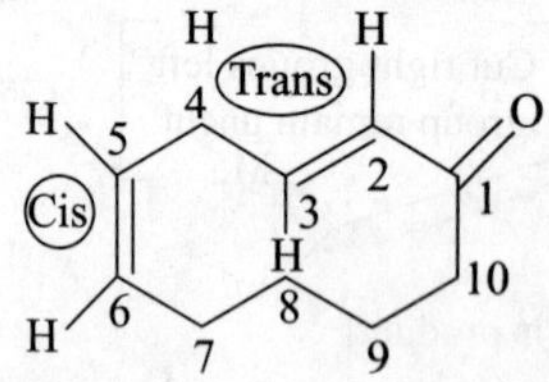

Ans: (b)

Passage 5

101. to 103.

1st pair

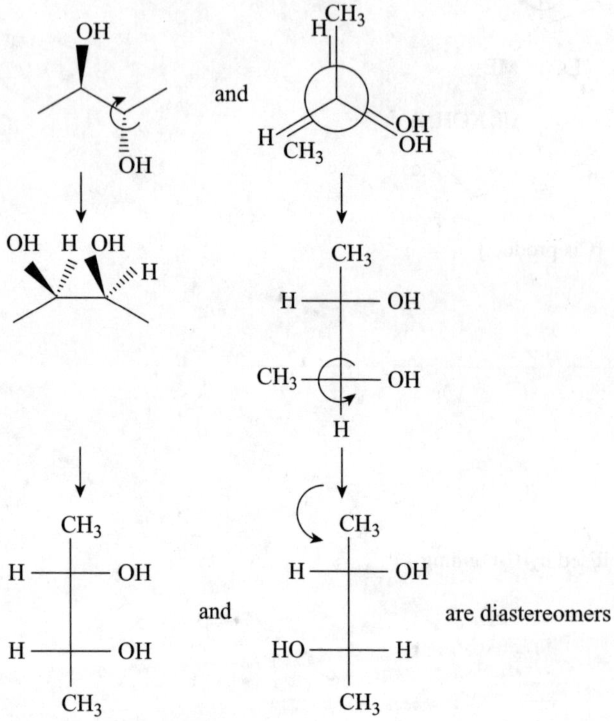

2nd pair: Enantiomers (mirror image of each other with absence of POS and COS)
3rd pair: Identical ['S_4' present]
4th pair: Identical (meso form of same compound)
5th pair: Identical (meso form of same compound)

Ans: 101 (c), 102 (c) and 103 (a)

Passage 6

104. (a)

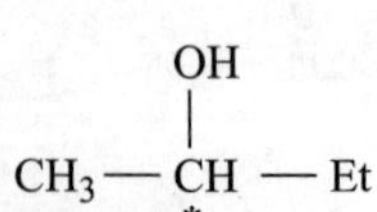

(b) No POS. and No COS. Therefore optically active.

(c) 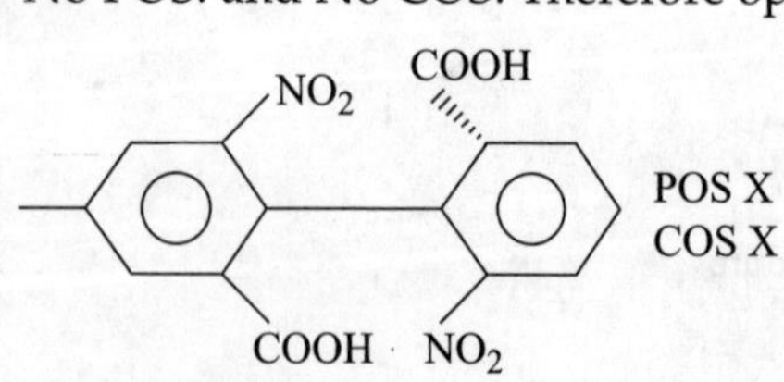

Therefore optically active.

(d) 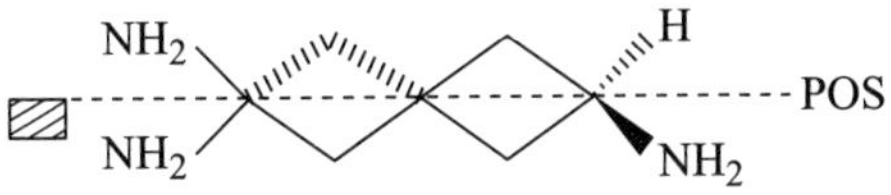

Cut right groups left group remain uncut

Therefore optically inactive. **Ans:** (d)

105. (b) Within plane POS is present **Ans:** (c)

106. to 108.

(i) 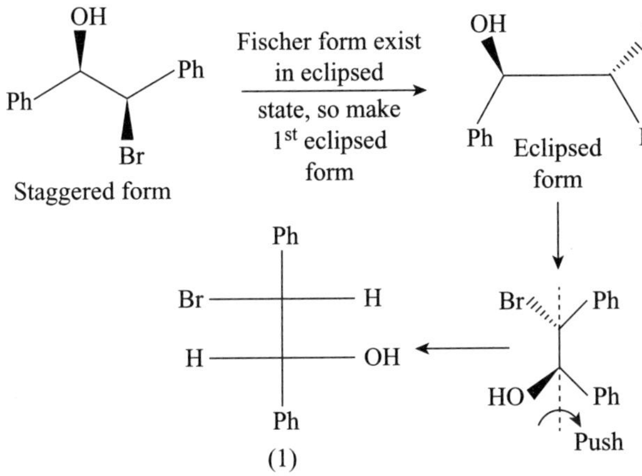

(iv) 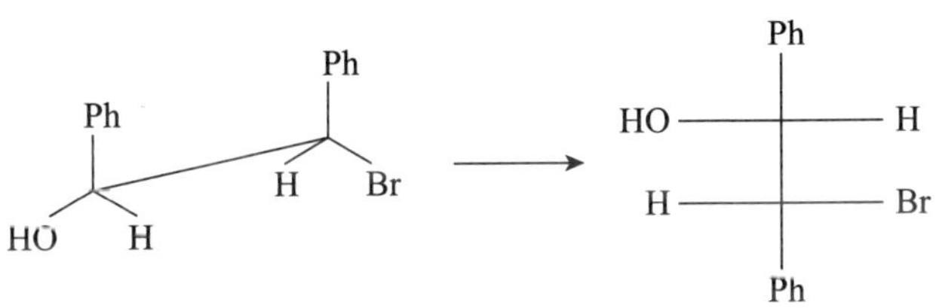

(v) 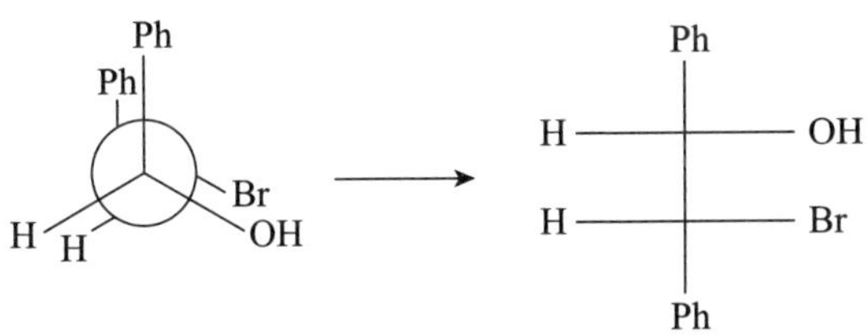

(vi) 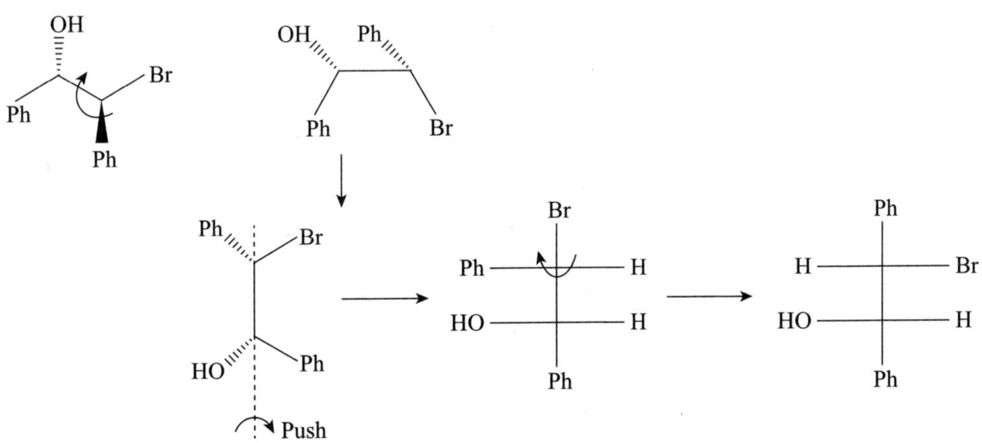

Ans: 106 (d), 107 (c) and 108 (a)

109.

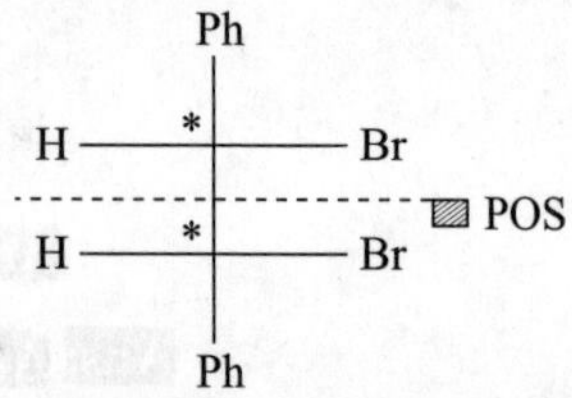

* MeSO compound
* DU = 8

Ans: (a)

Matrix Type

110. (a)

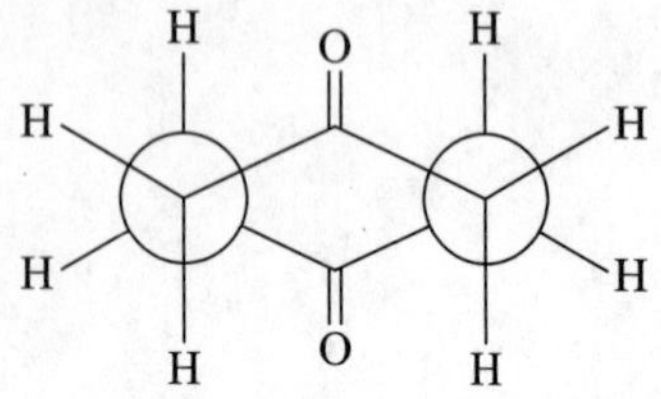

H = O Projection. This is Newman's chair form that is 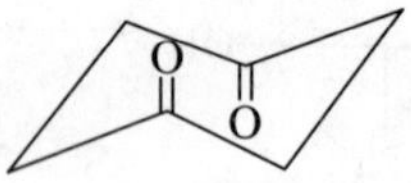 out of 1 mole compound, few may exist in boat form 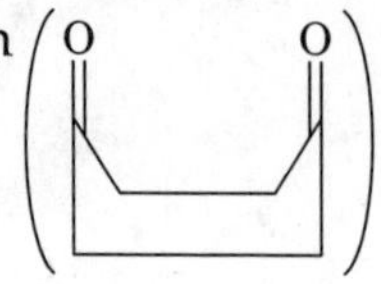 where M ≠ O. Therefore, net dipole moment of 1 mole of compound is non-zero.

DU = 3

Ans: (r, s and t)

(b) DU = 8. In 1 mole of compound few may exist in gauche form where $\mu \neq O$. So (μ_{net}) for 1 mole ≠ O.

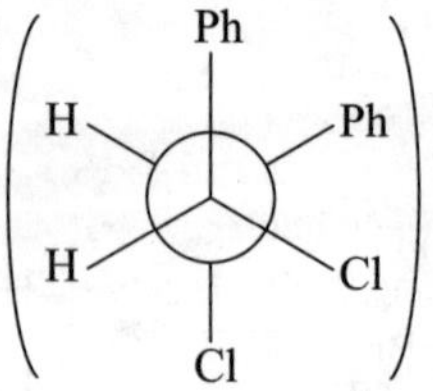

Ans: (q, r and s)

(c) DU = 4

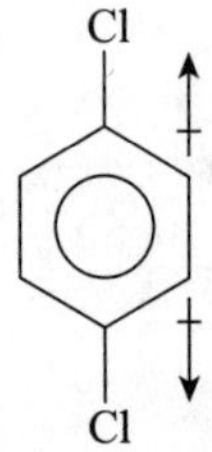

Ans: (p, q and r)

111. (b) $-\overset{O}{\overset{||}{C}}-$ is bivalent functional group on changing arrangement of carbon-atom both side of it, metamers form.

Ans: (r and t)

(d) 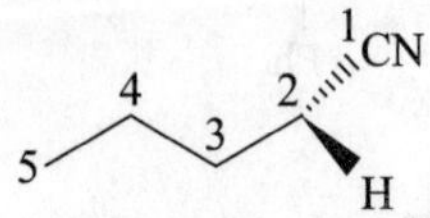 and 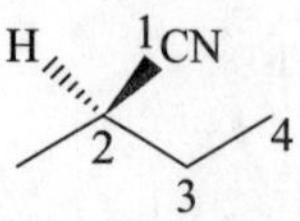 are chain ISO.

Ans: (s)

112. (a)

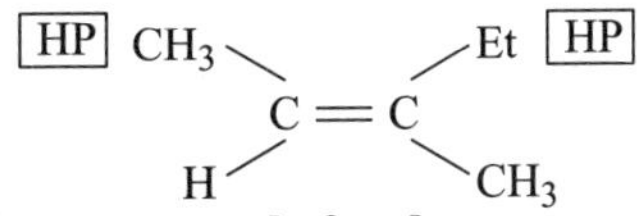

[z-form]

[*HP : Higher priority]

within plane, POS is present.

Ans: (q and t)

(b)

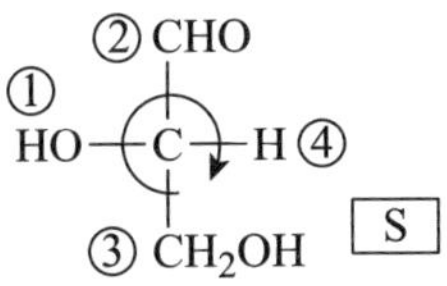

Ans: (s)

(c)

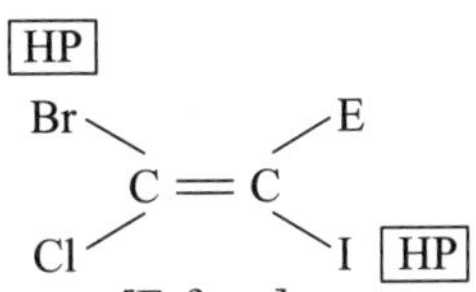

[E-form]

Planar molecule; within plane POS is present.

Ans: (p and t)

(d)

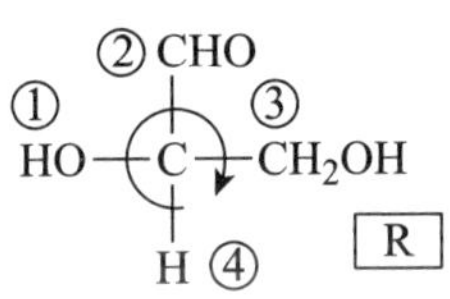

Ans: (r)

113. (a)

 C_3

About axis perpendicular to plane

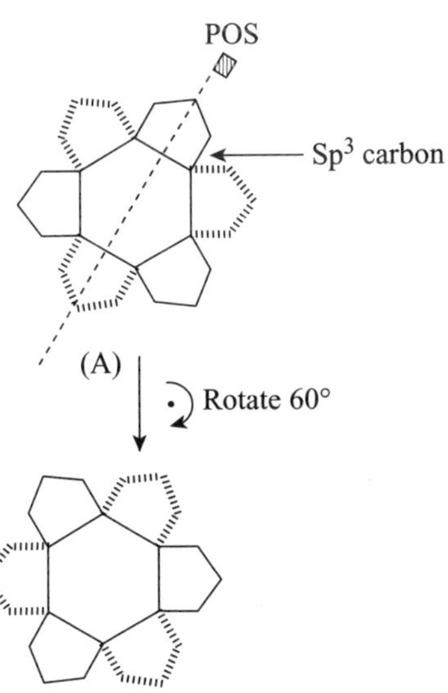

(B)

Ans: (r and s)

On rotating A, 60° A, we got B that is mirror image of A therefore S6 (AAOS) is present.

(b) ⤵ C_5 (AOS)

About axis perpendicular to plane of paper.

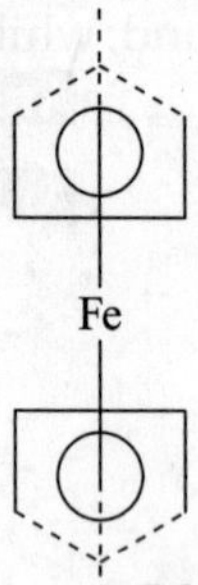

COS also present

▨ POS

Ans: (s and t)

(c) POS present (within plane POS)

→ ⤵ C_6 (AOS)

→ ⤵ S_6 (AAOS)

→ COS present

(d) ⤵ S_2 (AAOS)

Ans: (q, r and s)

114. (b) Alcohol and phenol are functional isomers. **Ans:** (s)

(d)

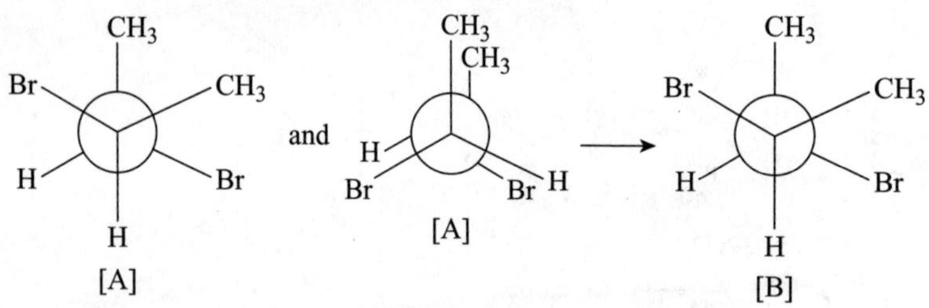

A and B are mirror image about top mirror, so enantiomers. **Ans:** (t)

115. (a)

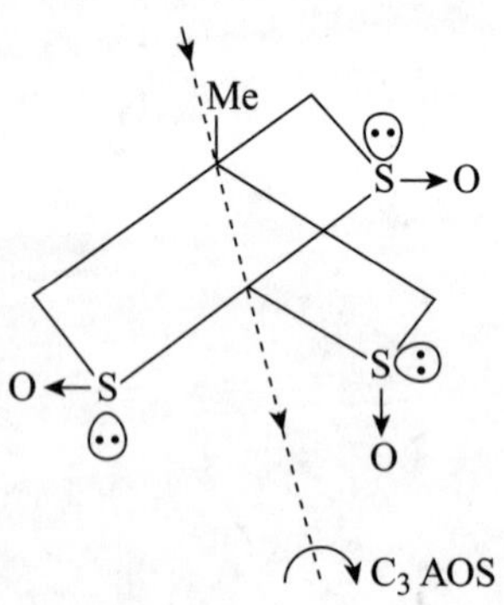

POS X
COS X

Ans: (q and s)

(c)

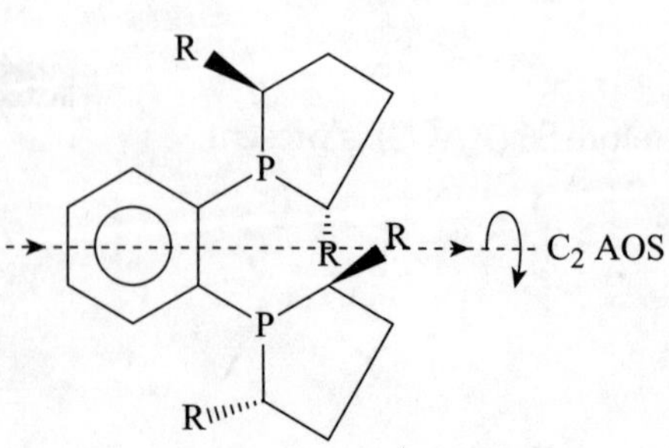

POS X
COS X

Ans: (q and s)

(d) POS X

COS X

AOS X optically active compound with absence of AOS called asymmetric compound; while optically active compound with presence of AOS called dissymmetric compound. **Ans:** (q and t)

116. (a)

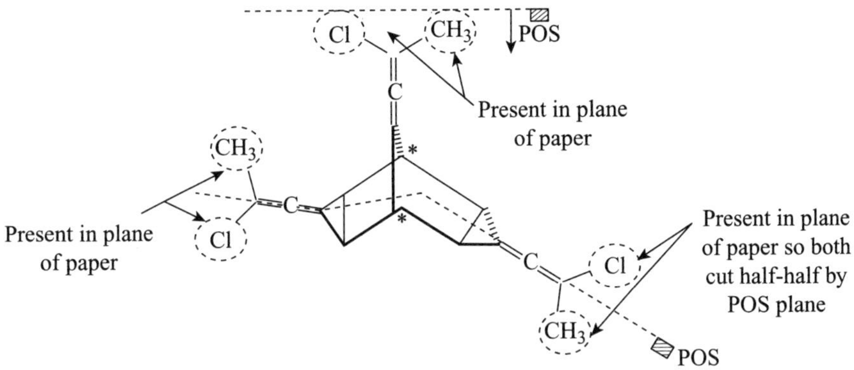

Each –Cl and –CH_3 cut half-half individually by POS plane.

2 Chiral centre are present with POS, so meso compound. **Ans:** (p and r)

(c)

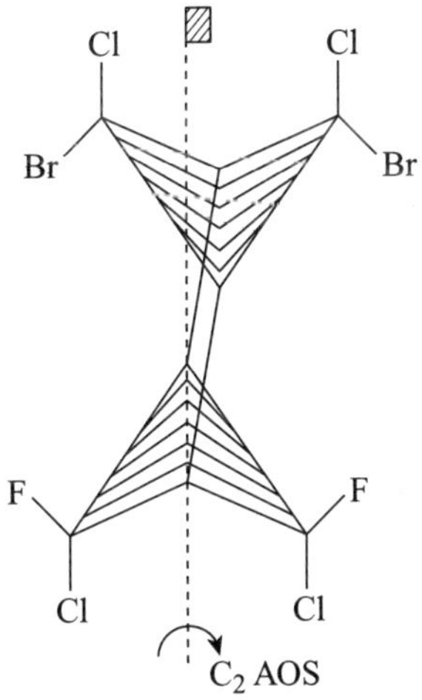

Chiral centre is not present so say no to Meso

Ans: (p and s)

(d)

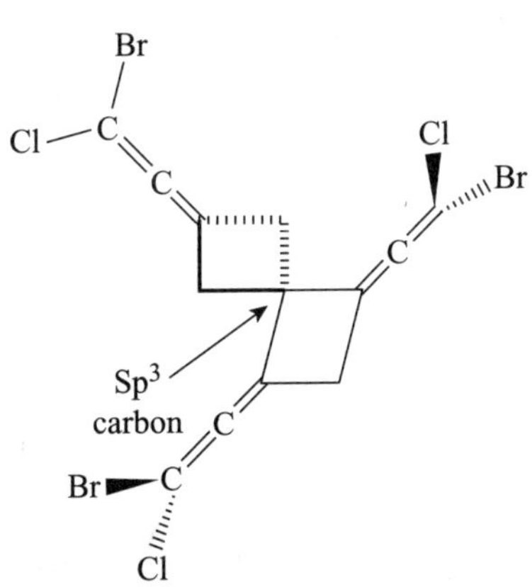

POS X
COS X
AOS X

Ans: (q and t)

117. (a), (b)

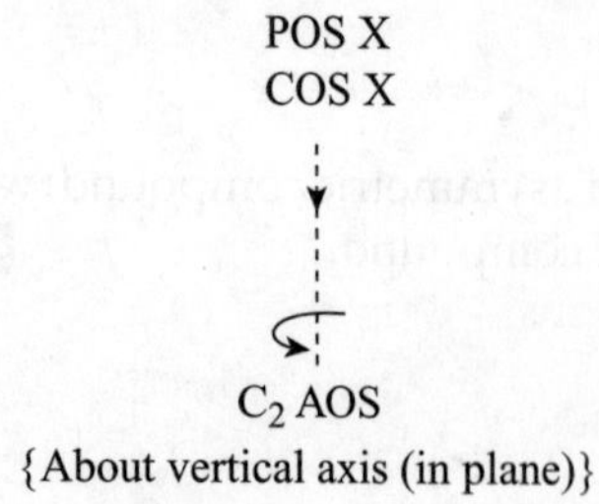

Ans: (a) and (b) (q, s and t)

(c)

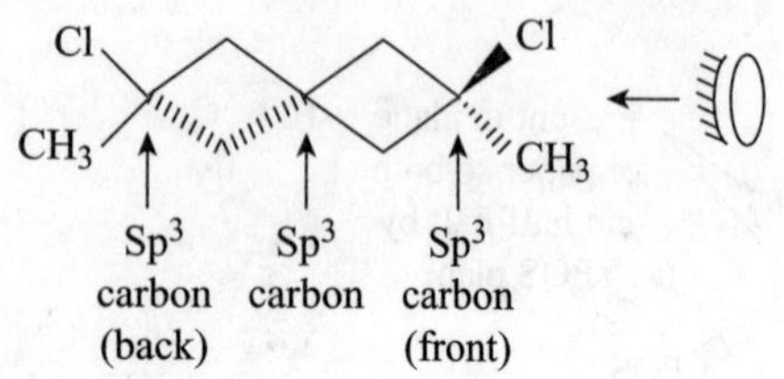

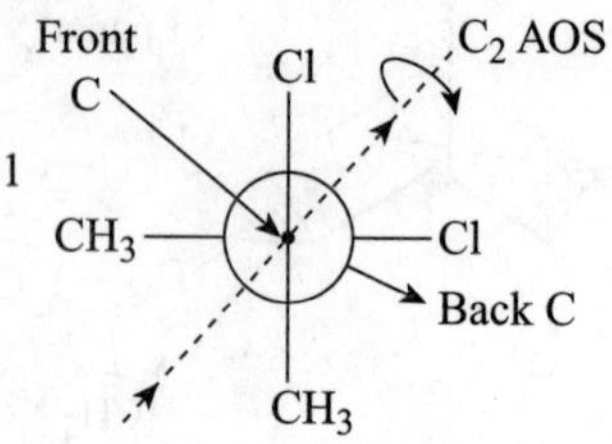

Ans: (q, s and t)

(d)

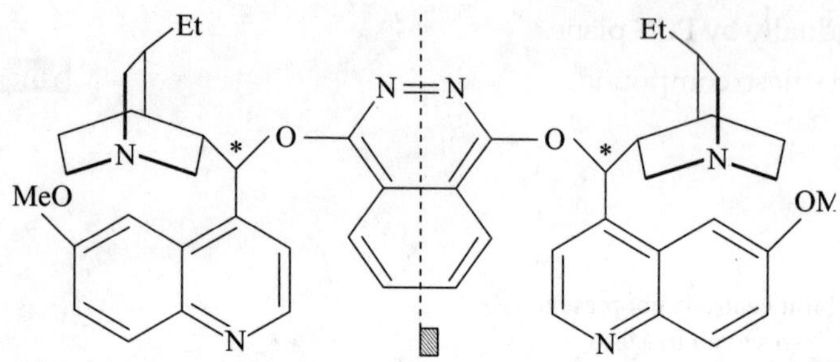

Ans: (p and r)

118. (a) POS X
COS X
AOS X

Ans: (q and t)

(b)

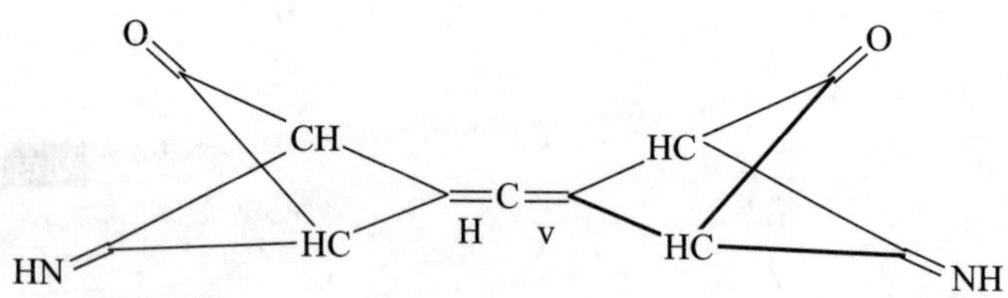

POS X
COS X
AOS X

Ans: (q and t)

(c)

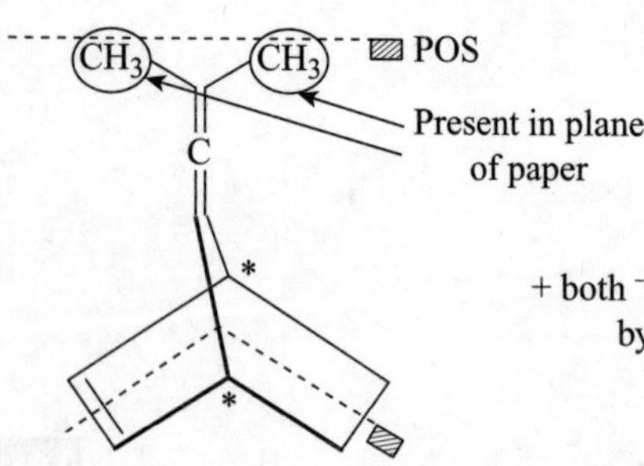

+ both $^-CH_3$ cut half-half by POS plane

Ans: (p and r)

(d)

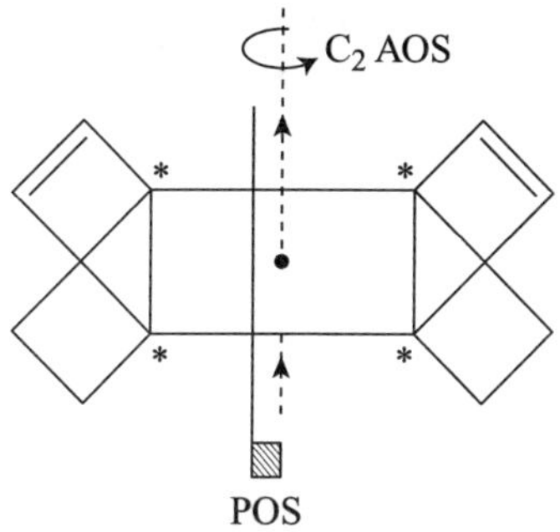

Ans: (p, r and s)

119. (a)

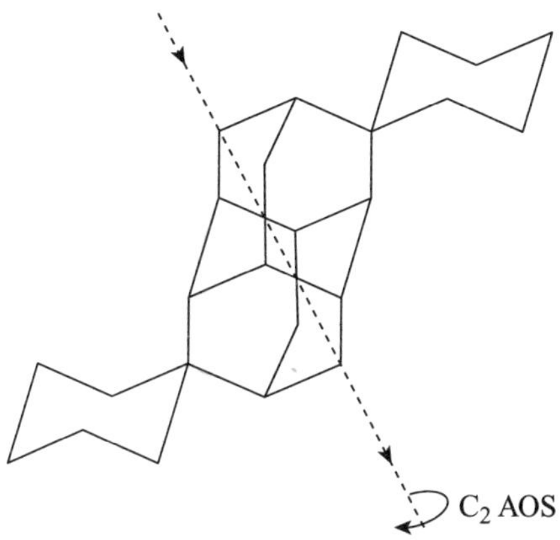

Chiral centre present

COS and So meso compound

Ans: (p, r and s)

(b)

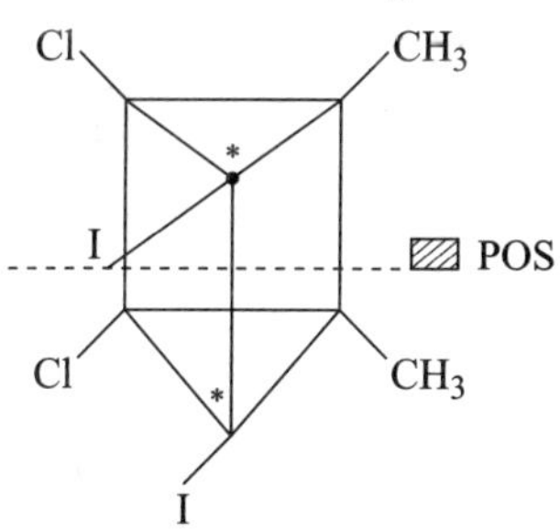

(c)

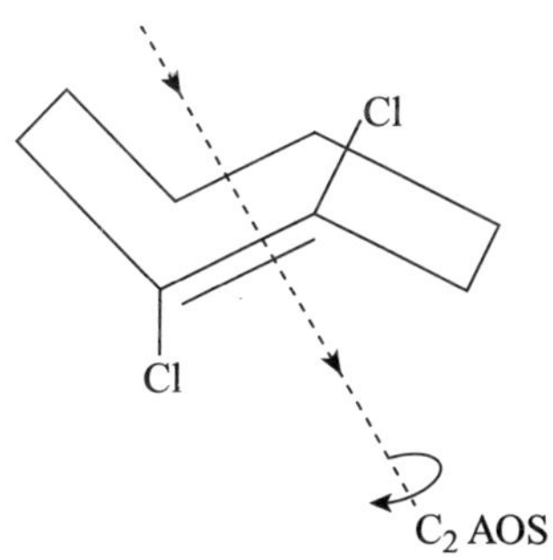

Ans: (p and r)

(d) POS X

COS X

AOS X

Ans: (q and t)

120. (a)

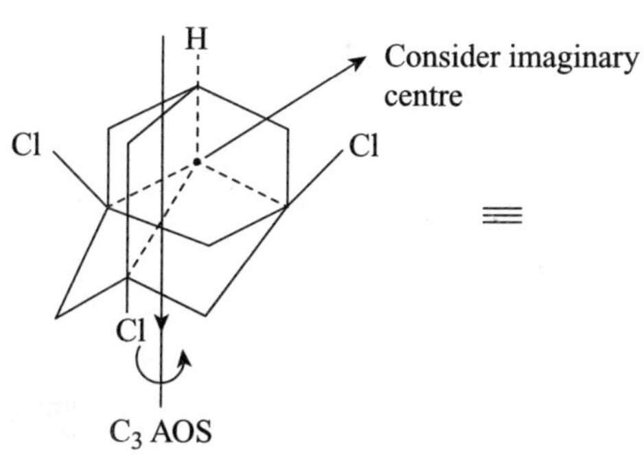

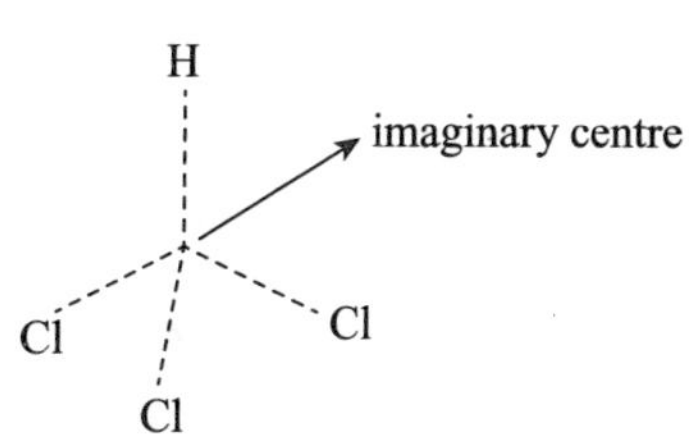

it has POS

Ans: (p and s)

(b)

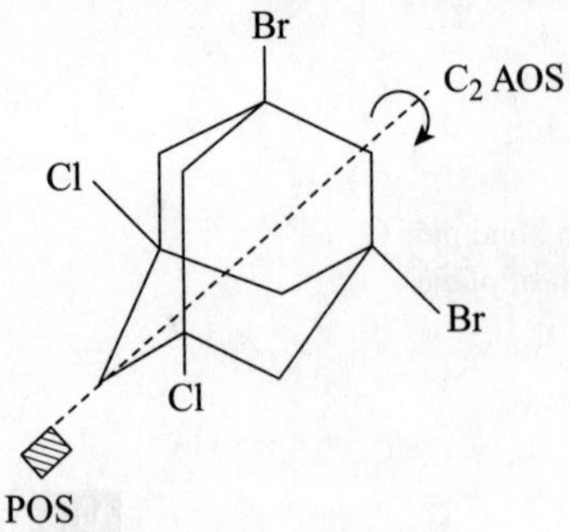

Ans: (p and s)

(c) POS present

Ans: (p)

(d) POS X
COS X
AOS X

Ans: (q and t)

121. (a) POS X DU = 3
COS X
AOS X

Ans: (q, r and t)

(b) (c) POS X DU = 5
COS X
AOS X

Ans: (q, r and t)

Ans: (q, r and t)

(d) POS X DU = 5
AOS X

Ans: (p and r)

122. (a)

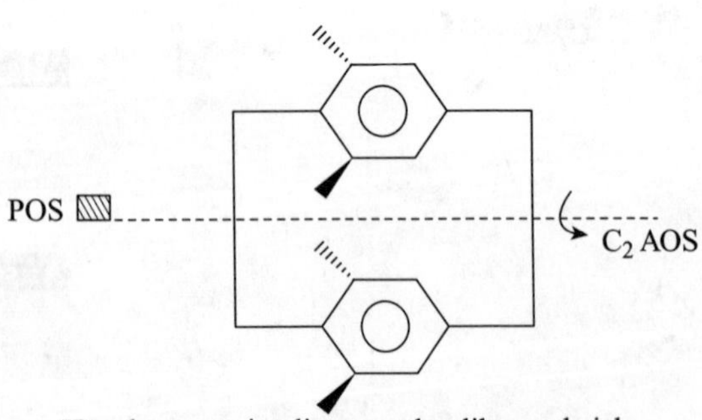

Here benzene ring lie over other like sandwich

Ans: (p and s)

(b) COS

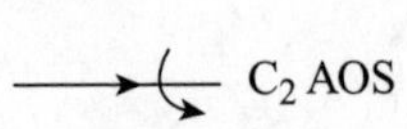

About horizontal axis (in plane)

Ans: (p and s)

(c)

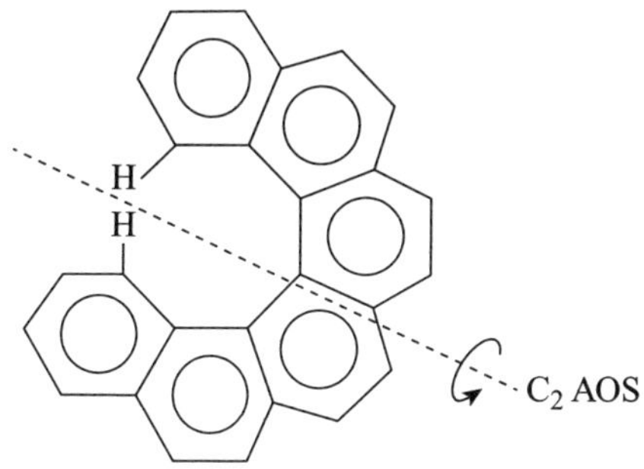

Due to repulsion between bond pair C has terminal ring goes out of plane therefore POS X

Ans: (q, r and s)

(d)

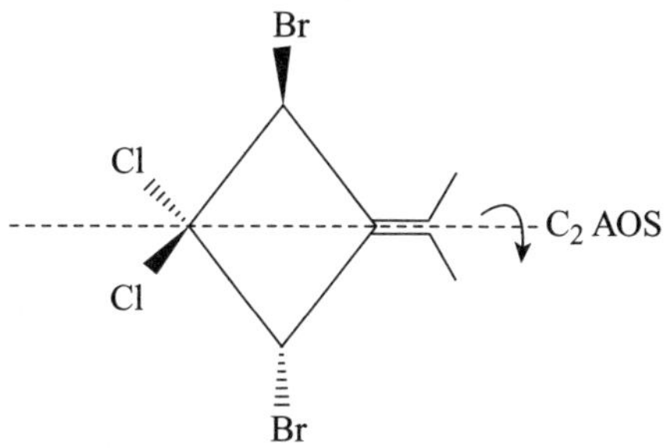

Ans: (q, r and s)

123. (a)

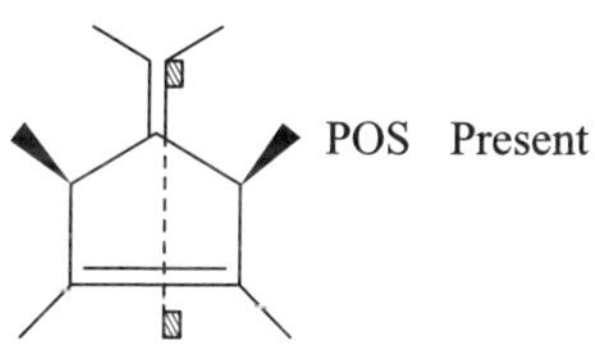

Ans: (p)

(b)

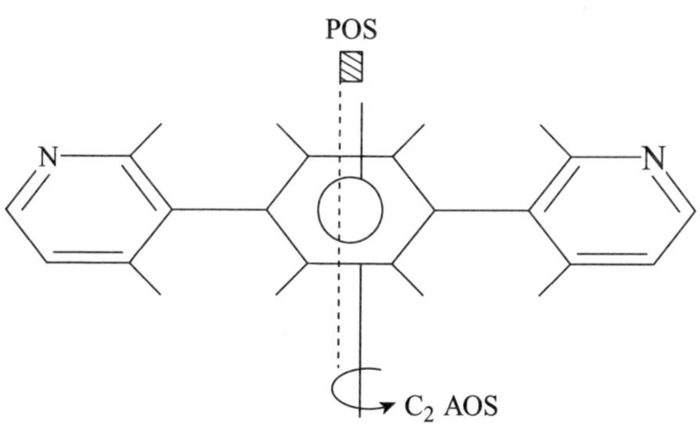

Ans: (p and s)

(c) POS X

COS X

AOS X

Ans: (q and t)

(d)

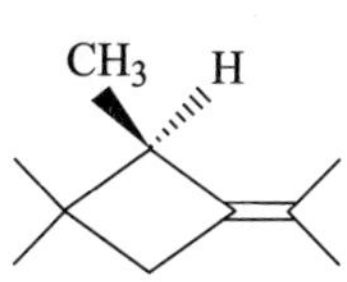

POS X
COS X
AOS X

Ans: (q and t)

124. (a)

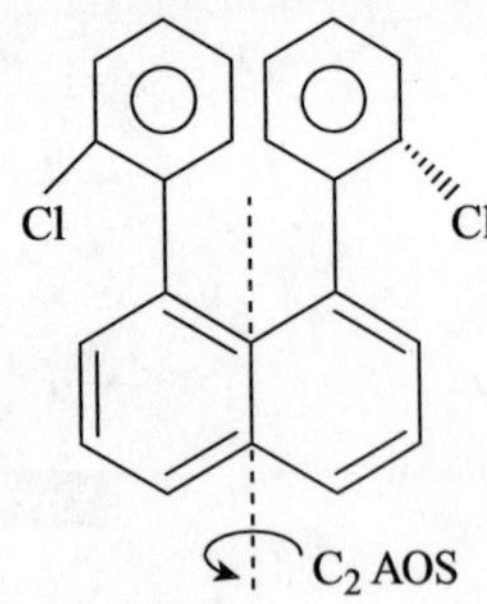

POS X
COS X

Ans: (q, r and s)

(c)

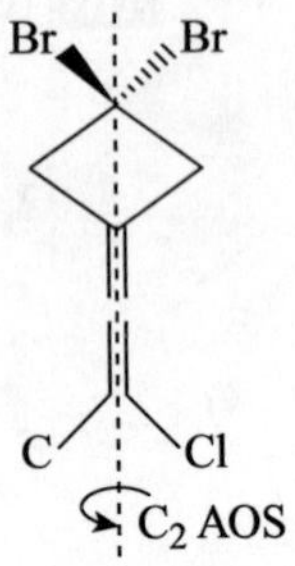

Within plane POS present.

Ans: (p and s)

(d) Refer Question Number 122 (d) **Ans:** (q, r and s)

125. (a) Chiral centre = 4
POS X **Ans:** (q and r)

(b) Chiral centre = 4
POS X **Ans:** (q and r)

(c) Chiral centre = 4
POS Y **Ans:** (r and s)

(d) POS X
COS X, so optically active. **Ans:** (p)

126. (a)

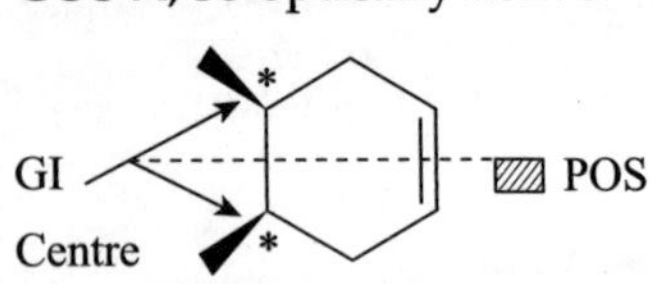

Ans: (q, r and s)

(b)

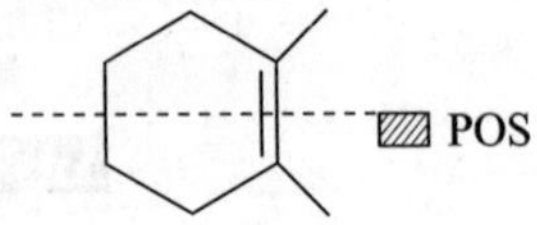

Ans: (q and s)

(c)

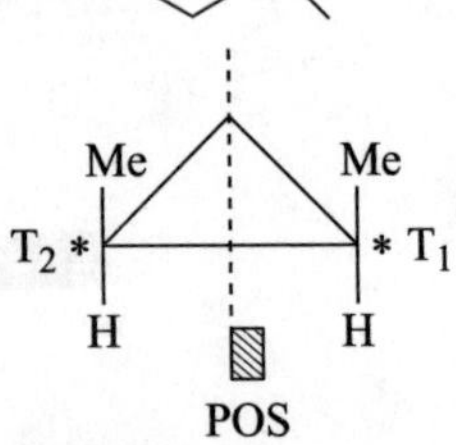

GI is possible about T_1 and T_2 terminal

Ans: (q, s and r)

(d)

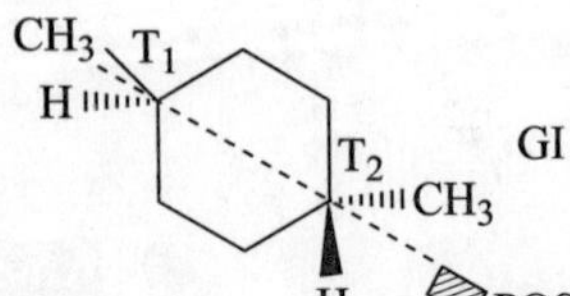

GI is possible about T_1 and T_2 terminal

Ans: (q, r and s)

127. (a)

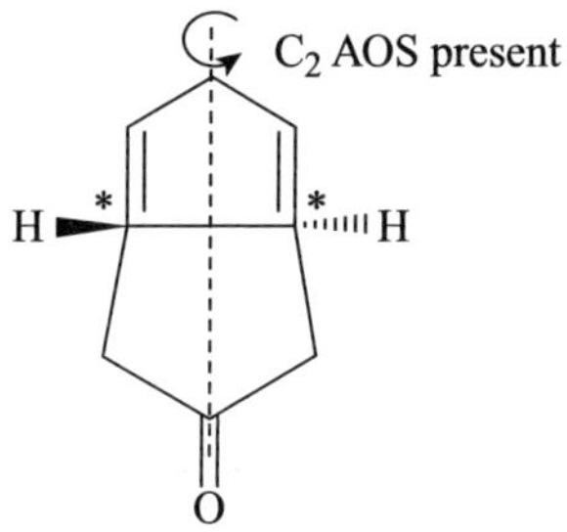

POS X
COS X
Optically active

Ans: (p and r)

(b)

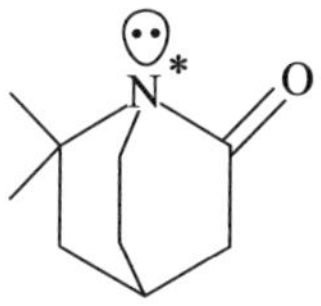

POS X
COS X ⇒ Optically active
AOS X

Ans: (p and q)

(c)

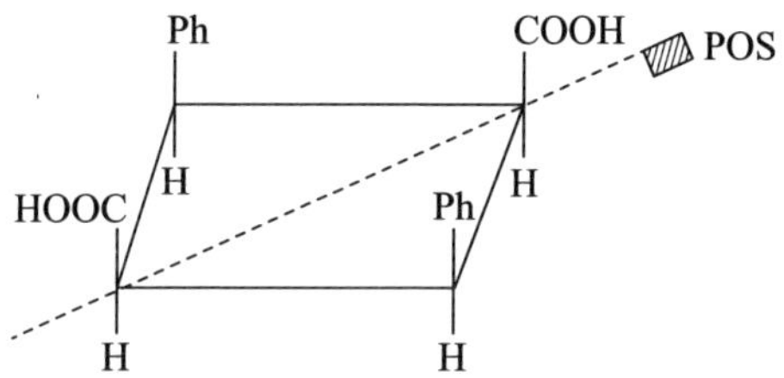

Ans: (s)

(d) POS X
COS X ⇒ Optically active

Ans: (q)

Integer Type

128. Unit $= 1 = \mu_A X_A + \mu_G X_G$

We know $\mu_A = 0$ — μ_A : dipole moment of anti-form

Therefore $1 = \mu_G X_G$ — μ_G : dipole moment of gauche-form

$1 = \mu_G(1 - X_A)$ — $X_A X_G$: mole fraction of anti and gauche form, respectively

$= \mu_G(1 - .82)$

$\mu_G = \dfrac{1}{18} = 5.5$

Ans: (5.55)

129. $CH_3—CH_2—CH_2—CH_2—NH_2$ ⇒ ①

$CH_3—CH_2—\overset{*}{C}H(NH_2)—CH_3$ ⇒ ②

$CH_3—CH(CH_3)—CH_2—NH_2$ ⇒ ①

$CH_3—C(NH_2)(CH_3)—CH_3$ ⇒ ①

Total : 5

Ans: (5)

130.

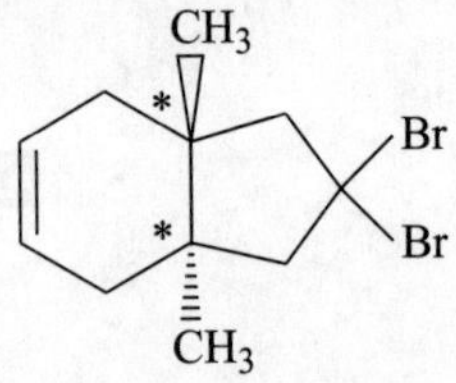

Ans: (2)

131.

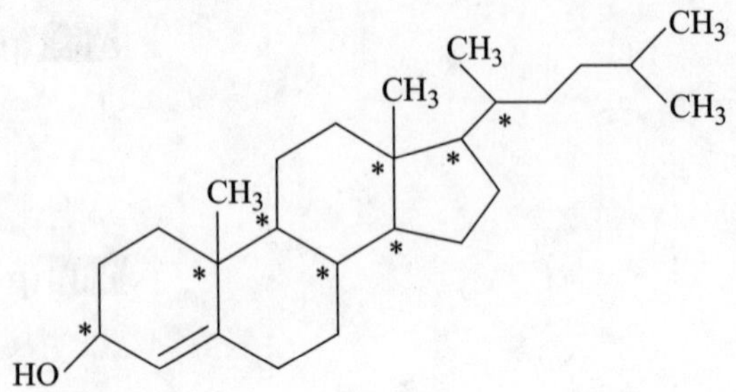

Ans: (8)

132.

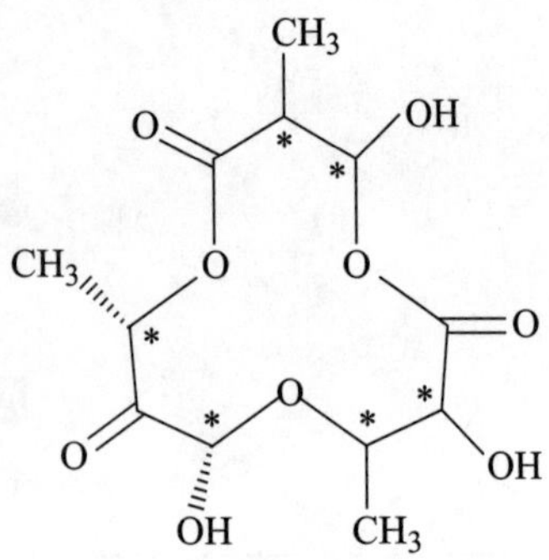

Asymmetric compound

Total stereoisomers
$= 2^n = 2^6 = 64$

Ans: (64)

133.

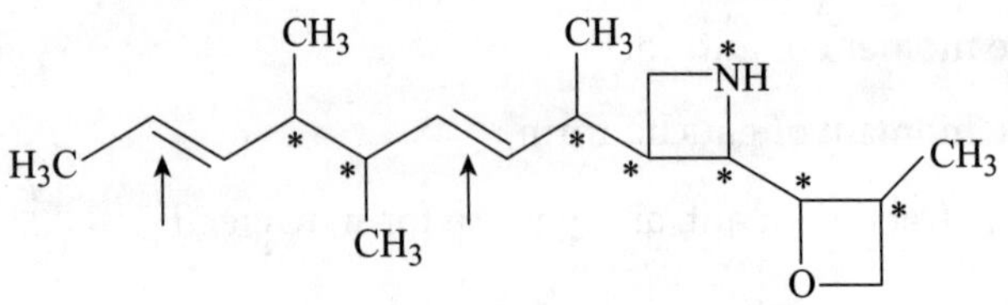

[Asymmetric compound]

Total stereogenic centre = 10 (including chiral centre and GI centre)
So, stereoisomer $= 2^n = 2^{10}$

Ans: (1024)

134.

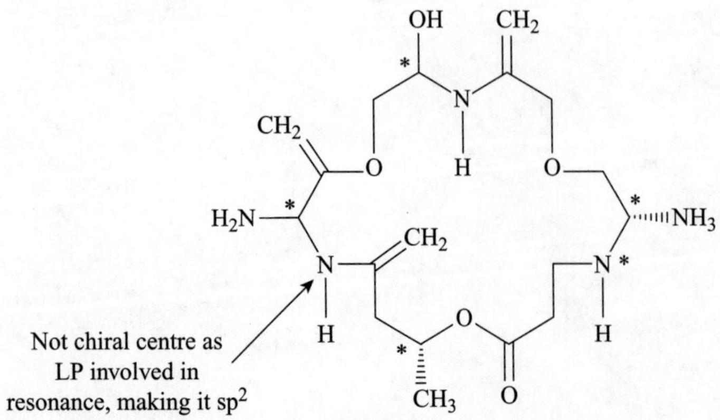

Stereoisomers $= 2^5$

Ans: (32)

135. Chiral centre = 3

Stereoisomers = 2^3

Ans: (8)

136. (a), (b): within plane POS present

(b) 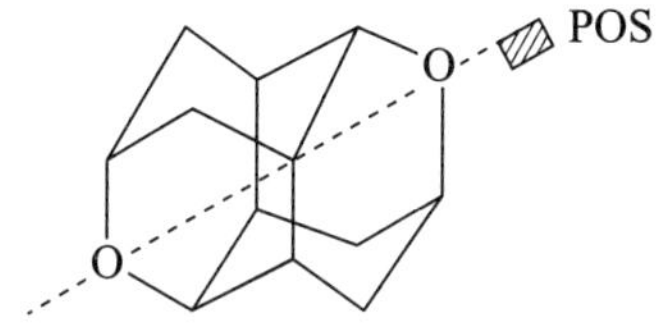

Optically active: c, d, e, f, g, i, j, k, l

Ans: (9)

137. 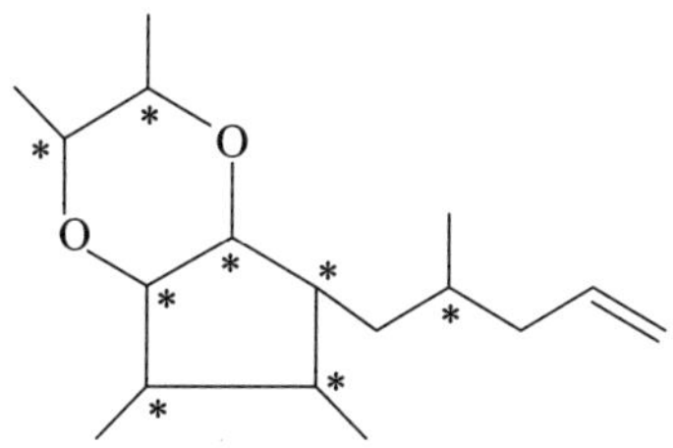

Stereocentre = 8

Ans: (8)

138. Optically inactive: a, d, g, k, m, n, q

Ans: (7)

139. 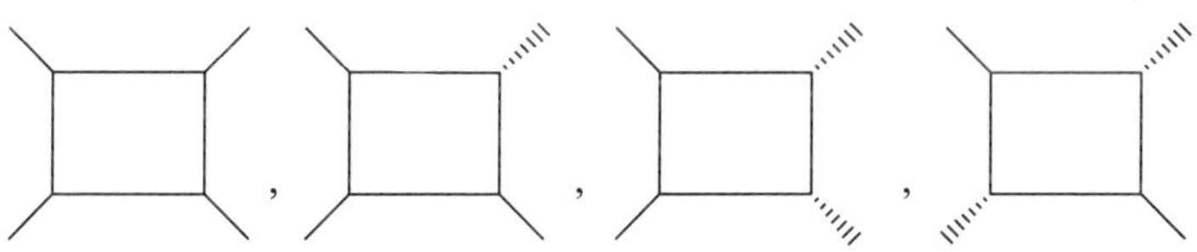

Ans: (4)

140. 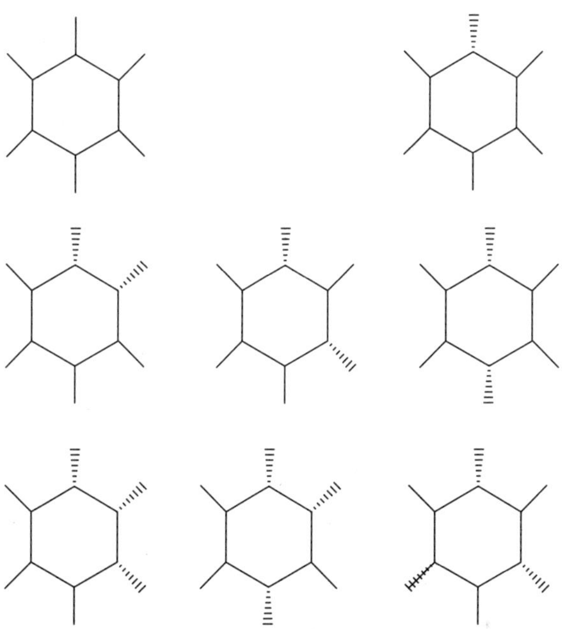

Ans: (8)

141.

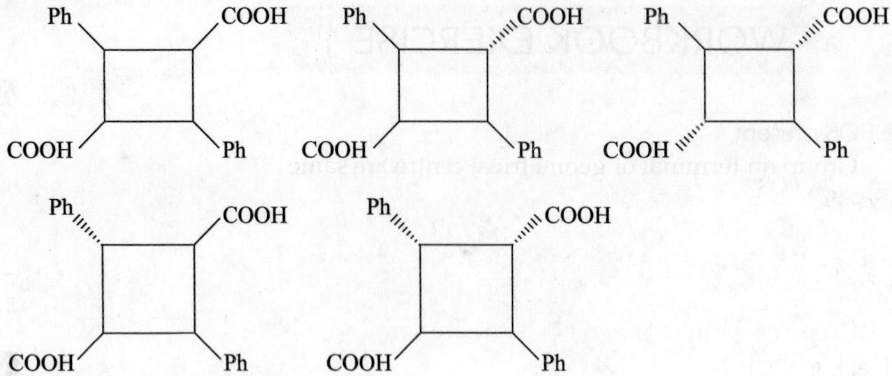

Ans: (5)

WORKBOOK EXERCISE 1

1. 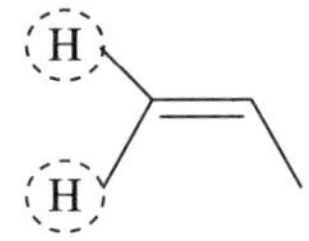 Group on terminal of geometrical centre are same.

2. 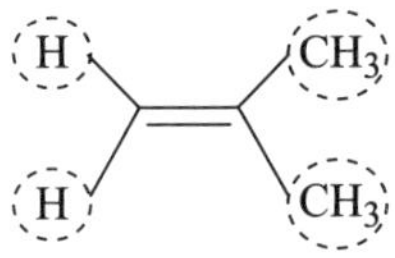 Group on both terminal of alkene are same.

3. 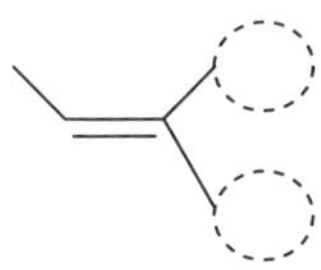 Two $-CH_3$ (identical group) are attached at one terminal of alkene.

4. 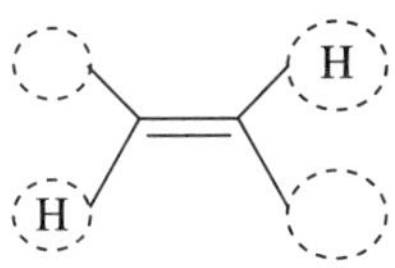 Both group on each terminal of alkene are different.

5. 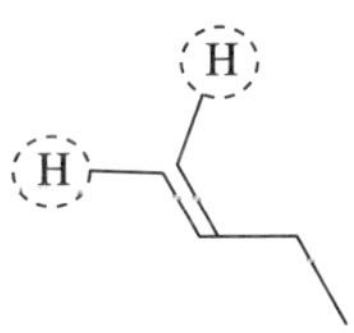 Group on one terminal of alkene are same.

6. 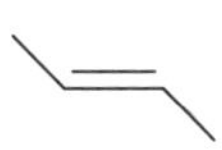 Same as 4th;  and 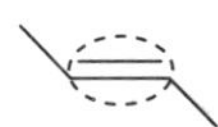 are same structure.

10. 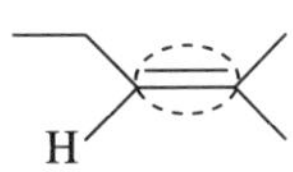 On right terminal of alkene two identical group ($-CH_3$ group) are attached.

12. 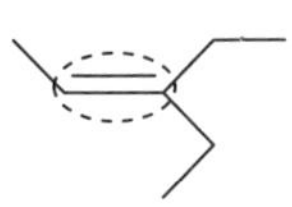 On right terminal of alkene two identical group ($-CH_2CH_3$ group) are attached.

18. 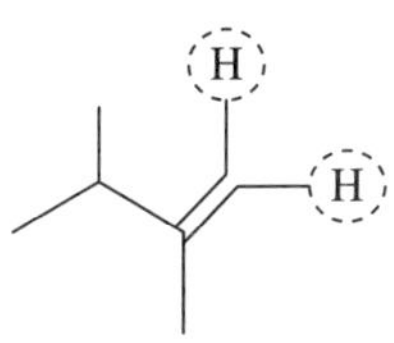 Two hydrogen are attached on one terminal of alkene.

23. 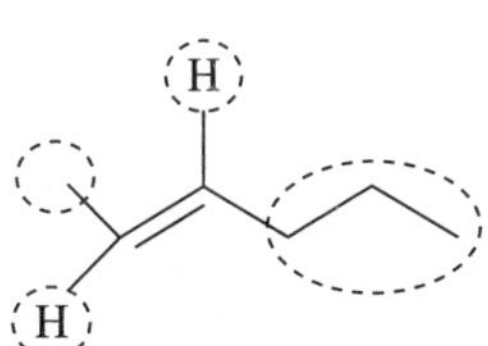 Both group on each terminal of alkene are different.

28. 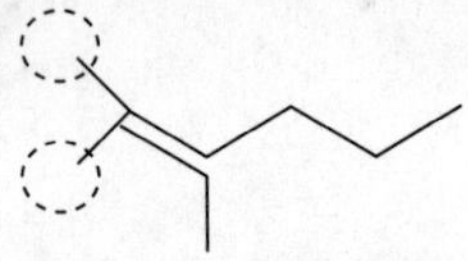

On left terminal of alkene two methyl group (identical group) are attached.

30. 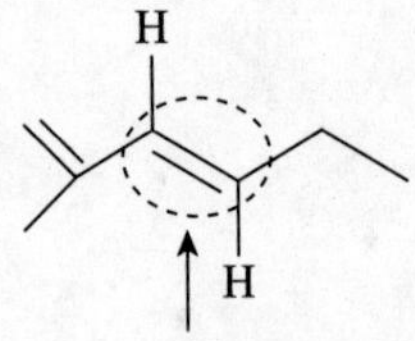

GI across this alkene

31. 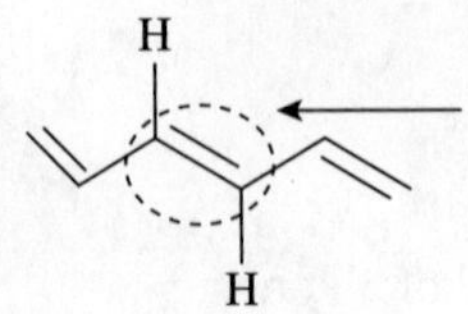

GI across this π bond

33. 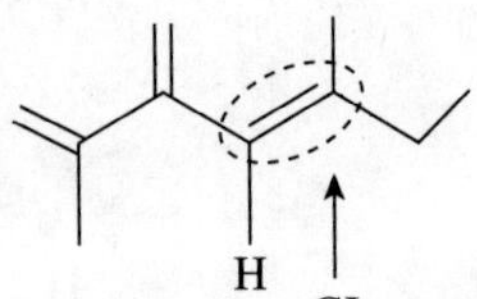

GI across this π bond

34. 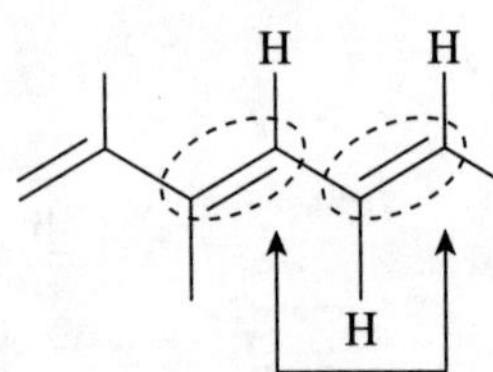

GI across these system

35. 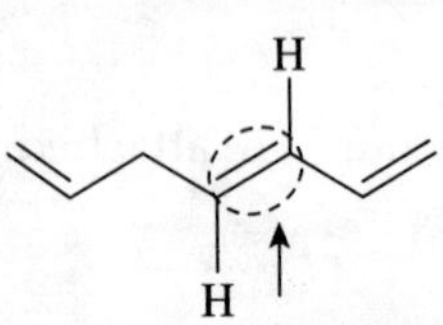

GI across this π bond

36. 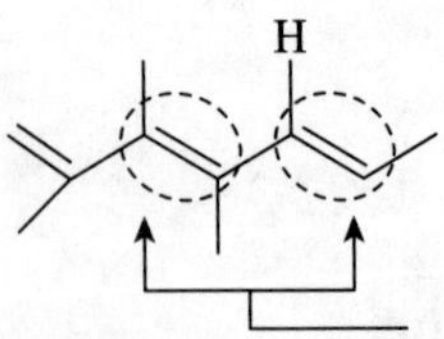

GI across these π bond

39. 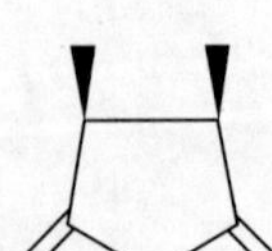 and 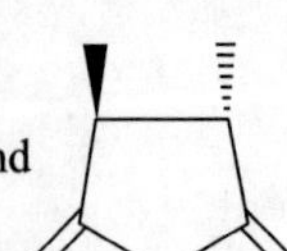

GI across ring system

40.

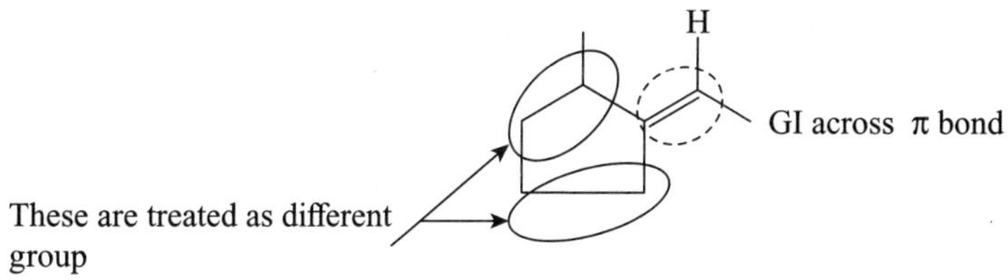

41.

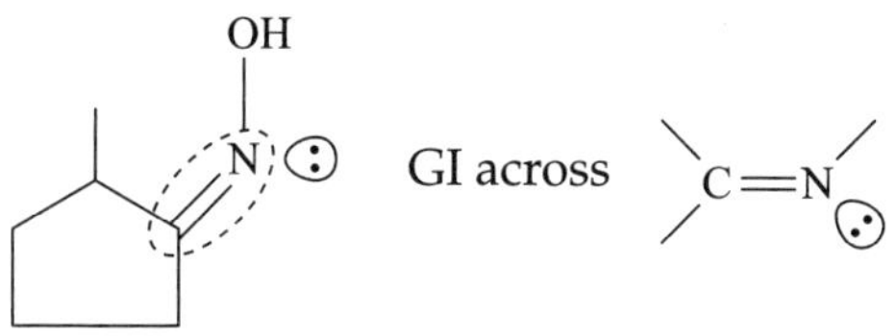

42.

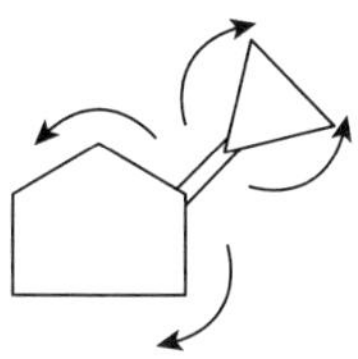

Identical group on each terminal of alkene.

44.

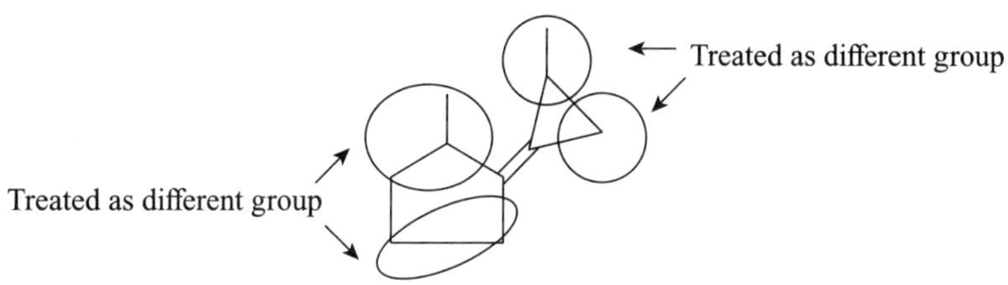

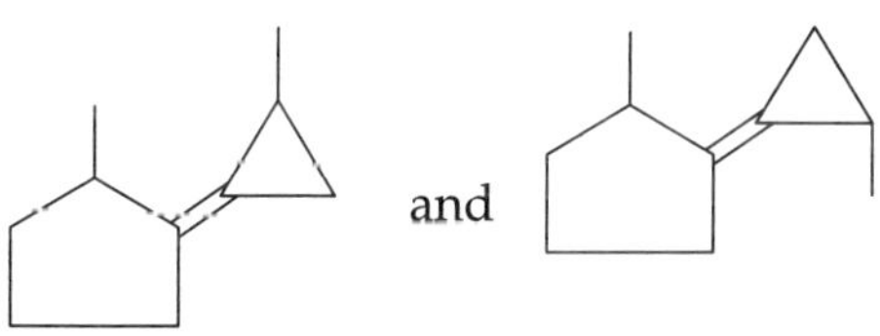

are GI; as distance between both $-CH_3$ changed across restricted system.

47.

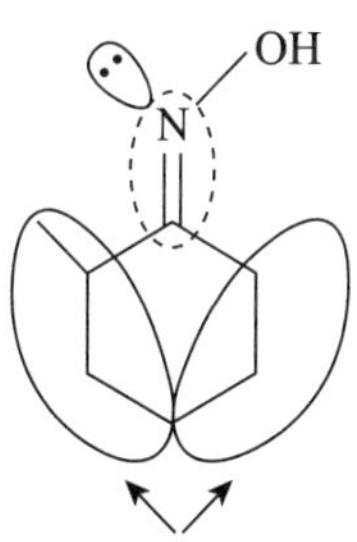

These are treated as different groups.

48.

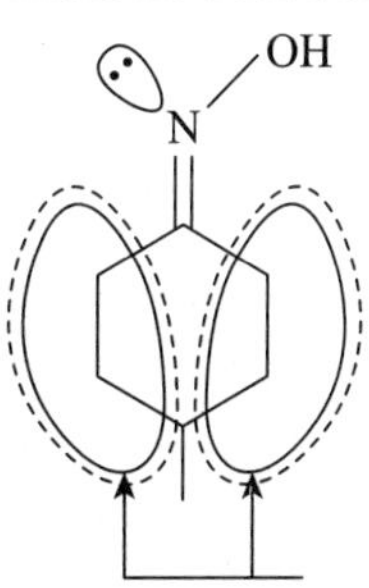

These both groups attached to sp^2 carbon are identical

49. Cyclic system having carbon less than 8 does not show GI across π bond.

50. 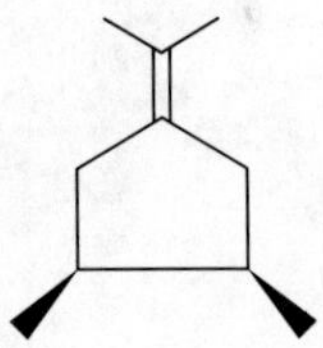 and 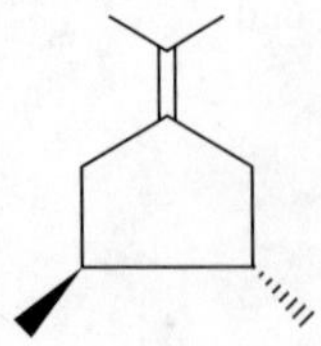 ; GI across ring

51. 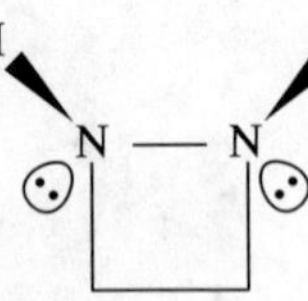 and 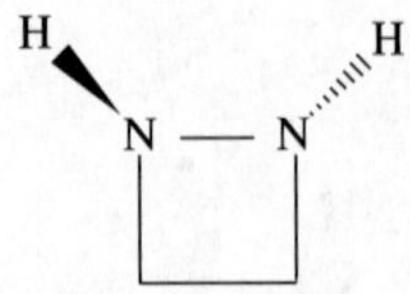

52. 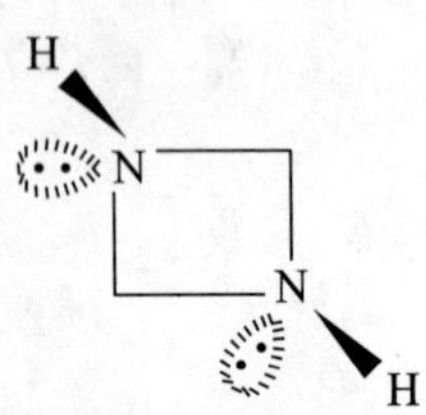 and 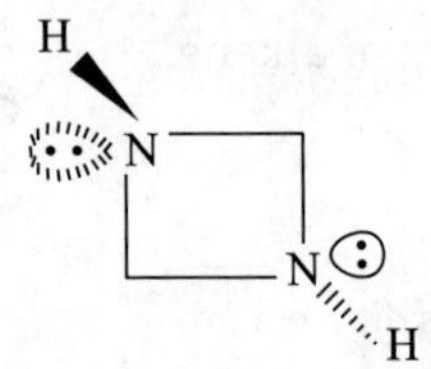

55. 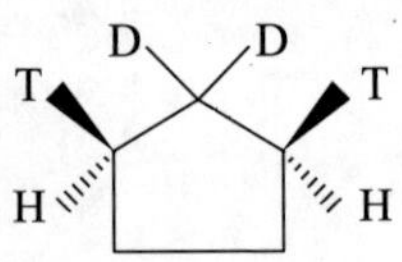 and 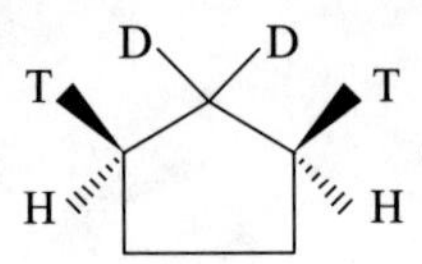 are GI

56. 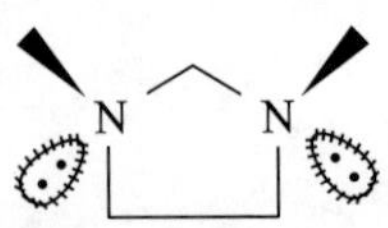 and 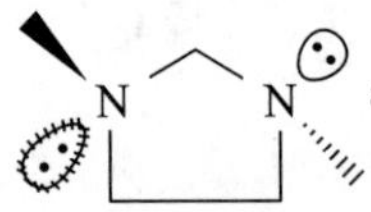 are GI

57. Lone-pair of Nitrogen are in conjugation with π bond. Nitrogen is in sp^2.

60. 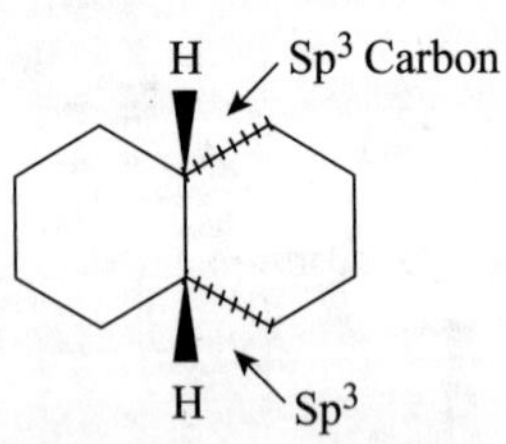

and 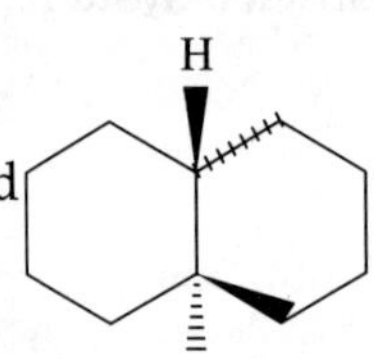

are GI; as distance between both –H got changed across restricted system.

61. 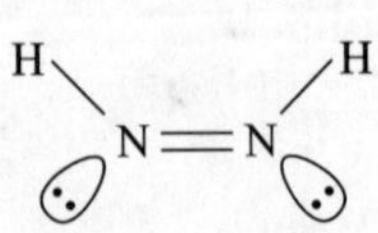 and 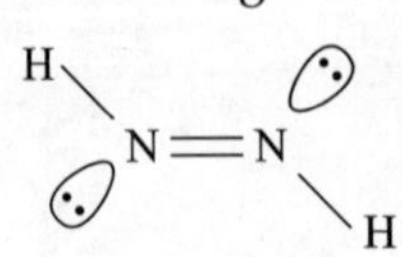 are GI

62. 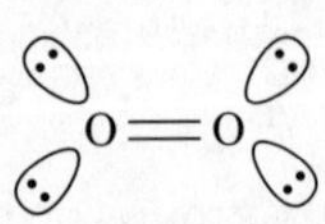

64. 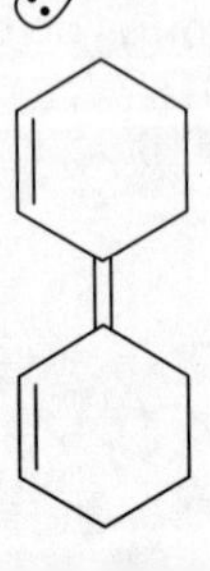 and 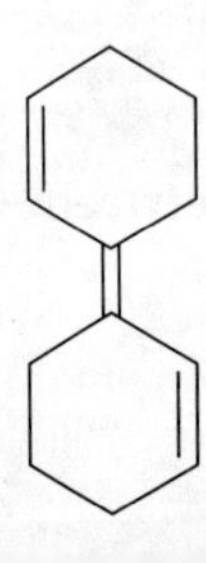 are GI

66. 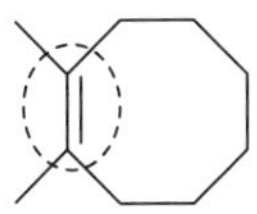 Cyclic system having carbon equal or greater than 8 show GI across π bond.

67. 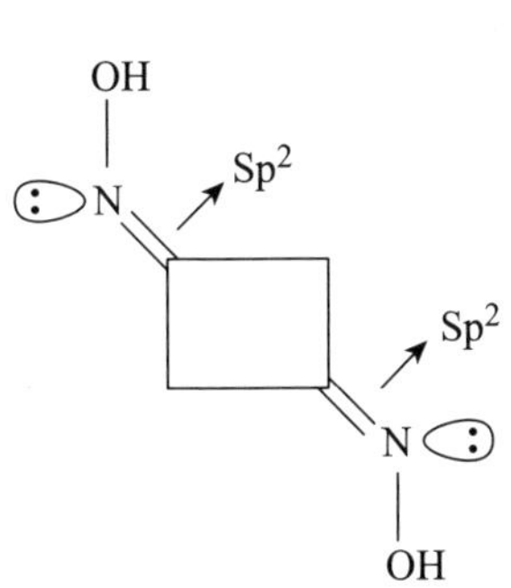 and 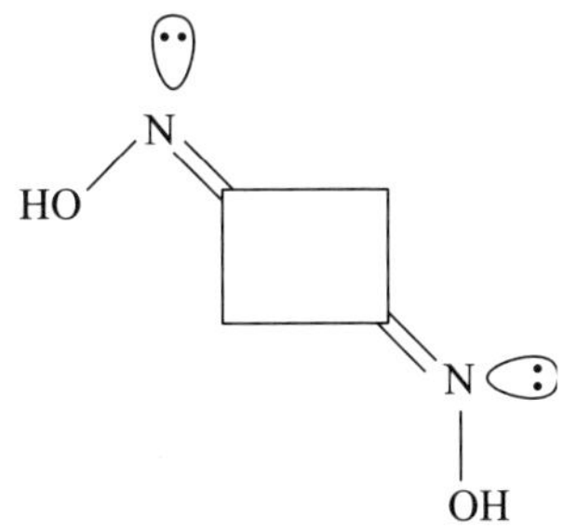 are GI

In the system of π-π bond interaction, both nitrogen in the ring are in the same plane.

71. 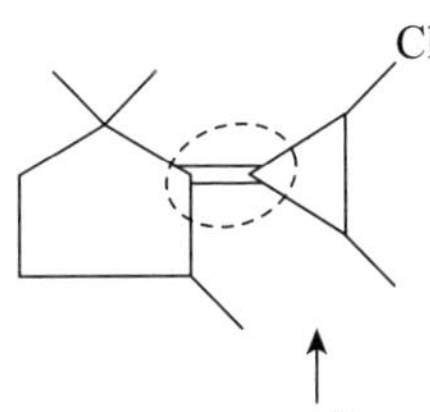 and 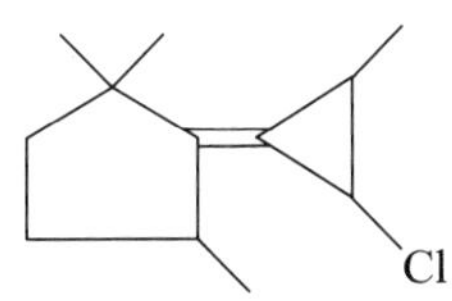 are GI

GI across this π bond

72. 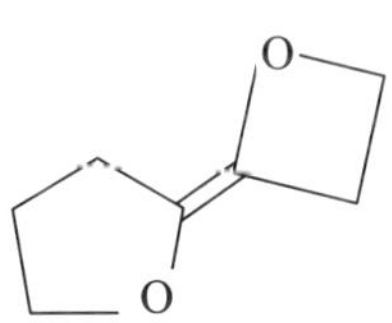 and 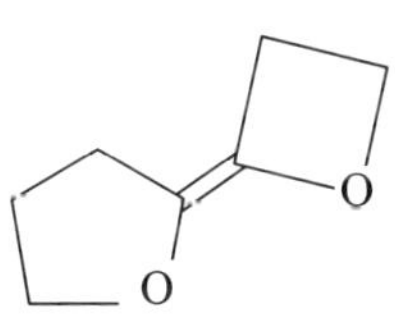 are GI

74. 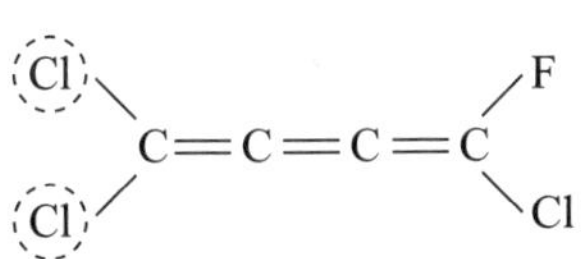 Group on left terminal of planer system are identical.

76. 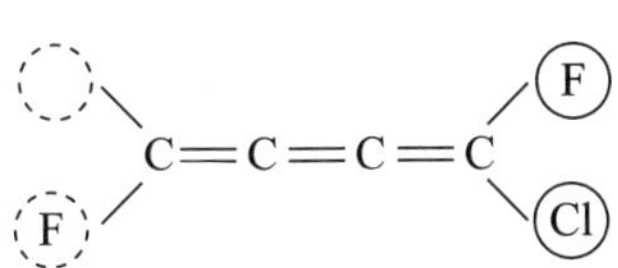 Group on each terminal of planer system are different.

80. In allene system having even number of π bond, both terminal are perpendicular to each other. So does not show GI.

82. In allene system having odd number of π bond, both terminals are in same plane. Both groups on each terminal are different, therefore shows GI.

83. Both rings become perpendicular to minimize steric repulsion between –COOH and –CH3.

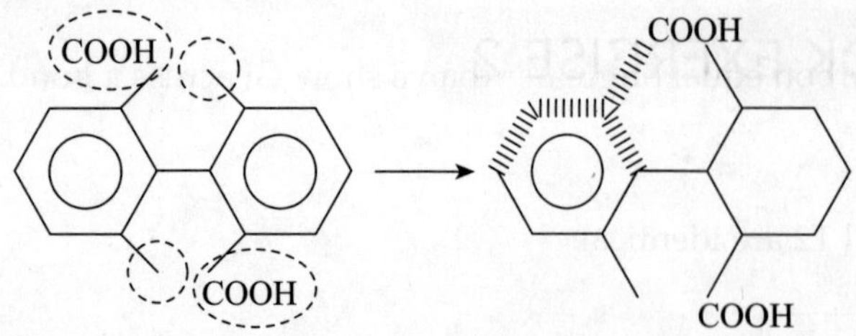

86. 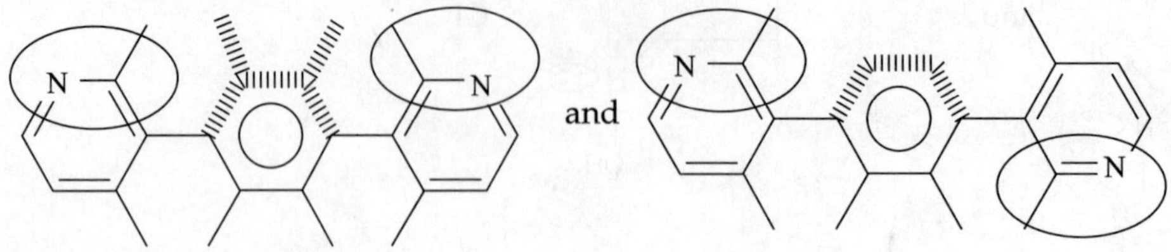

are GI

Both terminal rings are in same plane *i* where middle ring provide restricted rotation system.

87.

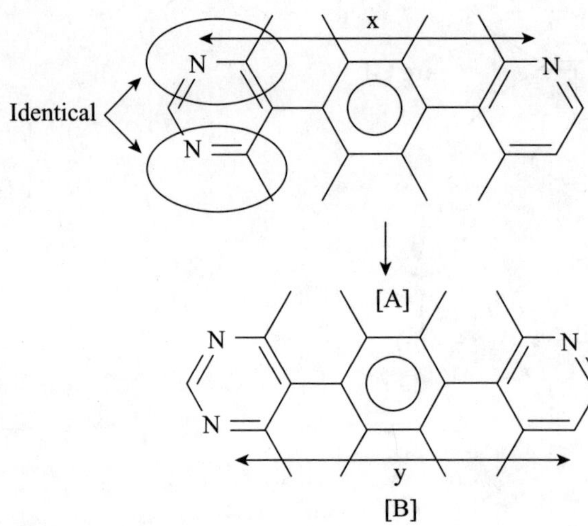

Distance x and y are same. Even after changing configuration distance between left and right nitrogen remains same or geometry remains same. Therefore [A] × [B] are not GI.

89. 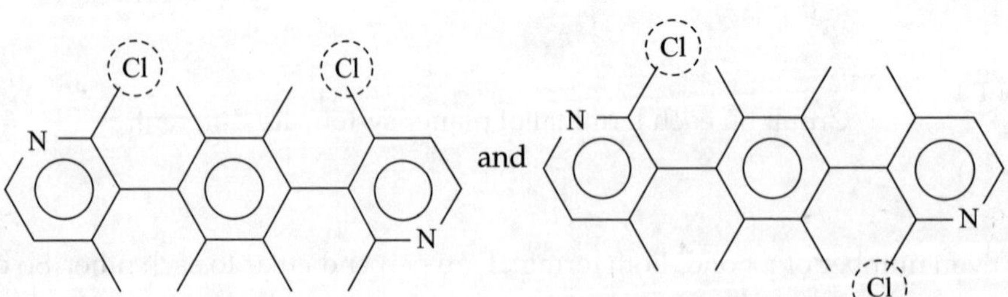

are GI distance between both chlorine in both system is different. Middle ring is responsible for restricted rotatory system.

WORKBOOK EXERCISE 2

4. 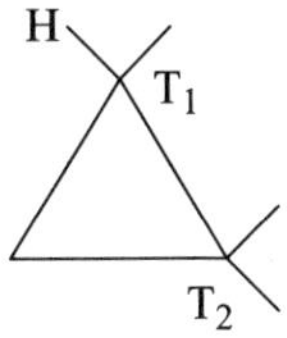

Both groups on terminal T2 are identical.

8. 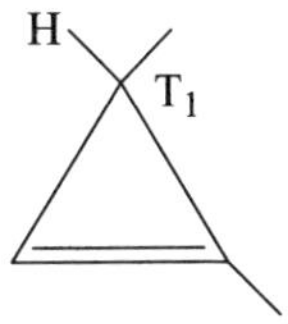

There is only 1 terminal having two different groups.

11. 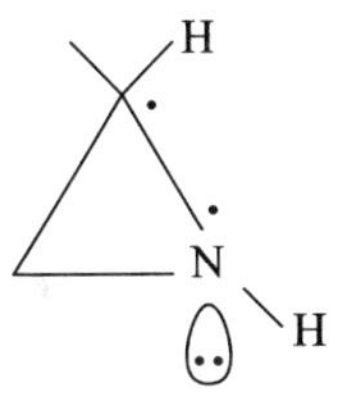

There are two stereocentre in ring system.

Note: Stereocentre are marked with dot (•)

12. 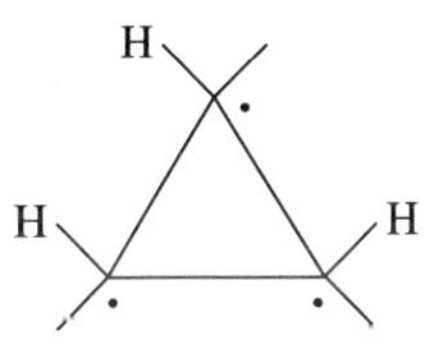

3 Stereocentre are there

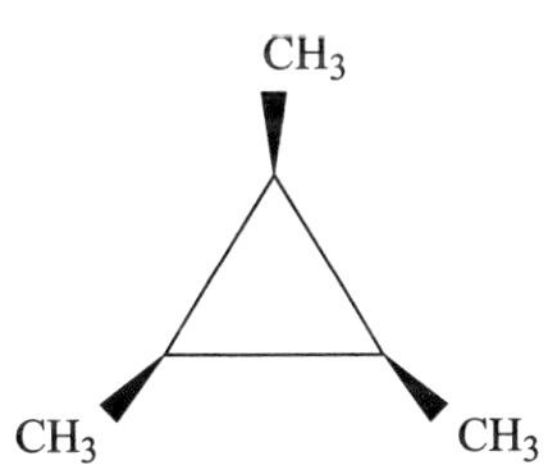

and

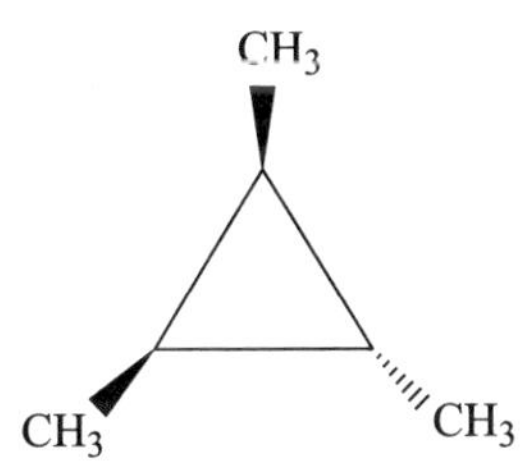

are GI

14. 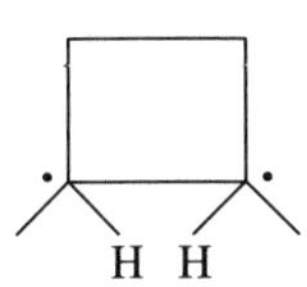

17. 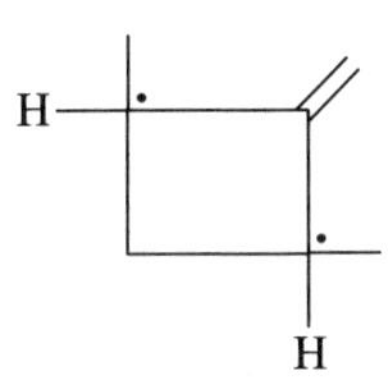

There are 2 stereocentre

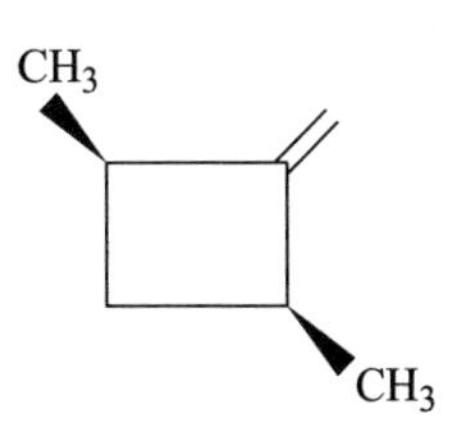

and

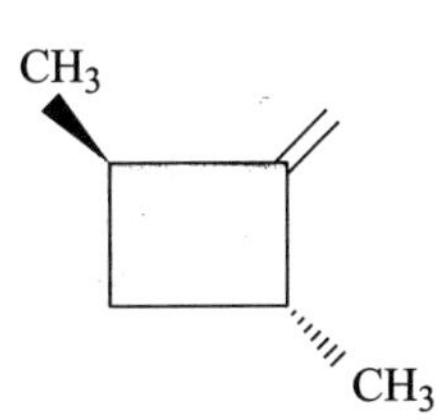

are G.I.

21. 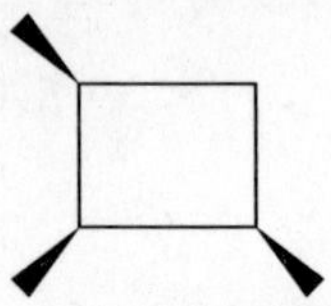 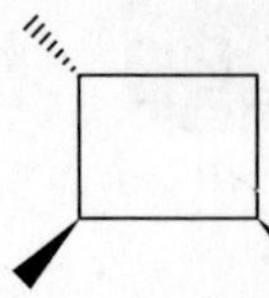 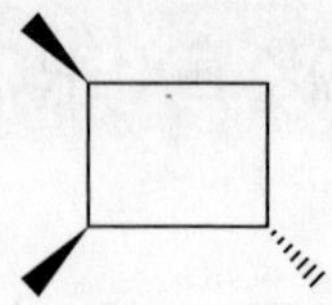 and 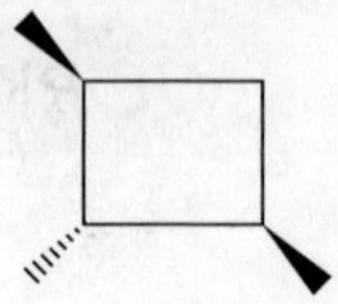 are GI of each other.

23. 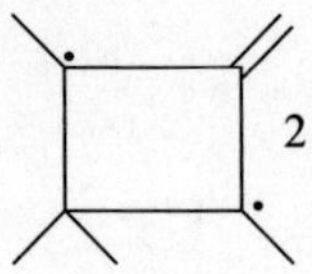 2 Stereocentre.

28. 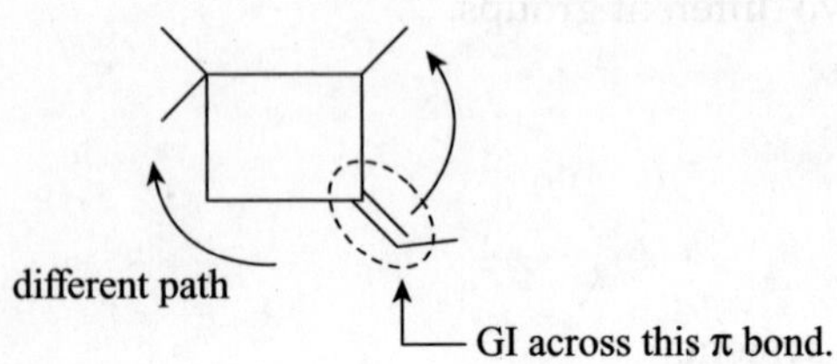

29. 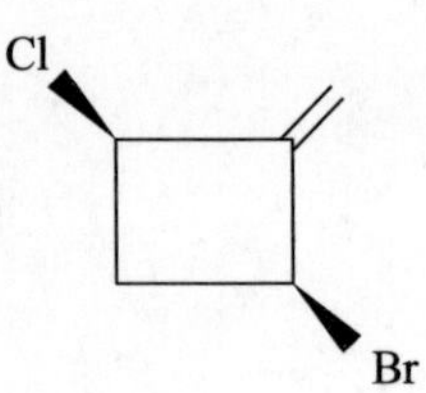

and 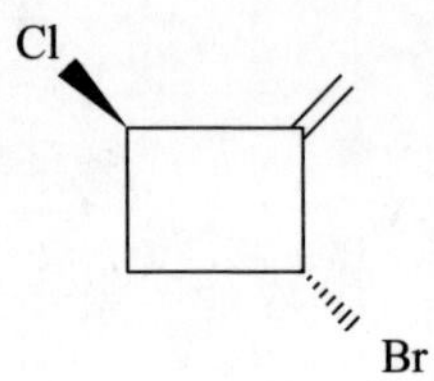

are GI.

31. 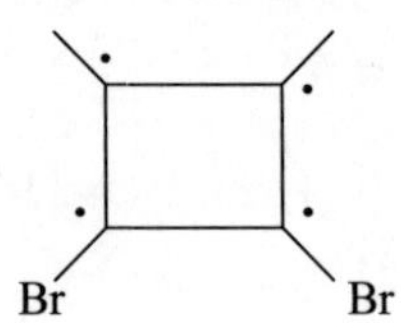

There are 4 stereocentre.

34. 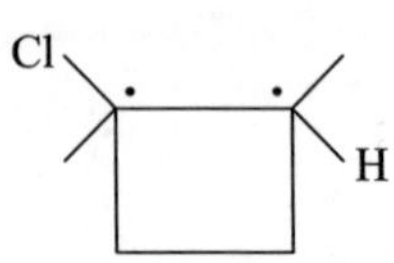

2 stereocentre are there.

42. 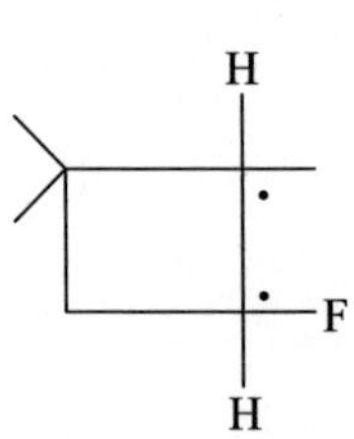

45. 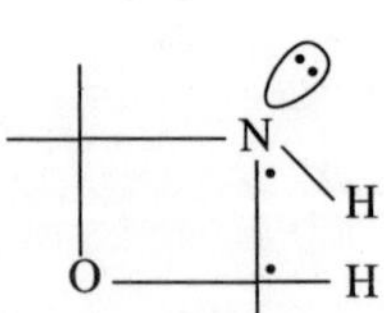

There are 2 stereocentre.

Note: Stereocentre is represented by dot (.)

46. 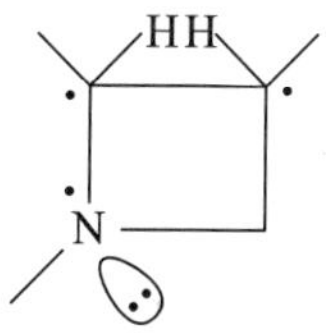 There are 3 stereocentre.

55. to 57. Cyclic system having carbon less than 8 does not show GI across π bond.

59. 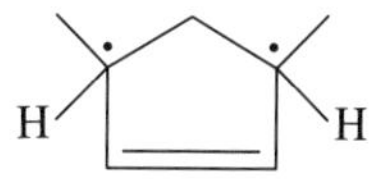 2 Stereocentre

70. 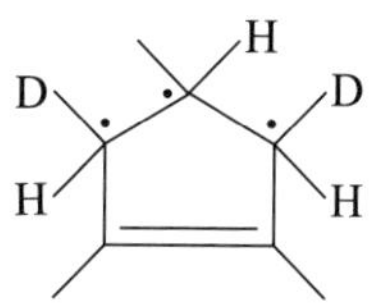 3 Stereocentre

75. 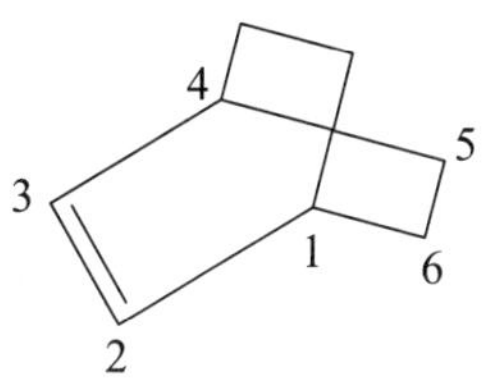 Cyclic system having π bond with 6 carbon does not show GI across π bond.

78. 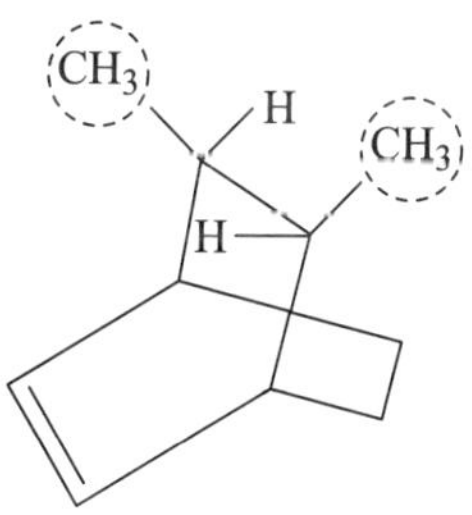 and 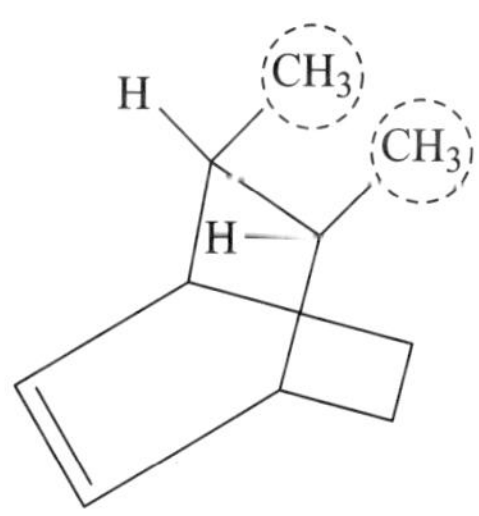 are GI

X>Y

79. 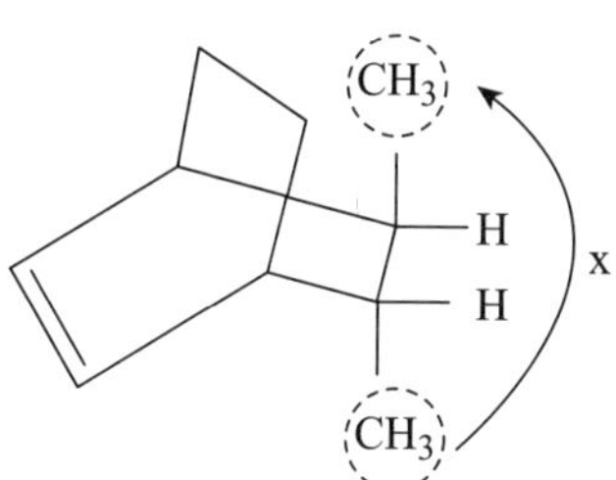 and 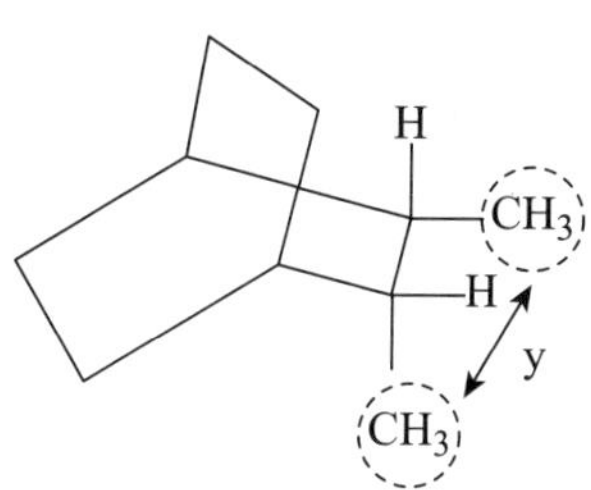 are GI

X>Y

CHAPTER 3

Hydrocarbons

LEVEL 1

1. 'Br_2' addition is anti-addition and on trans alkene it gives erythro product.

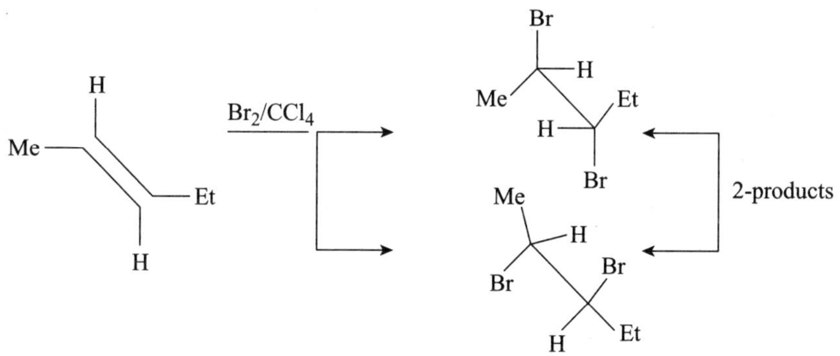

Ans: (b)

2. –F, $-\overset{\oplus}{N}R_3$, $-\overset{\oplus}{S}R_2$ groups give Hofmann's elimination product by formation of partial anion in T.S. Therefore the correct option is 'A'. **Ans:** (a)

3. Addition of Br–Cl is anti and on reaction with cis alkene it gives 'threo' product.

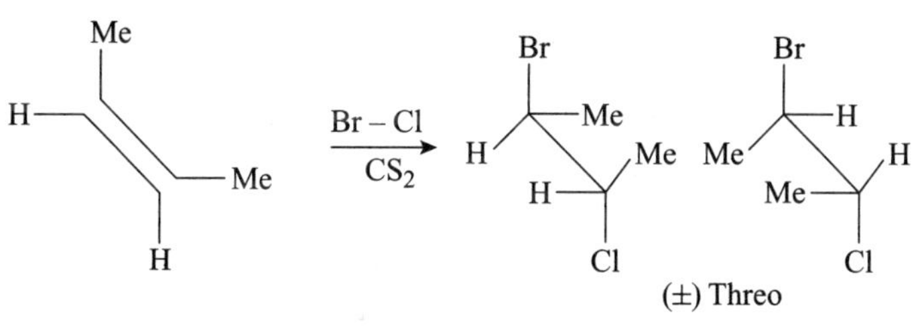

Ans: (b)

5. Hydroboration oxidation is syn addition.

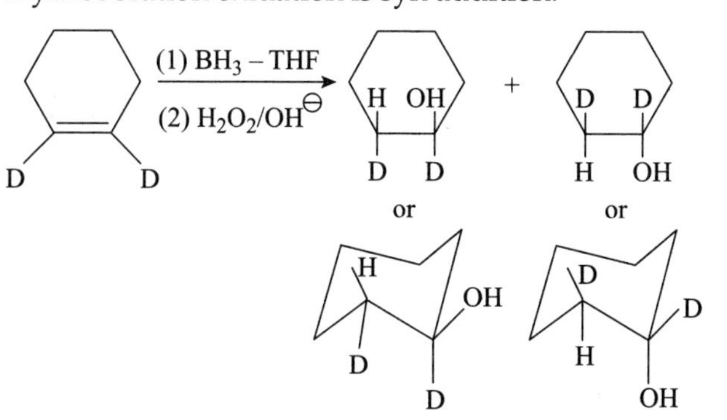

Ans: (c)

6.

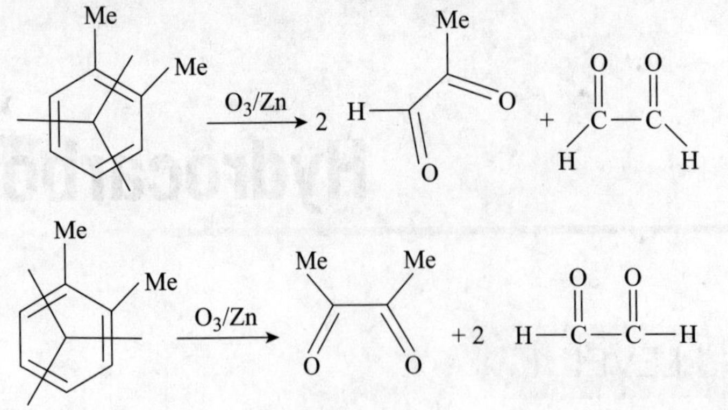

Ans: (d)

7.

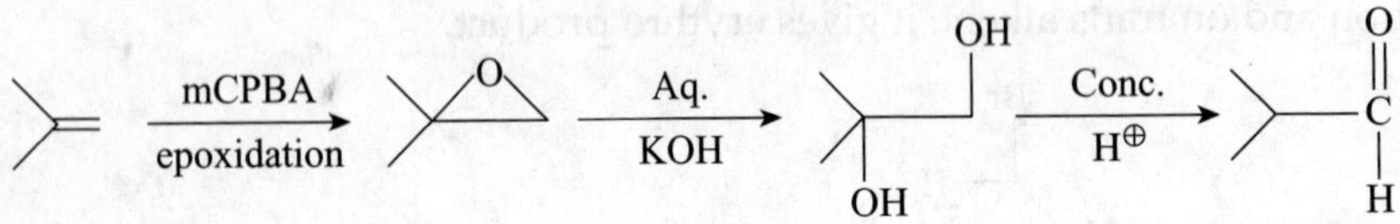

Ans: (a)

8.

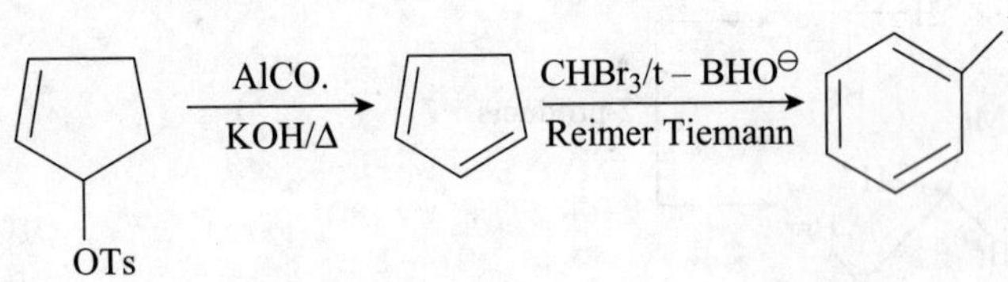

Ans: (c)

9.

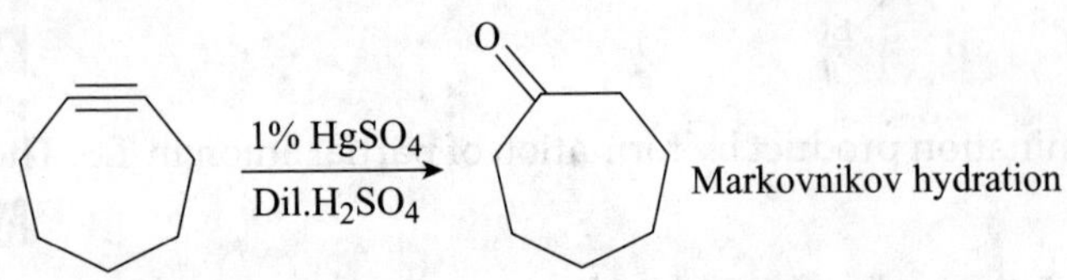

Ans: (d)

10.

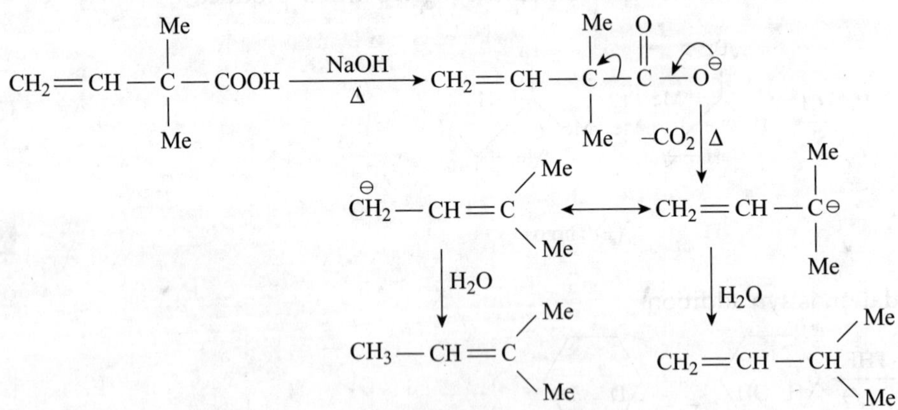

Ans: (a) and (b)

11.

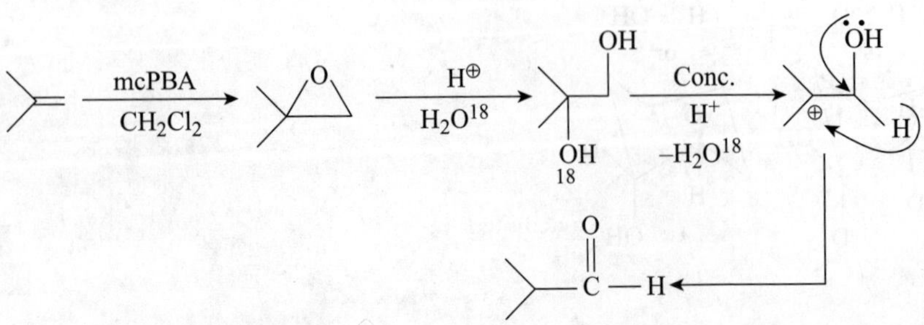

Ans: (a)

12.

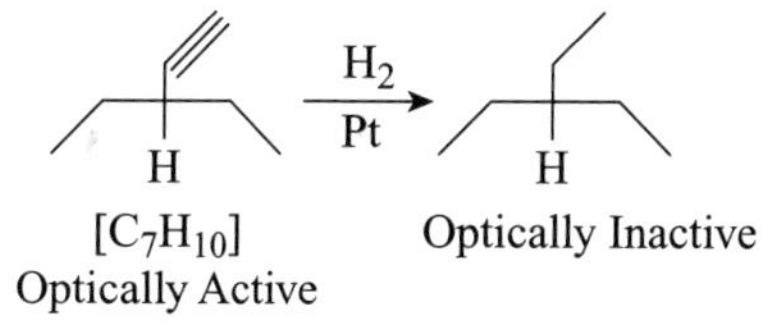

Ans: (c)

13.

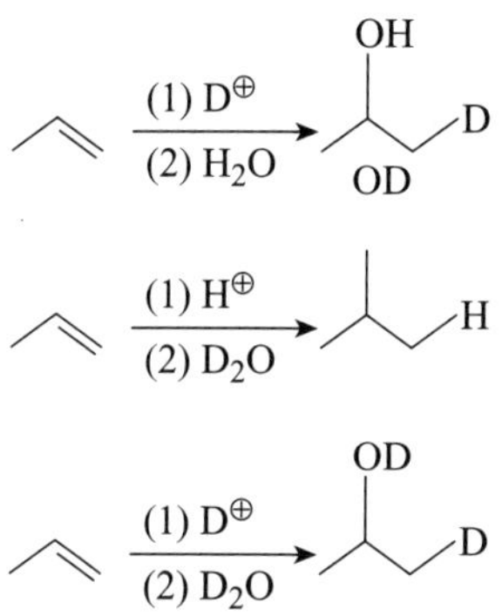

Ans: (b)

14.

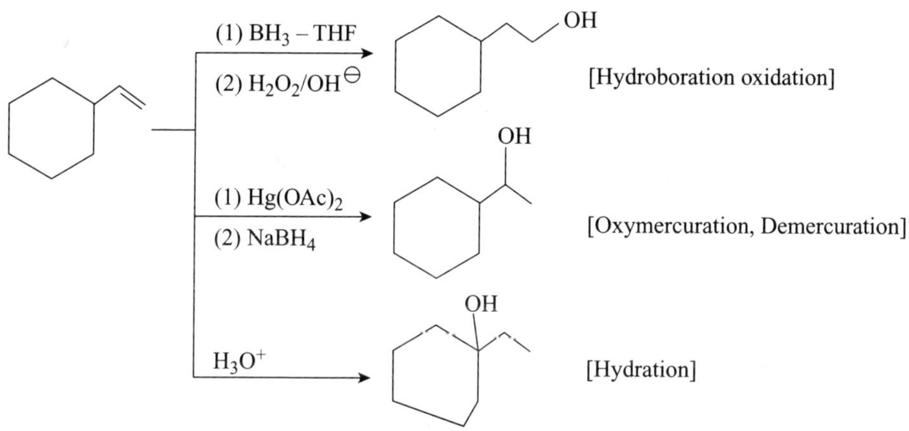

Ans: (b)

15.

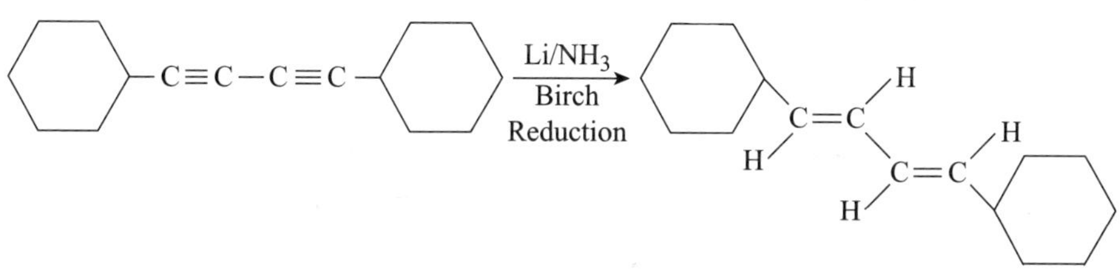

Ans: (b)

16. Rate of dehydration depends on stability of carbocation. **Ans:** (c)

17.

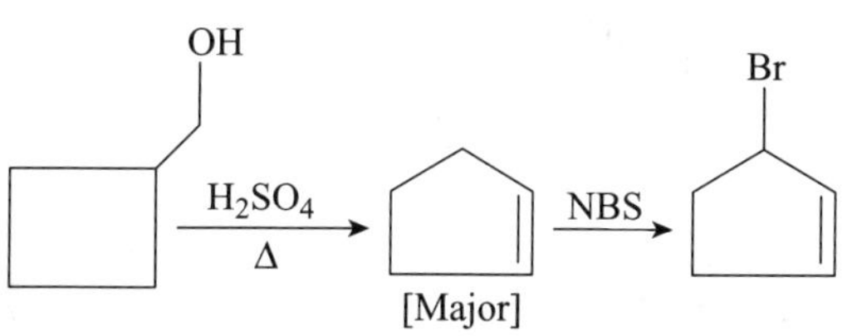

Ans: (c)

18.

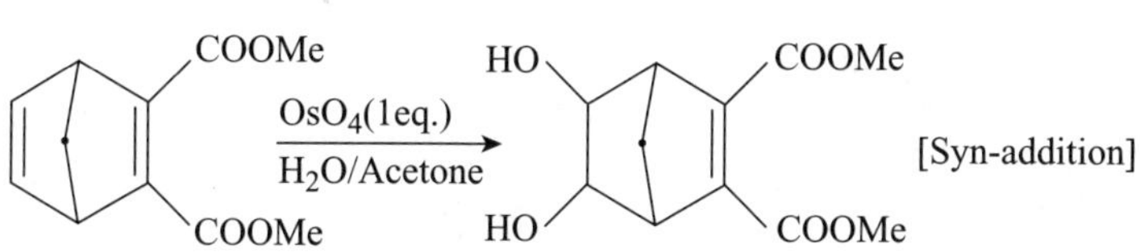

Ans: (b)

19. 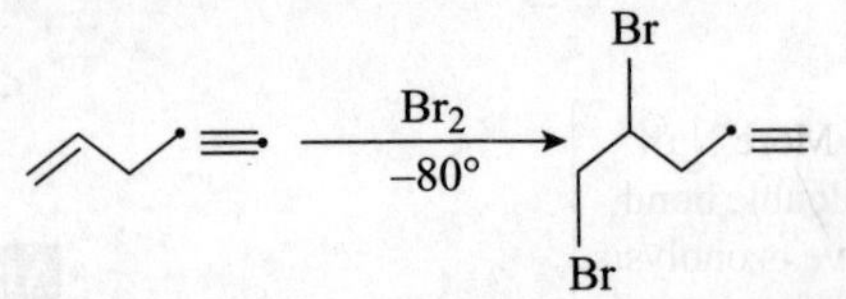

Alkene is more nucleophillic than alkene, so more reactive towards $E^{\oplus}$ addition. **Ans:** (d)

20. Terminal and internal alkynes are differentiated by (a) Ozonolysis (b) Ammonical Cu_2Cl_2 (3) Conc. $KMnO_4$.

Ans: (a)

21. 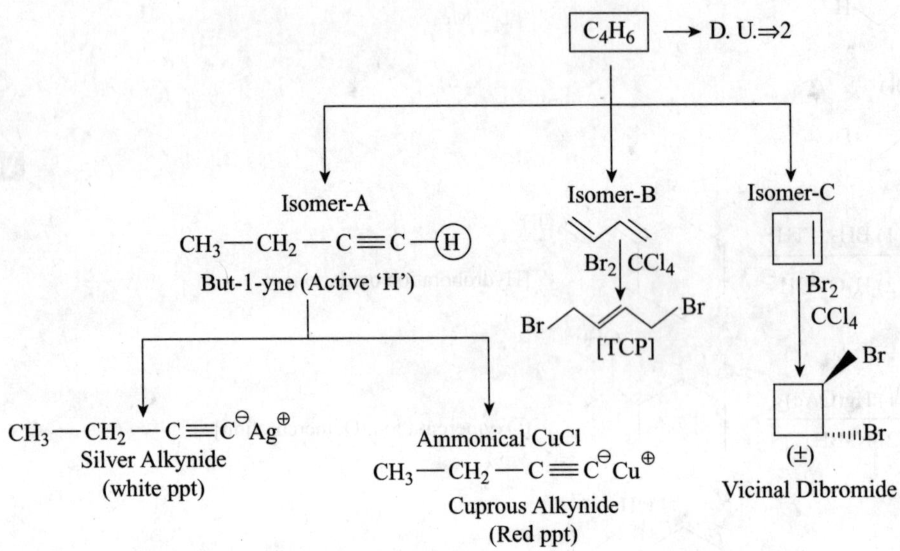

Ans: (a)

22. 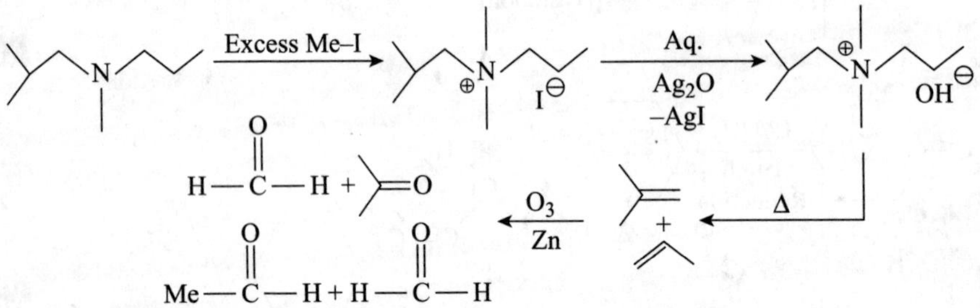

Ans: (d)

23. 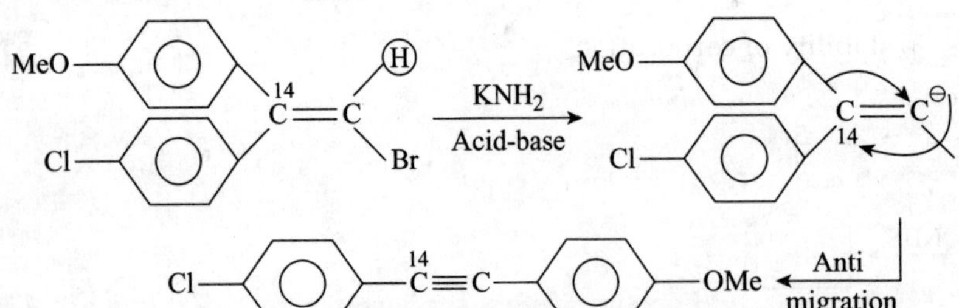

Ans: (d)

24. 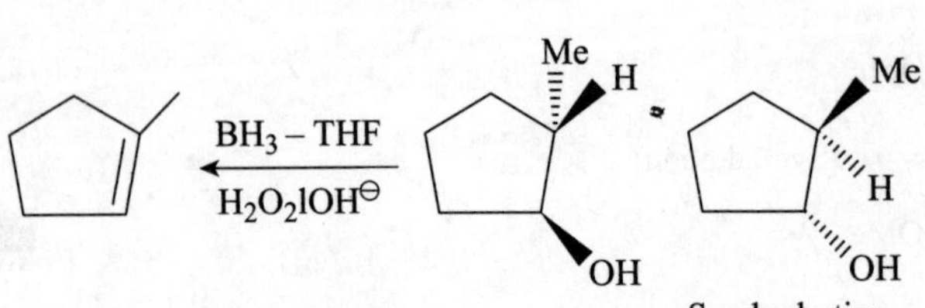

Ans: (c)

25.

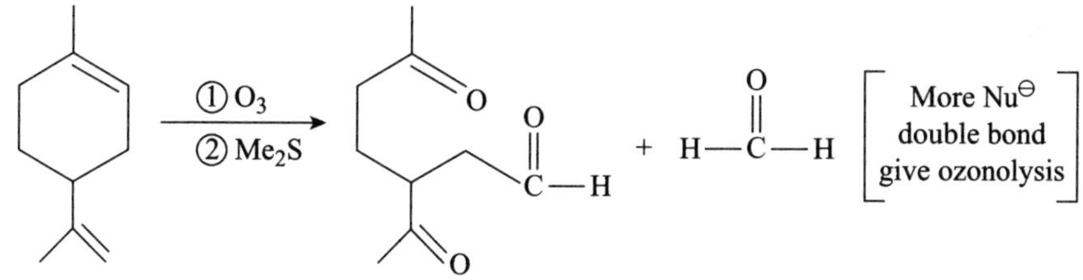

Ans: (b)

26.

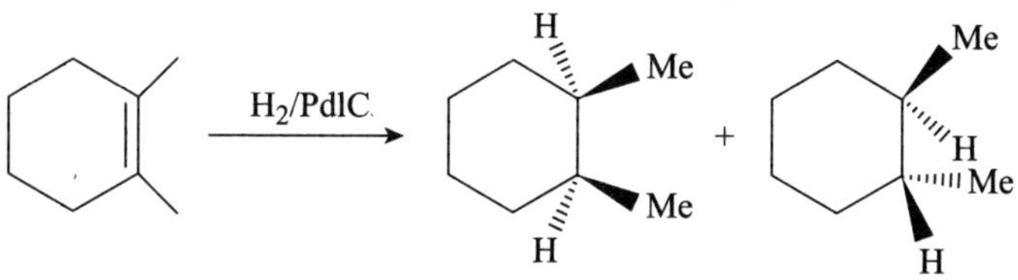

Syn-addition 80% Anti-addition 20%

Ans: (d)

27.

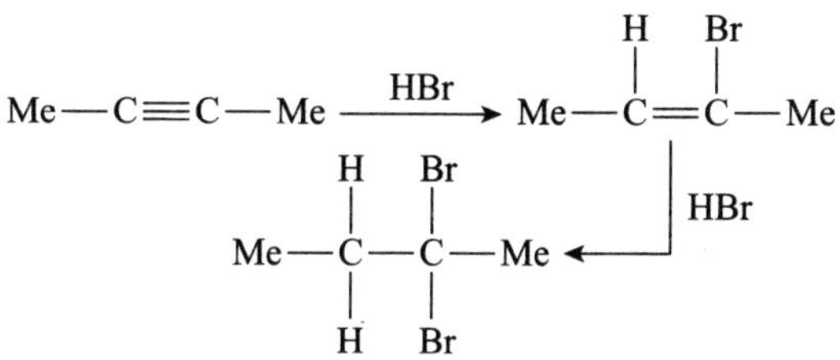

[Gem dihalide]

Ans: (c) and (d)

28.

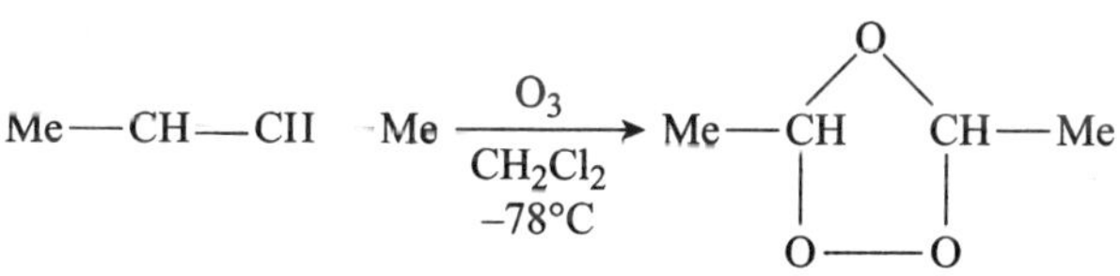

Ozonide

Ans: (d)

29.

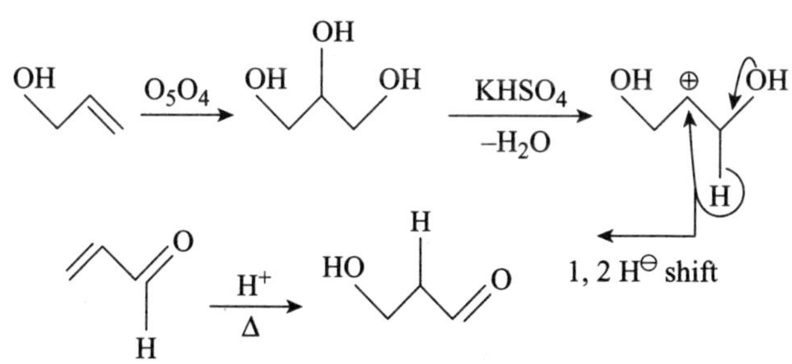

Acrylaldehyde Dehydration

Ans: (c)

30.

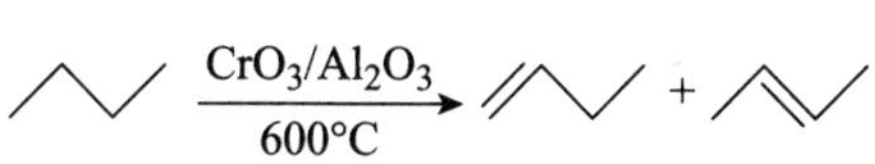

It is dehydrogenation reaction

Ans: (a)

31. **Free radical halogenation/**Anti-Markovnikov addition is enthalpy supported reaction. If one of the step is highly endothermic then reaction does not occur.

Ans: (c)

33. $\underset{\text{Base}}{[Ag(NH_3)_2]OH} + \underset{\text{Acid}}{R{-}C{\equiv}C{-}H} \rightleftharpoons \underset{\text{Salt}}{R{-}C{\equiv}C^{\ominus}Ag^{\oplus}} + NH_4OH + NH_3$

Alkyne behave as acid

Ans: (d)

34. Reactivity order fluorination > chlorination > bromination. **Ans:** (b)

35.

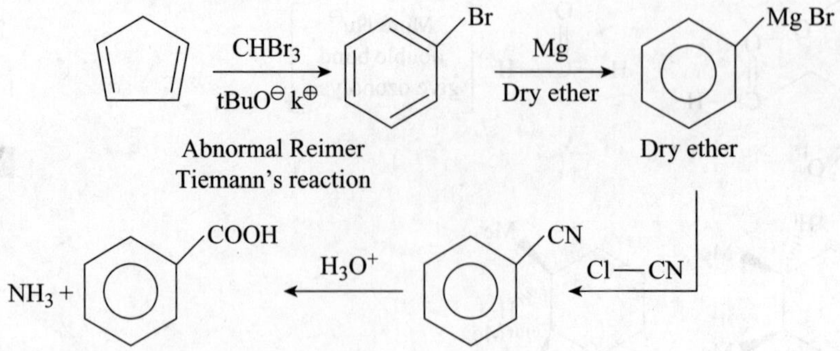

Ans: (c)

36.

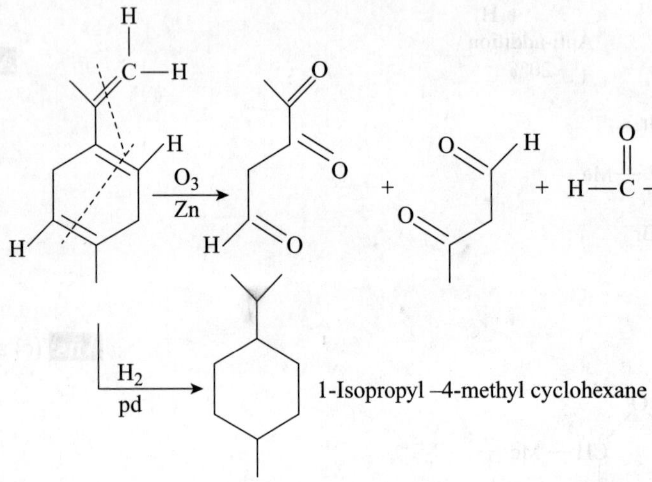

Ans: (d)

37.

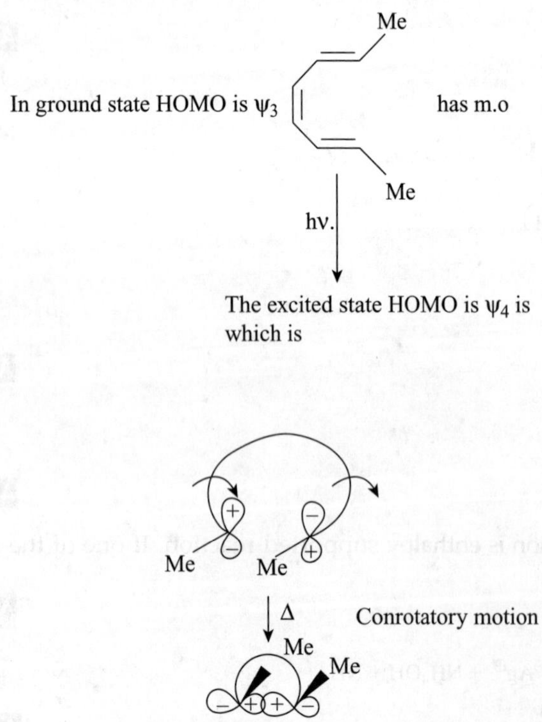

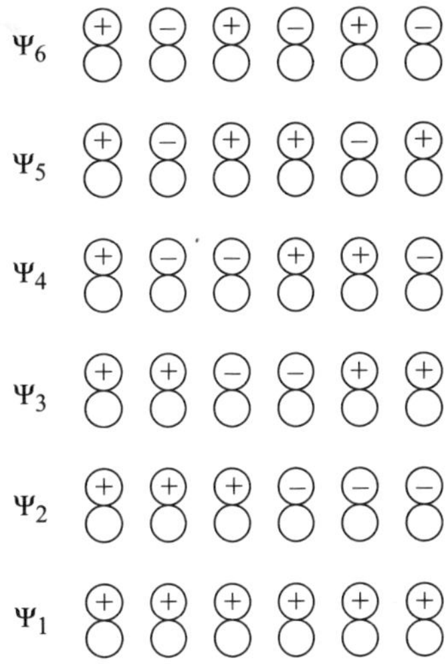

Ans: (c)

38.

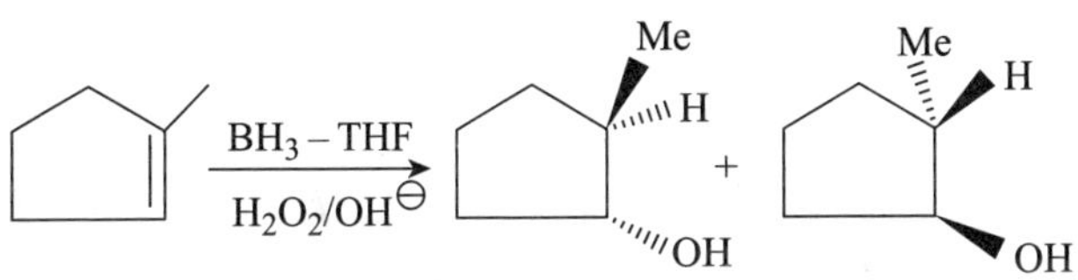

Ans: (a)

39.

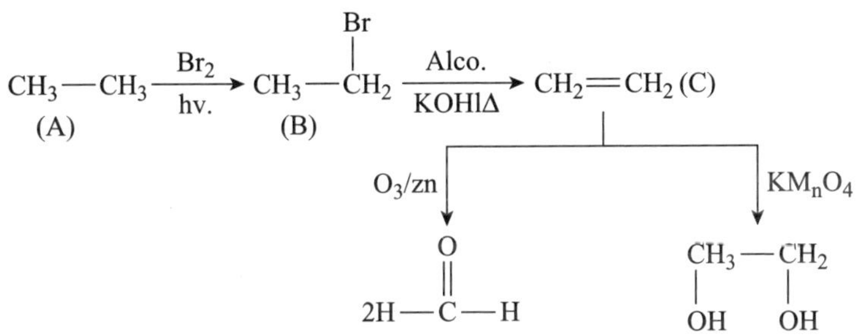

Ans: (c)

40.

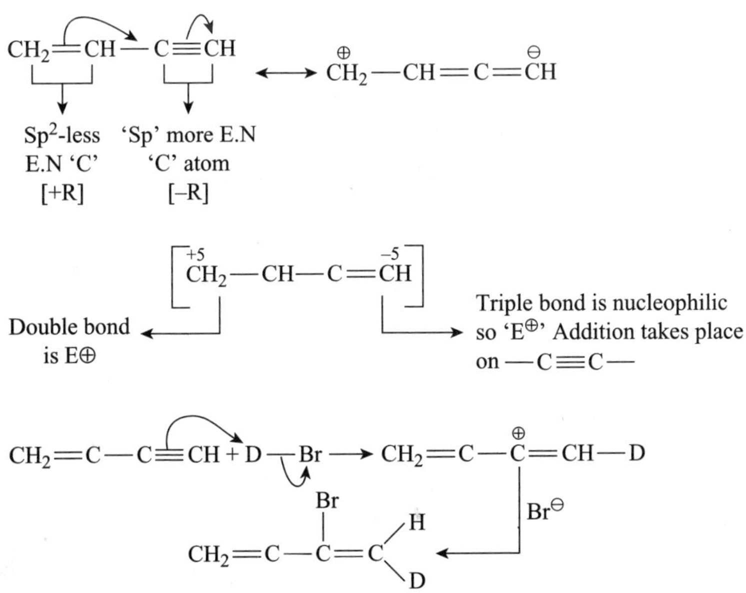

Ans: (a)

41.

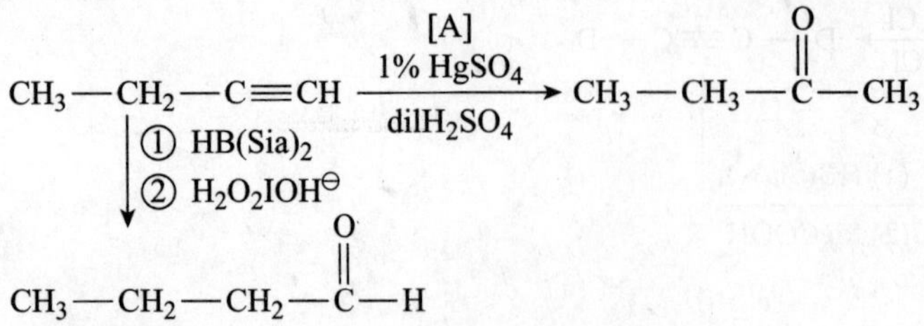

Ans: (b)

42.

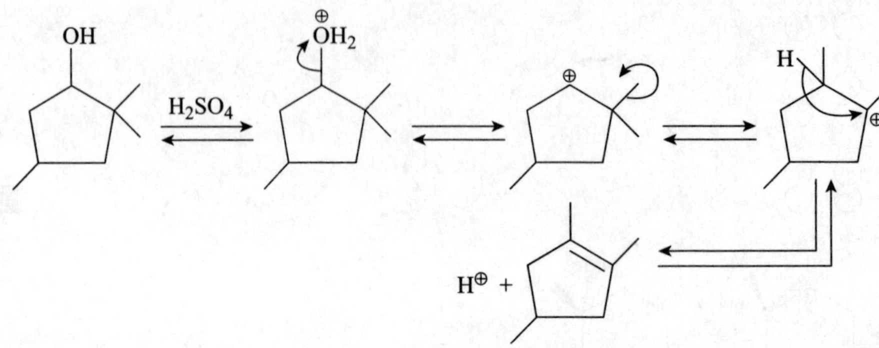

Major product

Ans: (b)

43. During Kolbe electrolysis reactions involve or

At anode:

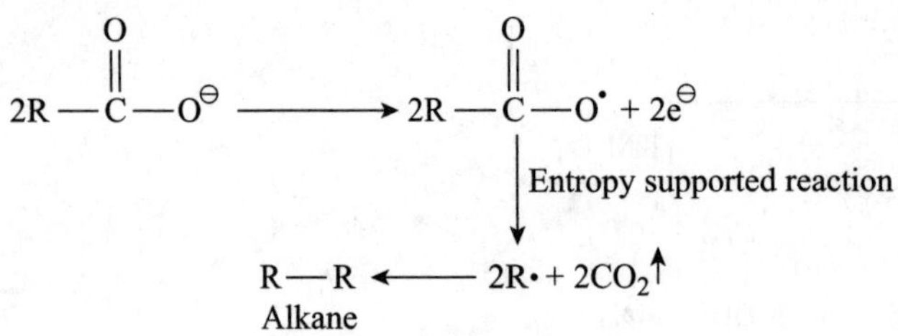

At cathode:

$H_2O \rightleftharpoons H^{\oplus} + OH^{\ominus}$

Reduction potential of $H^{\oplus} > Na^{\oplus}$. So $H^{\oplus}$ from H_2O migrate towards cathode.

$2H^{\oplus} + 2e^{\ominus} \longrightarrow 2H\cdot \longrightarrow H_2\uparrow$ and 'NaOH' is formed as side product 'so' during reaction pH of medium increases.

Ans: (b)

44.

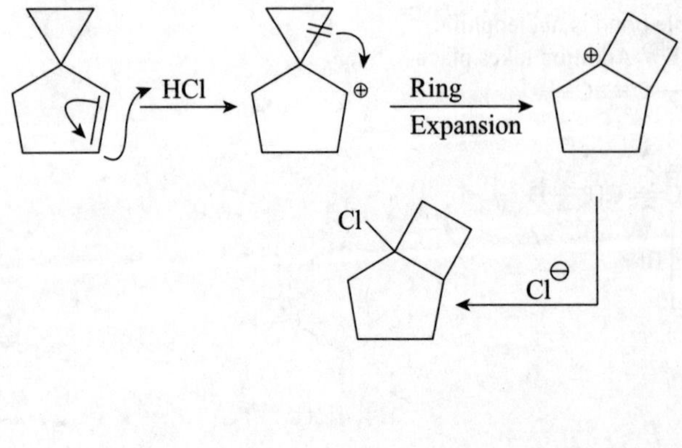

Ans: (c)

45.

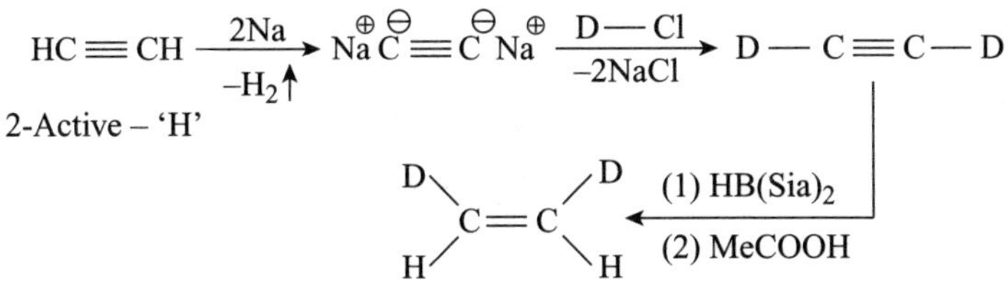

Ans: (c)

46.

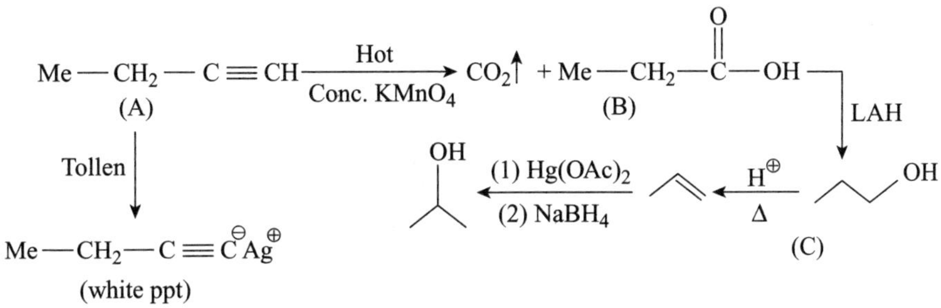

Ans: (b)

47. 1. H — Br Peroxide give Anti-Markovnikov addition.

2. Attacking species is Br.

Mechanism

1. Chain initiation

$$R—O—O—R \xrightarrow{hv.} RO\bullet$$

$$RO\bullet + H—Br \longrightarrow RO—H + Br\bullet \ \Delta r = -ve$$

2. Chain propagation

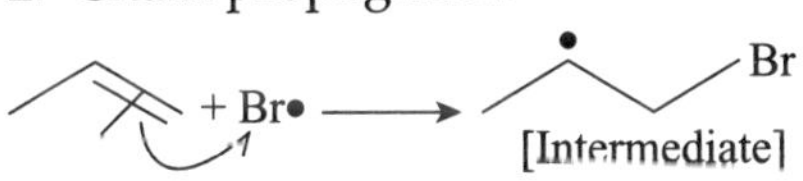

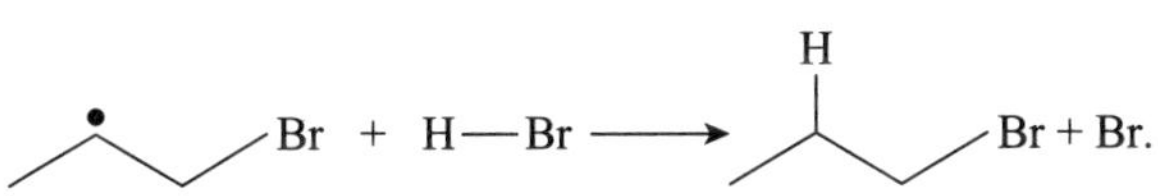

Ans: (b)

48.

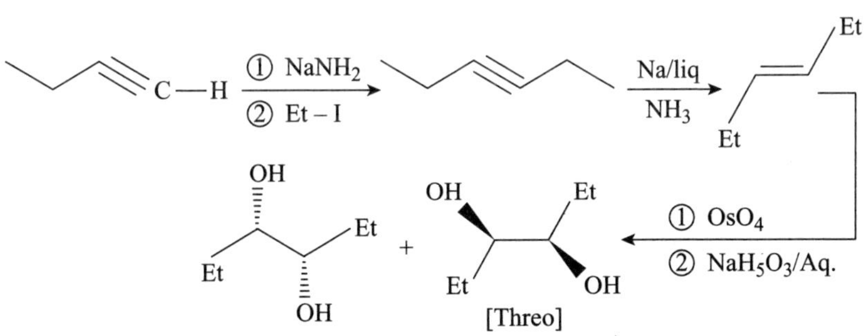

Ans: (c)

49.

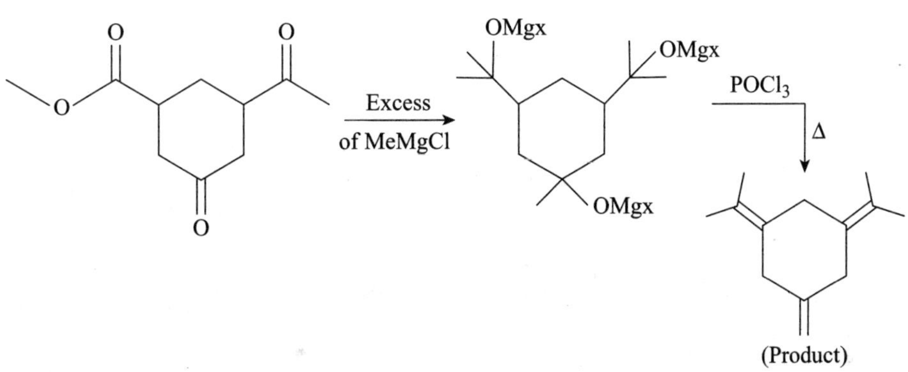

Ans: (b)

50.

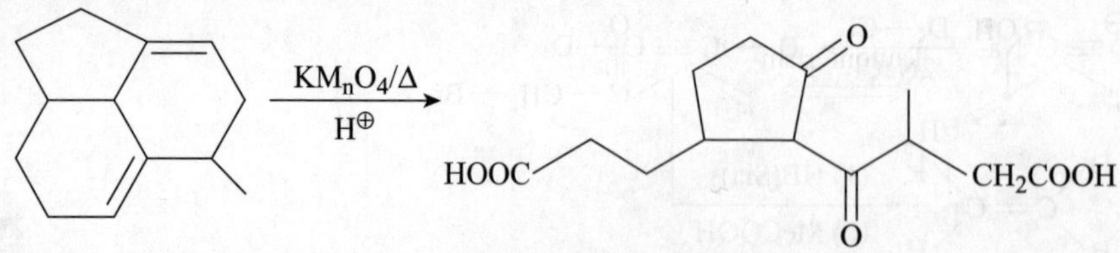

Ans: (d)

51.

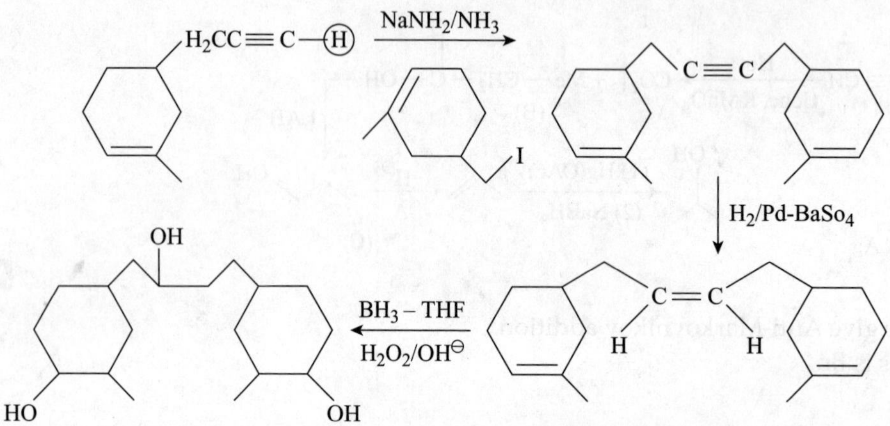

Ans: (c)

52.

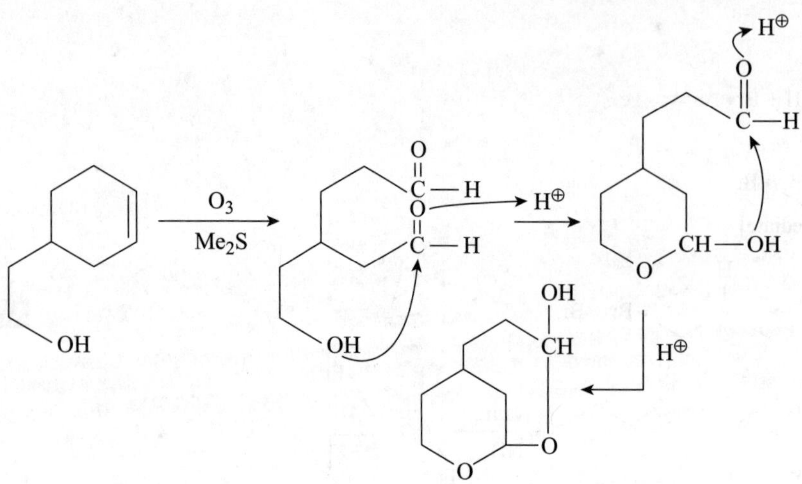

Ans: (b) and (d)

53.

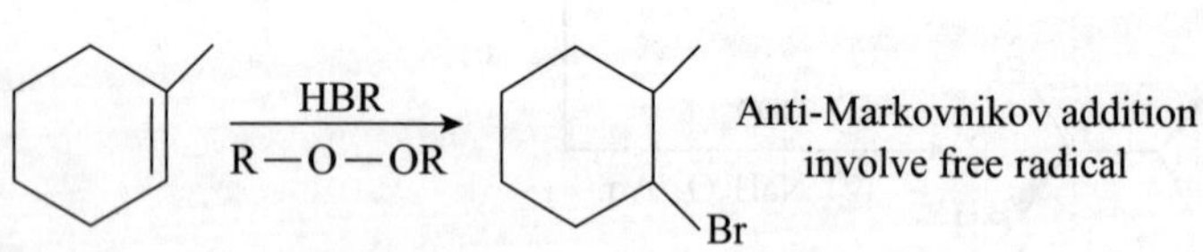

Ans: (c)

54.

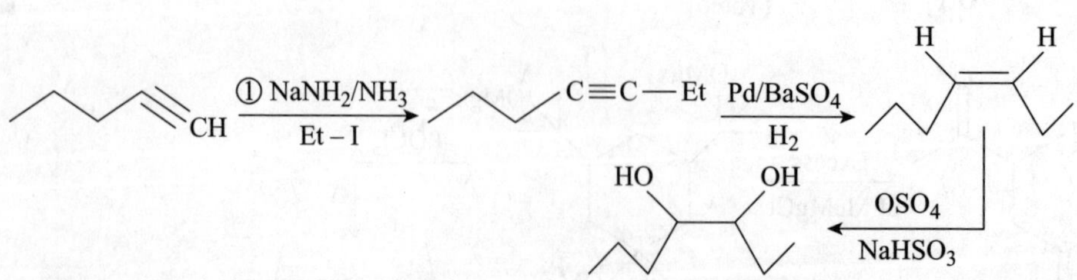

Ans: (b)

55.

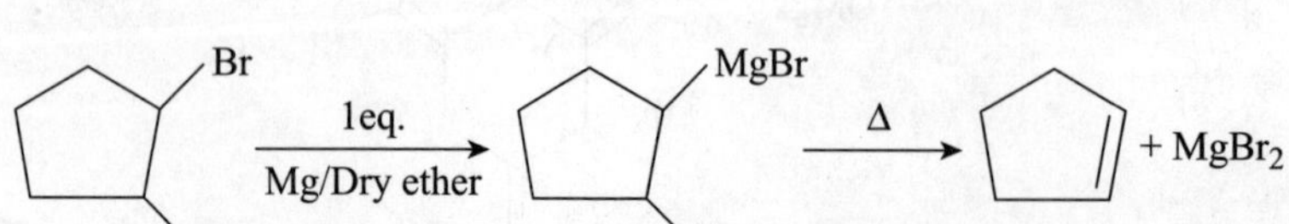

Ans: (c)

56. 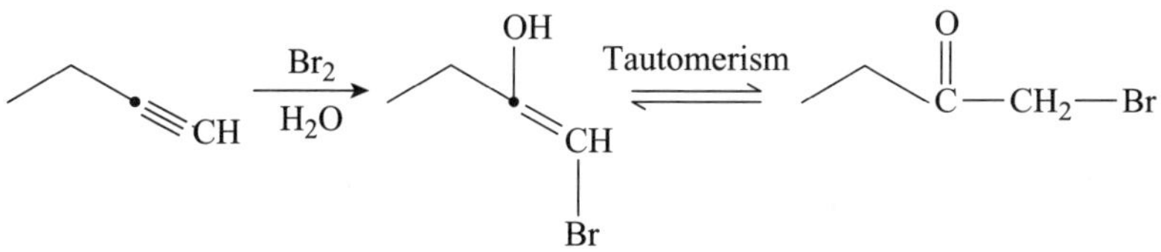

Ans: (d)

57. 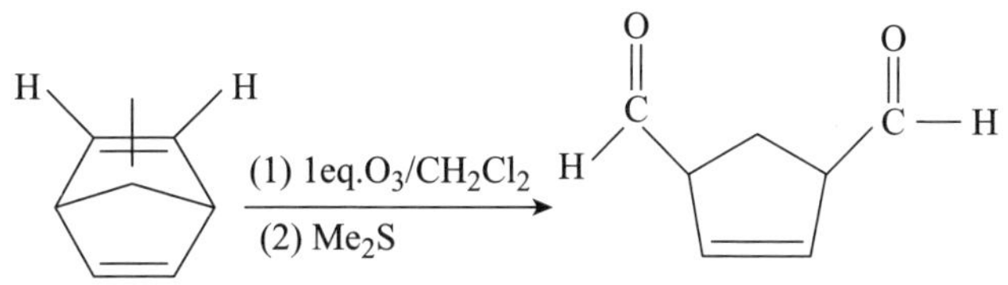

Ans: (b)

58. 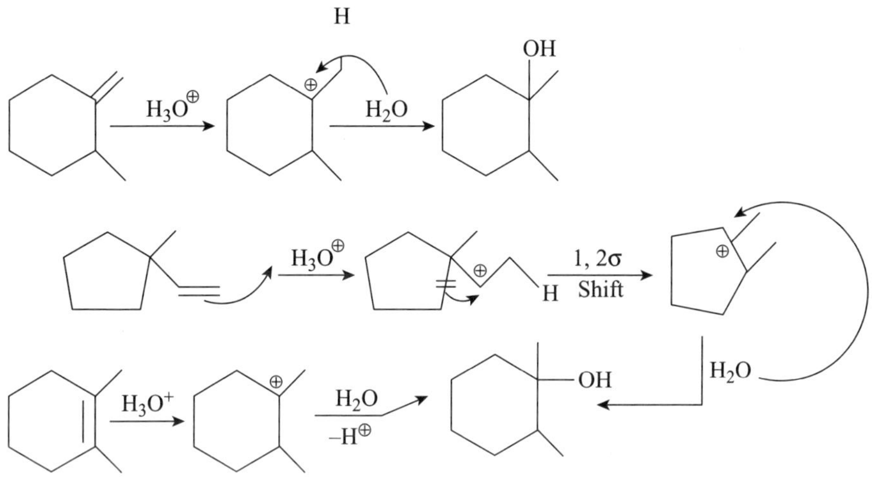

Ans: (a), (c) and (d)

59. 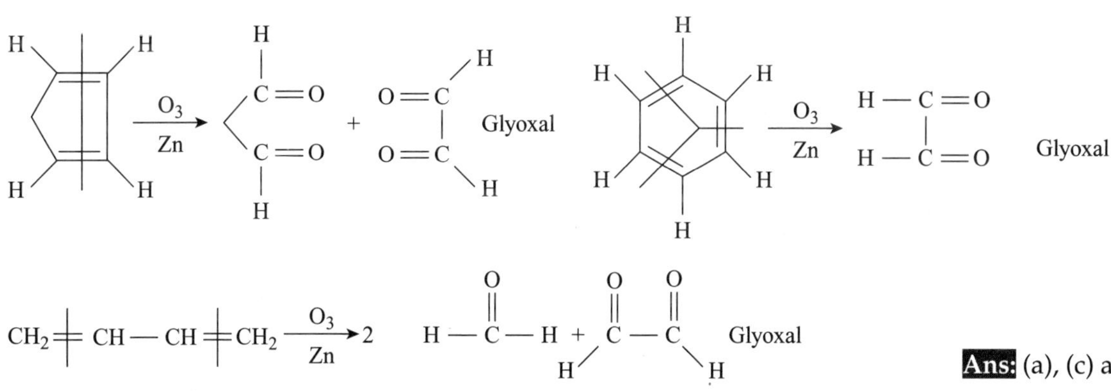

Ans: (a), (c) and (d)

60. 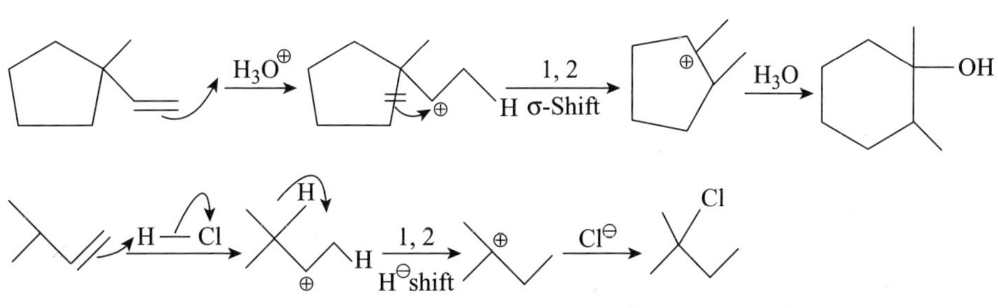

Ans: (a) and (b)

61. 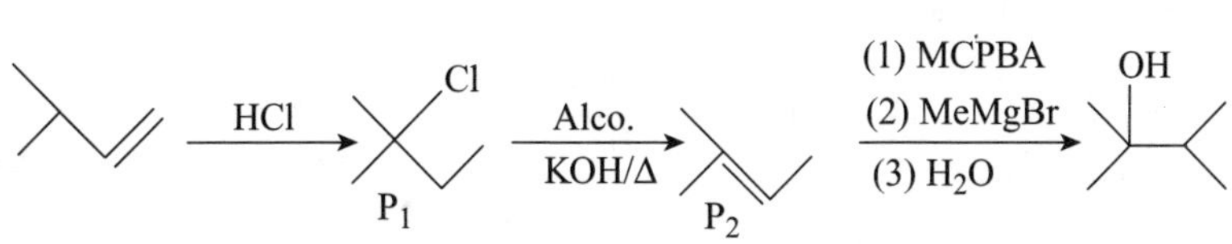

Ans: (b)

62.

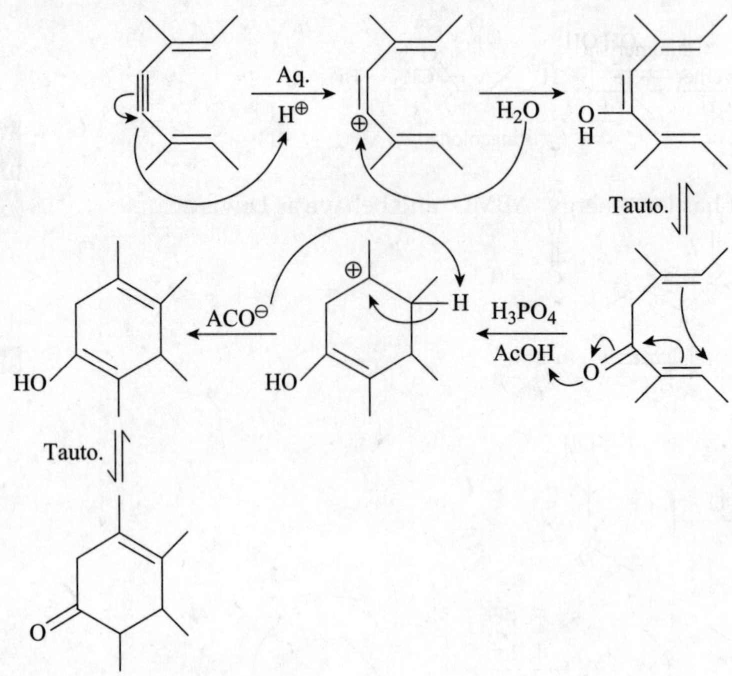

Ans: (d)

63.

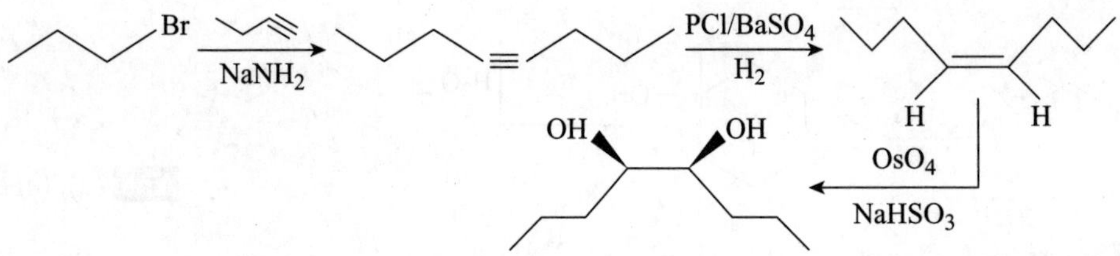

Ans: (d)

64. When carbocation stabilises by resonance it cannot form non classical carbocation (NCC) and give 'syn' and 'anti' both product.

Ans: (d)

65.

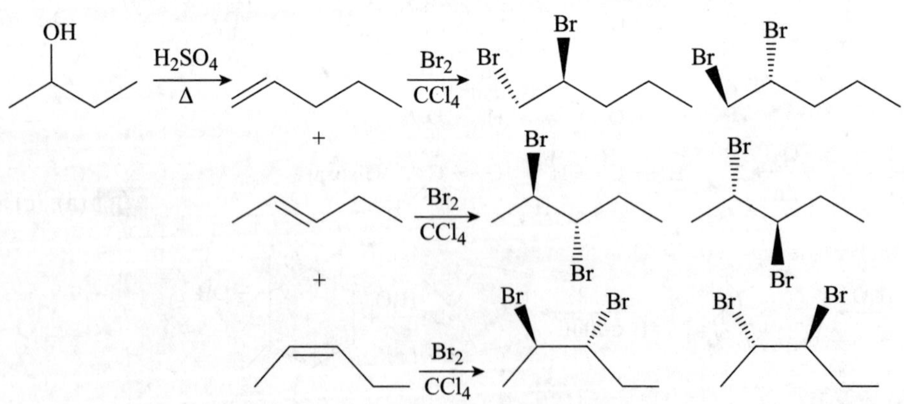

Ans: (c)

66.

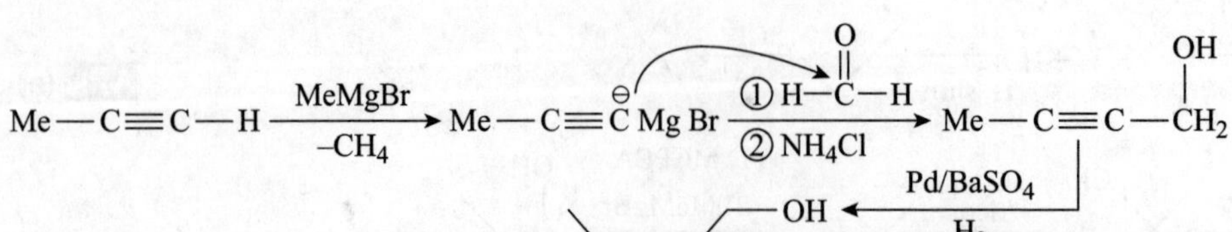

Ans: (d)

67.

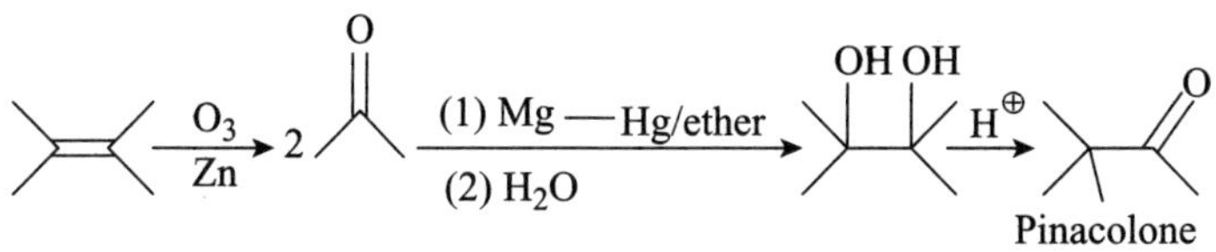

Ans: (b)

68. $:\ddot{\text{Br}}-\ddot{\text{Br}}:$ has weak 'σ' bond so it has low energy 'ABMO' and behave as Lewis acid. Ans: (d)

69.

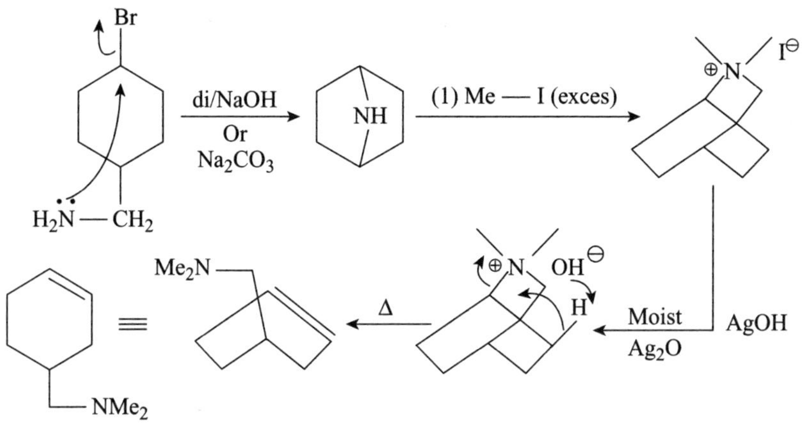

Ans: (c)

70.

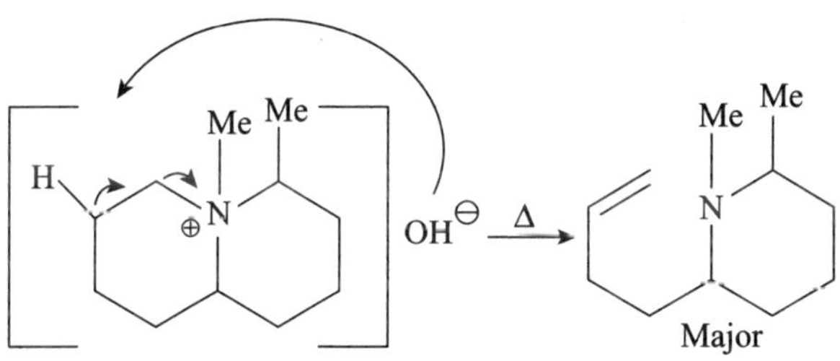

Ans: (c)

71.

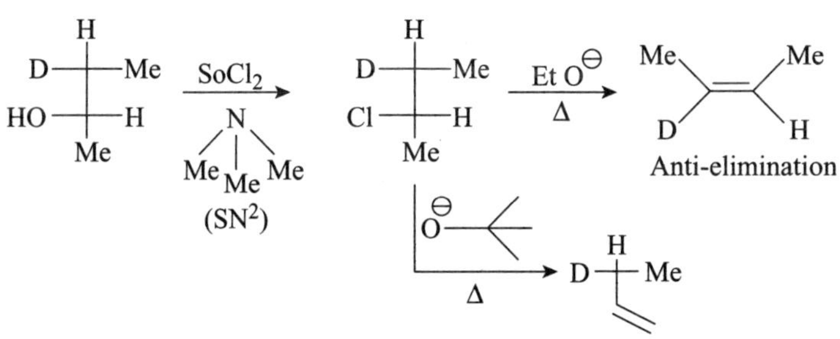

[Bulky base prefer Hofmann's elimination]

Ans: (d)

72.

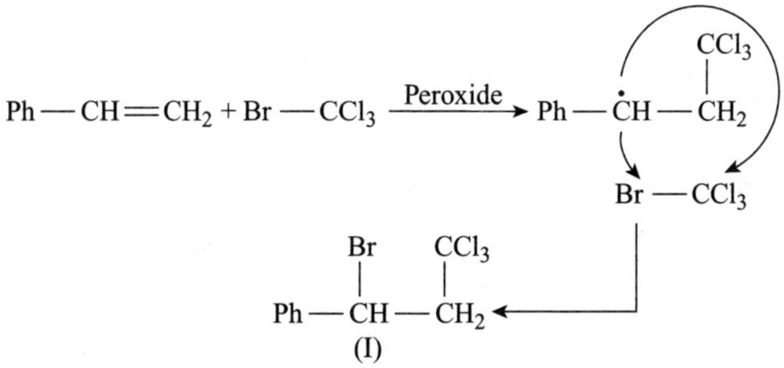

Ans: (a) and (c)

73.

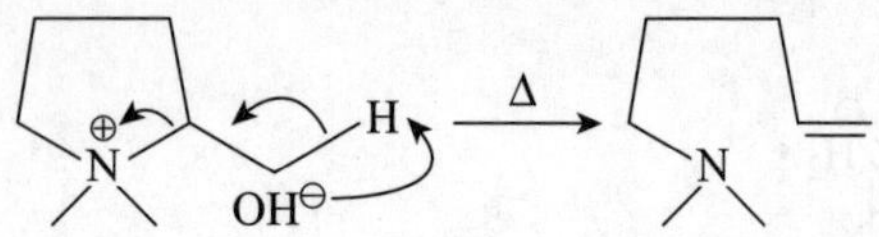

Ans: (b)

74. (a)

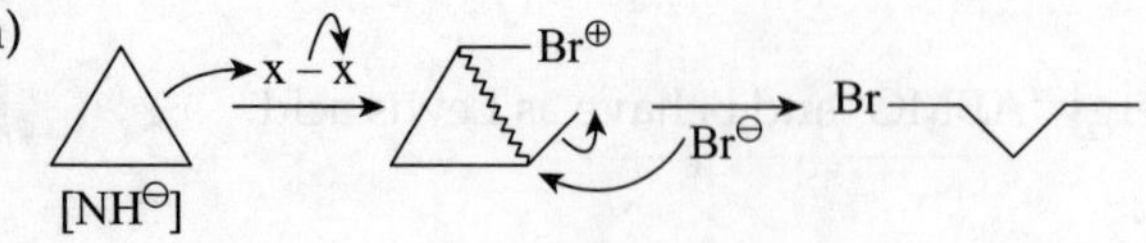

Prefer addition

(b) Selectivity for bromination

	3°	: 2°	: 1°
	1600	: 82	: 1
For chlorination	3°	: 2°	: 1°
	5°	: 3.8°	: 1

(c)

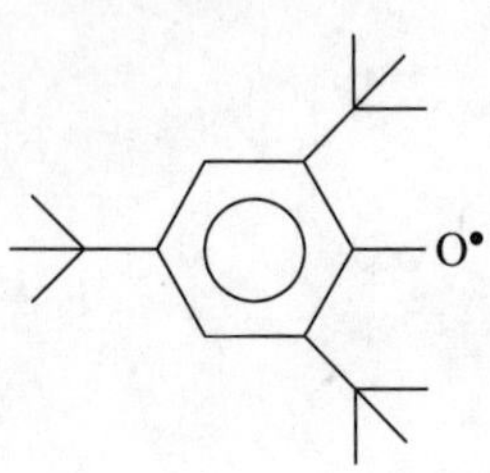

Sterically hindered so cannot give dimerisation reaction

Ans: (a), (b) and (c)

75. Terminal alkyne give white ppt. with Tollen.

Ans: (d)

76.

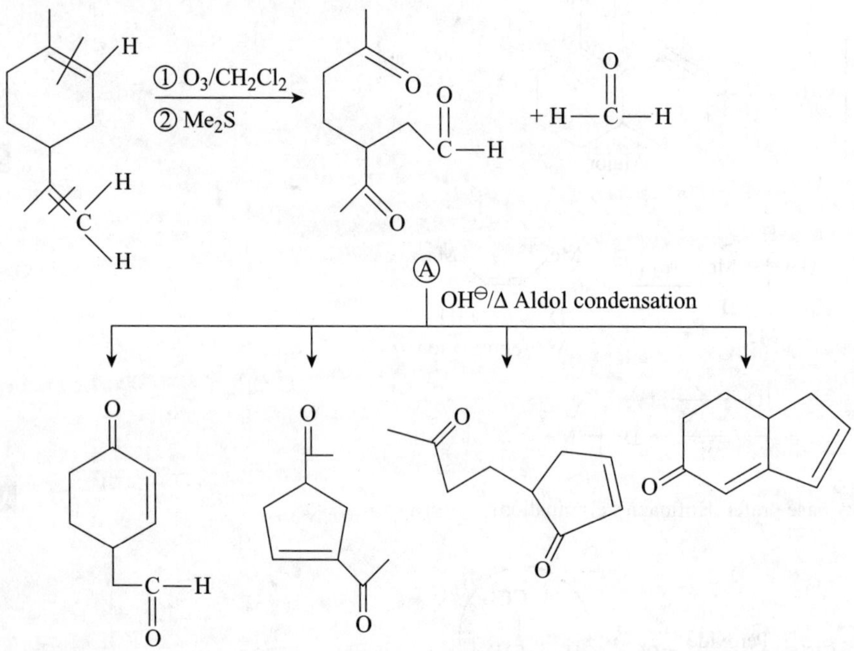

Ans: (a), (b), (c) and (d)

77. 'ABCD' all are correct option because Hofmann's elimination takes place from 'B' position where carbanion is more stable.

Ans: (a), (b), (c) and (d)

78.

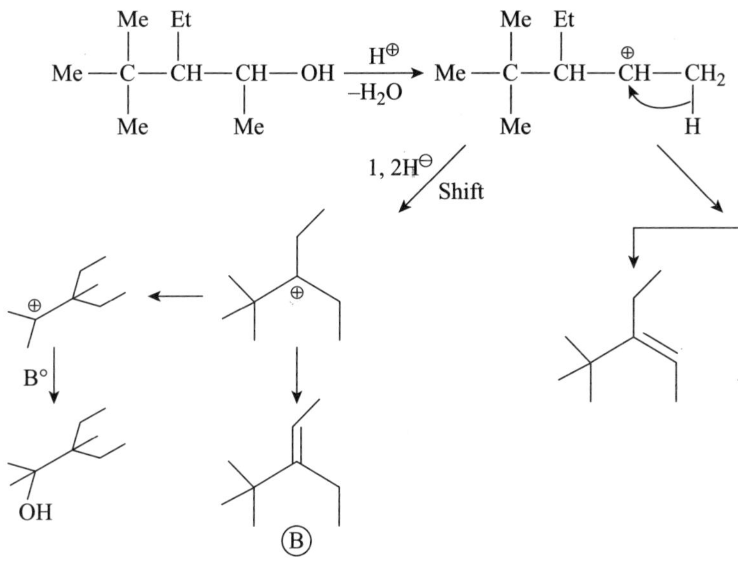

Ans: (b) and (d)

79. **Ans:** (a) Oxymercuration demercuration.

(b) Hydroboration oxidation.

(c) Anti-Markovnikov H-Br addition.

80.

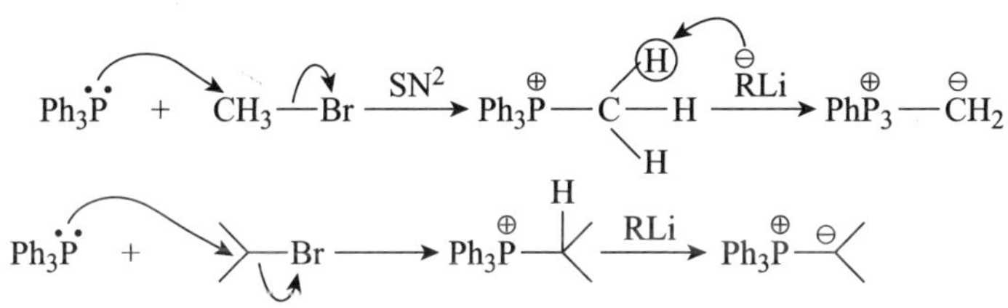

Ans: (b) and (c)

81.

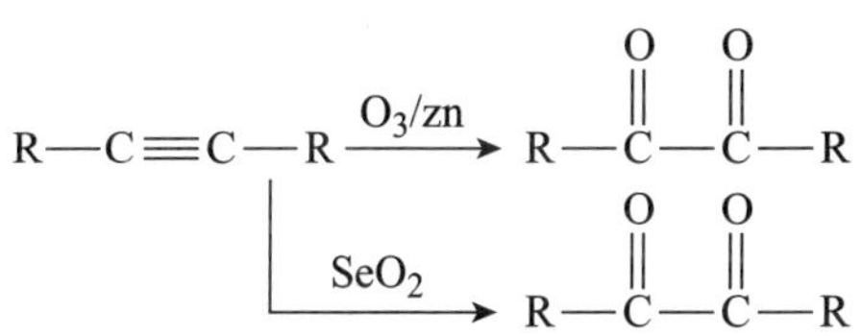

Ans: (a)

82.

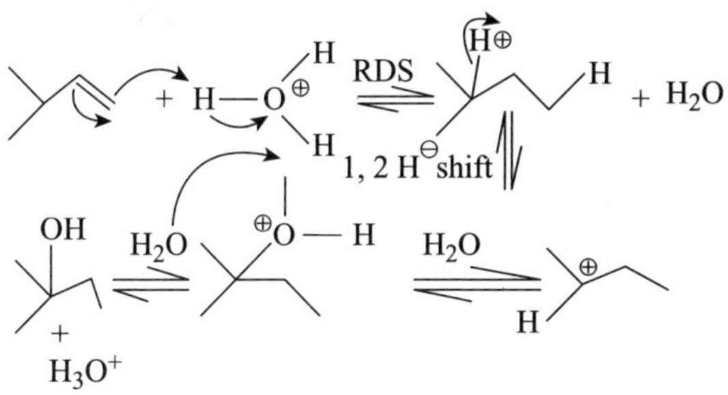

Ans: (a)

83. (A)

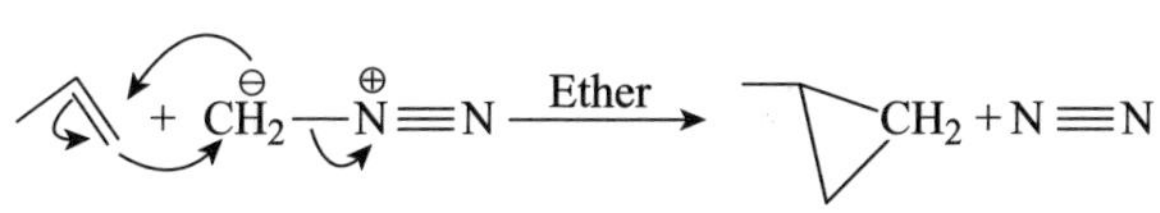

(C) $I—CH_2—I + Zn \longrightarrow I—CH_2—Zn—I$

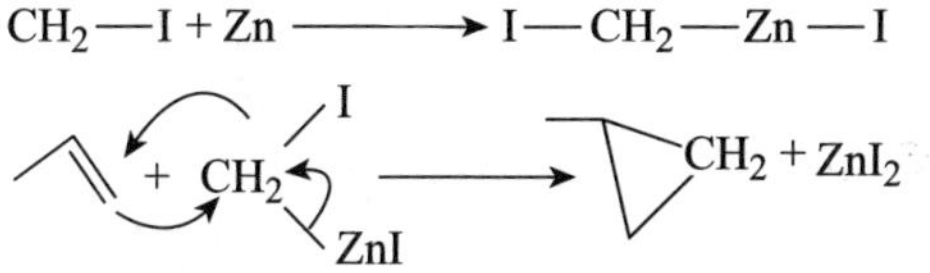

Ans: (a) and (c)

84. (A)

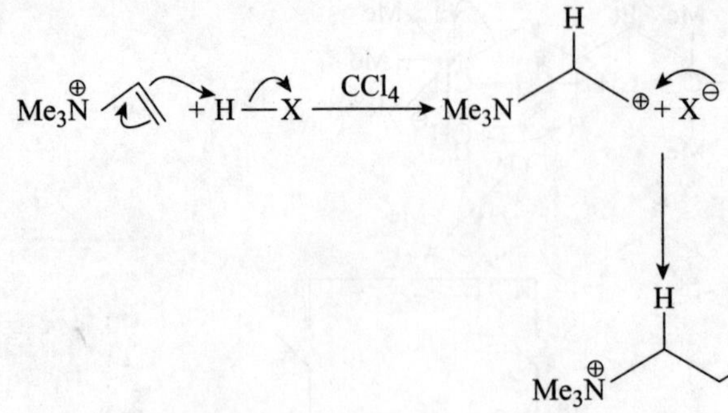

(B)

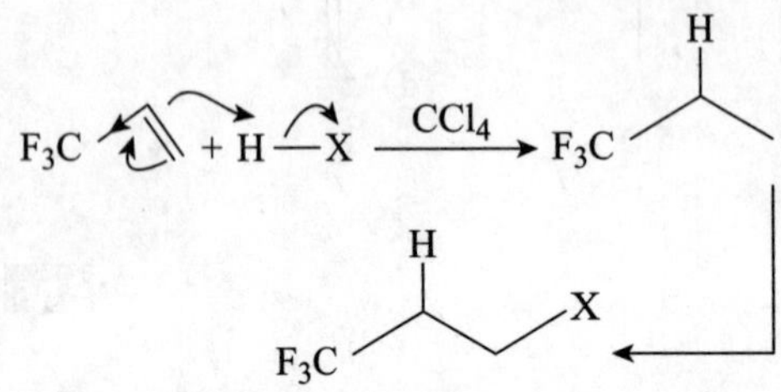

Ans: (a) and (b)

85. (A)

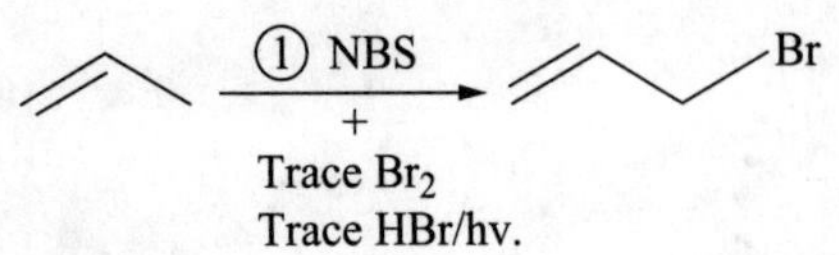

(B)

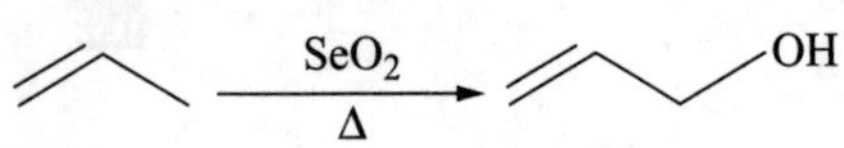

(C)

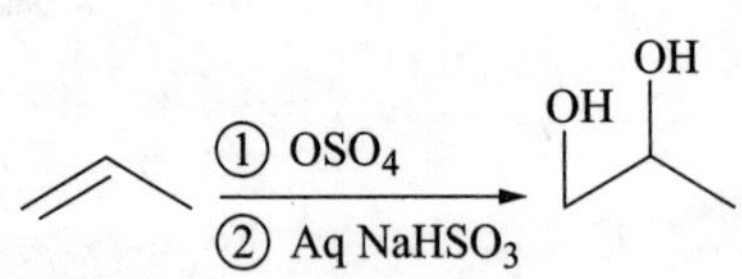

(D)

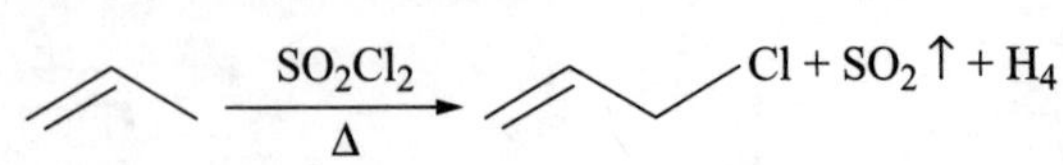

Ans: (a), (b) and (d)

86. 'σ'-bond of cyclopropane behave as π-bond so it gives addition reaction.

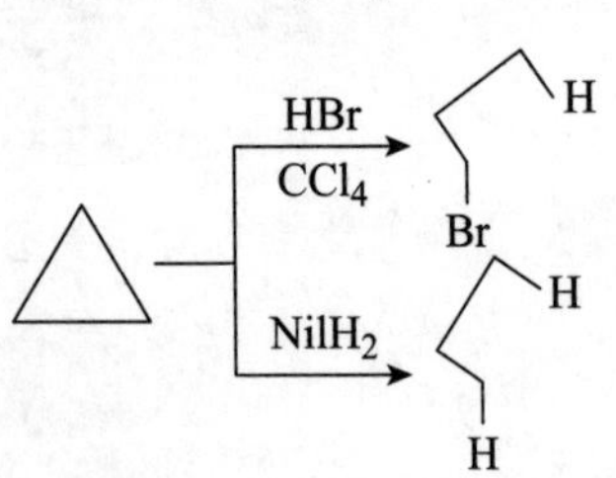

Ans: (b) and (c)

87. (B)

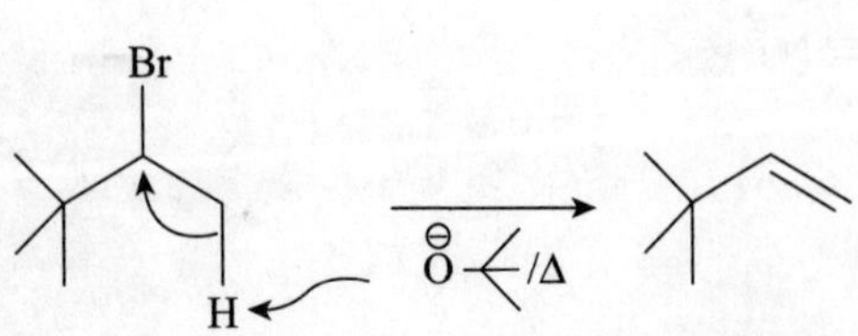

(C) 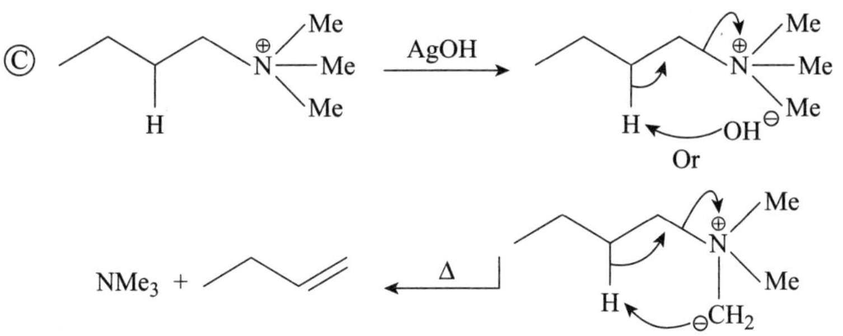

Ans: (b) and (c)

88. 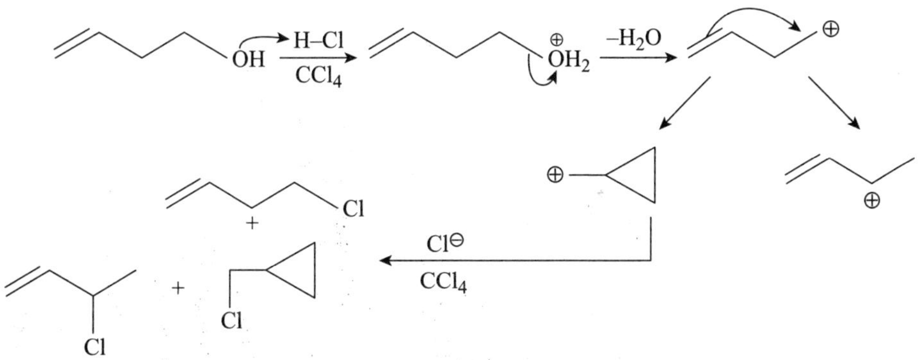

Ans: (a), (b), (c) and (d)

89. Allylic and benzylic bromination occur with NBS. Ans: (a) and (d)

90. Terminal alkyne have active/Acyclic 'H'. Ans: (a), (b) and (c)

91. 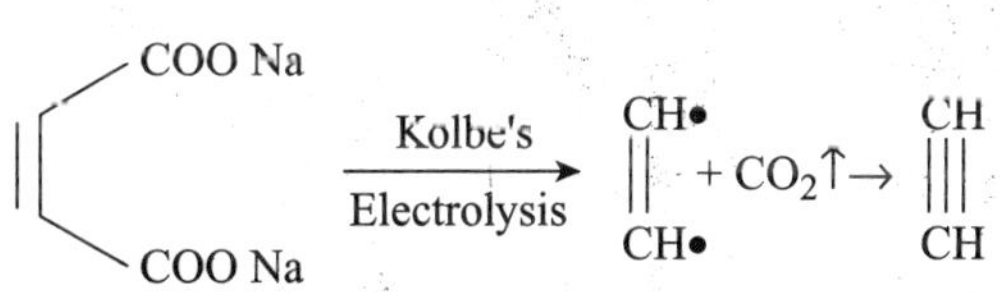

Socl.maleate

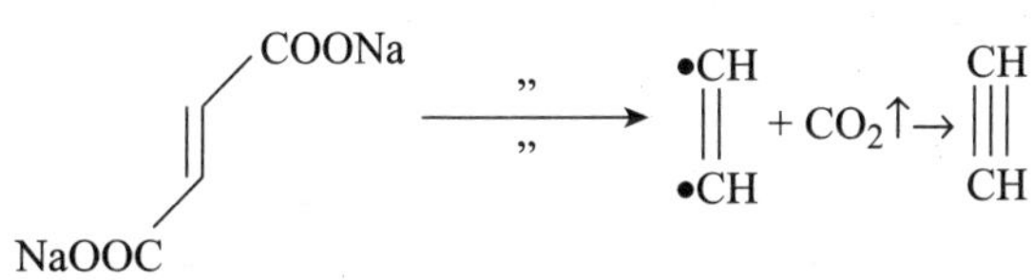

Socl.fumerate

Ans: (a) and (b)

92. 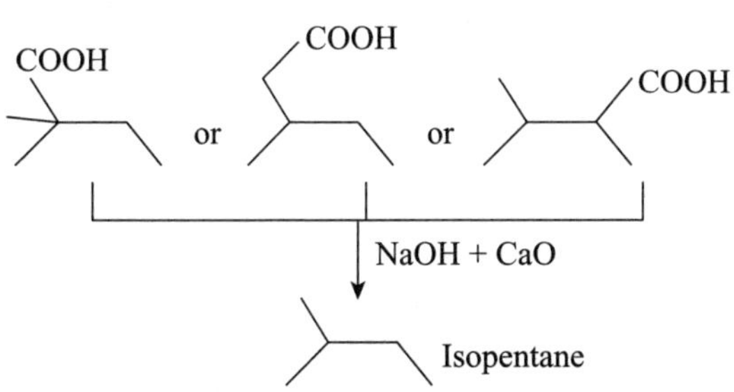

Ans: (a), (b) and (c)

93. Odd number of 'C' containing alkane cannot prepared as good yield in Wurtz reaction.

Ans: (a) and (c)

94.

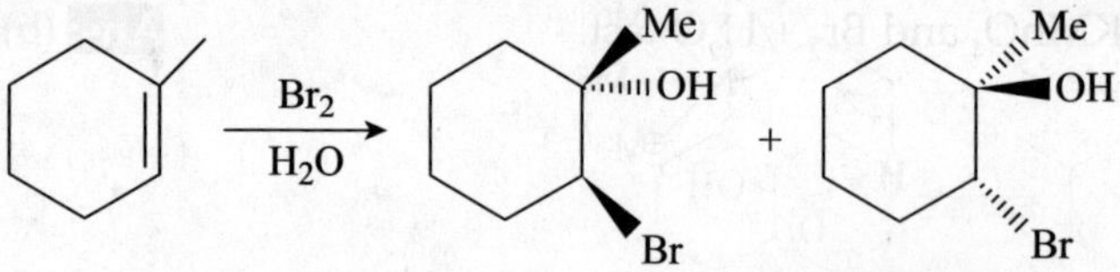

[Anti-addition]

Ans: (a)

95.

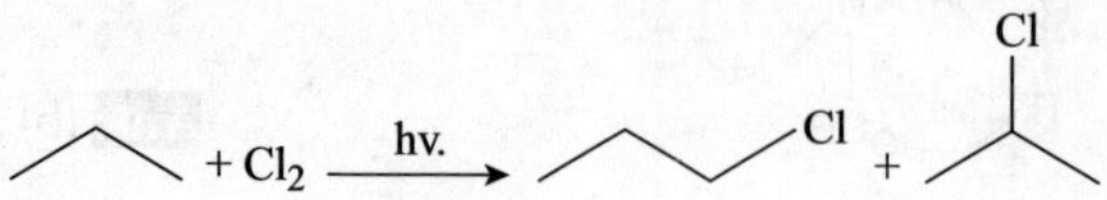

% yield = 45% : 55%

Number of H = 6 : 2

Per 'H' atom selectivity $= \frac{45}{6} : \frac{55}{2} = 1 : 3.6$ (1° : 3°)

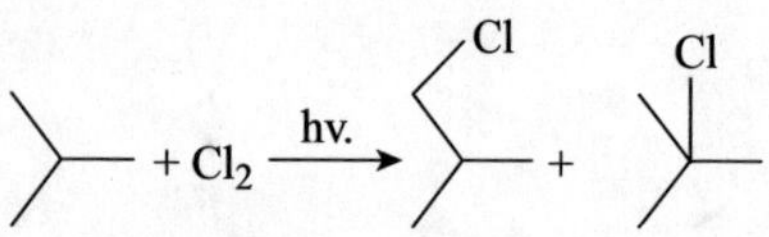

% yield = 65% : 35%

Number of H = 9 : 1

Per 'H' atom selectivity $= \frac{65}{9} : \frac{35}{1} = 1 : 4.8$ (1° : 3°)

Per 'H' atom selectivity = 1°: 2° : 3°

1 : 3.6 : 4.8

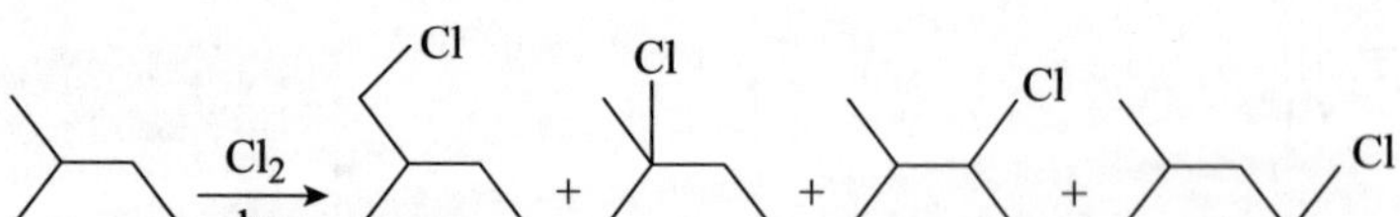

	1°		3°		2°		1°		
Selectivity	1	:	4.8	:	3.6	:	1		
Number of H =	6	:	1	:	2	:	3		Total
Product ratio =	6	:	4.8	:	7.2	:	3	=	21

% of 2 product$=\frac{7.2}{21}\times 100 = 34.28\%$

Ans: (d)

96.

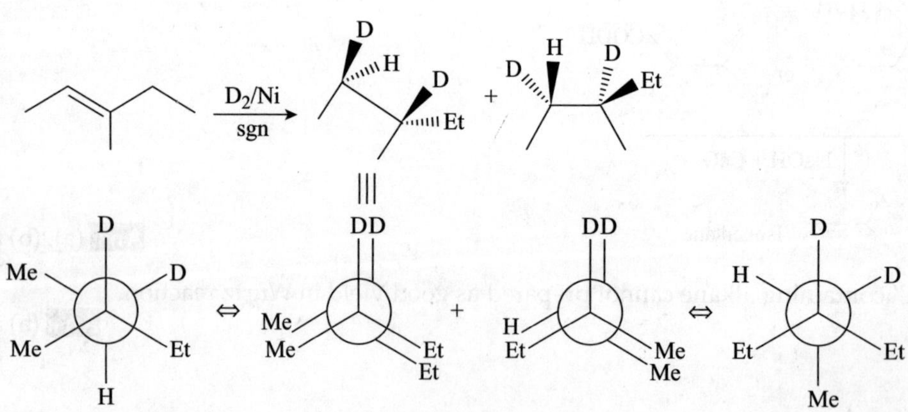

Ans: (d)

97. Test for unsaturation is 1% alkaline cold $KMnO_4$ and $Br_2 + H_2O$ test. **Ans:** (b) and (d)

98. (a), (b), (c)

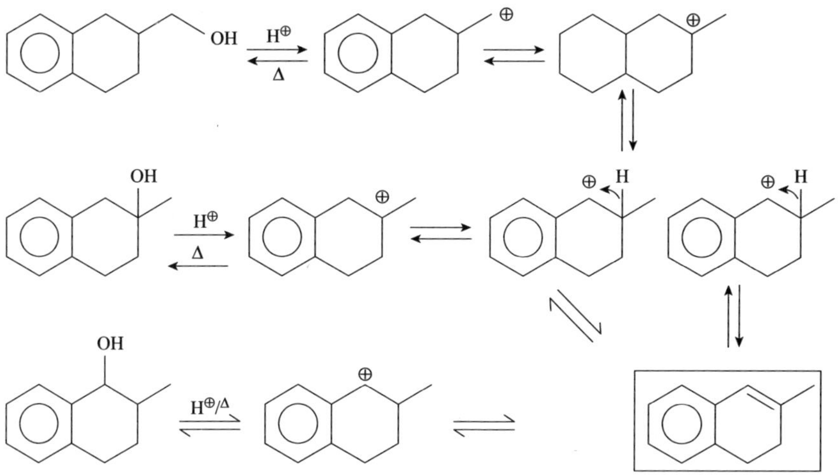

Ans: (a), (b) and (c)

99. (a)

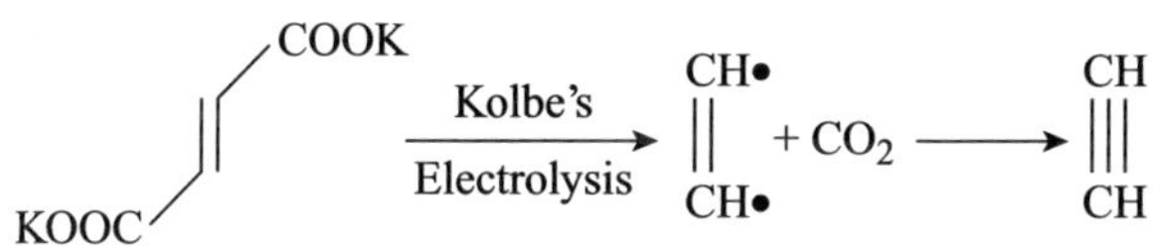

(b)

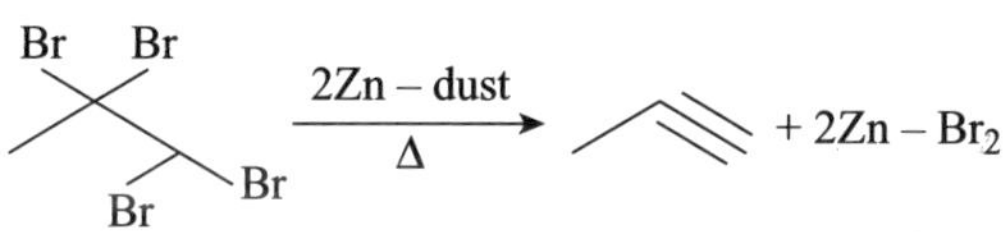

(c)

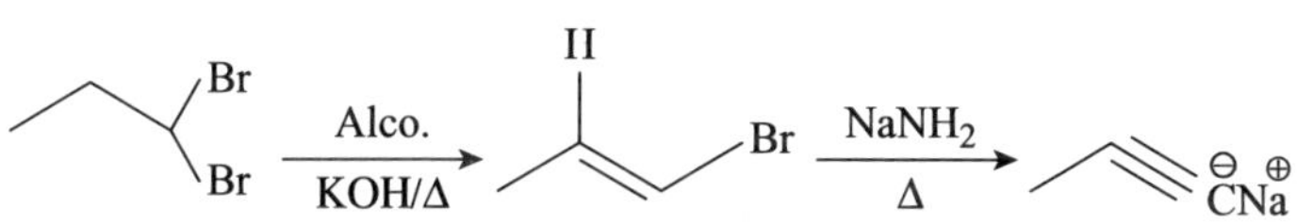

(d)

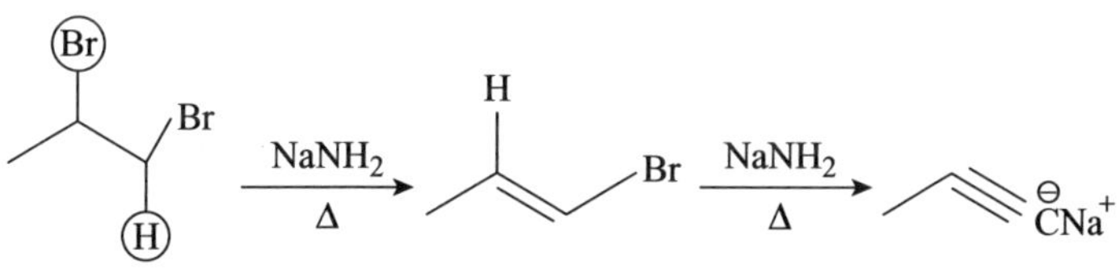

Ans: (a), (b), (c) and (d)

100.

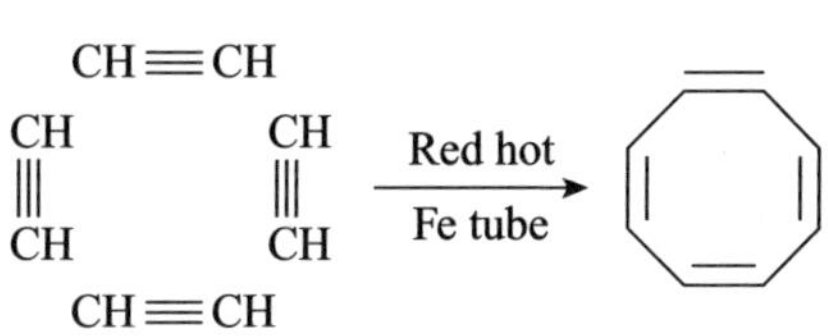

Ans: (b)

LEVEL 2

1. 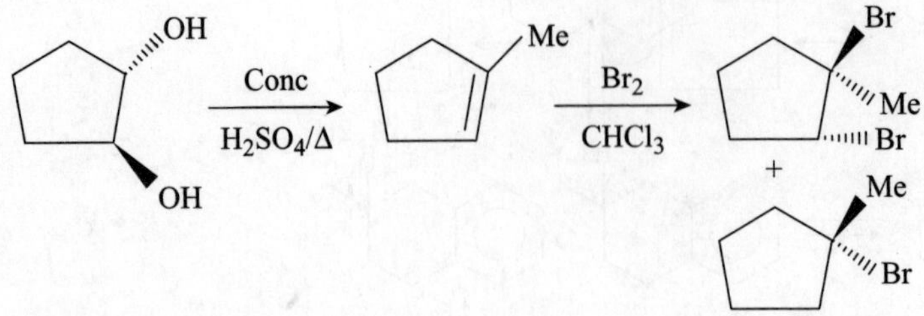

Ans: (b)

2. 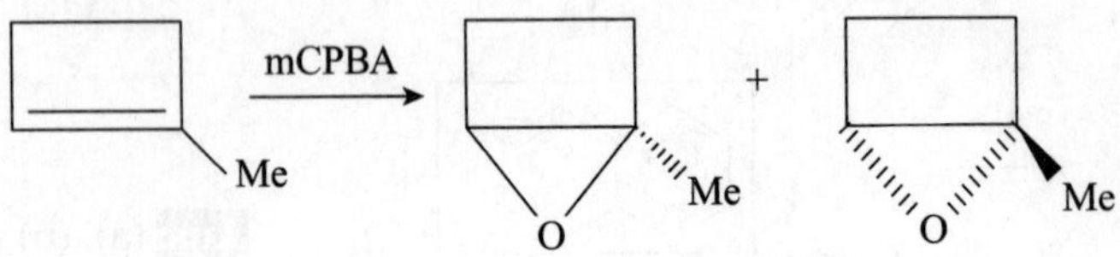

Ans: (a) and (c)

3. 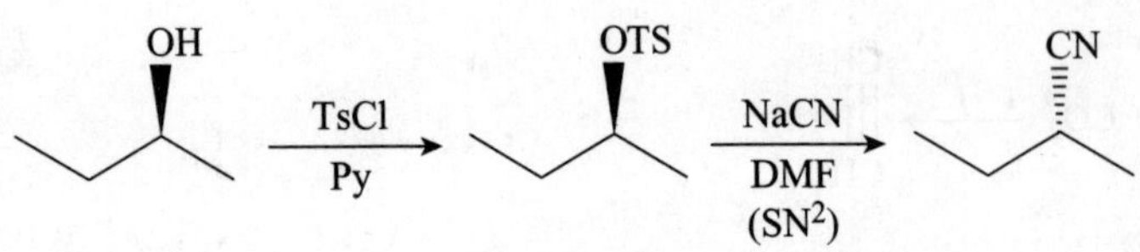

Ans: (c)

4. 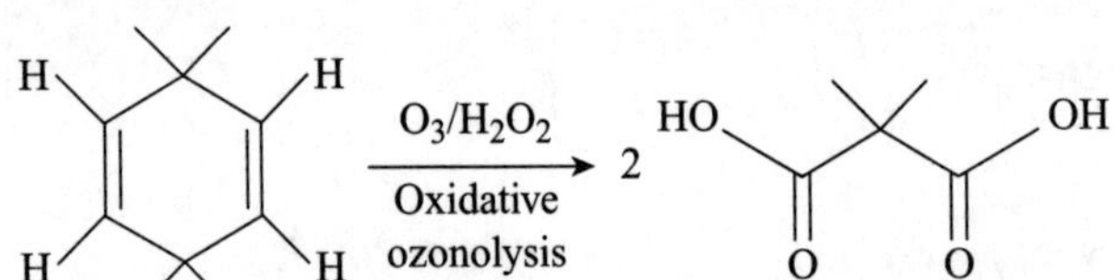

Ans: (d)

5. 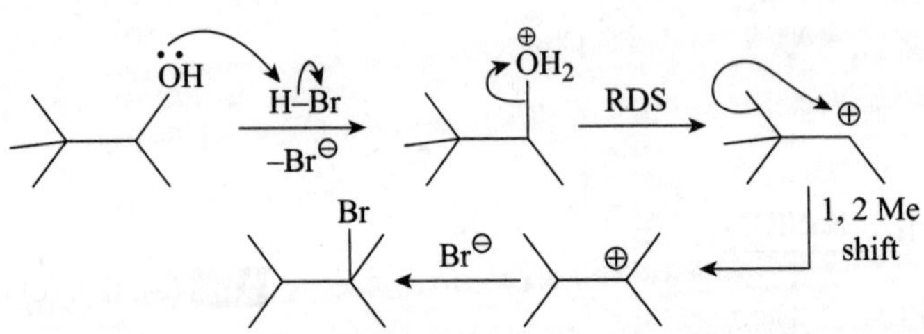

Ans: (d)

6. 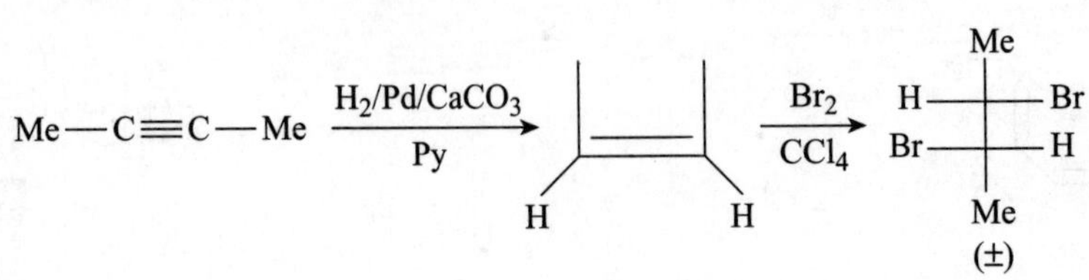

Ans: (d)

7. 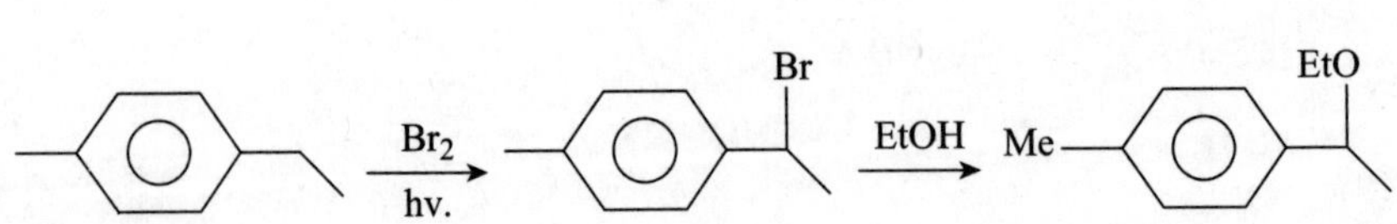

Ans: (a)

8.

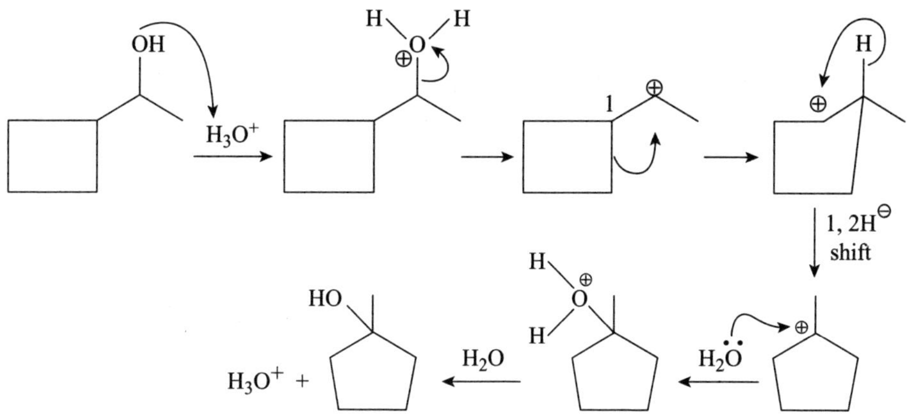

Ans: (a) and (b)

9.

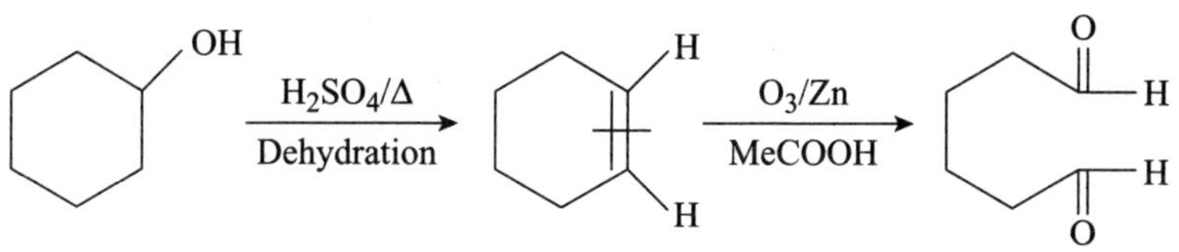

Ans: (b)

10. (c)

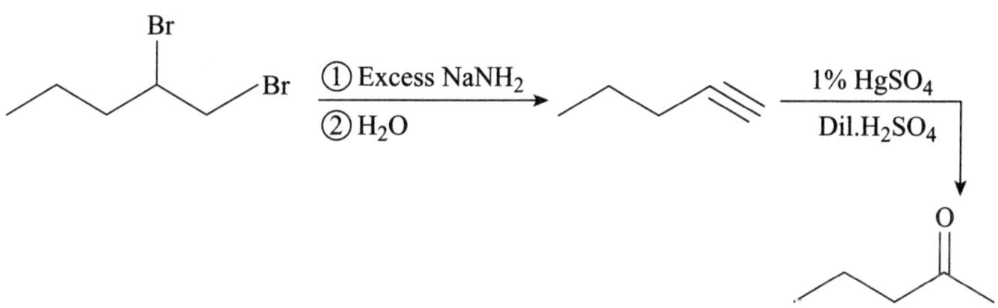

(d)

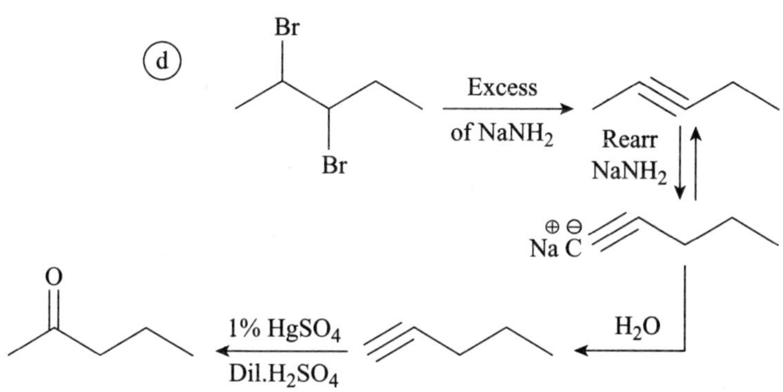

Ans: (c) and (d)

11.

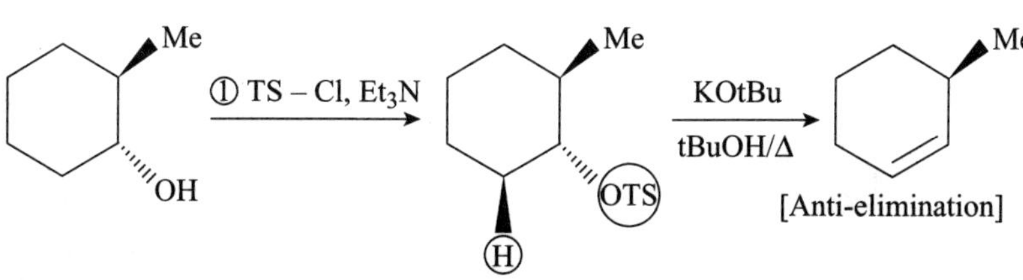

Ans: (a) and (b)

12.

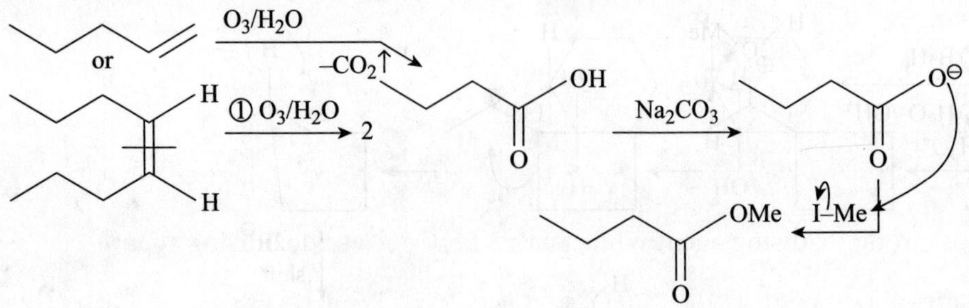

Ans: (b) and (a)

13.

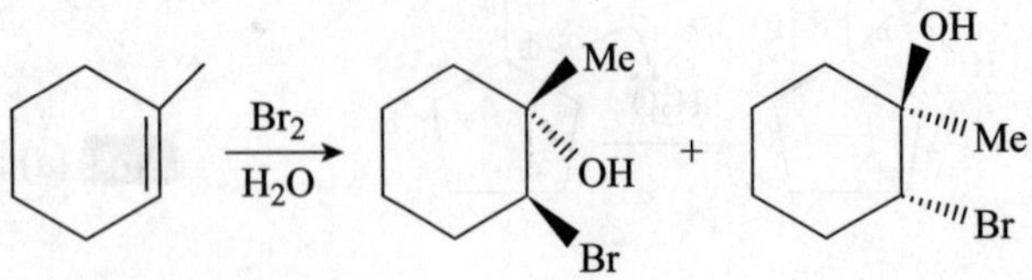

Ans: (d)

14.

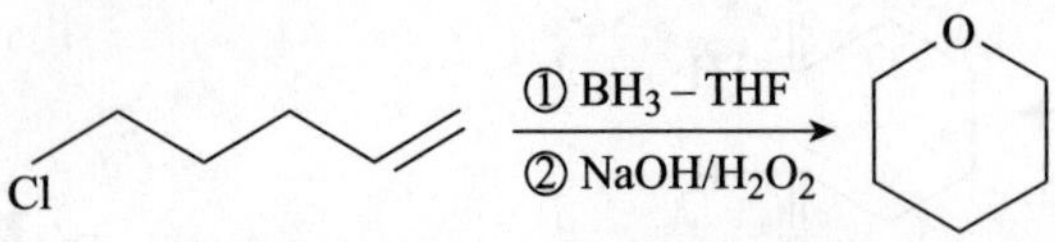

Ans: (d)

15.

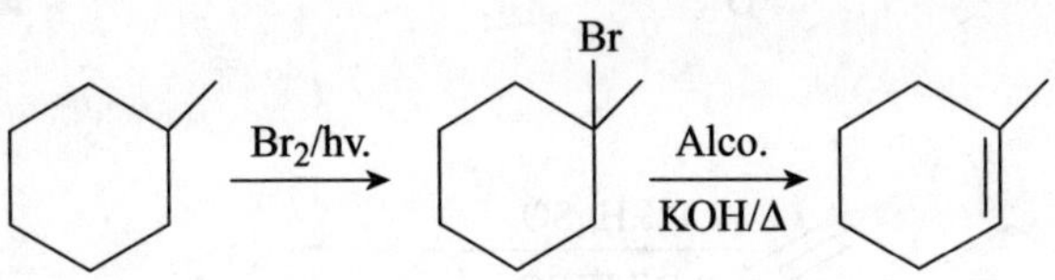

Ans: (d)

16.

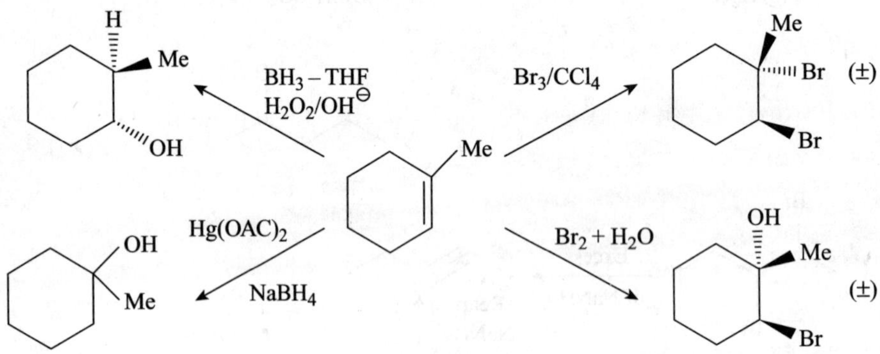

Ans: (a), (b) and (d)

17.

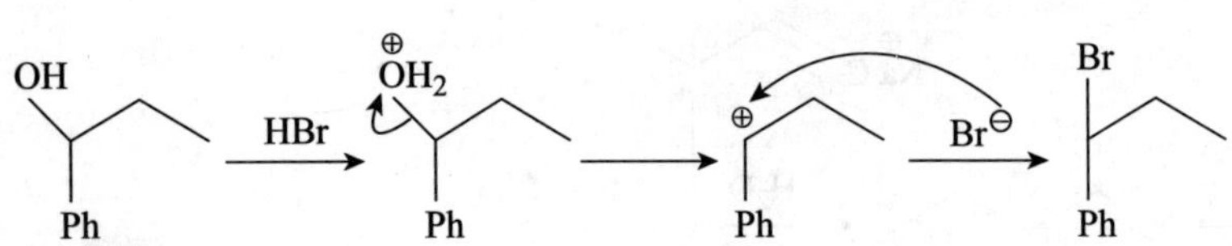

Ans: (b)

18.

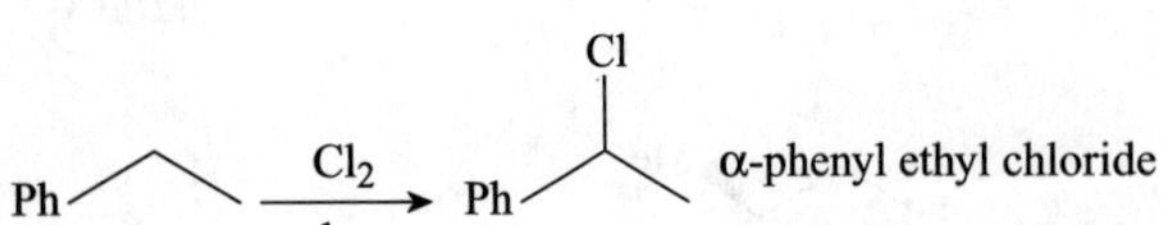

Ans: (b)

19.

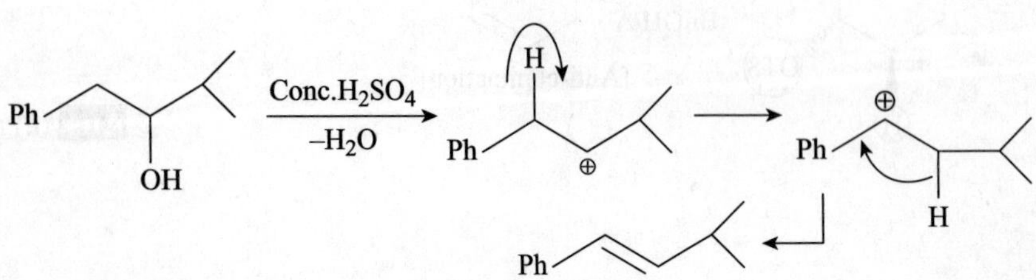

Ans: (b)

20. 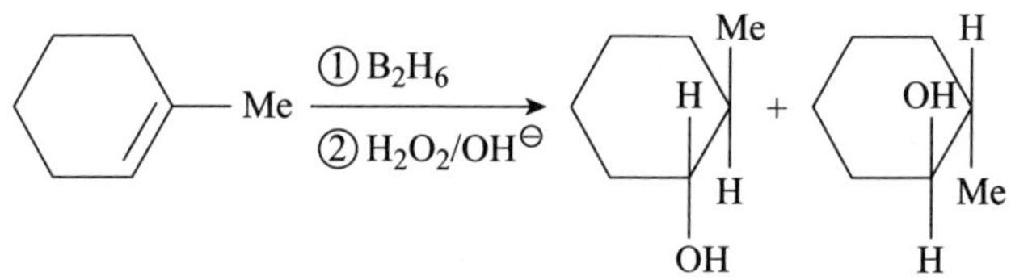

Ans: (b)

21. (a) Acidify $KMnO_4$ is strong oxidising agent while basic $KMnO_4$ is weak oxidising agent.

(b) $2KMnO_4 + 3H_2SO_4 \longrightarrow 2MnSO_4+K_2SO_4+3H_2O+5[O]$

while

[Neutral/Basic medium] $2KMnO_4 + H_2O \longrightarrow 2MnO_2+ 2KO_4+3[O]$

(c) 

(d) 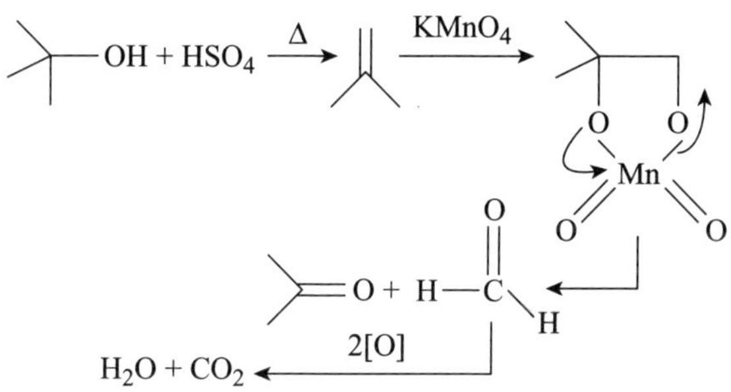

Ans: (c)

22. Functional groups having oxygen atom at α-position give reaction with $HIO_4.2H_2O$.

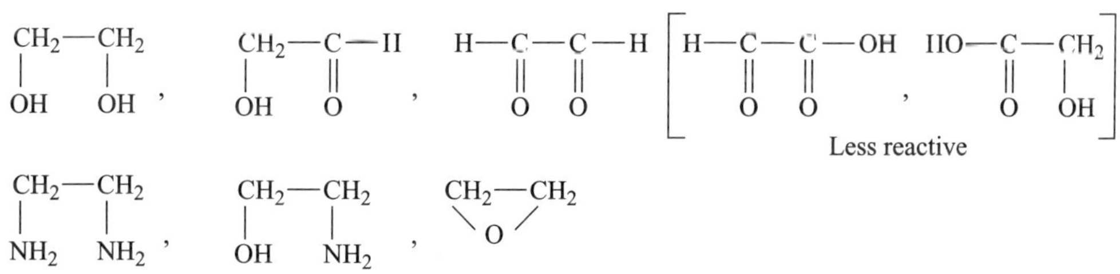

Ans: (d)

23. 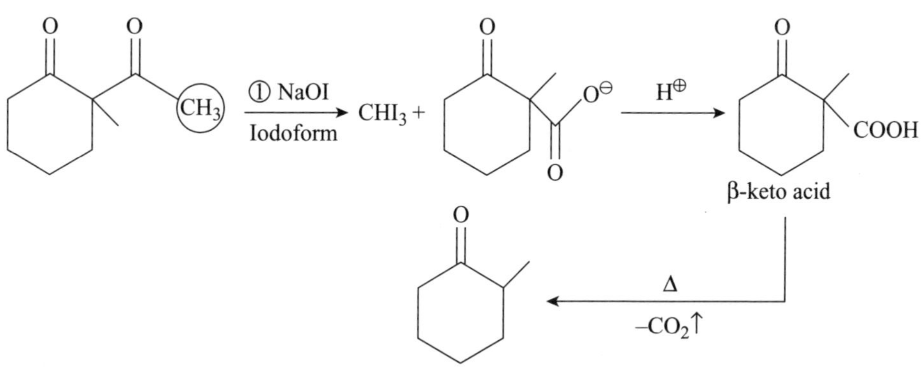

Ans: (c)

24. H_2SO_4 is strong dehydrating agent and form various hydrate $[H_2SO_4 . nH_2O]$

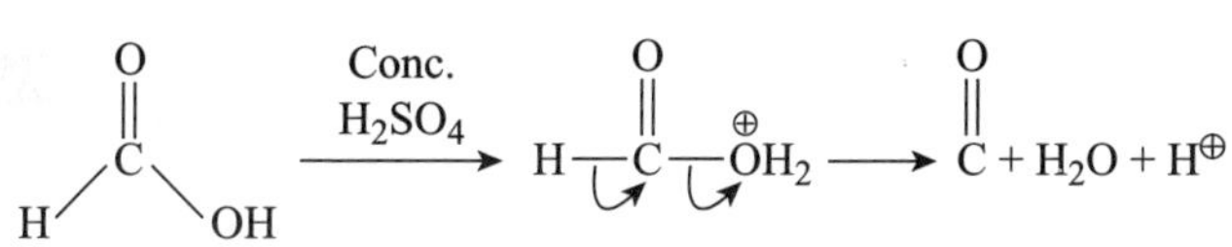

Ans: (a)

25.

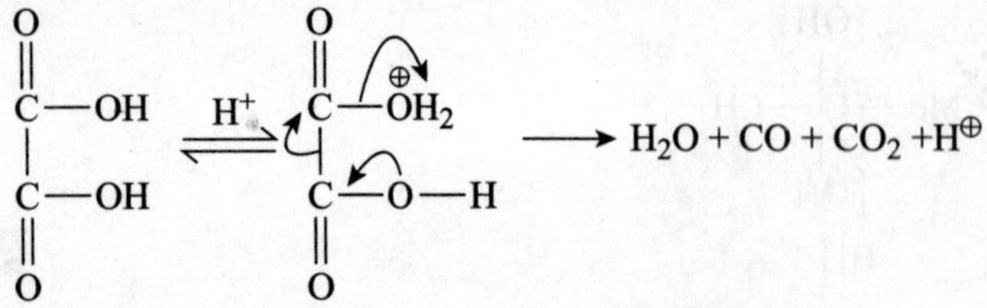

Ans: (d)

26.

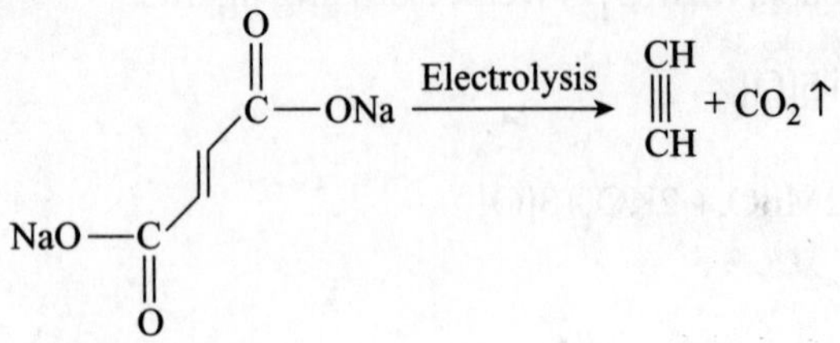

Ans: (b)

27.

$$HO{-}H + H{-}\overset{O}{\overset{\|}{C}}{-}O^{\ominus}Na \xrightarrow{\Delta} CO_2\uparrow + H_2 + NaOH$$

Ans: (b)

28.

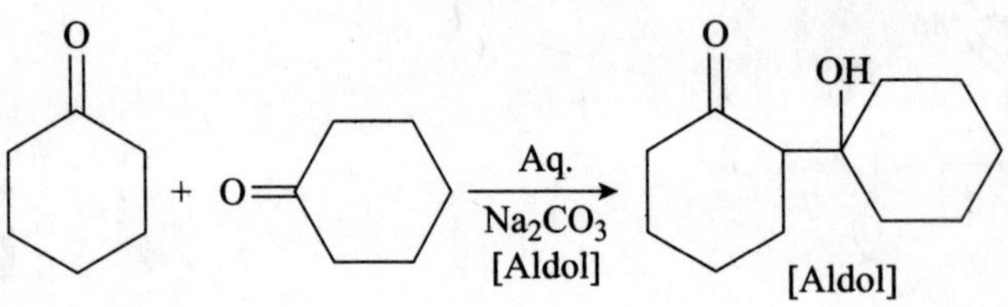

Ans: (c)

29.

$$Et{-}C{\equiv}C{-}H + BH_3 \longrightarrow (Et{-}CH{=}CH)_3{-}B \xrightarrow{H_2O_2/OH^{\ominus}}$$

$$Et{-}CH_2{-}\overset{O}{\overset{\|}{C}}{-}H \xrightleftharpoons{\text{Tauto.}} \underset{\text{enol}}{Et{-}CH_2{-}CH{=}CH{-}OH}$$

Ans: (a)

31.

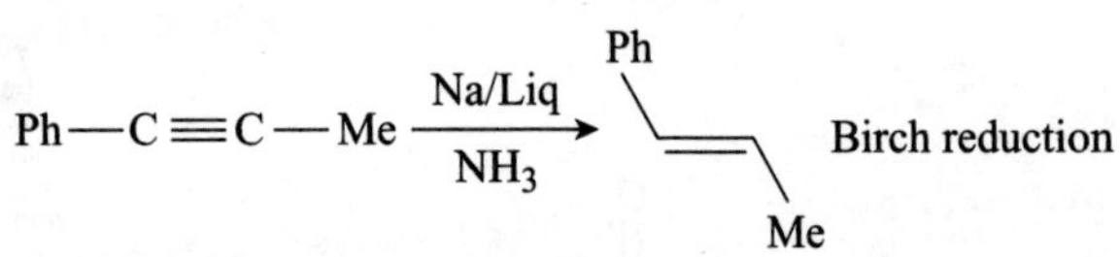

Ans: (c)

32.

$$Et{-}C{\equiv}C{-}Me \xrightarrow[\text{Dil.}H_2SO_4]{1\%\ HgSO_4} Et{-}\overset{O}{\overset{\|}{C}}{-}CH_2{-}CH_3 + Et{-}CH_2{-}\overset{O}{\overset{\|}{C}}{-}CH_3$$

Ans: (d)

33.

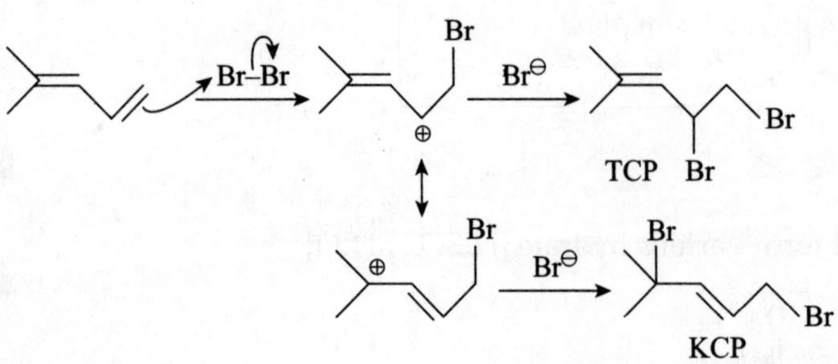

Ans: (b)

34.

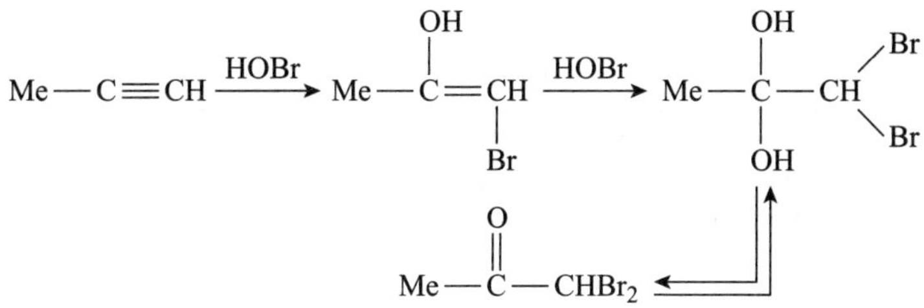

Ans: (b)

37.

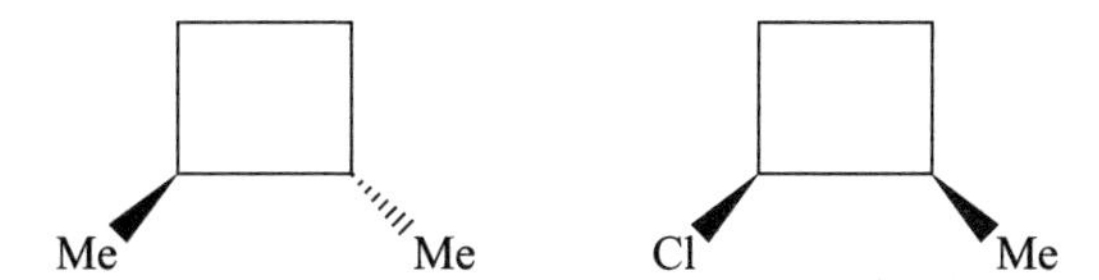

Plane of symmetry absent
Centre of symmetry absent
Optically active

Plane of symmetry absent
Centre of symmetry absent
Optically active

Ans: (a)

38. Boiling point of hydrocarbon α-surface area lowest bp is expected for 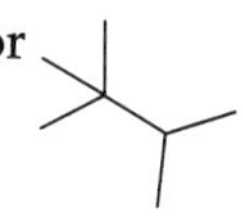

Ans: (c)

39.

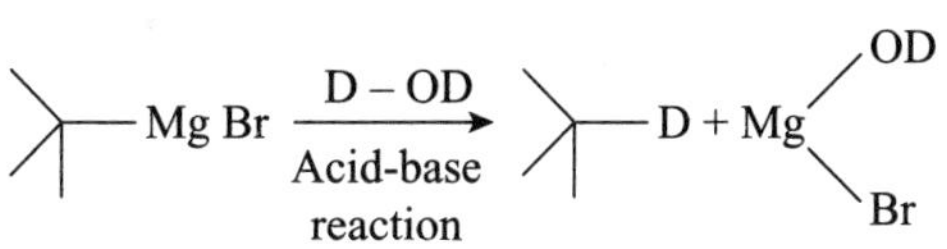

Ans: (b)

40.

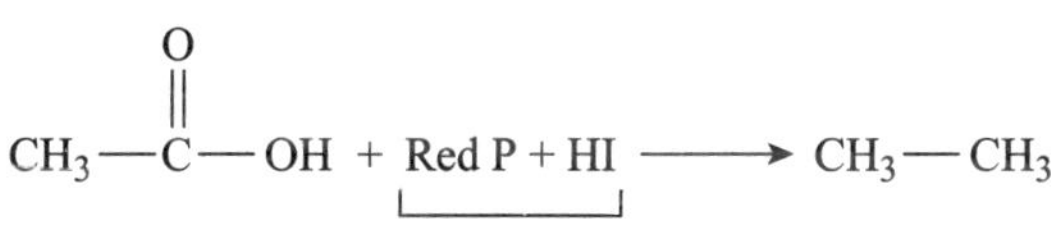

Ans: (b)

41.

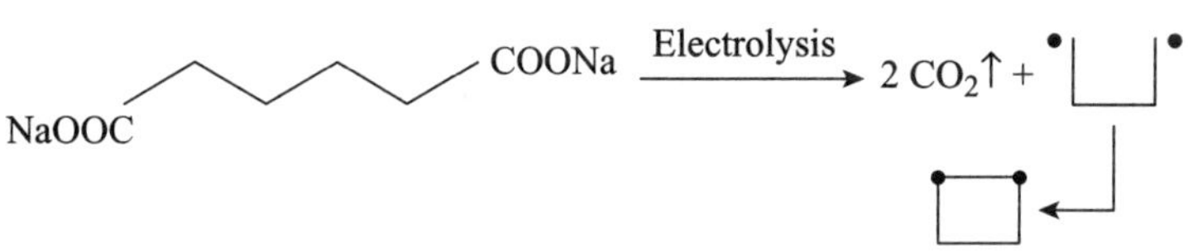

Ans: (a)

42.

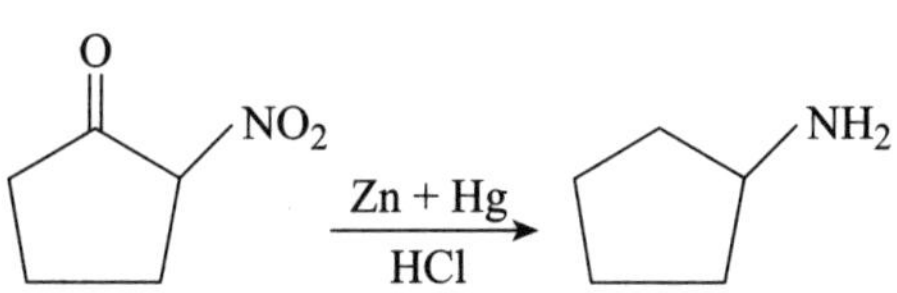

Reduction >C=O as well as –NO_2 both group takes place.

Ans: (b)

43.

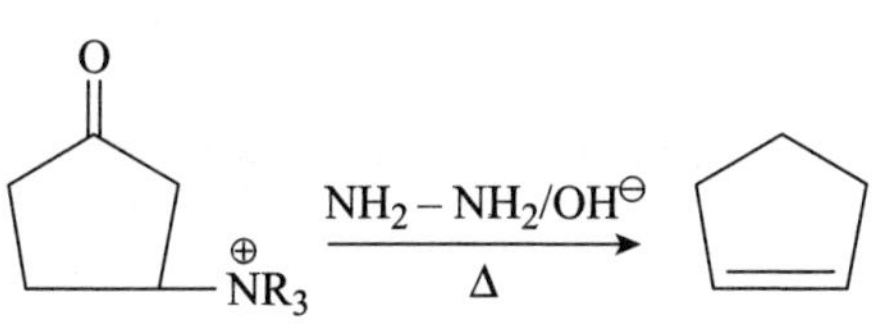

Ans: (a)

44. 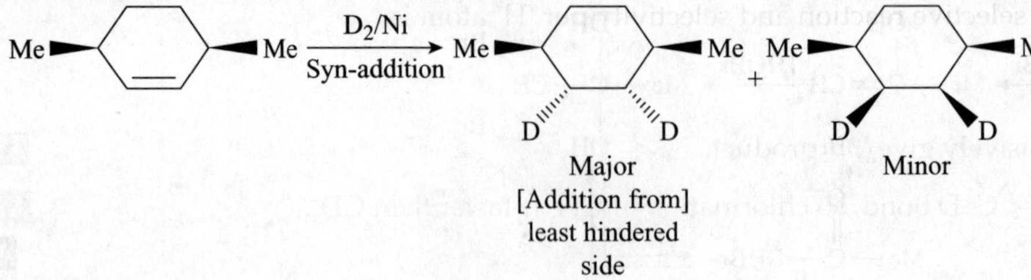

Ans: (d)

45. 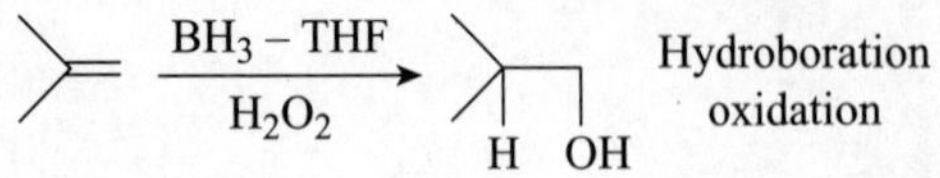

Ans: (b)

46. 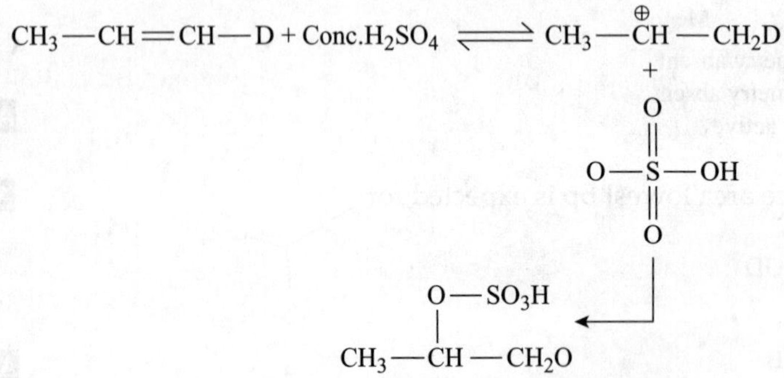

Ans: (b)

47. Heat of hydrogenation α unstability of alkene $\alpha \dfrac{1}{\text{Number of } \alpha \text{ 'H'}}$

Ans: (a)

48. Stability order

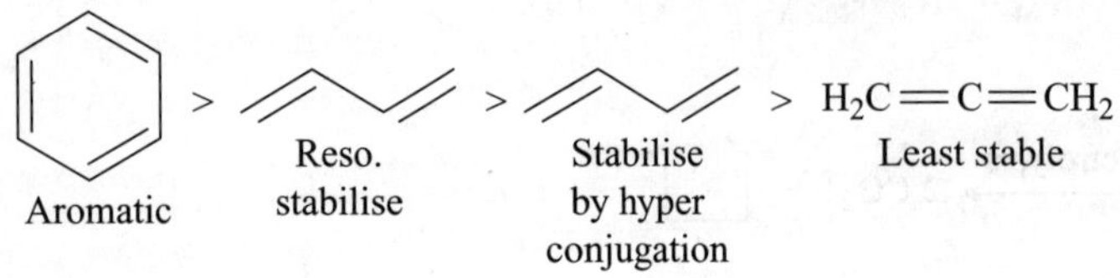

Ans: (b)

49. Frankland reaction.

Ans: (a)

50. 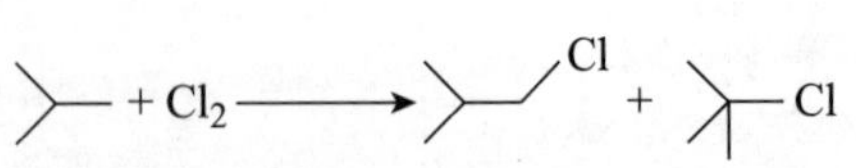

	1°	:	3°
Selectivity per 'H' atom	1	:	5
Number of 'H' atoms	9	:	1
Product ratio	9	:	5

So, major product is Isobutyl Chloride.

Ans: (b)

51. By Frankland reaction.

Ans: (b)

52. If regioselectivity is absent then product formation depends on number of 'H' atoms and 'H' atom ratio in n-pentane is 3:2:1 for 1°, 2° symm. 2°.

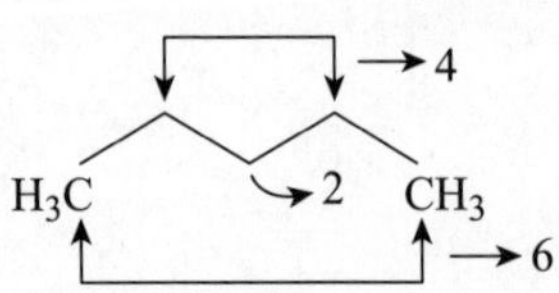

Ans: (b)

53. Bromination is most selective reaction and selectivity per 'H' atom is
3° : 2° : 1°
1600 : 82 : 1
So bromination exclusively give '3°' product. **Ans:** (a)

54. Bond energy of C–H < C–D bond, so chlorination of CH_4 is faster then CD_4. **Ans:** (b)

55.

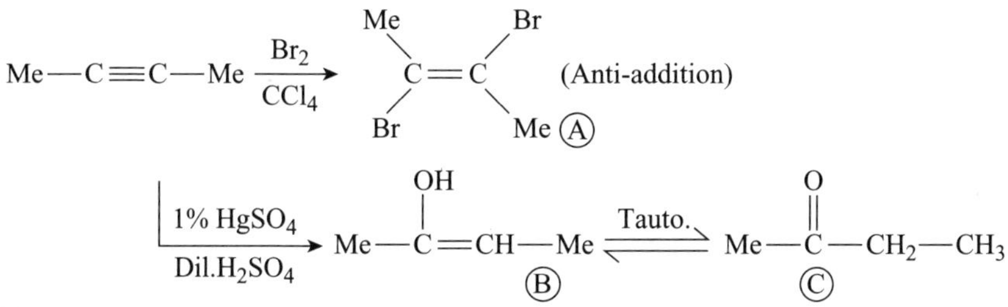

Ans: (b) and (c)

56.

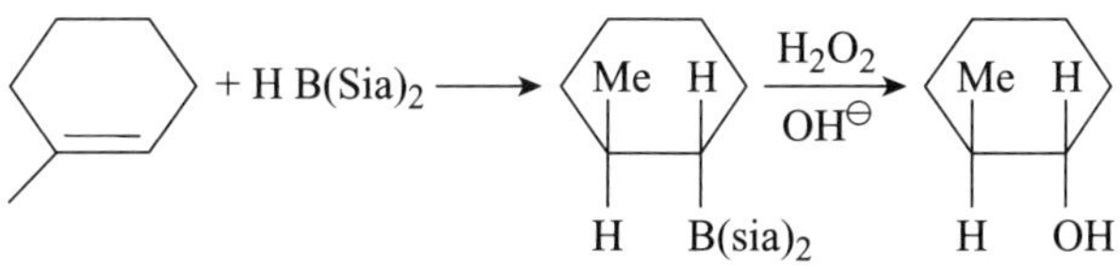

Ans: (c)

57.

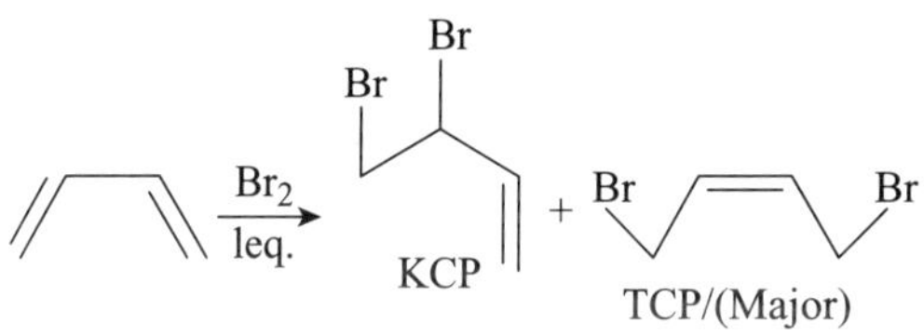

Ans: (a) and (b)

58.

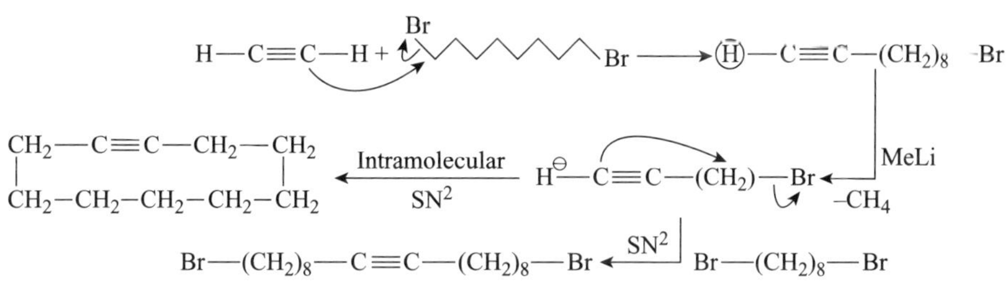

Ans: (a), (c) and (d)

59.

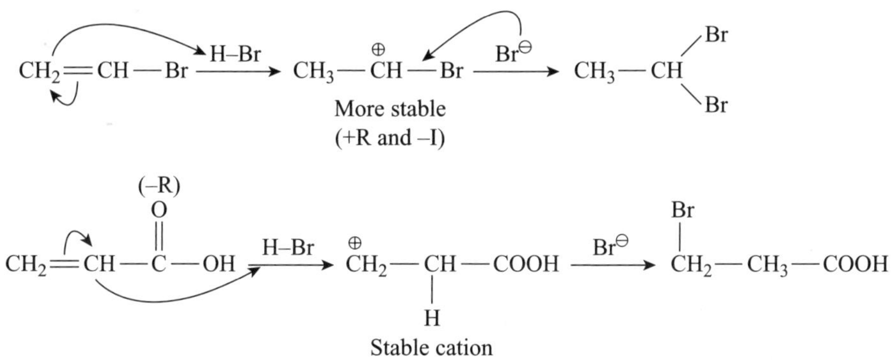

Ans: (b) and (d)

60.

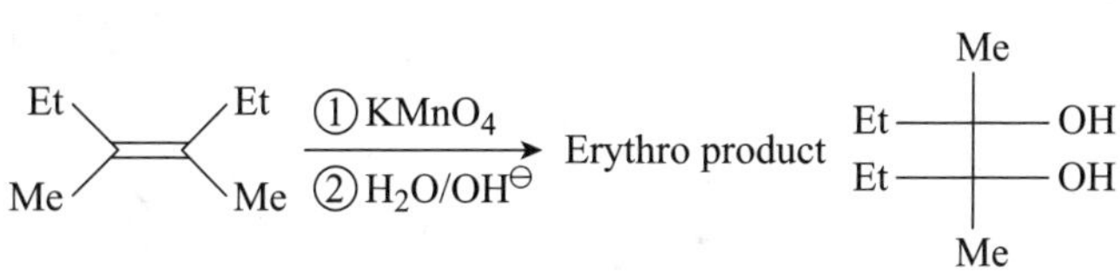

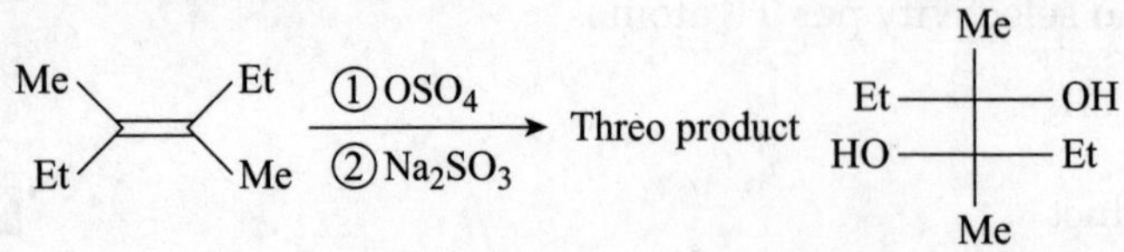

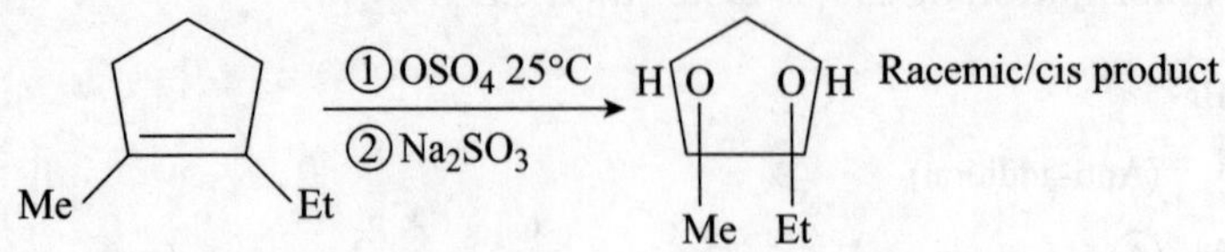

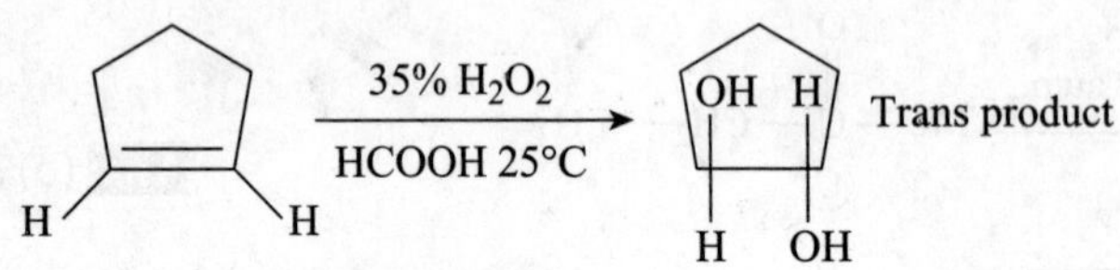

Ans: (c)

61.

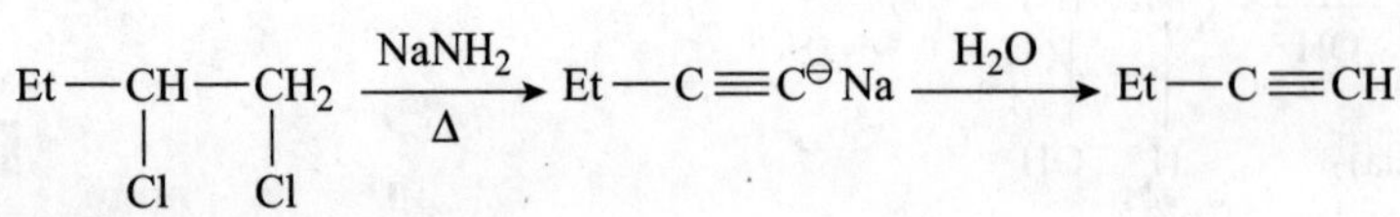

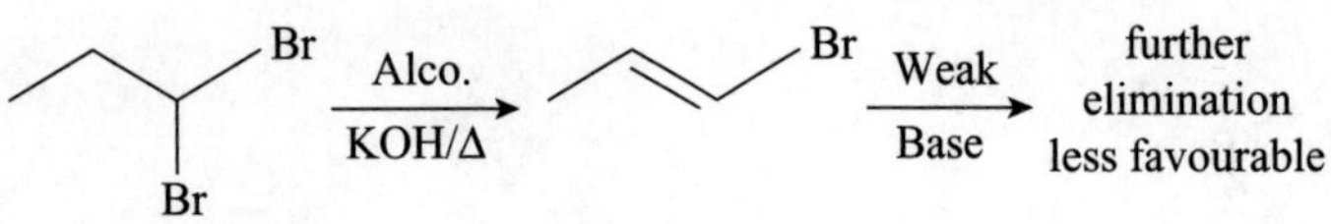

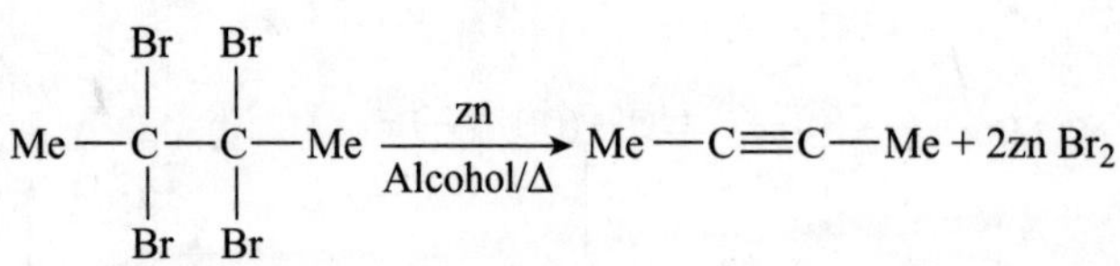

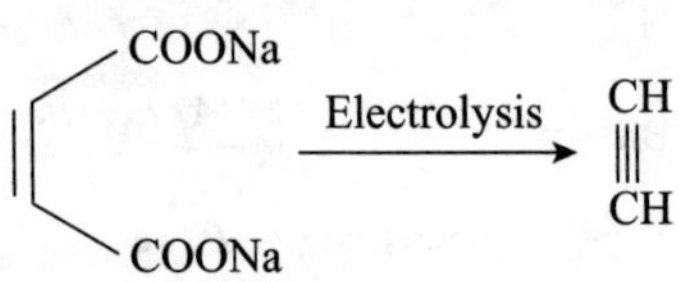

Ans: (a), (c) and (d)

62.

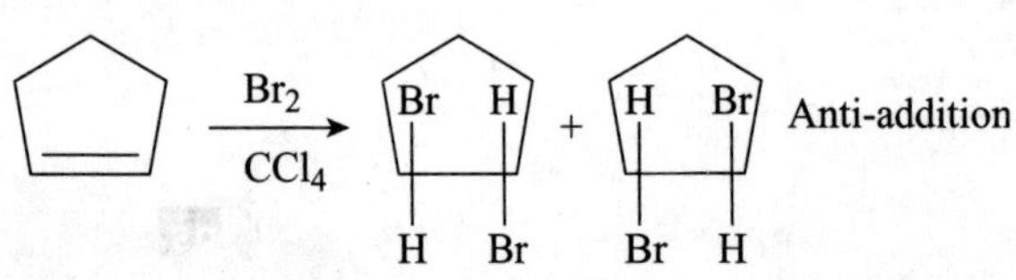

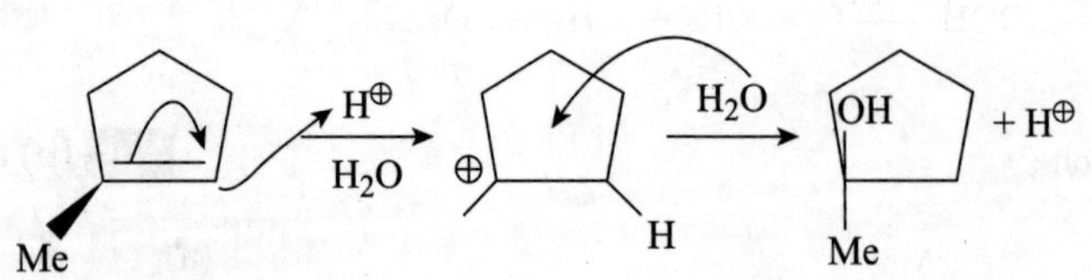

Ans: (a) and (c)

63.

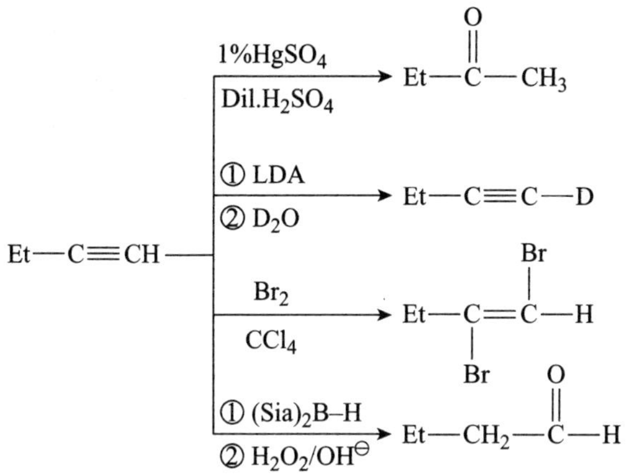

Ans: (a), (b) and (d)

64.

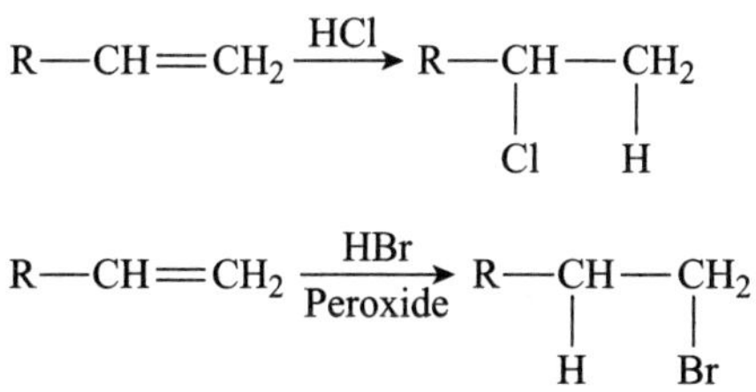

Ans: (a) and (c)

65.

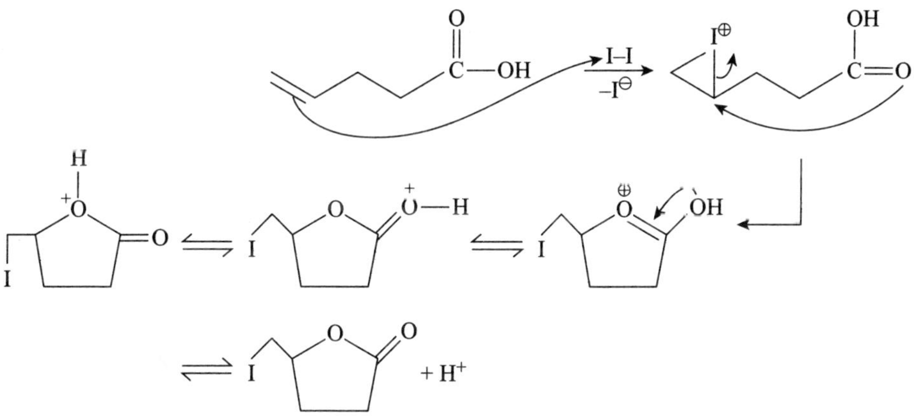

Ans: (a) and (c)

66.

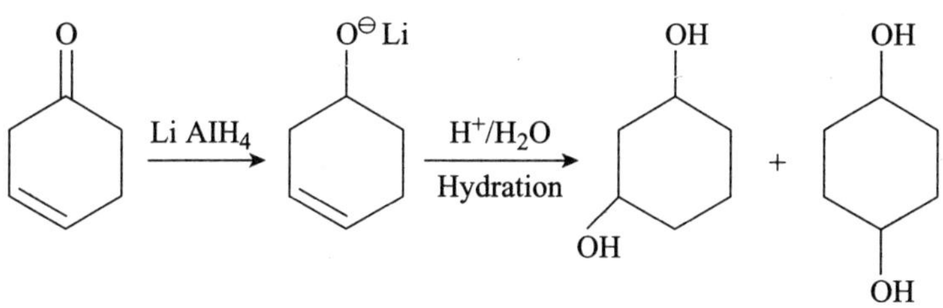

Ans: (b), (c) and (d)

67.

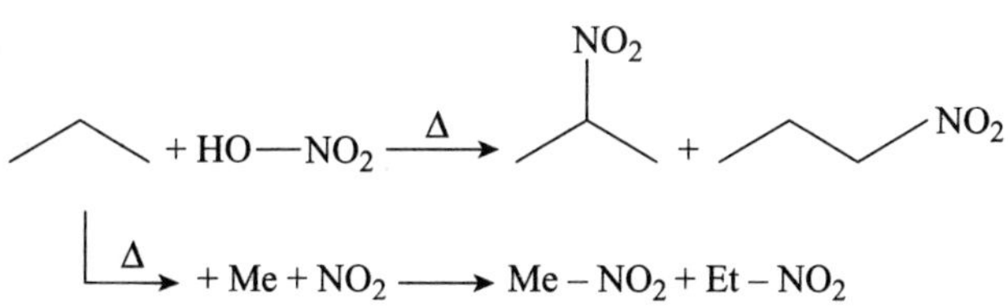

Ans: (a), (b), (c) and (d)

68.

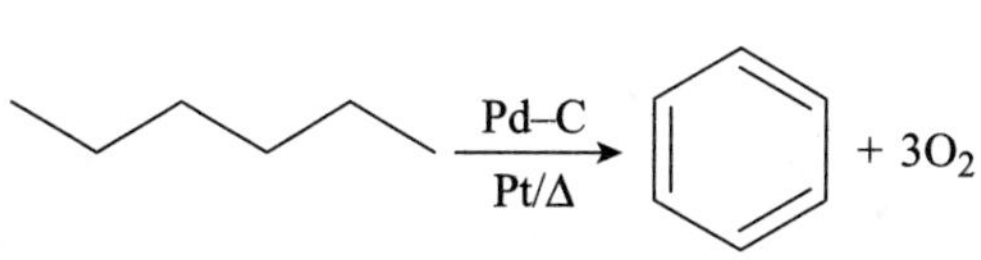

Ans: (b)

69. Iodination of alkane is carried out in presence of oxidising agents HIO_3 and HNO_3. **Ans:** (b) and (c)

70. Cracking involve homolytic fission and free radical intermediates. **Ans:** (a) and (b)

71. 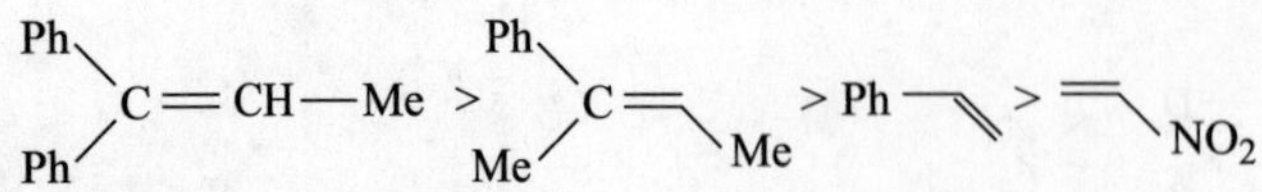

Ans: (b)

72. $E^{\oplus}$ Attack on nucleophillic π-bond. **Ans:** (a)

73. 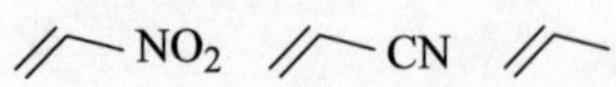

give anti-Markovnikov's addition **Ans:** (a)

74. 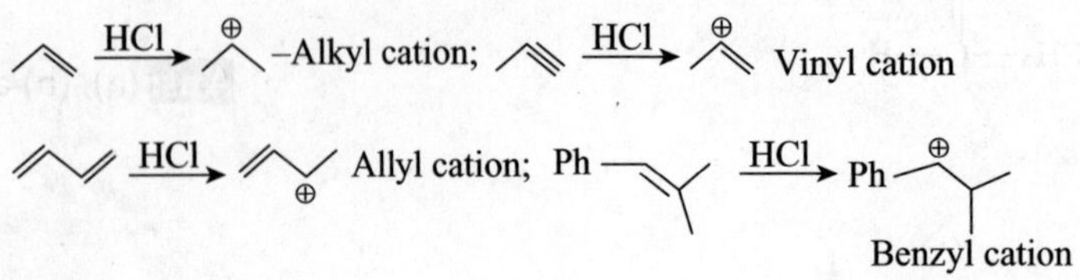

Ans: (c)

75. Statement –1, 2, 3 is correct. **Ans:** (d)

76. Statement –1 is correct. **Ans:** (a)

77. Statement –1, 2, 3 are correct. **Ans:** (a)

78. Dil. $KMnO_4$; $OsO_4/NaHSO_3/OH^{\ominus}$; Ni/H_2 give syn addition. **Ans:** (c)

82. Fehling's solution does not react with $Ph-\overset{\overset{\displaystyle O}{||}}{C}-H$ but easily react with aliphatic aldehyde. **Ans:** (a)

83. 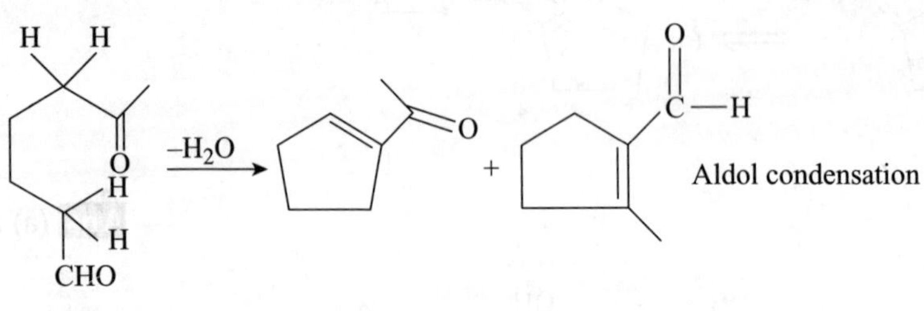

Ans: (d)

85. 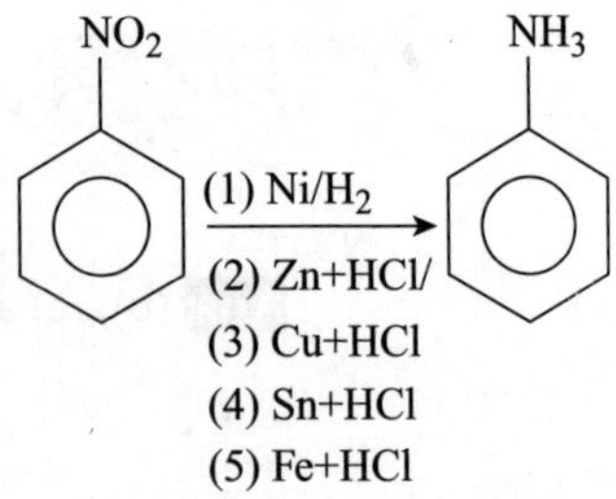

Ans: (b)

87. 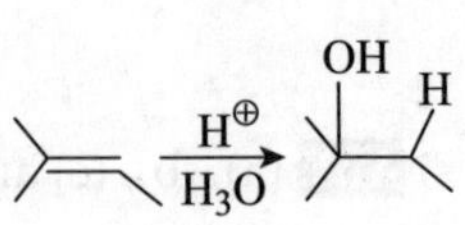

Ans: (c)

88. $$\text{Me}-\text{C}\equiv\text{C}-\text{Me} \underset{\text{Alco. KOH}}{\overset{\text{NaNH}_2\text{ Alkane + Solvent}}{\rightleftarrows}} \text{Me}-\text{CH}_3-\text{C}\equiv\overset{\ominus}{\text{C}}\,\text{Na}^{+}$$

Ans: (c)

89. Rate of dehydration increases with increase in carbocation stability. **Ans:** (c)

91.

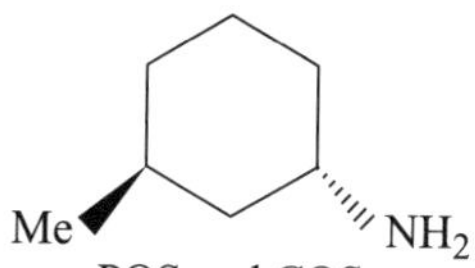

POS and COS absent

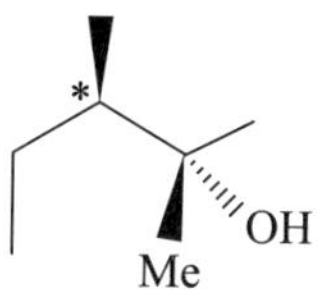

POS and COS absent

Ans: (c)

92. 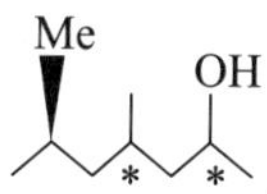 $2^2 = 4$ Isomers **Ans:** (c)

94. $H^{\oplus}/H_3O^+$ in hydration reaction act as catalyst. **Ans:** (c)

Comprehension Type

Passage 1

96.

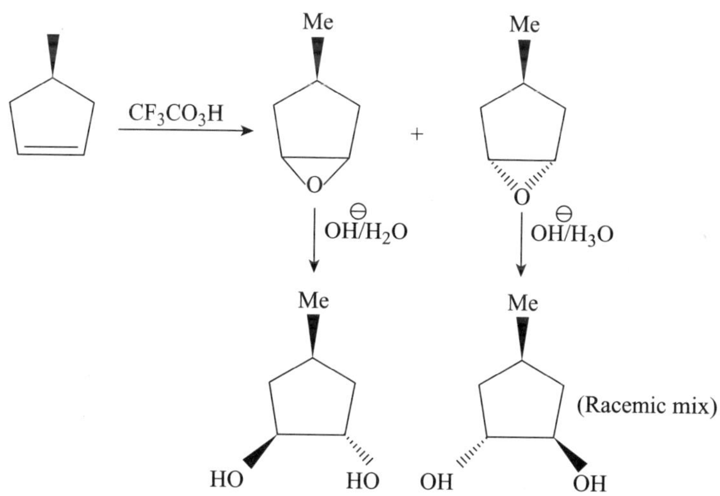

Ans: (d)

97.

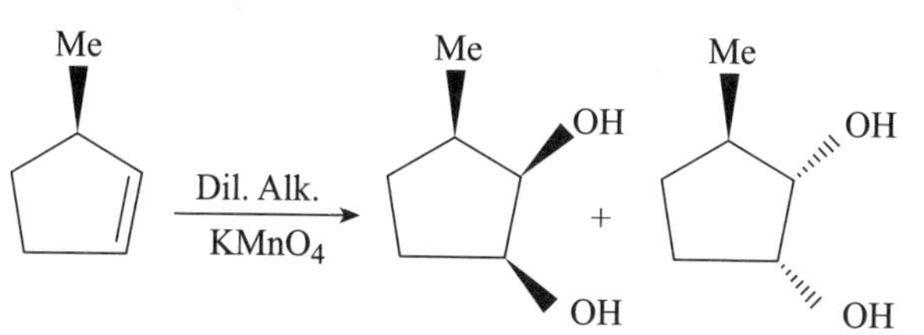

Diastereo isomers

Ans: (b)

Passage 2

99.

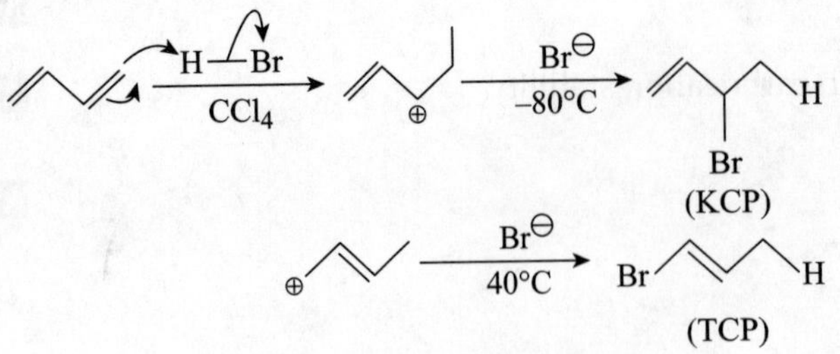

Ans: (b)

Passage 3

102.

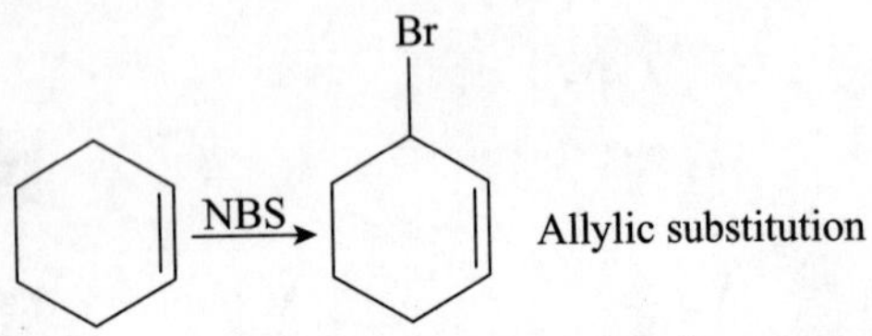

Ans: (a)

103.

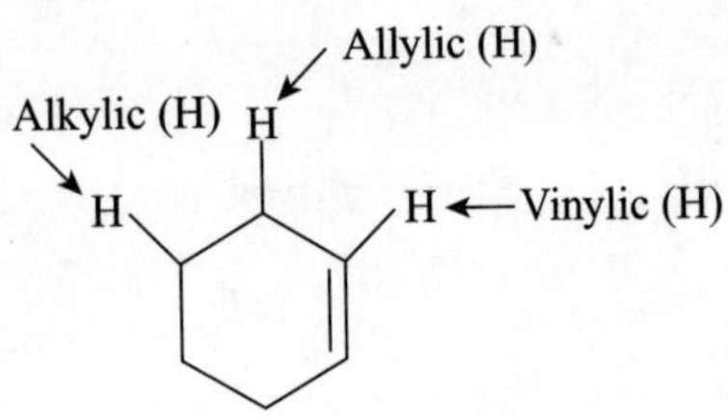

Ans: (d)

104.

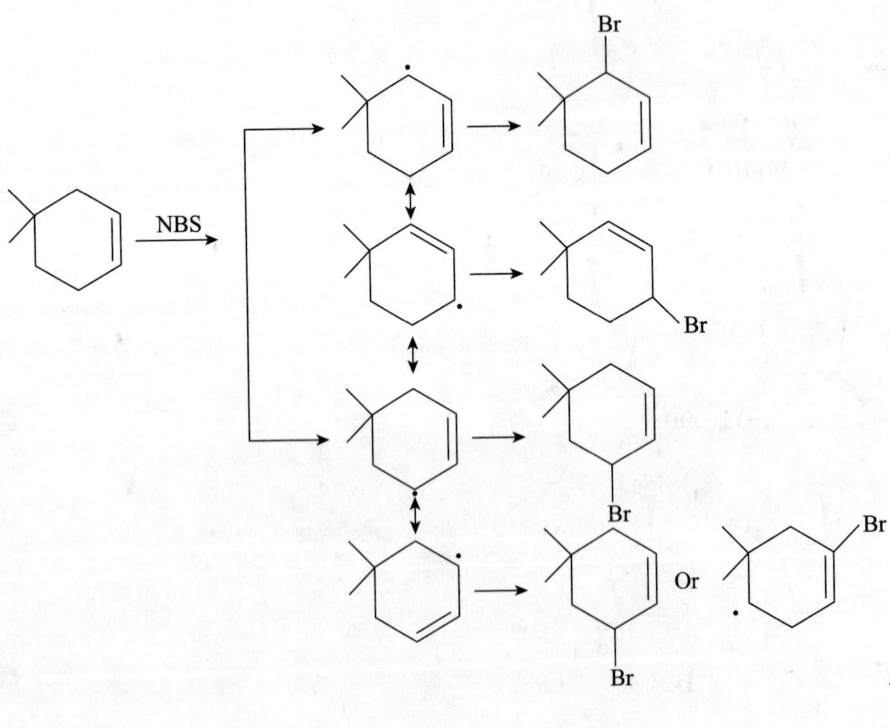

Ans: (c)

Passage 4

105. 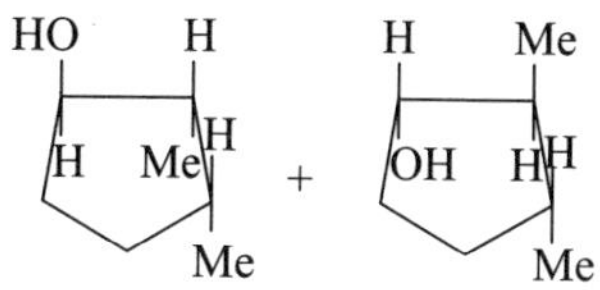 Product are formed

Ans: (c)

106. 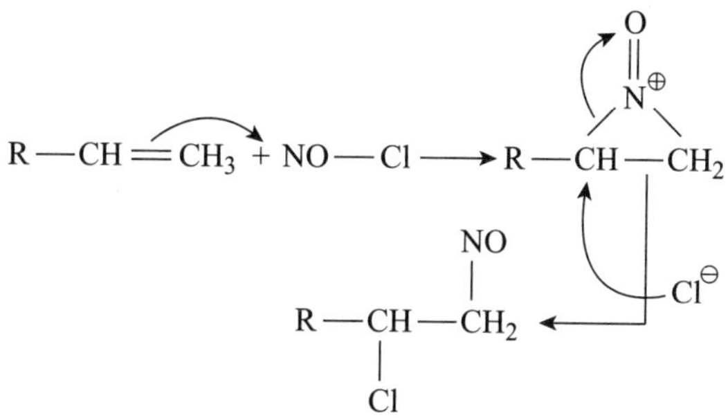

Ans: (d)

Passage 5

110. 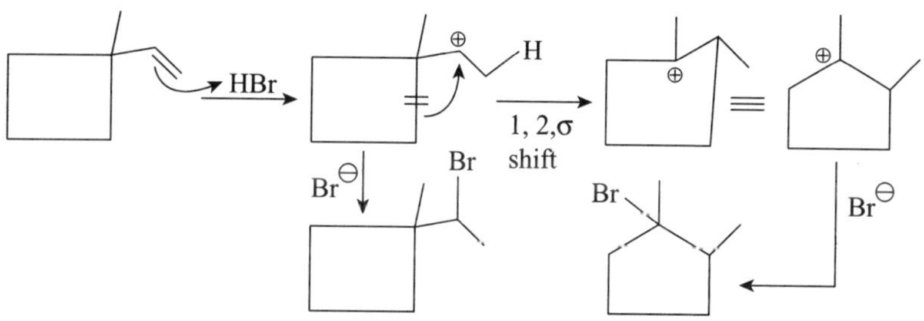

Ans: (d)

Passage 6

111. 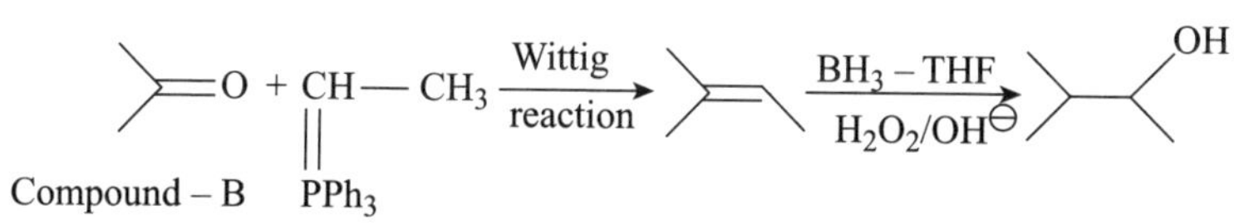

Ans: (d)

112. 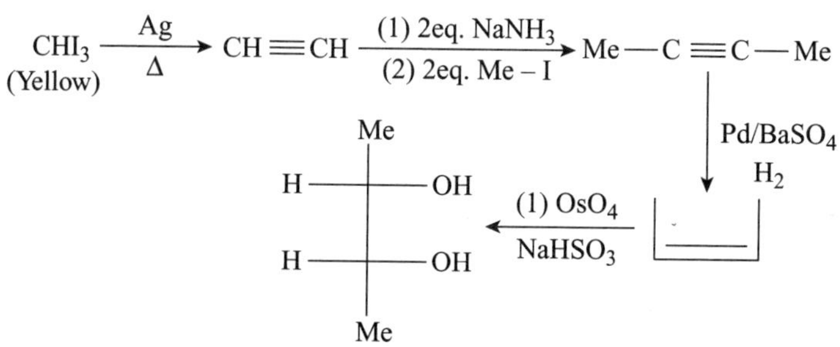

Ans: (a)

Passage 7

113. On ozonolysis α-farnesene give Acetone.

Ans: (c)

114. 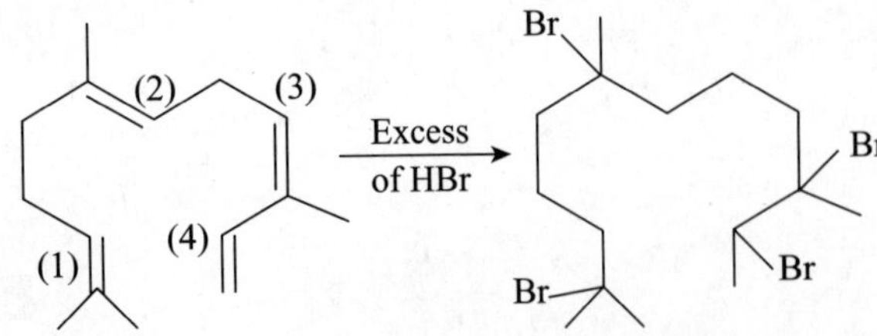

Ans: (b)

115. After first conjugate addition '4'th double bond is least reactive. Ans: (d)

116. 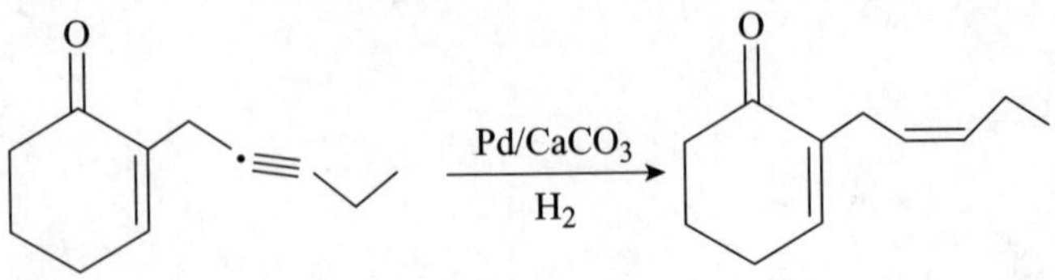

Ans: (a)

117. 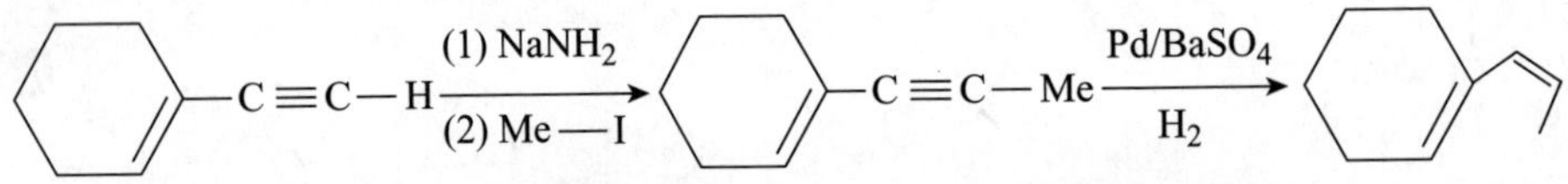

Ans: (c)

119. Alkyne for carbonyl product on reaction with 1% $HgSO_4$ dil. H_2SO_4. Ans: (b) and (c)

121. $Me-C\equiv C-H \xrightarrow{NaNH_2} Me-C\equiv \overset{\ominus}{C}\,\overset{\oplus}{Na} + NH_3$ Ans: (c)

122. 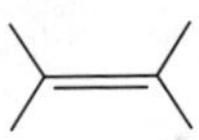 has '12' α-H and is most stable. Ans: (a)

125. 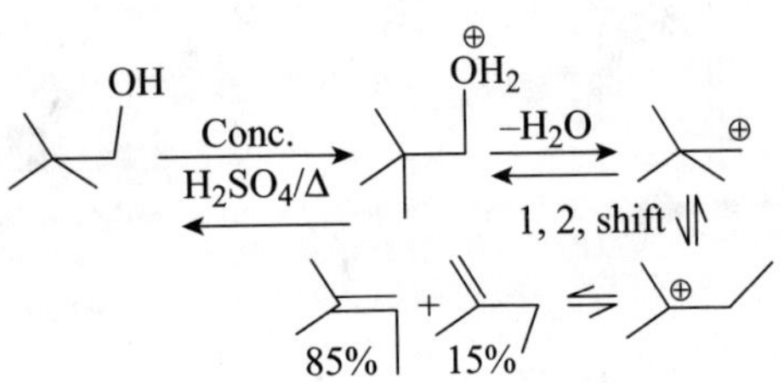

Ans: (c)

126. 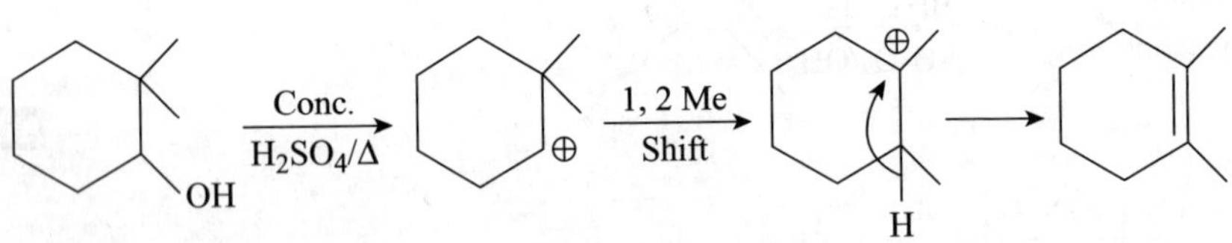

Ans: (a)

127. 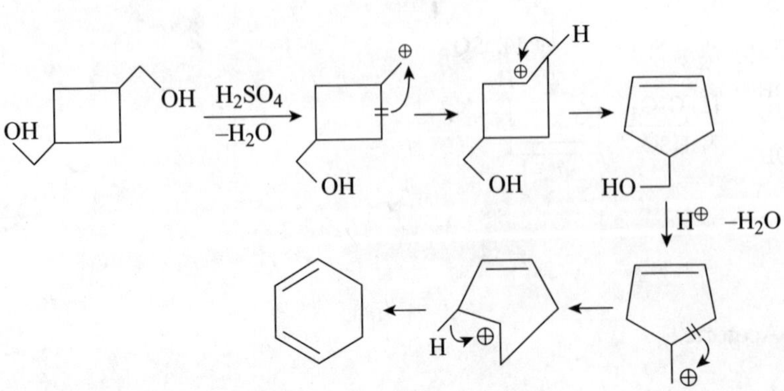

Ans: (a)

128.

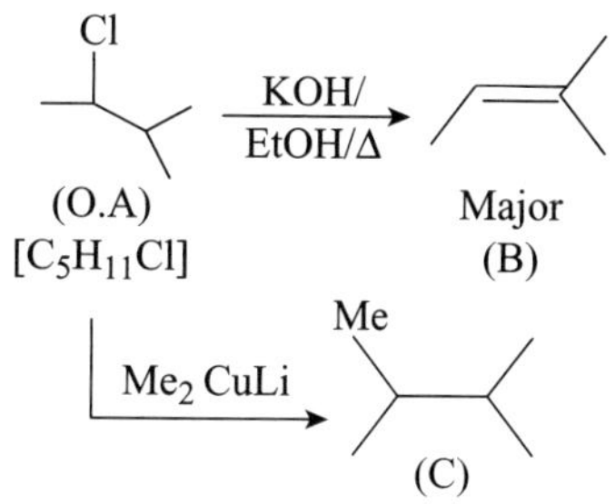

Ans: (c)

Passage 13

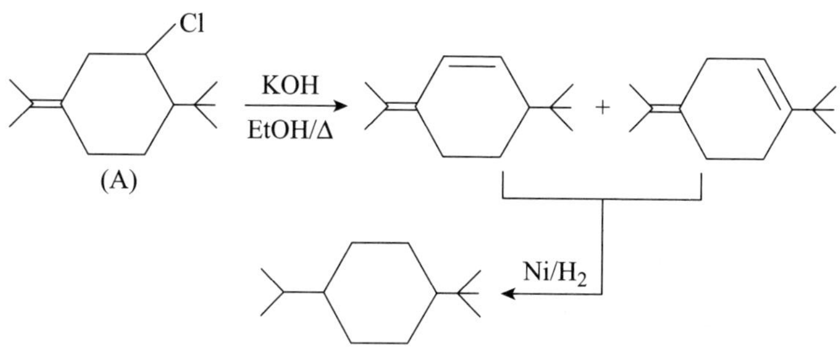

Passage 14

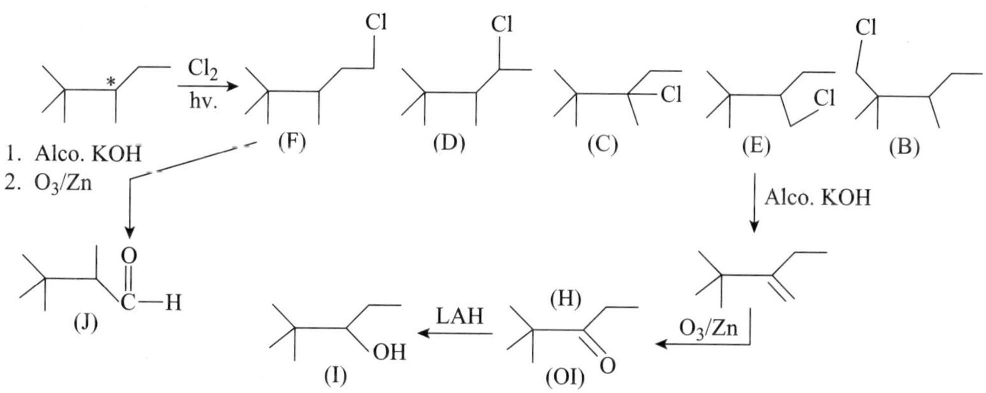

Passage 15

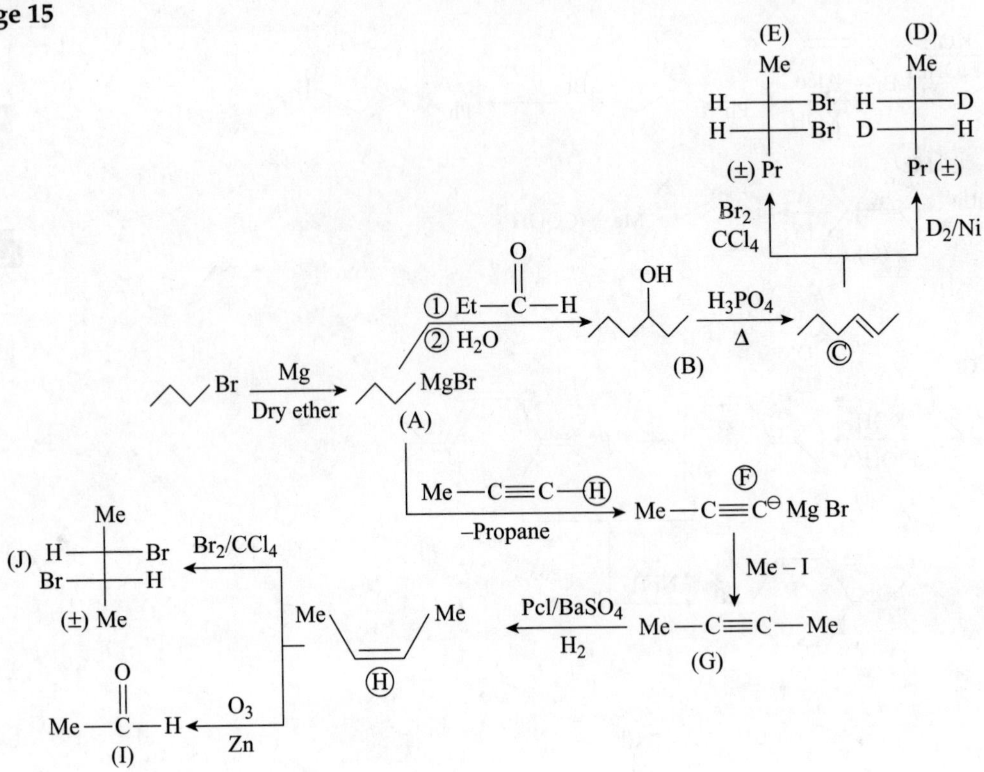

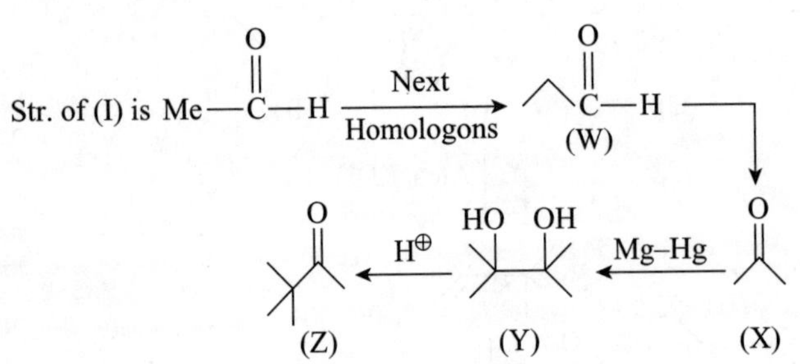

Passage 16

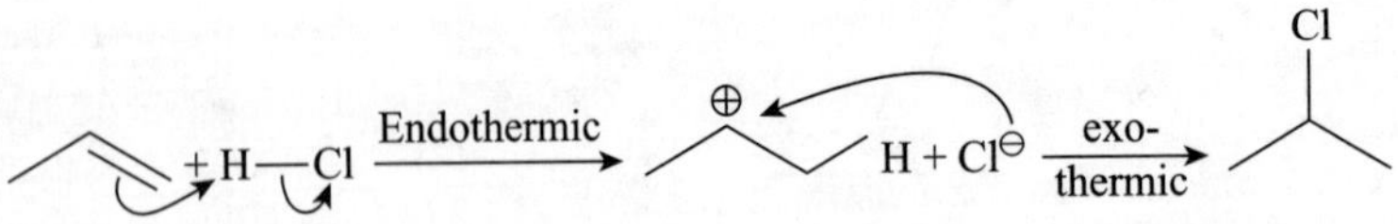

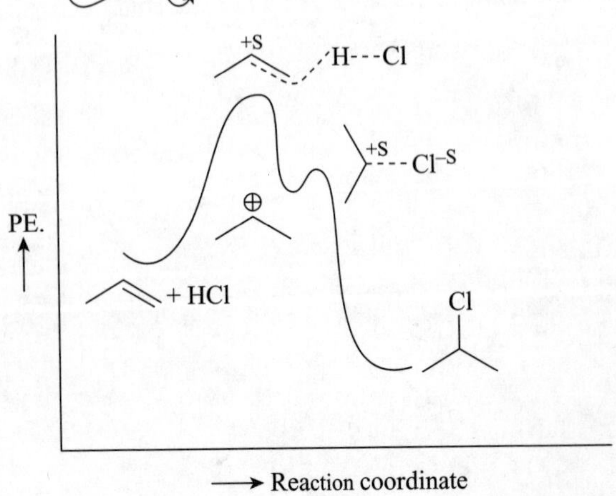

Matrix Type

146. (a)

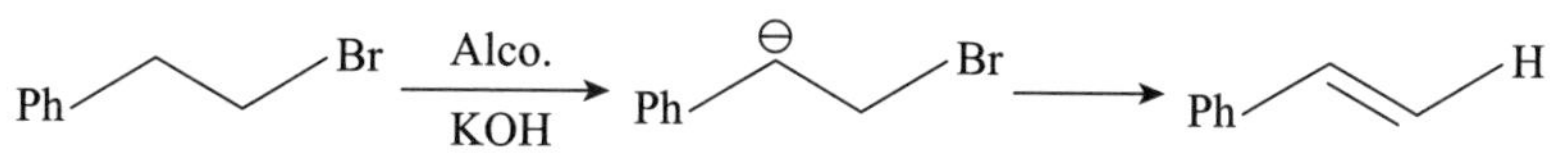

Ans: (r)

(b)

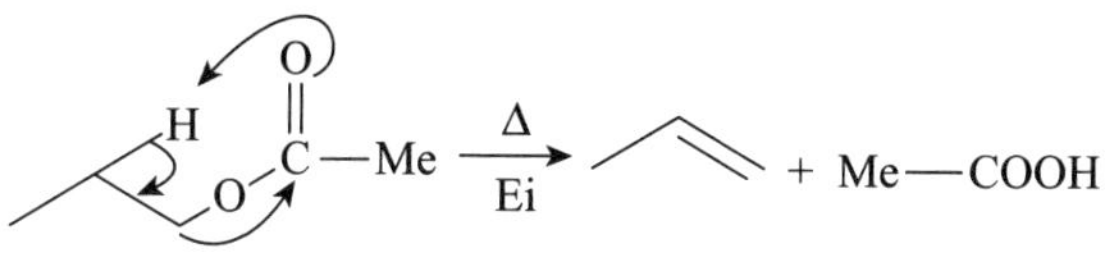

Ans: (s)

(c)

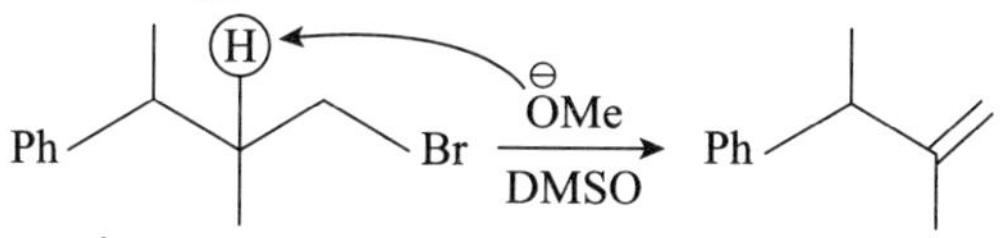

Ans: (r)

(d)

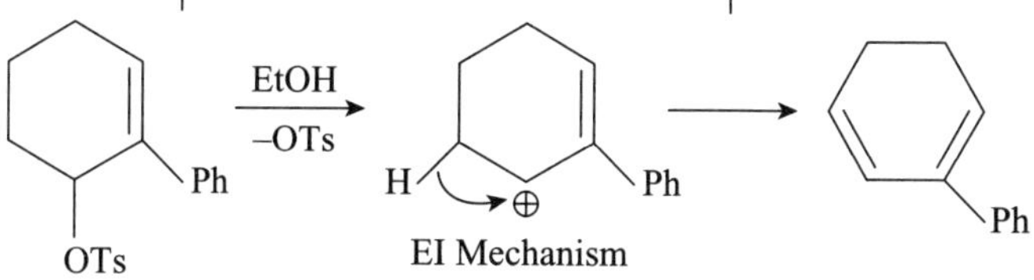

Ans: (p)

147. (a) 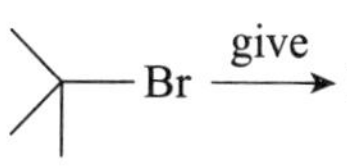 give → E_1, E_2, SN^1 reactions and No^+ give SN^2 EICB

Ans: (q), (r) and (s)

(b) 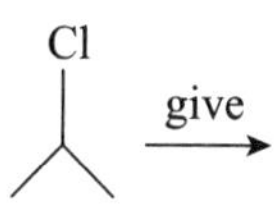 give → E_1, E_2, SN^1, SN^2 reactions and No^+ give EICB

Ans: (q), (r) and (s)

(c) 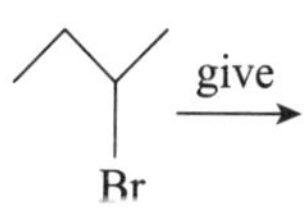 give → E_1, E_2, SN^1, SN^2 reactions and No^+ give EICB

Ans: (q), (r) and (s)

(d) 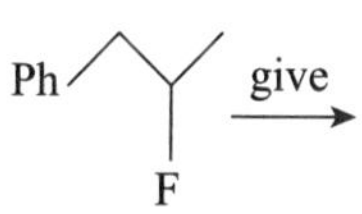 give → E_1, CB reaction. [Unimolecular reaction having 2nd order mechanism]

Ans: (p)

Integer Type

159.

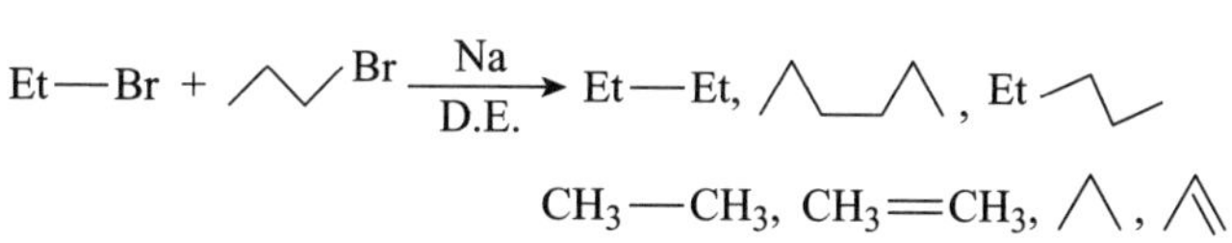

Ans: 7

160.

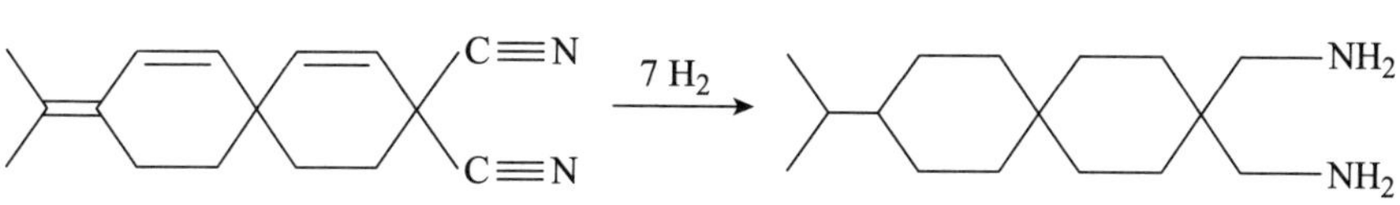

Ans: 7

161.

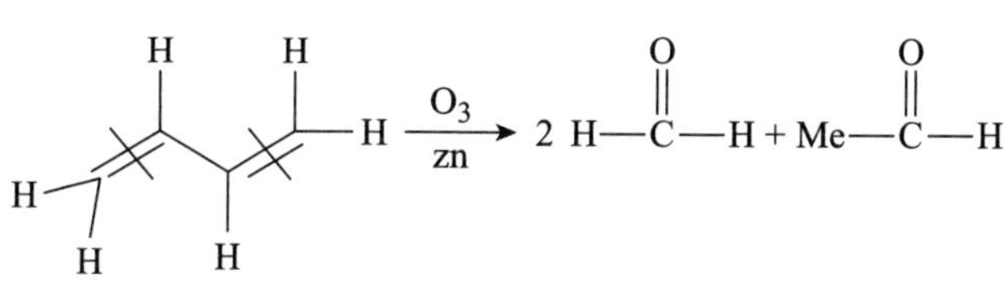

Ans: 2-different products and '3' moles of product.

162.

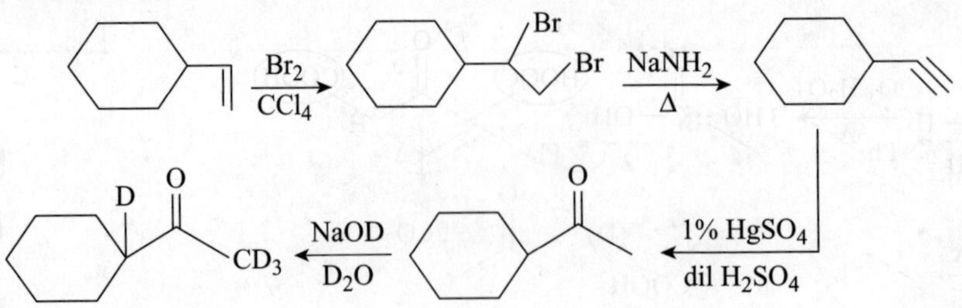

Ans: '4'-Deuterium exchange occur.

163.

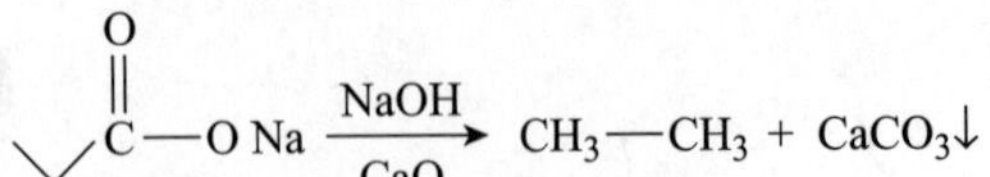

1 mole   1 mole

96 gm   22.4 (Lat NTP)

96 gm of 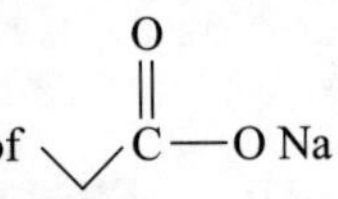 react and give 22.4 L $CH_3—CH_3$

$$\therefore 1\text{gm} = \underline{\qquad\qquad} = \frac{22.4}{96}$$

$$\therefore 38\text{ gm} = \underline{\qquad\qquad} = \frac{22.4}{96} \times 38 = 10\text{ L}$$

Ans: (10)

164. 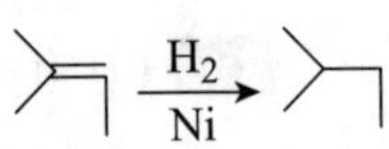 Number optical isomer form.

Ans: (0)

165.

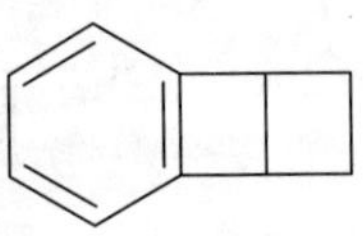 ⟶ 3 ring + 3 π-bond so DU is = 6

Ans: (6)

166.

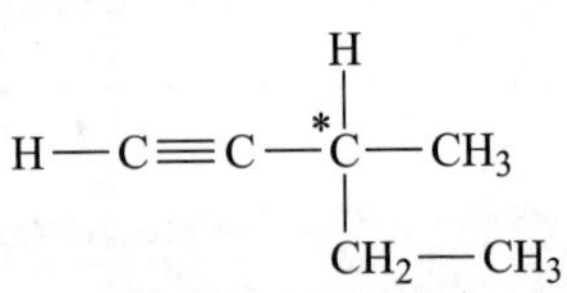

Ans: 6

167.

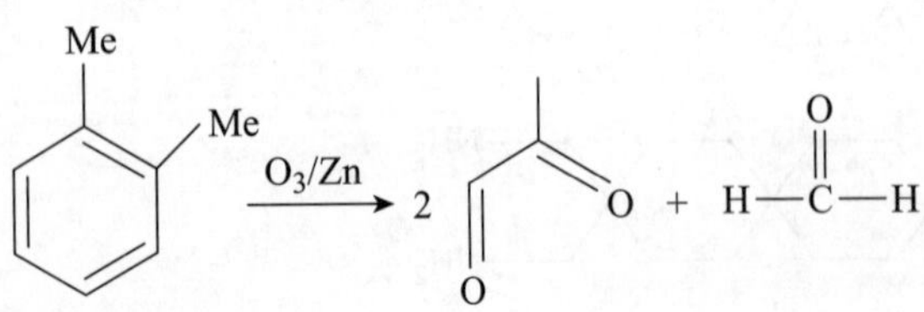

or

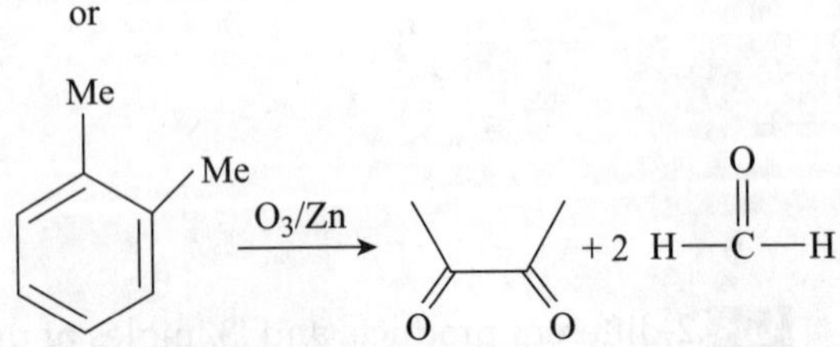

Total '3' different products.

Ans: (3)

168.

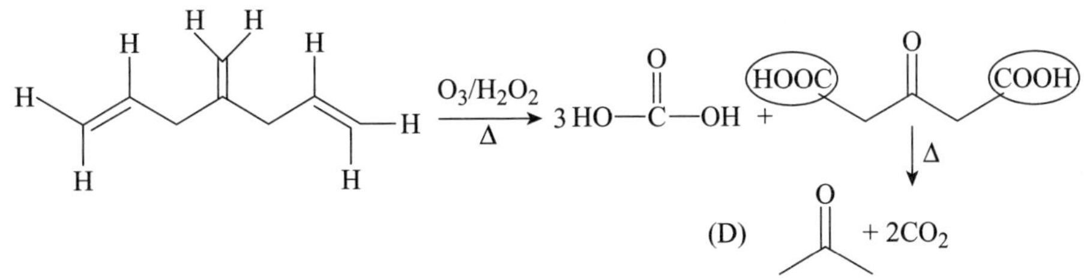

Total 5 mole of CO_2 liberate on heating

x = 5

x = 6

z = 11

Ans: 11

169. (Number of α-H = 9)

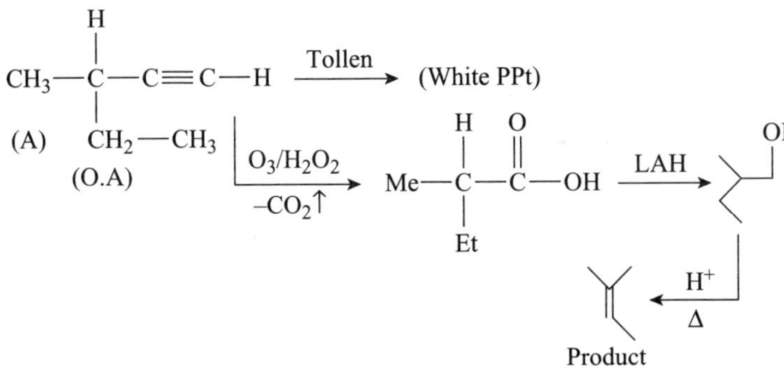

Ans: (9)

170.

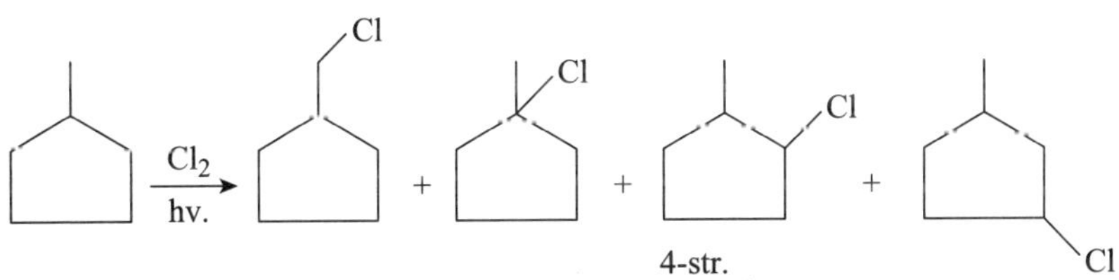

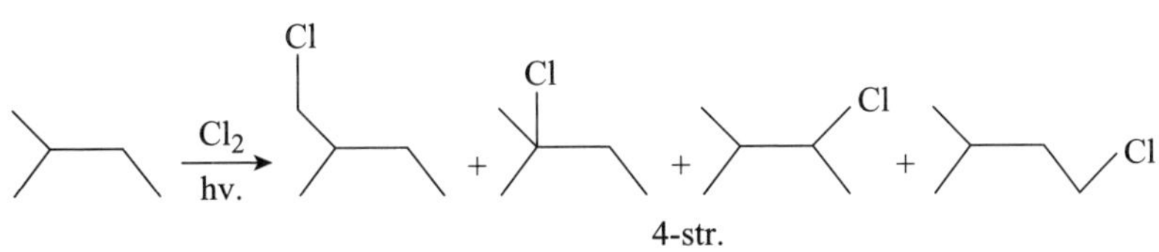

x + y = 8

Ans: (8)

CHAPTER 4

Alkyl Halides, Alcohols and Ethers

LEVEL 1

1. In SN_2, inversion will take place.

Ans: (b)

2. Reactivity ∝ stability of δ-formed on carbon during TS.

Ans: (b)

3.

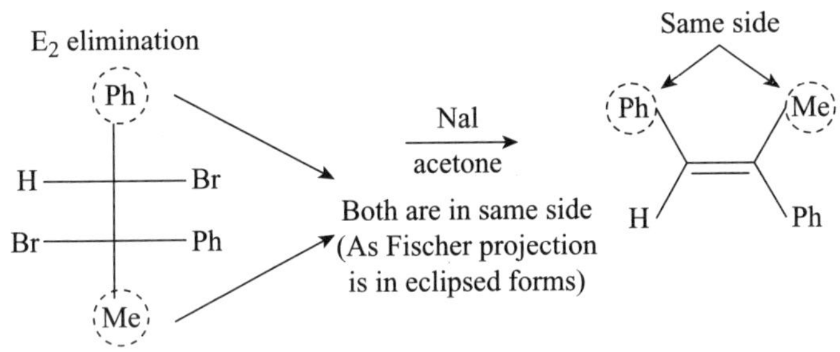

Ans: (c)

4. –OH is bad LG (in absence of $H^{\oplus}$); while –OTs is good LG.

Weak acid HCN is unable to protonate ROH.

Ans: (c)

5.

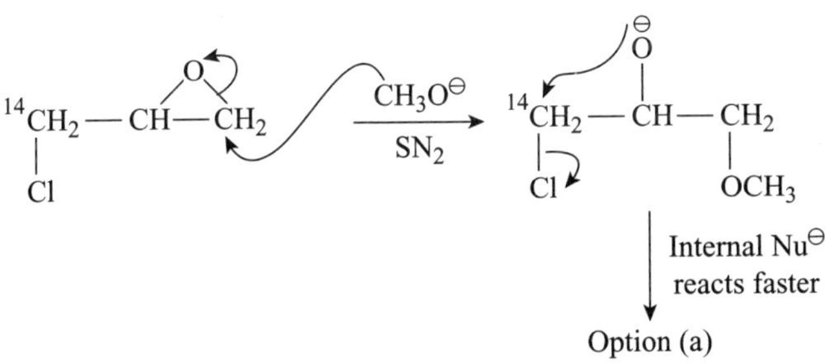

Ans: (a)

6. (a)

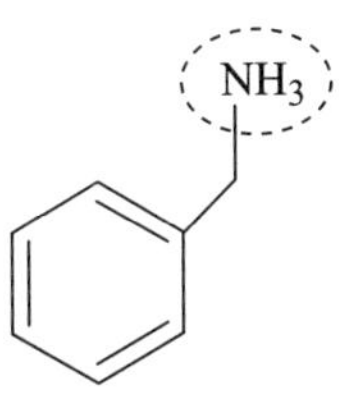

(b)

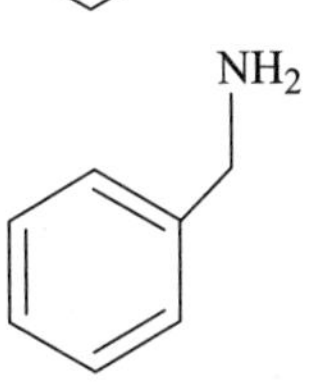

(c) 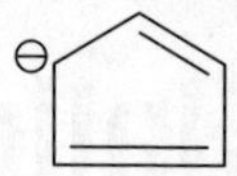

Aromatic

Ans: (d)

7. Rate $\propto$ stability of carbocation. **Ans:** (c)

8. 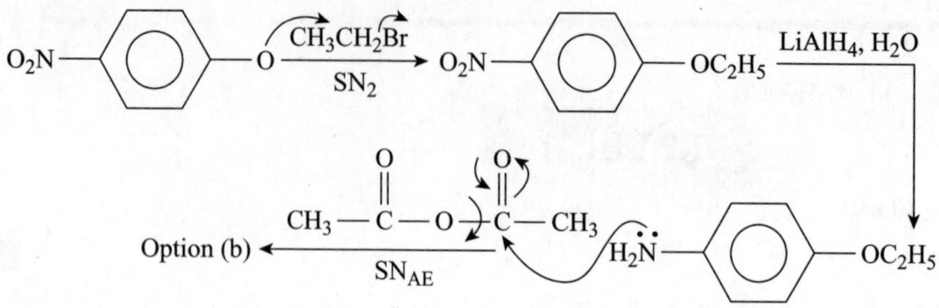

Ans: (b)

9. 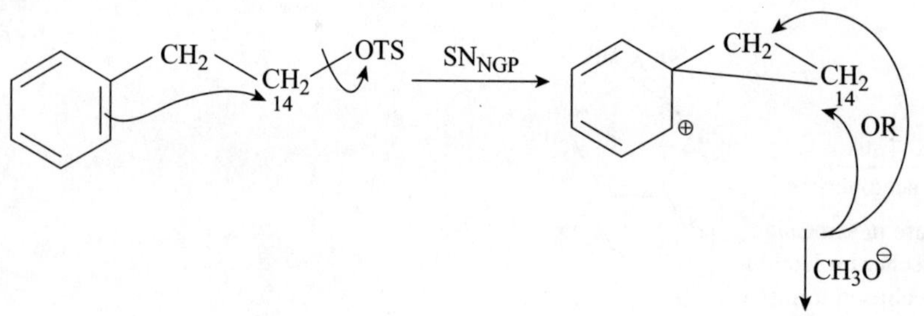

Ans: (d)

10. 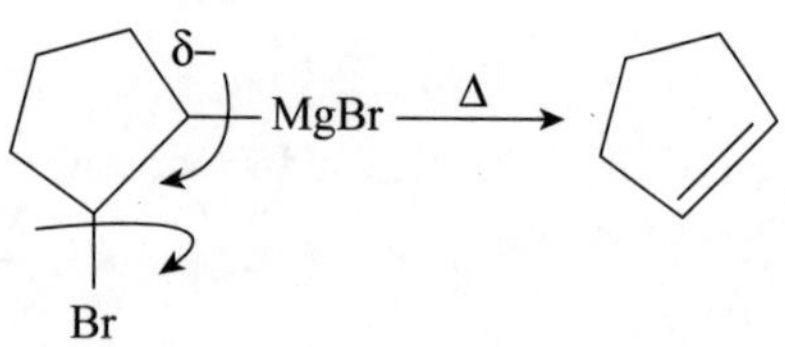

; heat favours elimination.

Ans: (c)

12. Alcoholic AgCN (Polar protic solvent); favours SN_1.

Ans: (b)

13. 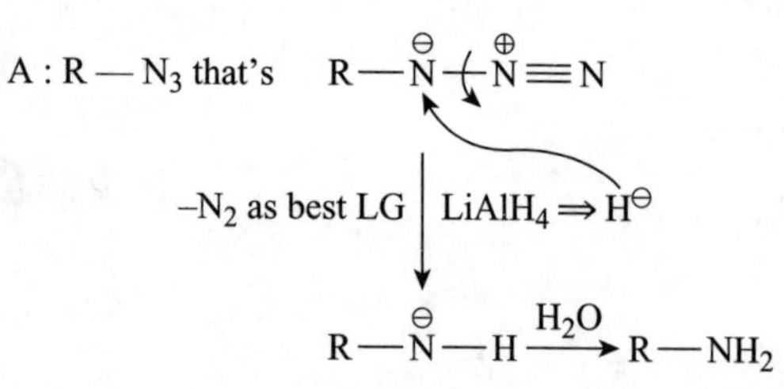

Ans: (c)

14.

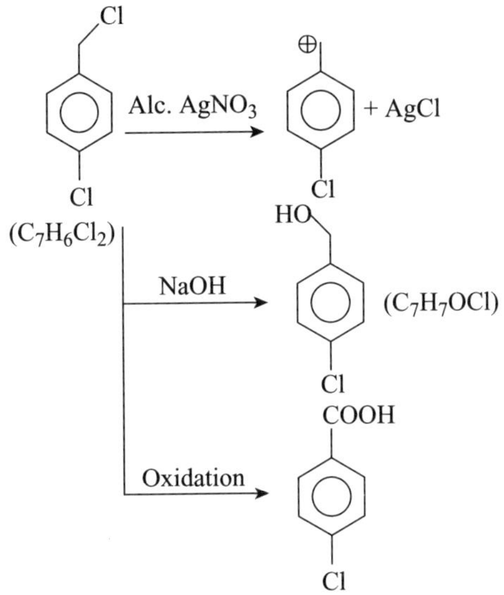

Ans: (a)

19. (i)

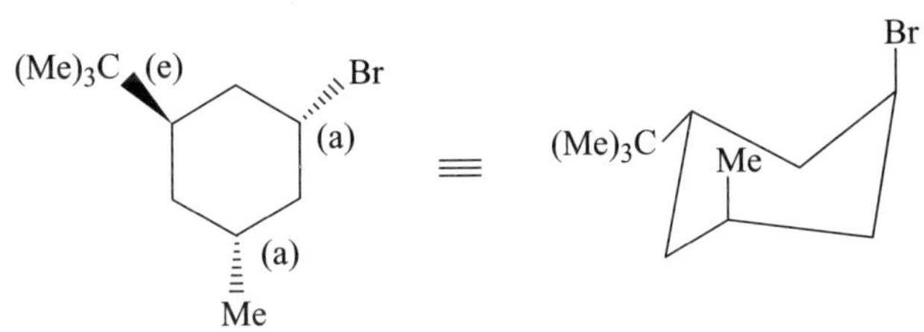

(ii)

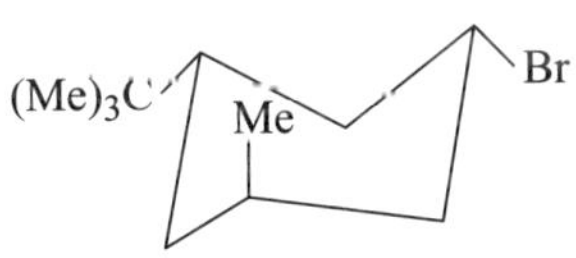

Both (i) and (ii) make cation as intermediate of equal stability. Therefore, unstability of reactant decides rate of reaction (activation energy) unstable: i > ii.

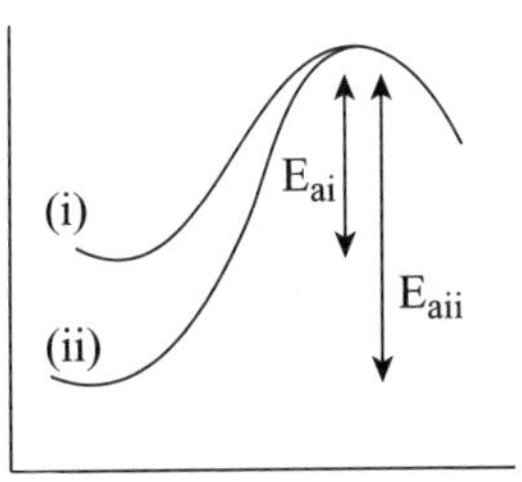

Activation energy: Ea(ii) > Ea(i). So, reactivity: i > ii

Ans: (d)

22. (a)

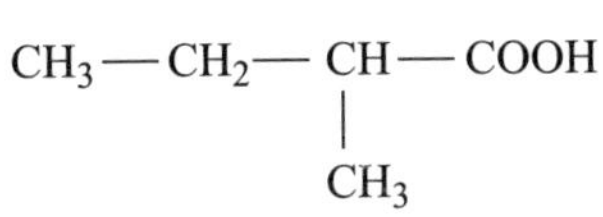

(b)

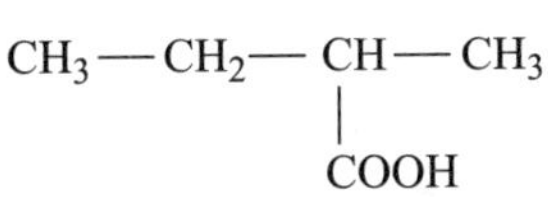

(d)

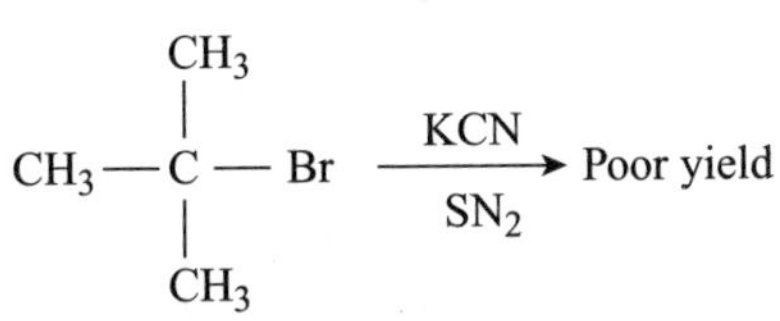

Ans: (c)

23. LG ability $\propto \dfrac{1}{\text{base strength}}$

Ans: (d)

24.

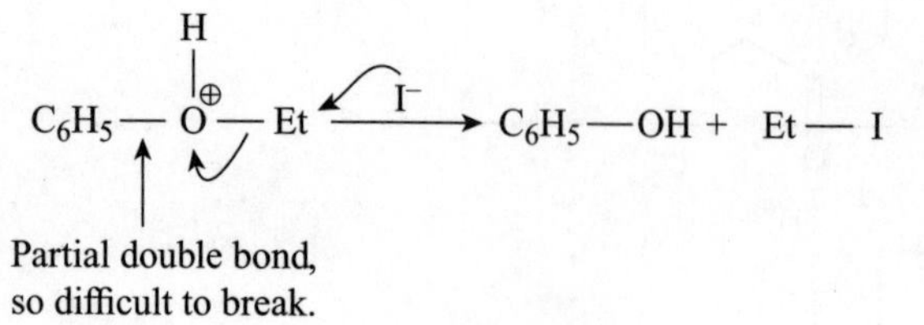

Ans: (b)

26.

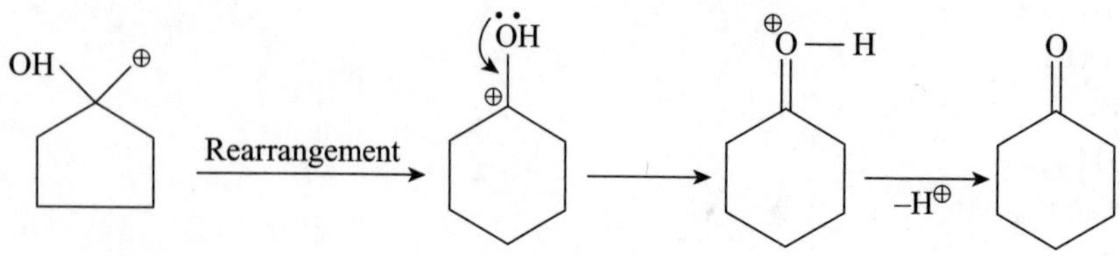

Ans: (b)

27. Rate for esterification $\propto \dfrac{1}{\text{base strength}}$

Ans: (c)

28.

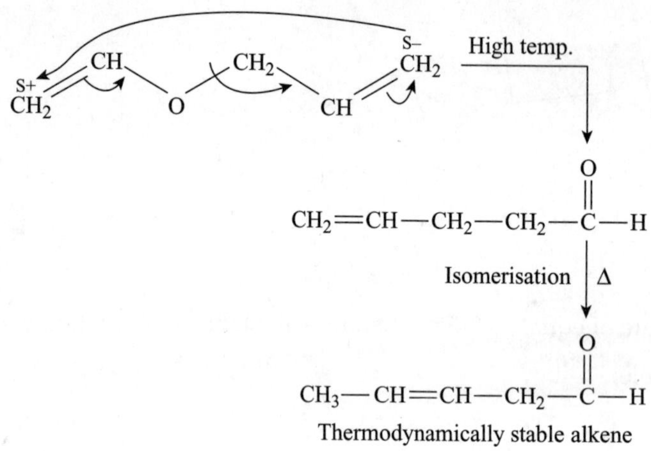

Ans: (d)

30.

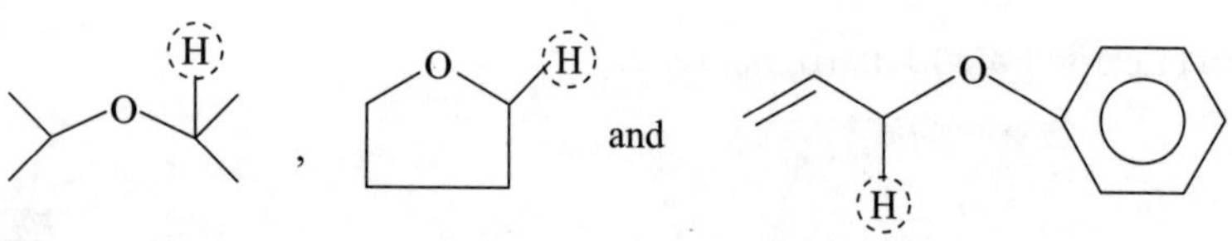

have oxidisable H. The "H" attach to C that's directly linked to O is oxidizable.

Ans: (d)

31. SN_1 mechanism.

Ans: (b)

32.

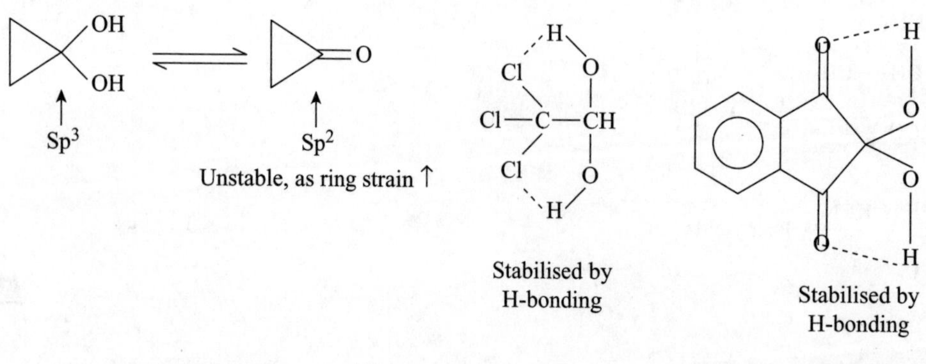

Ans: (d)

33. 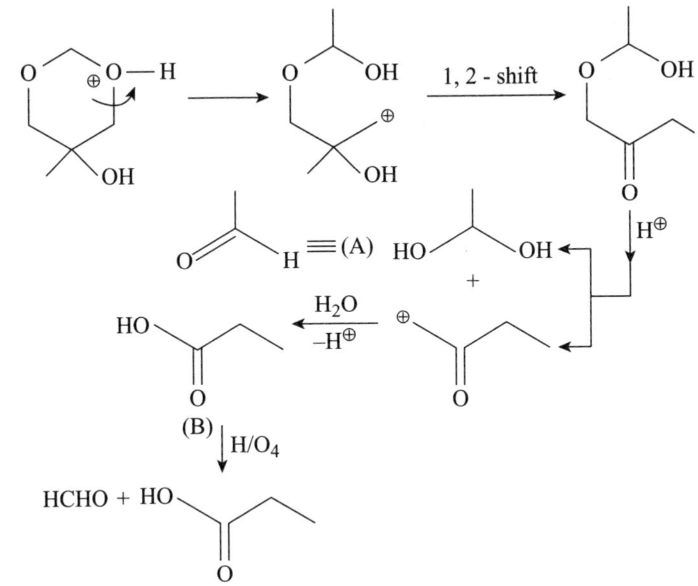

Ans: (d)

34. 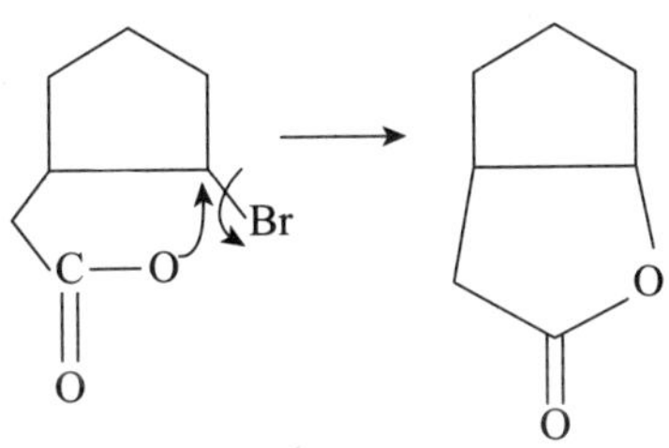

Ans: (b)

35. (a) 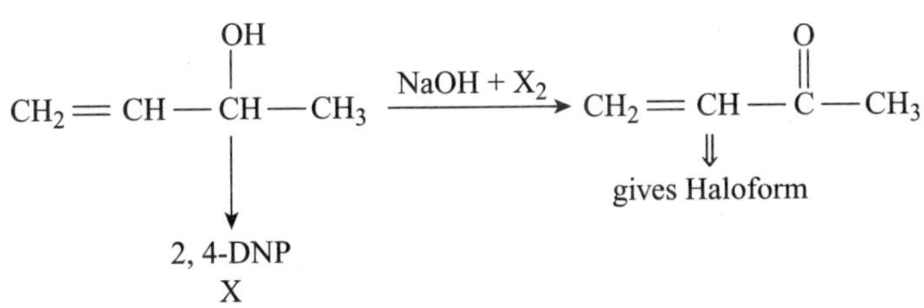

(d) 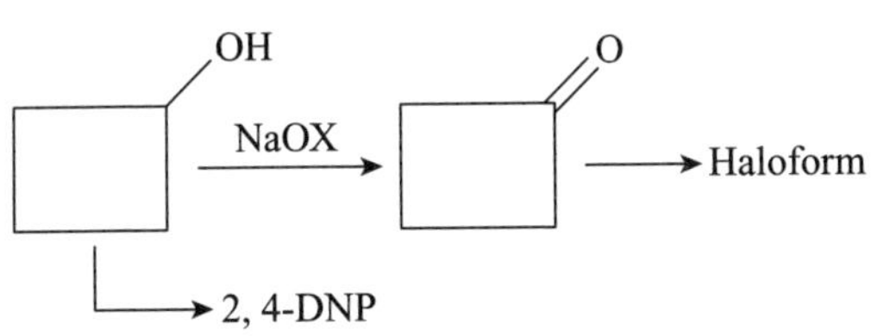

Ans: (a)

36. 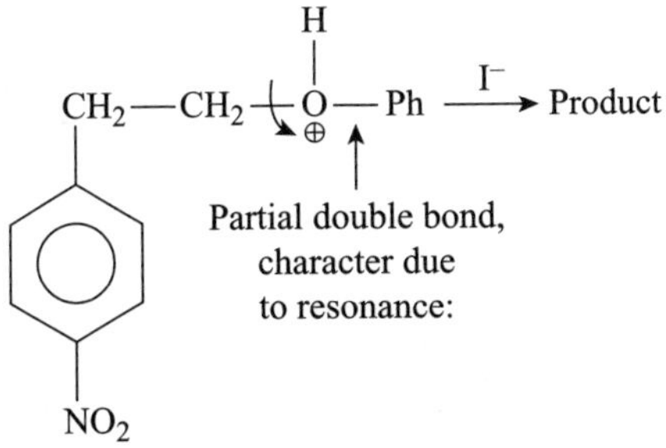

Ans: (d)

37.

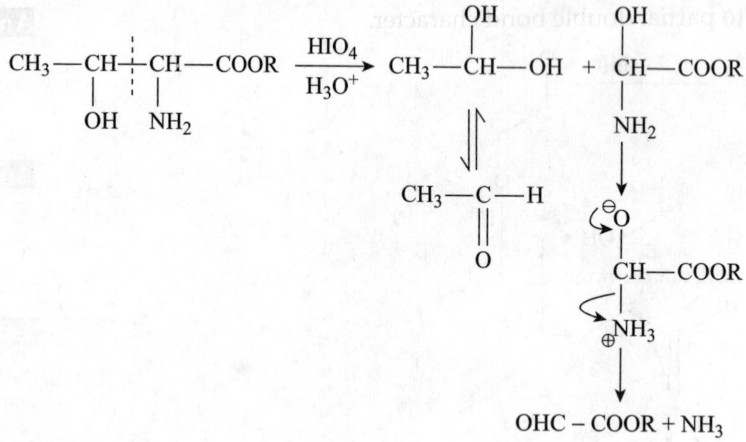

Ans: (c)

40. (a) H_3O^+ used in small amount to increase rate of reaction.

(b) H_3O regenerates at the end of reaction. Ans: (b)

41. E_1 mechanism Ans: (a)

42. H_2O, neutral gives very poor yield, as $H^{\oplus}$ is needed in catalytic amount for better rate of reaction.

Ans: (b)

50. SN_1 mechanism Ans: (a)

51. Rate : Addition of nu$^{\ominus}$ on polar π-bond > Nucleophilic substitution

$S_1+ > S_2+$

∴ NH_3 (nucleophile) attacks 1st at S_1+

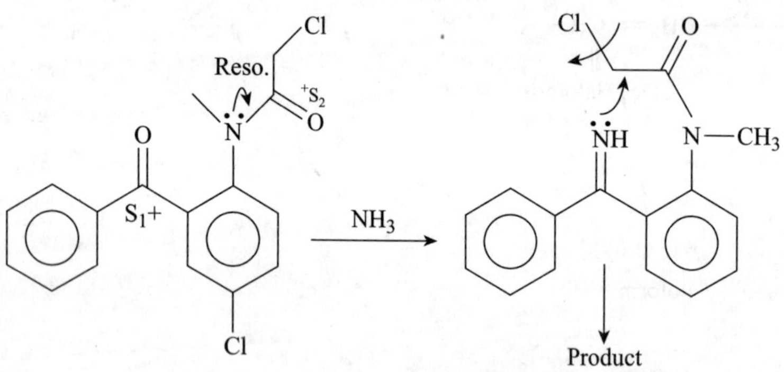

Ans: (d)

53. $Ph - CH_2 - CH_2 - NH_2$ Ans: (b)

54.

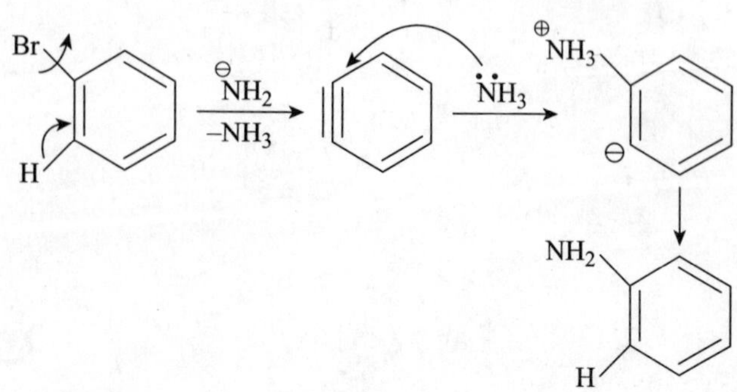

Ans: (d)

55. Aryl halide does not give SN_2 due to partial double bond character. **Ans:** (a)

56.

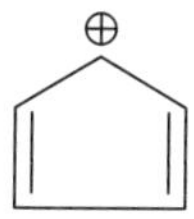

: Anti-aromatic

Ans: (c)

57.

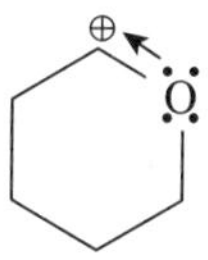

Stable as octet completion via resonance

Ans: (a)

60. Stable

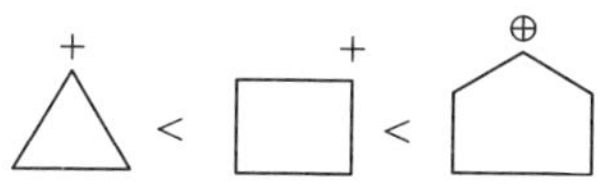

Ans: (d)

61.

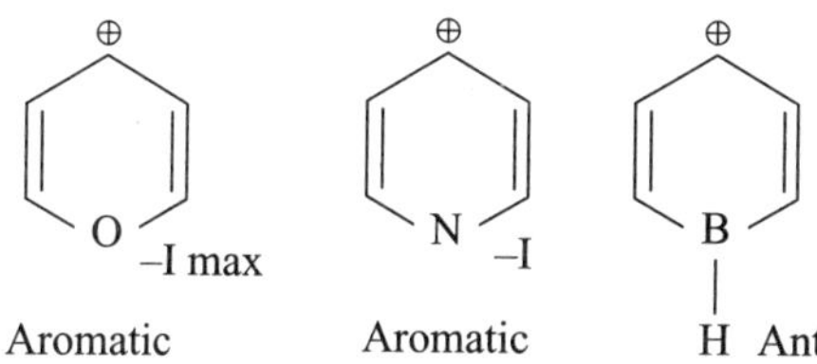

Aromatic   Aromatic   Anti-aromatic

Ans: (d)

62. (i)

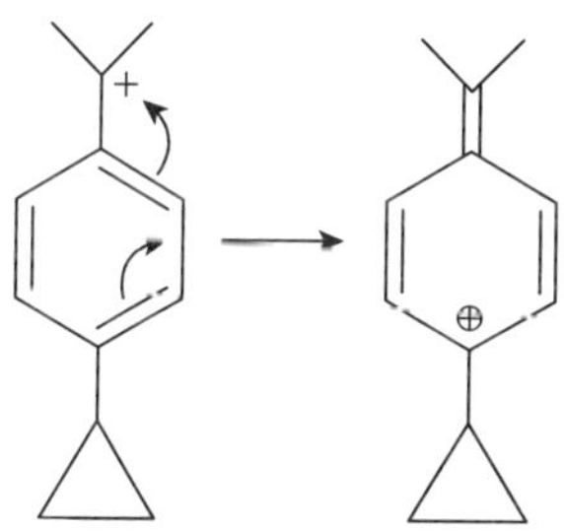

dancing resonance

(ii) Restricted dancing resonance due to SIR. **Ans:** (a)

64. Positive on bridgehead carbon in a compound having less number of carbon in ring is more unstable.

Ans: (b)

67. (iii)

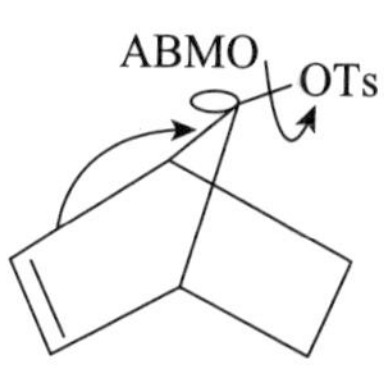

π-bond act as internal nucleophile. So LG leaves faster.

(ii)

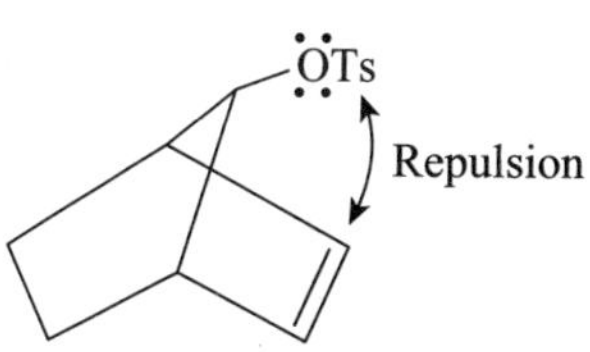

Unstable: ii > i more unstable, lesser activation energy needed
(For details, refer Question Number 19)

Ans: (b)

70. Reactivity $\propto$ Leaving group ability. **Ans:** (c)

71. to 74. Reactivity $\propto$ Stability of carbocation.

78. to 80. Reactivity $\propto$ Stability of S-formed on carbon during TS.

81. Reactivity ∝ unstability of reactant (for details : refer Question Number 19). **Ans:** (d)

82. Stable:

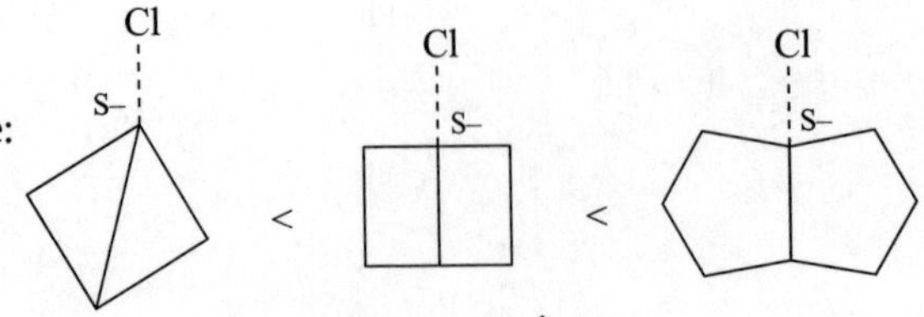

(TS of SN^2 has Sp^2 hybridisation).

Stability of TS ↑ SC

Ans: (d)

84. to 86. Rate of reaction ∝ Stability of product.

87. and 90. ROR : C–H > C–D > C–T for E_2 (ROR—rate of reaction).

89. E_2: Anti-elimination

(i) 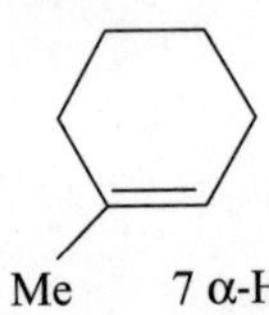 7 α-Hydrogen

(ii) 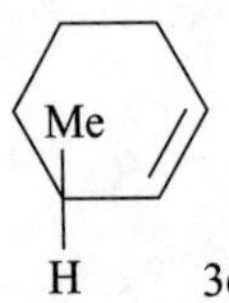 3α H

(iii) Very poor yield

Ans: (a)

91. to 93. Bredt's rule: double bond or cation at bridgehead will start to become stable if number of carbon in ring > 8.

92.

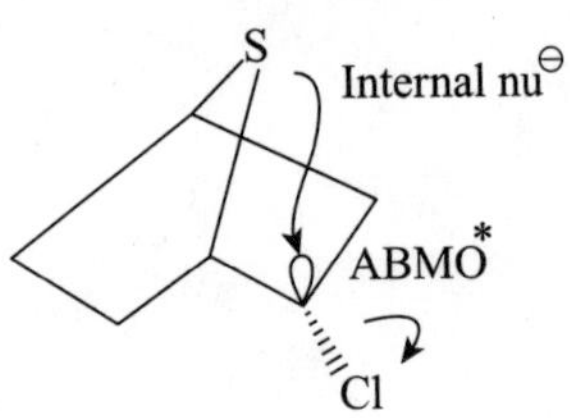

Ans: (a)

94.

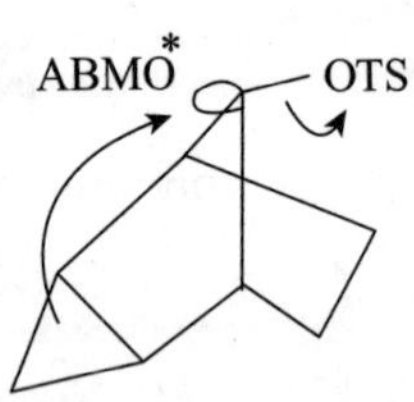

3-Mem. ring (equivalent to π-bond) act as internal Nu⊖

Ans: (b)

95.

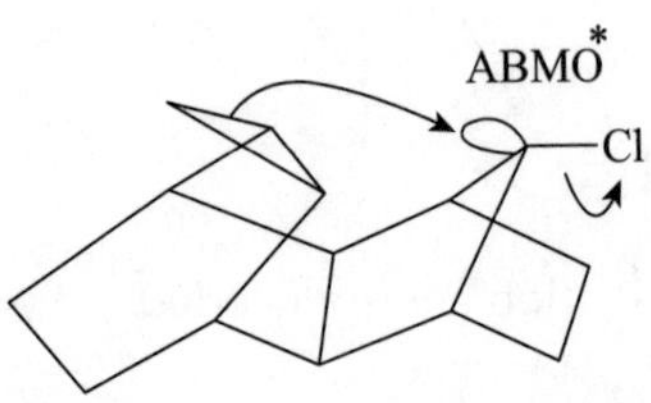

Ans: (c)

96. π-bond acts as internal nucleophile. Nucleophilicity of π-bond: i < ii < iii. **Ans:** (a)

98.

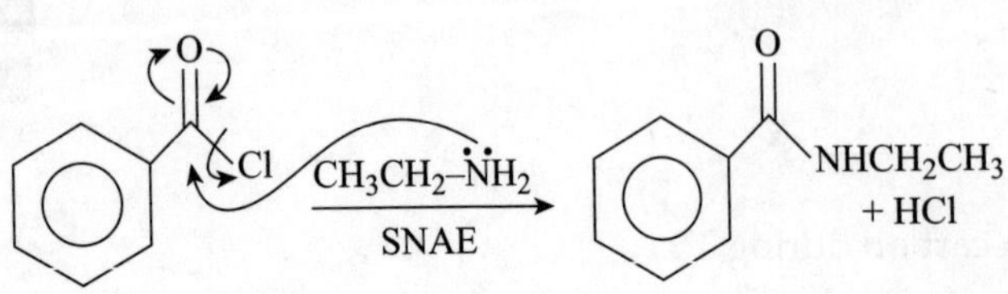

Ans: (b)

99.

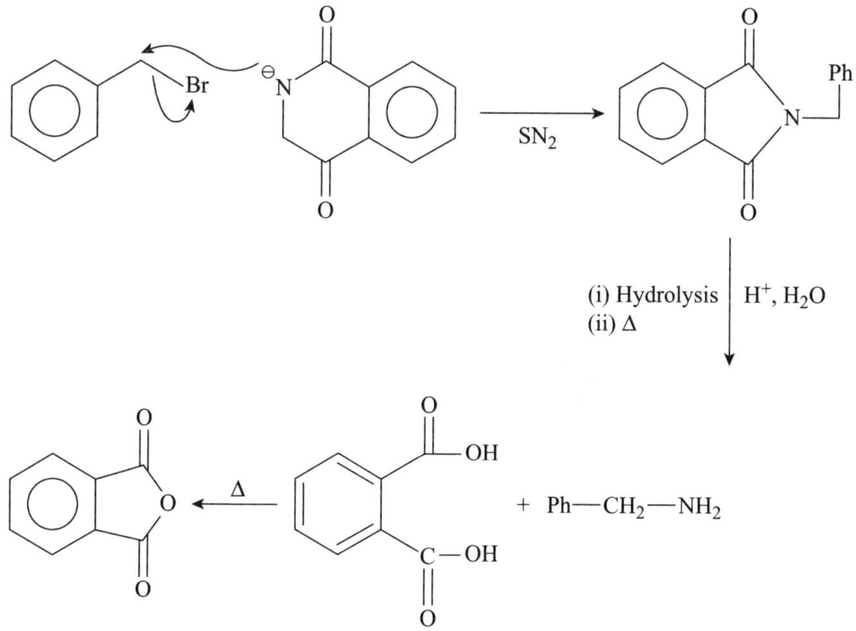

Ans: (b)

100.

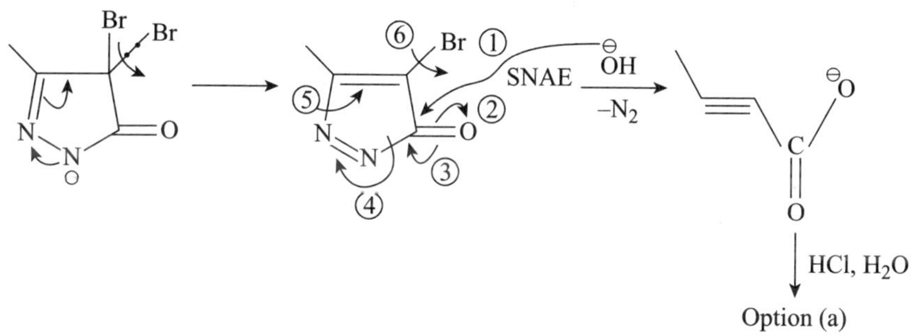

1, 2, ... 6 indicate sequence of e⁻ displacement.

Ans: (a)

101.

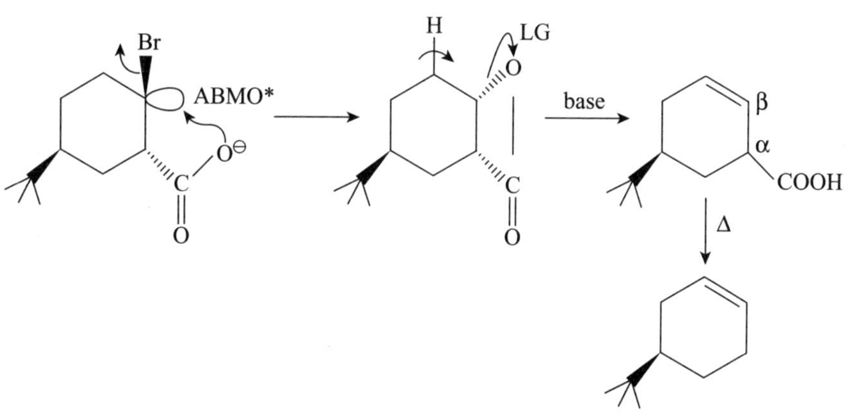

Ans: (b)

102.

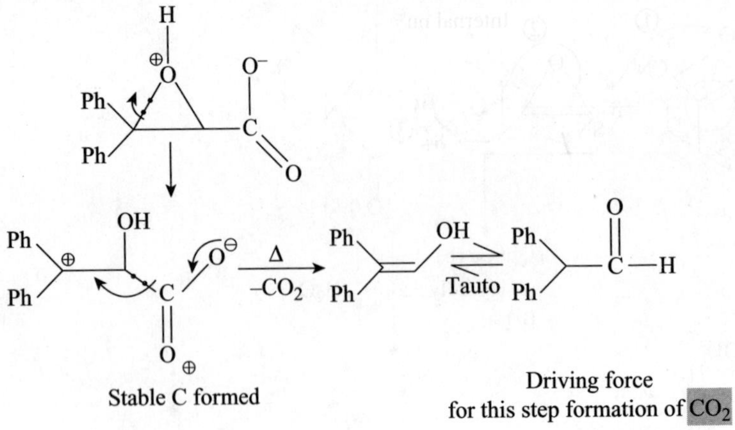

Ans: (b)

103.

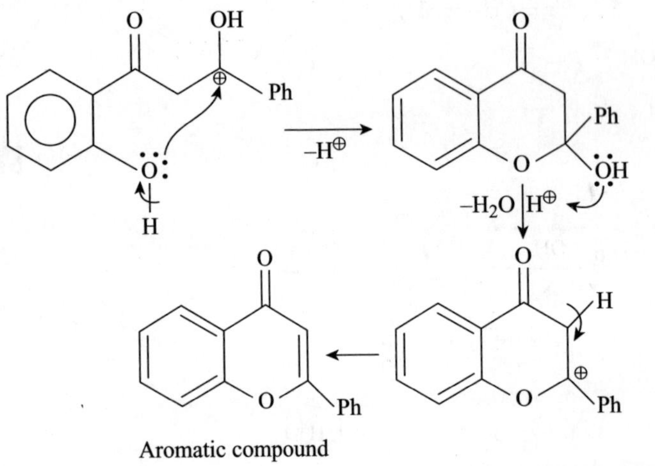

Ans: (b)

104.

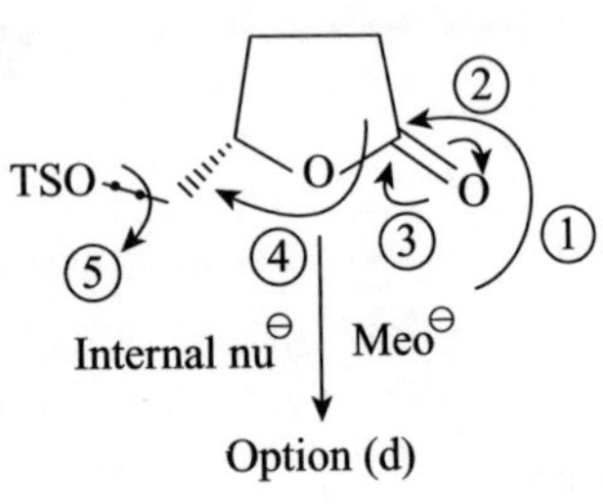

Rate of reaction addition on polar π-bond Nu ⊖ Substitution.

Ans: (d)

105.

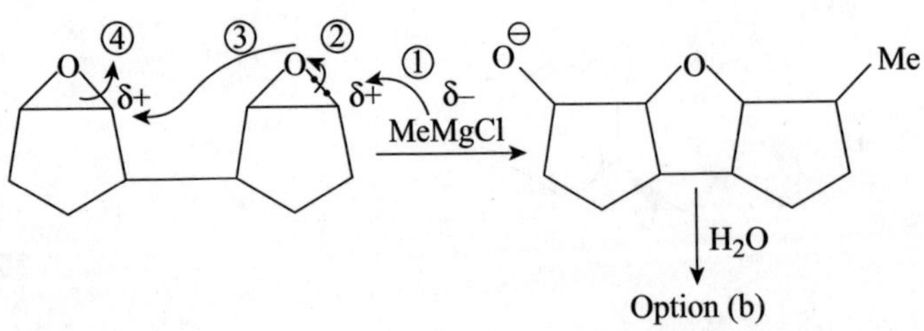

1, 2, 4 indicate sequence of electronic attack.

Ans: (b)

106.

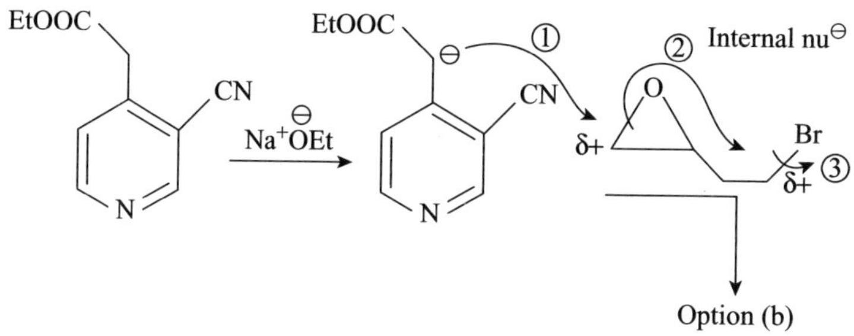

Ans: (b)

107.

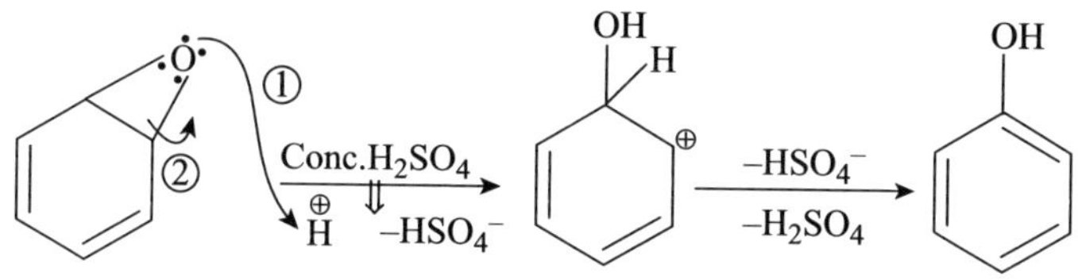

Ans: (d)

108.

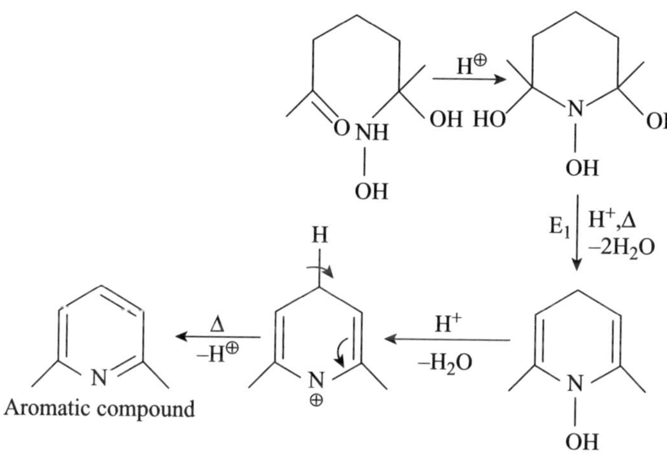

Ans: (a)

109.

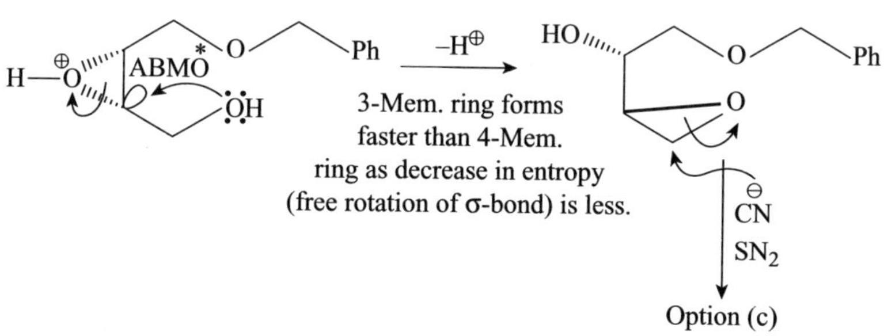

Ans: (c)

110. Reactivity order towards nucleophile:

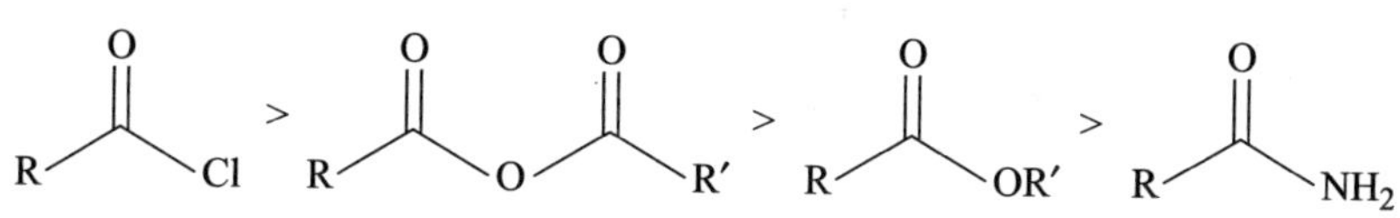

Ans: (d)

111. (a) 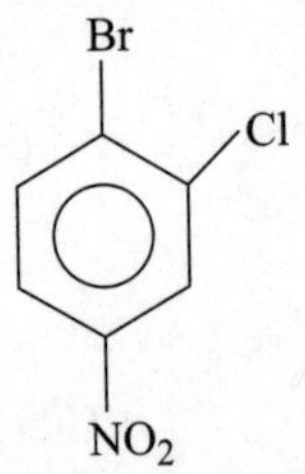 (b) 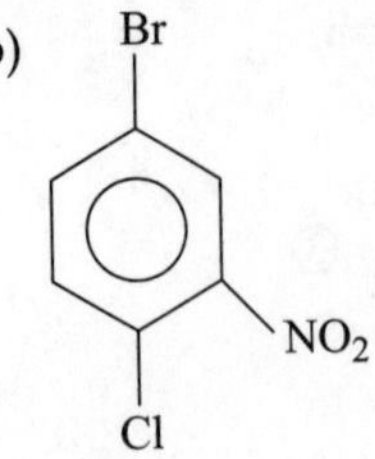 (c) 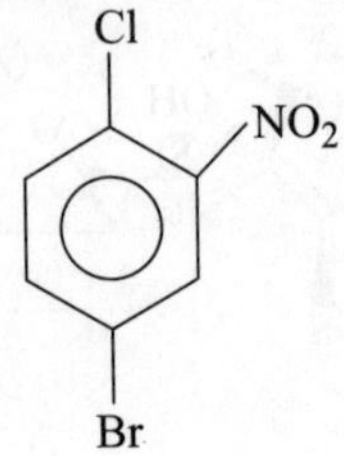

Ans: (c)

112. Nucleophilicity (in DMF) : $CH_3O^- > CH_3S^-$ **Ans:** (d)

113. Polar protic solvent at 3° carbon favours SN_1 LG ability: –Br > –Cl **Ans:** (c)

116. 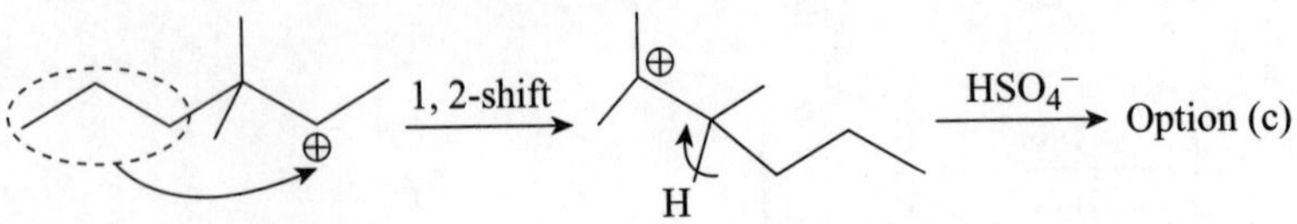

Ans: (c)

117. (a) 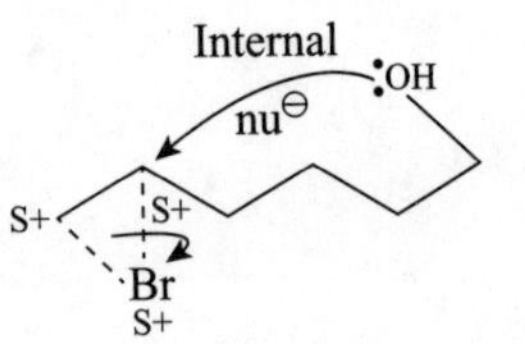 (c) 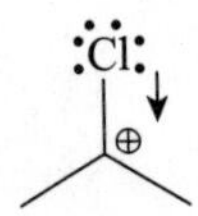 Stable carbocation via resonance

Ans: (c) and (d)

118. In step 1, BH_3 act as Lewis acid. **Ans:** (a)

121. 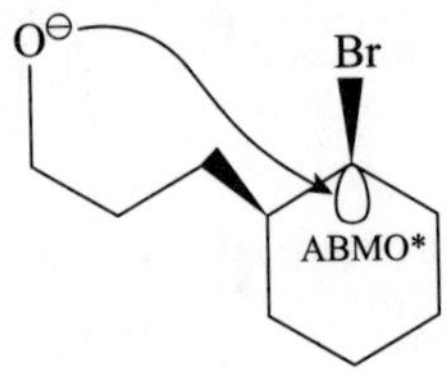

Ans: (d)

122. (a) (E or Z) 3-hexene both are possible.

(d) (E) - 2-hexene formed as final product. **Ans:** (b)

123. i, iii: Syn-addition

ii, iv: Anti-addition **Ans:** (d)

124. (d) 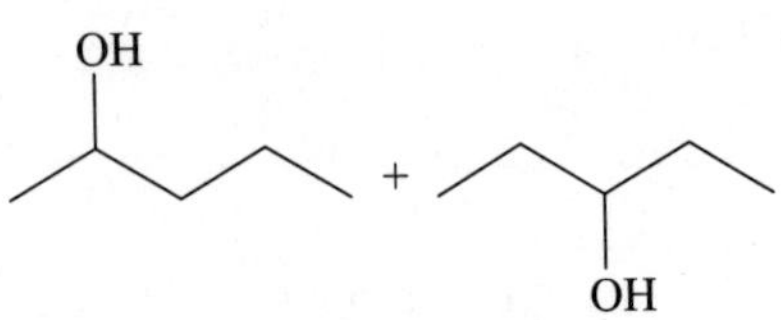 (Mixure of products)

Ans: (a)

125. (a) 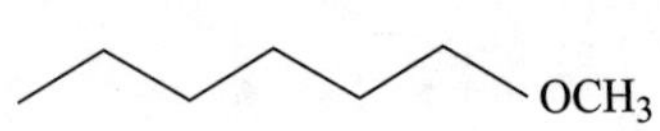

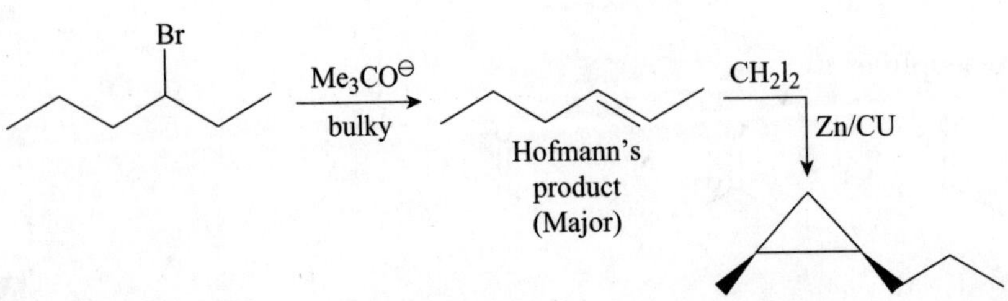

Ans: (b)

126.

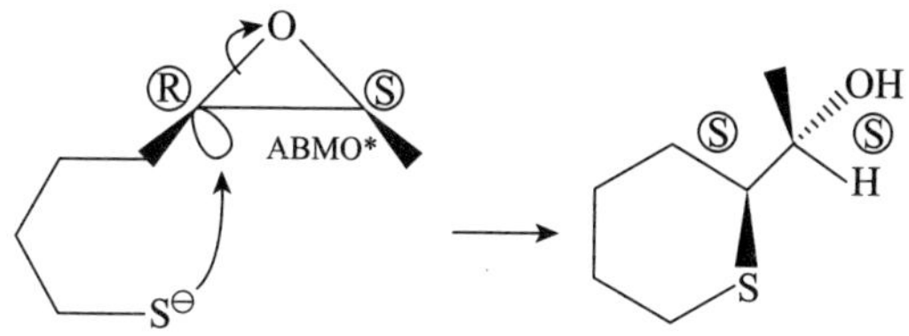

Ans: (b)

128. C.I. $Bu_3Sn–H + ln\bullet \rightarrow Bu_3Sn\bullet + Hln$

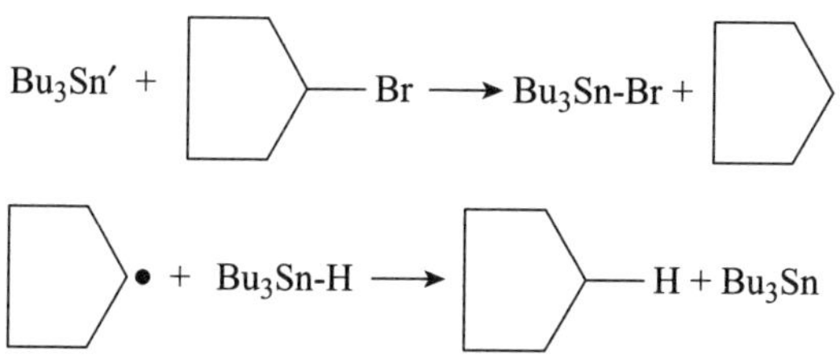

here ln• act as chain initiator.

Ans: (b)

130. First attack of $H^{\oplus}$ will be on >C=O, as >$\overset{\oplus}{C}$—OH is stable carbocation via octet completion.

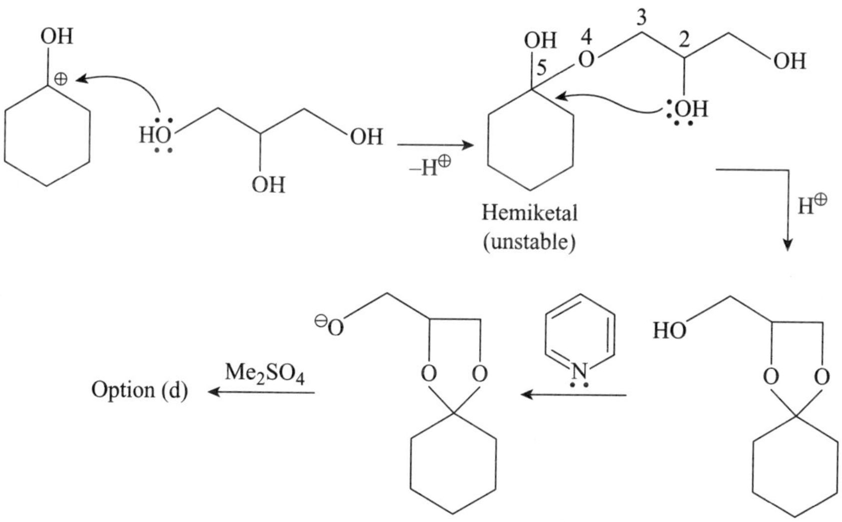

Ans: (d)

LEVEL 2

1. Rate $\propto \dfrac{1}{\text{Steric hindrance}}$

$\propto$ Stability of S- on carbon

Ans: (b) and (d)

2.

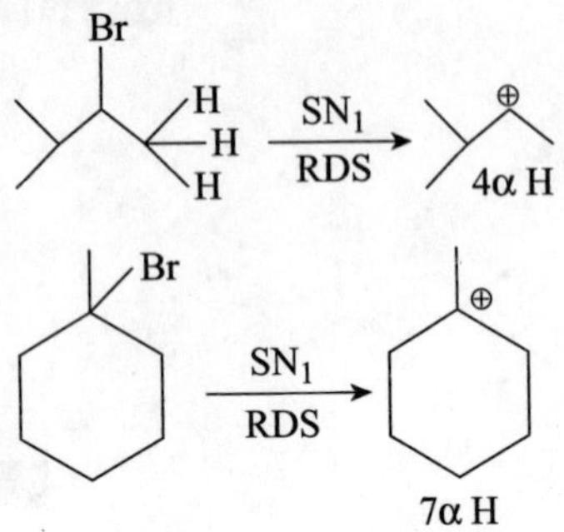

Ans: (a)

3. A:

B:

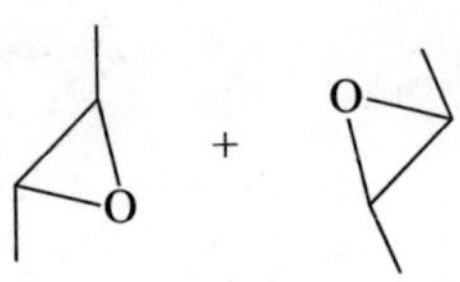

(Enantiomers)

Ans: (a)

4.

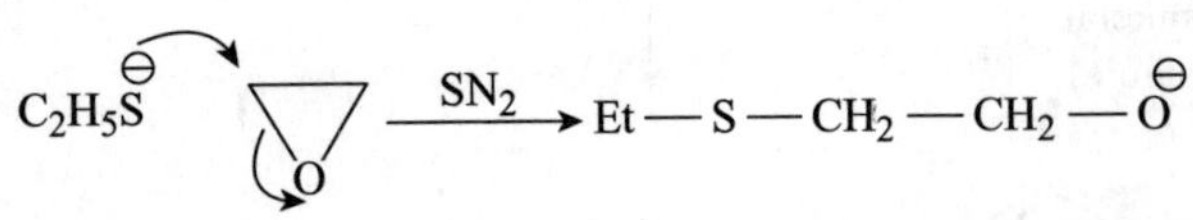

Ans: (a)

5. (a) SN_1 : Pseudo 1st order

(b) SN_2

Ans: (d)

6. (a)

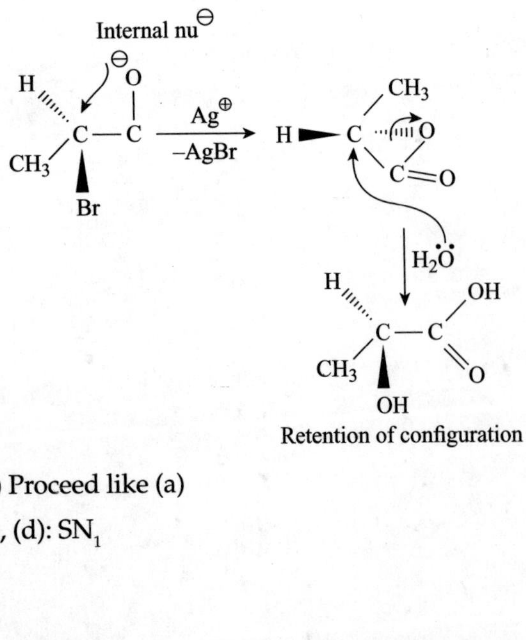

Retention of configuration

(b) Proceed like (a)

(c), (d): SN_1

Ans: (a) and (b)

8.

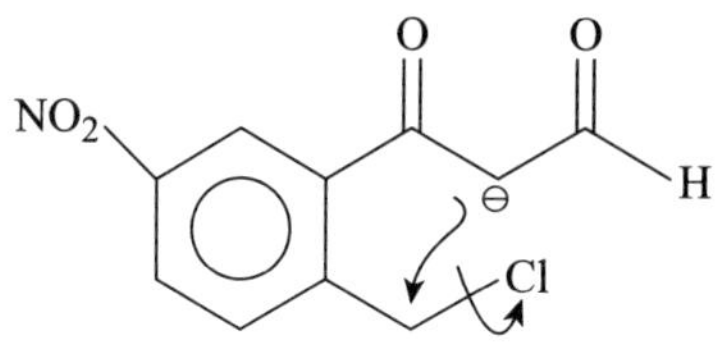

Ans: (a)

9. (a), (c): Internal nucleophile forces LG to leave better.

Ans: (a) and (c)

10 (a)

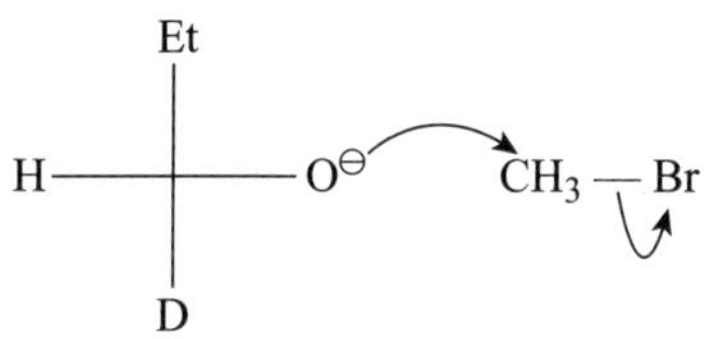

(b)

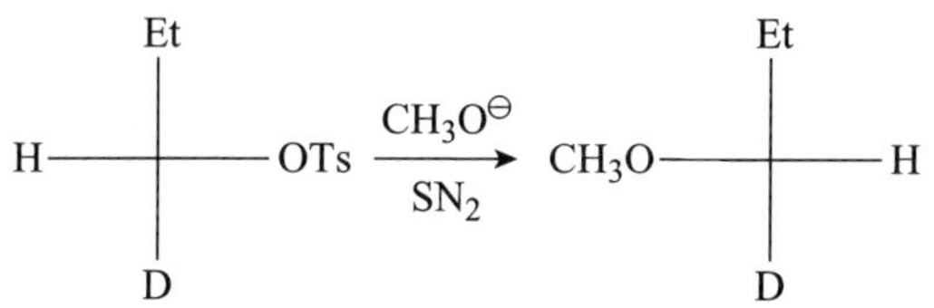

(c)

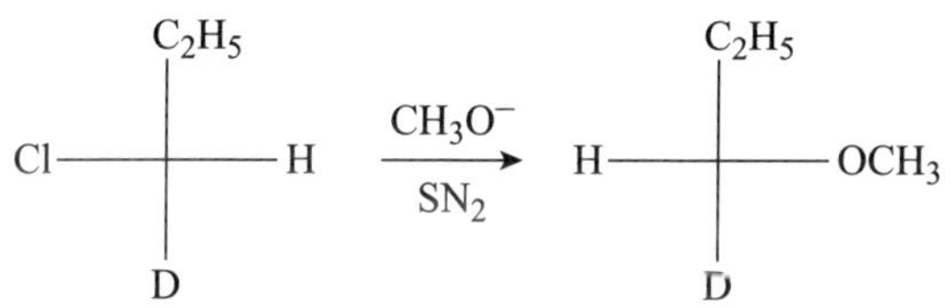

(d)

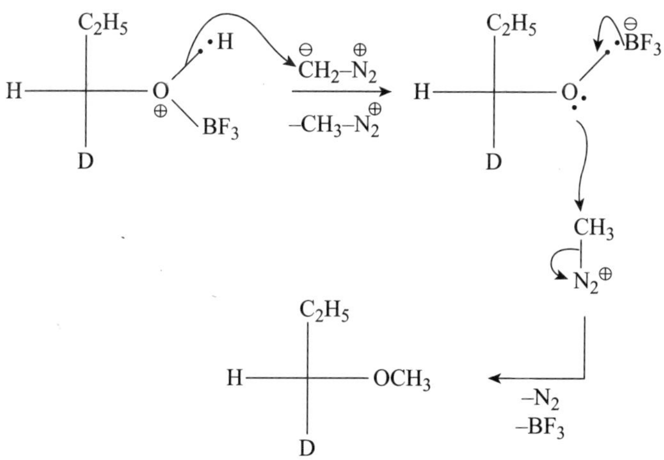

Ans: (a), (c) and (d)

11. A :

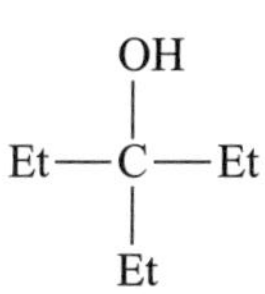

B :

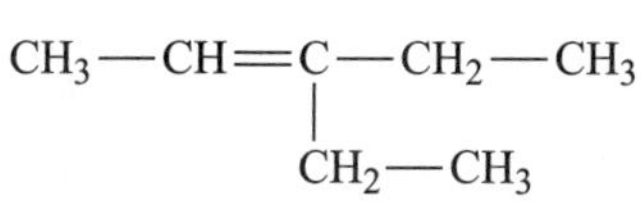

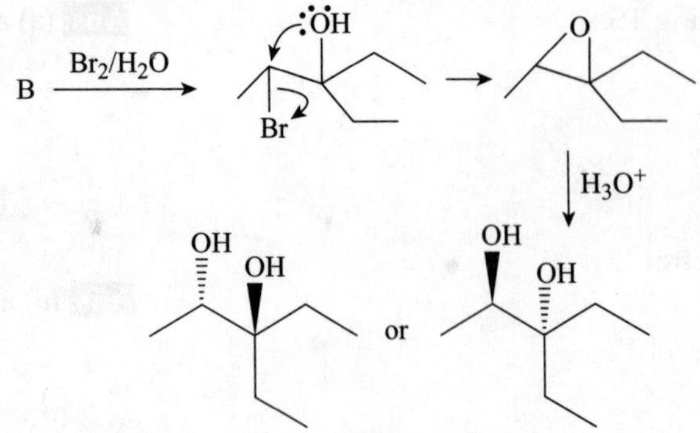

(Trans product)

Ans: (a) and (d)

12. (c) E and Z product

(d) Stability: $CH_3-CH_2-\ddot{O}-\overset{\oplus}{C}H_2 > CH_3-\overset{\oplus}{C}(CH_3)-CH_3$

Ans: (a), (b) and (c)

13. 3° alkyl halide in presence of strong base favours elimination. **Ans:** (b), (c) and (d)

14. SN_1 or SN_2 are not possible at bridgehead carbon. **Ans:** (a), (b) and (c)

15. (a)

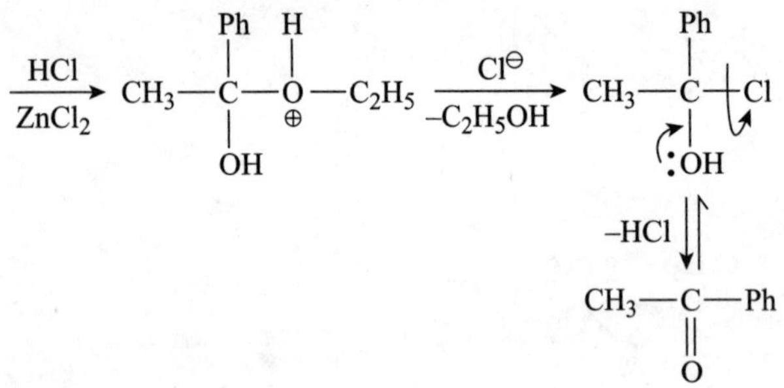

(d) Turbidity appears after very long time. **Ans:** (b) and (c)

16.

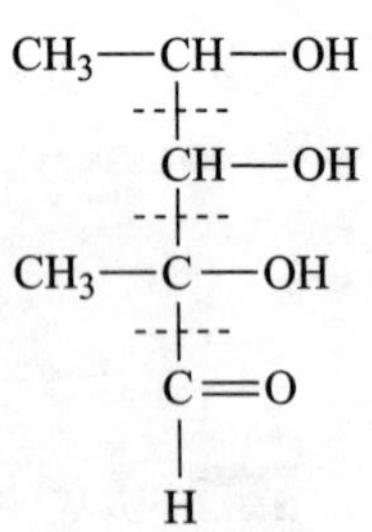

Ans: (d)

18. (a)

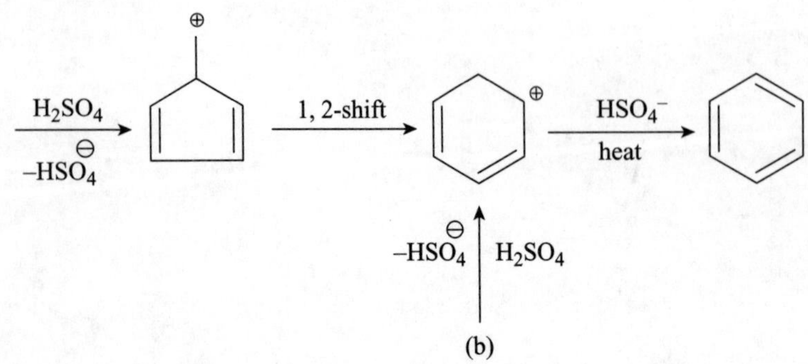

Ans: (a) and (b)

19. Reactivity ∝ Stability of δ-formed on carbon during TS. **Ans:** (a) and (c)

21. to 27.

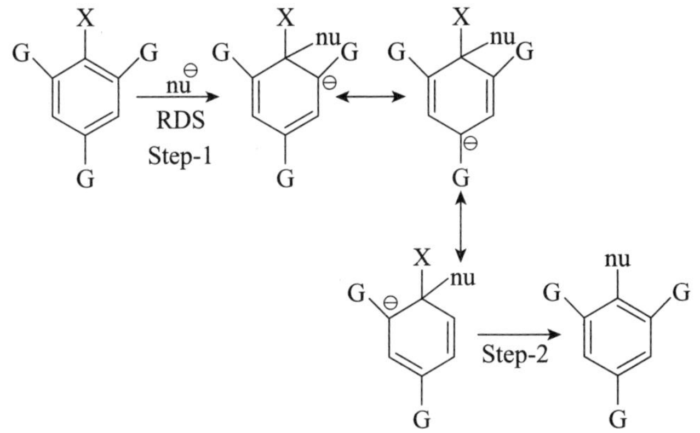

RDS is addition of nucleophile not step-2, so reactivity rate α stability of intermediate formed in RDS.

EWG nature of "G" and "X" stabilise –ve charge to increase rate.

28. to 34. Nucleophilicity $\propto \dfrac{1}{\text{Steric factor}} \propto$ e^- donating tendency (if steric factor same)

35. to 36. Rate ∝ Stability of carbocation.

37.

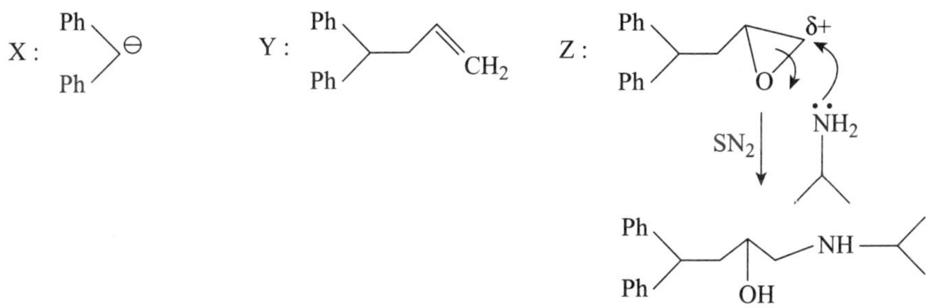

Ans: (b)

38.

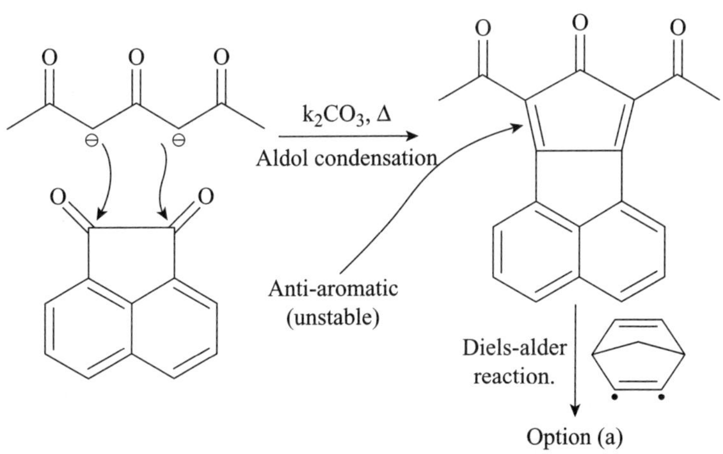

Ans: (a)

39.

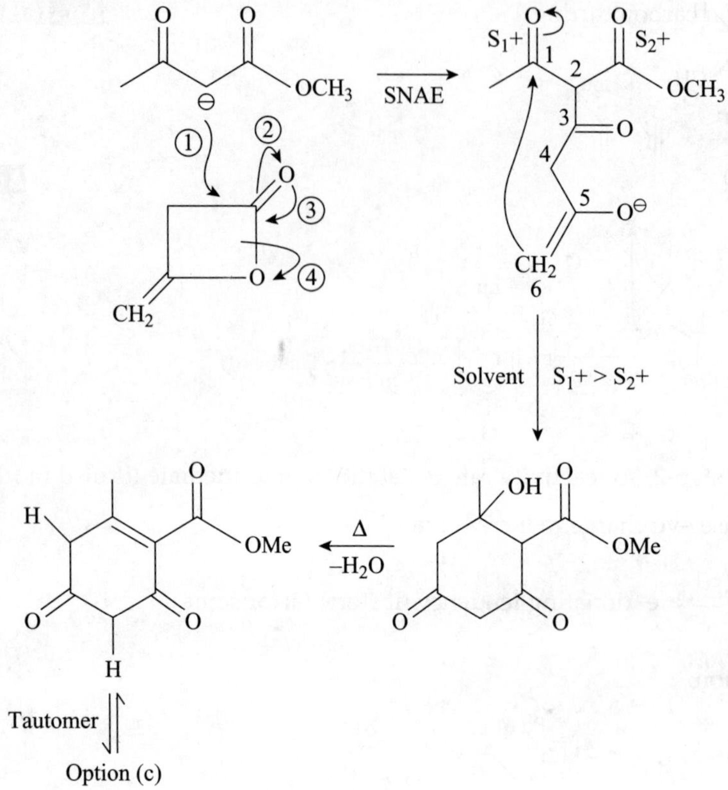

Ans: (c)

42. Aldol condensation

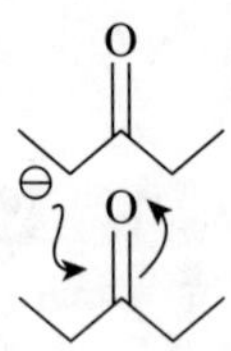

Ans: (b)

43.

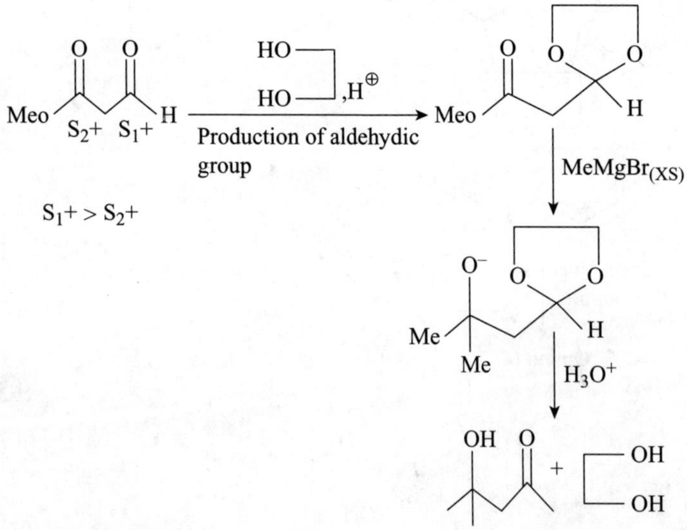

Ans: (c)

44.

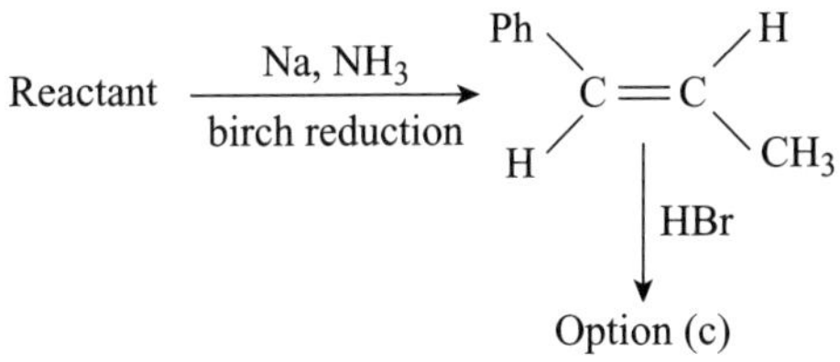

Ans: (c)

45.

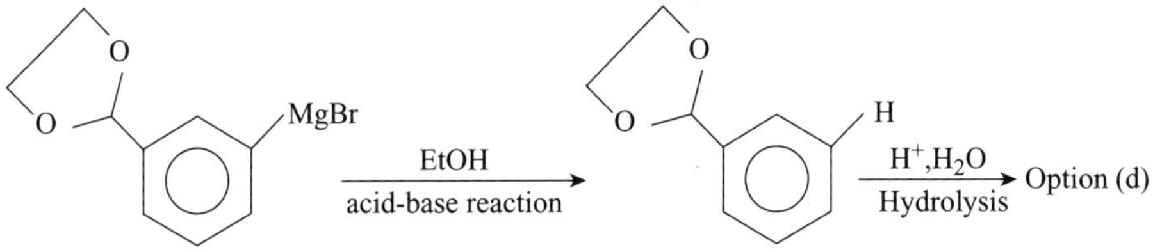

Ans: (d)

46.

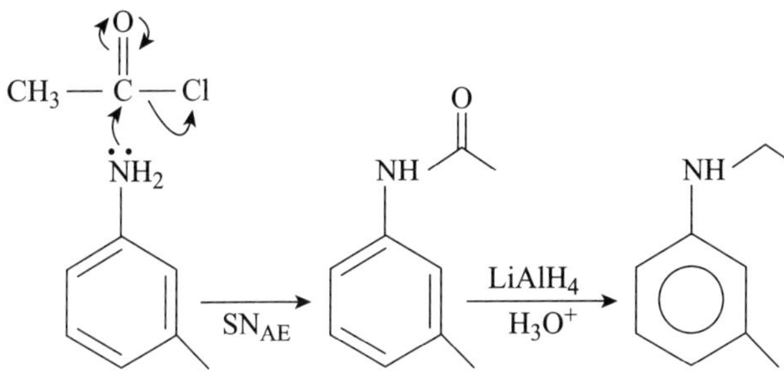

Ans: (d)

47.

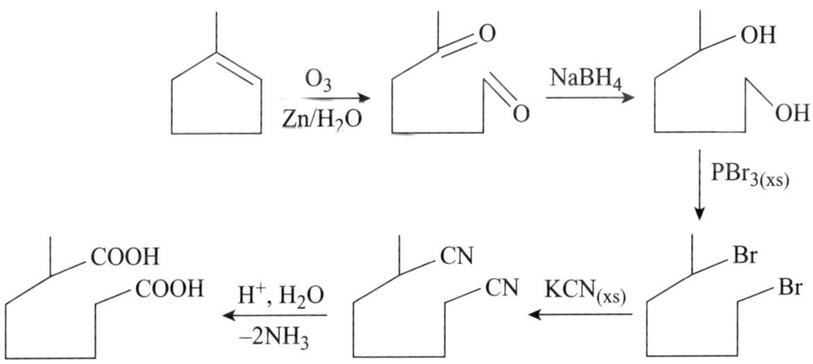

Ans: (b)

48.

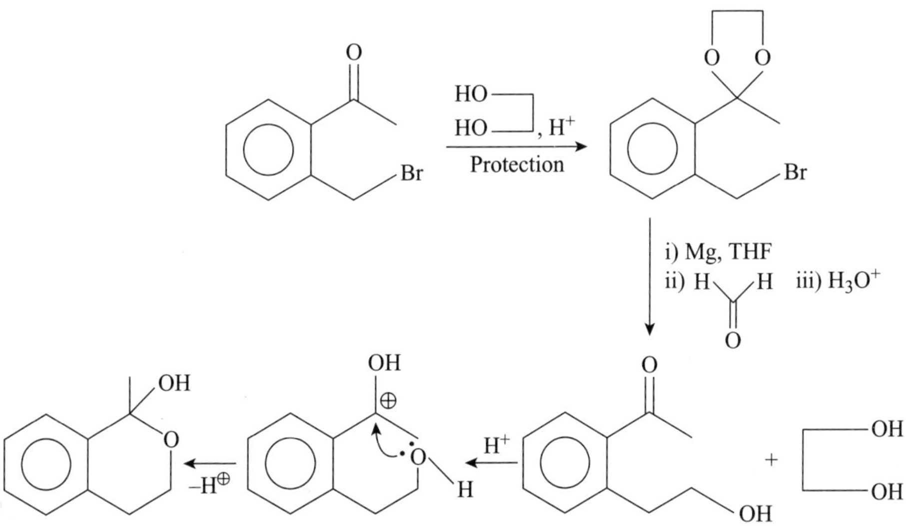

Ans: (a)

49. 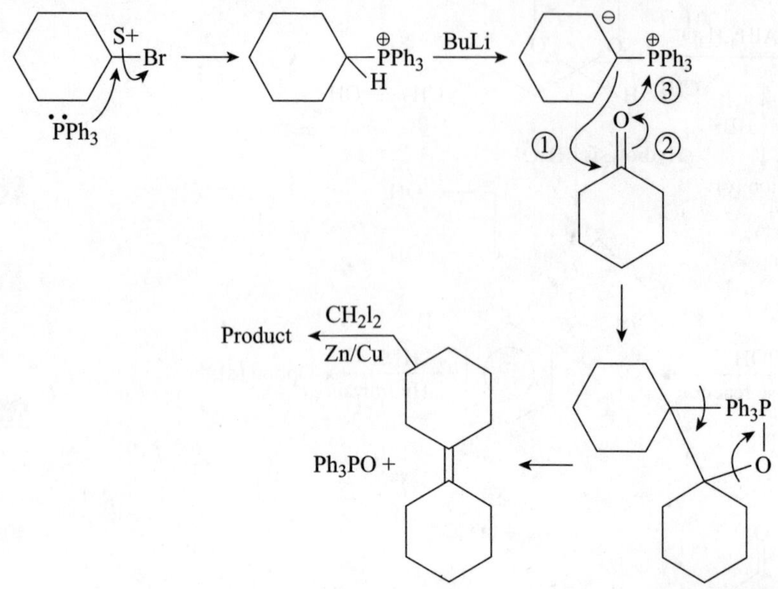

(Unstable)

Ans: (c)

50. 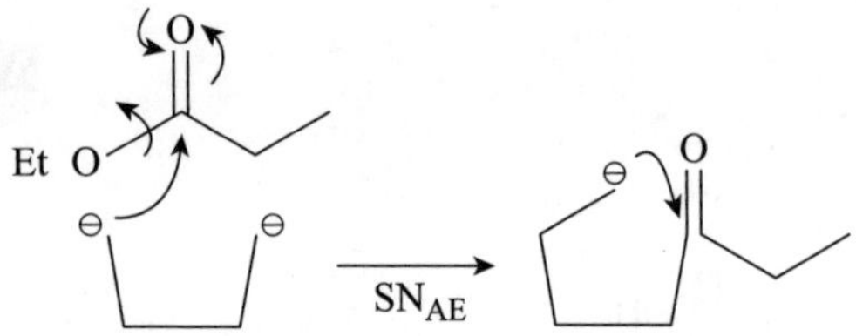

Ans: (c)

51. Trans alkene $\xrightarrow{\text{Anti-addition}}$ Erythro products

Ans: (d)

52. $LiN(i\text{-}Pr)_2$: act as base

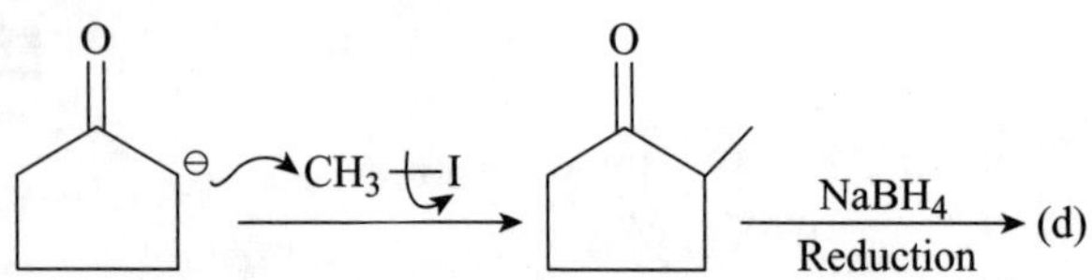

Ans: (d)

53. 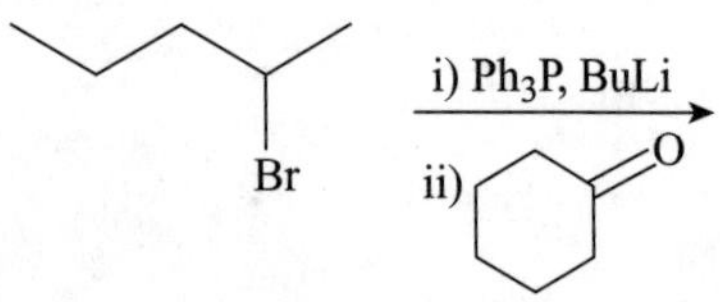 Similar path as in Question Number 49.

Ans: (a)

54. (i) Friedel-craft acylation.

(ii) Clemmensen reduction.

(iii) Friedel-craft alkylation.

Ans: (c)

55.

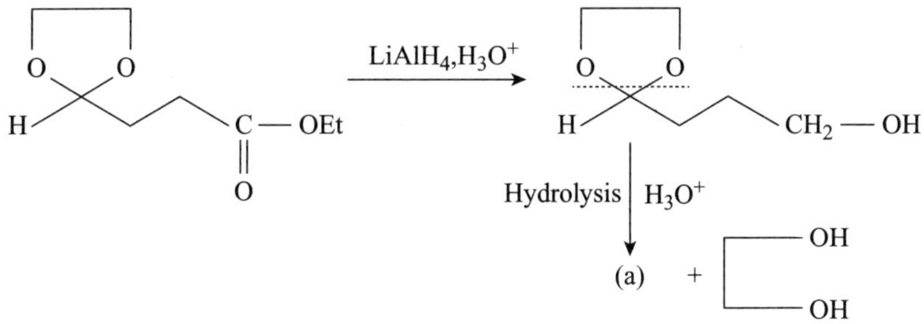

Ans: (a)

56.

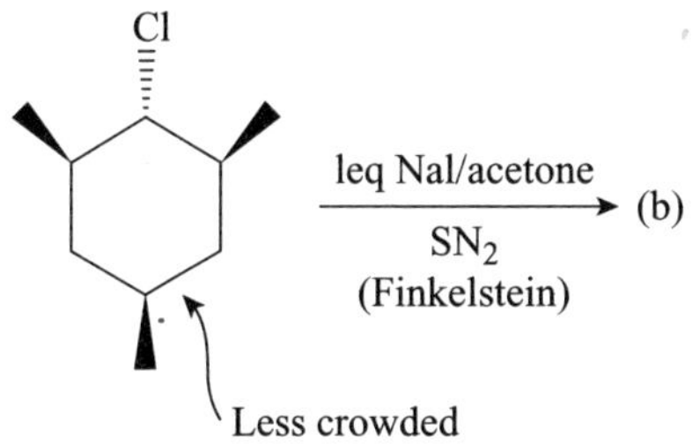

Ans: (b)

57.

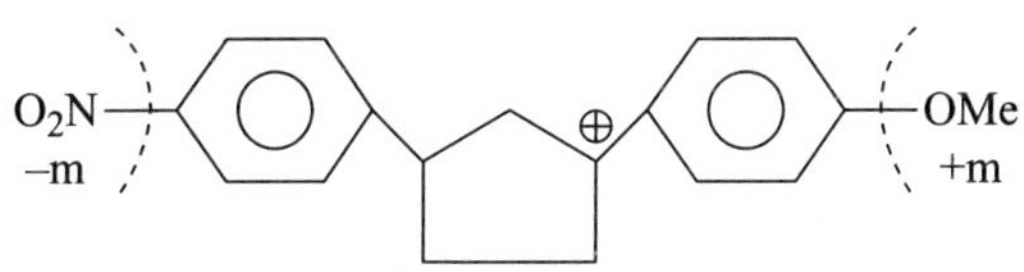

Stable C$^{\oplus}$ after rearrangement

Ans: (d)

58.

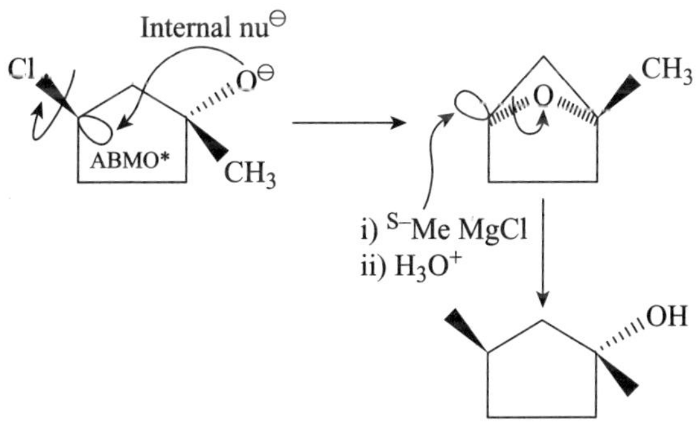

Ans: (d)

59.

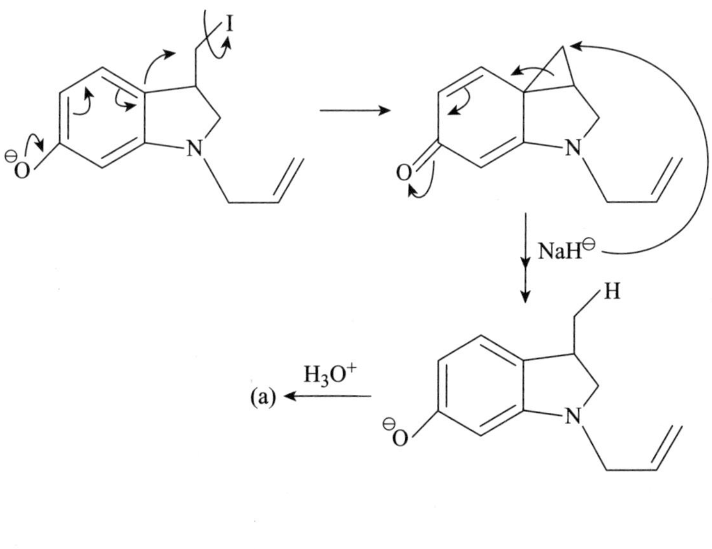

Ans: (a)

60.

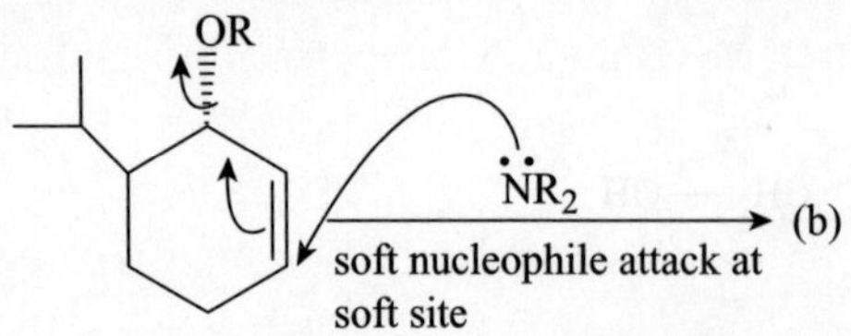

Ans: (b)

61.

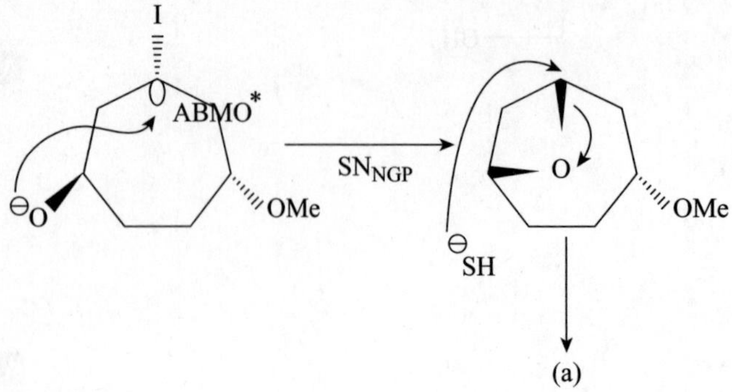

Ans: (a)

62.

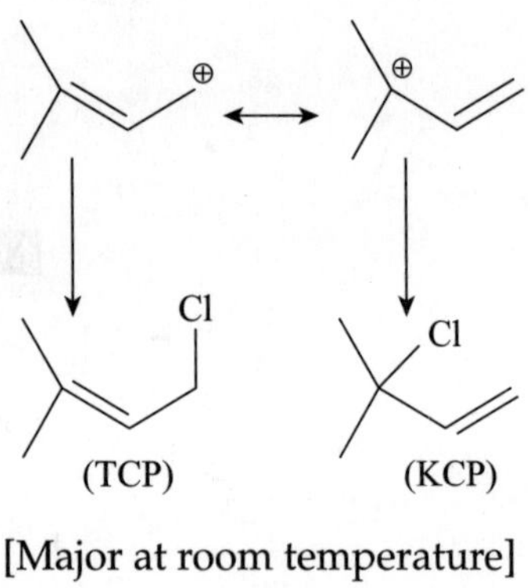

Ans: (a)

63. A :

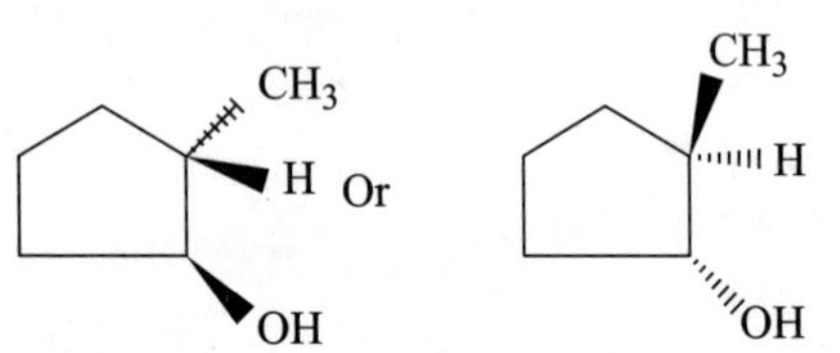

B :

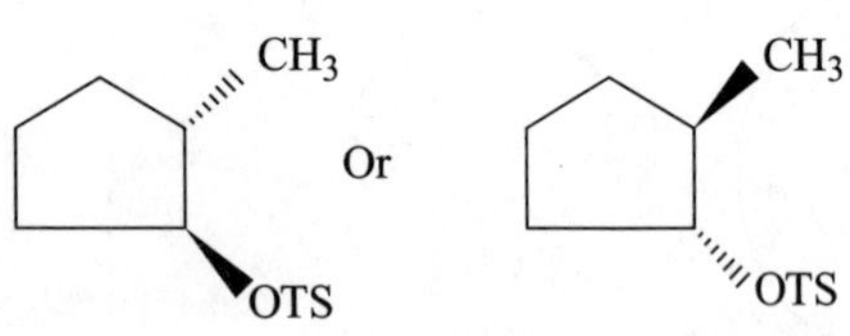

C :

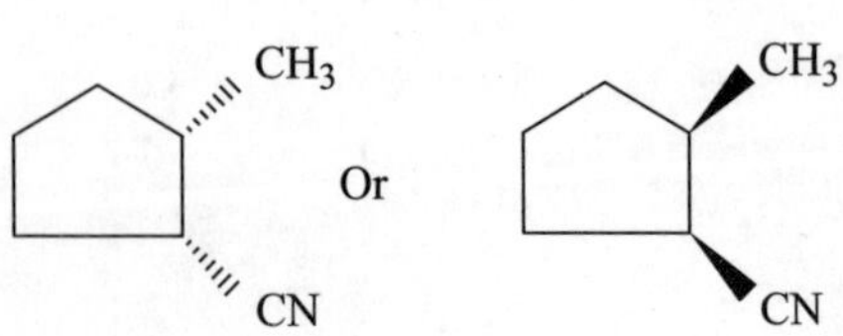

Ans: (a)

67. –F, $-NR_3$ are poor LG, gives Hofmann's product on elimination. Removal of H will take place from carbon where S– is more stable.

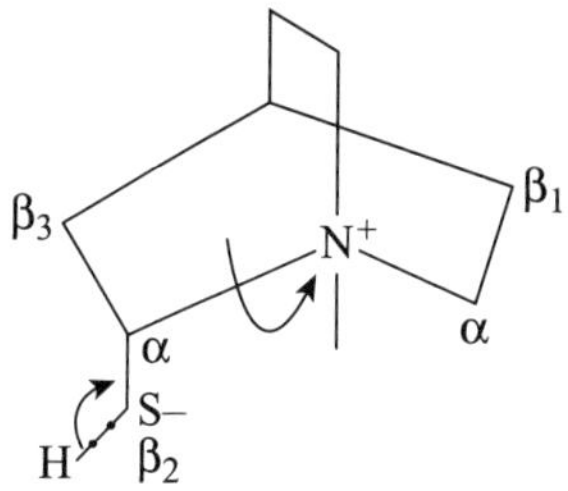

S– at β_2 position will be the most stable.

Ans: (a), (c) and (d)

70. (b) E_1CB **Ans:** (a) (b) and (d)

71. (a) E_2 ;

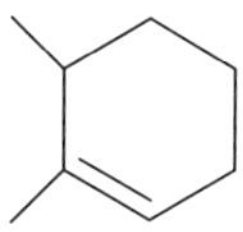

(Major product)

(c) Hofmann's product (d) E_2 ;

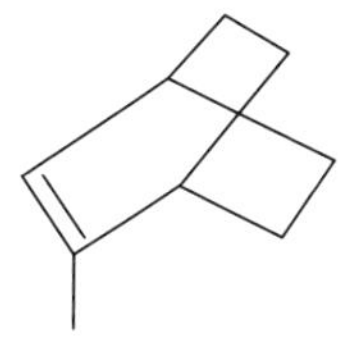

(Major product)

Ans: (b) and (c)

72. (a), (b), (c): SN_1

(d) E_2 **Ans:** (a) and (d)

74. Vinyllic, benzyllic and bridgehead alkyl halide does not give SN reaction. **Ans:** (d)

75. ArSN (for details : Refer Question Number 21). **Ans:** (c)

76. (d) Solvation of TS takes place, as S+ charge on TS, that leads to stablisation of TS; thus, making decrease in activation energy proceed to faster ROR (Rate of Reaction). **Ans:** (d)

77.

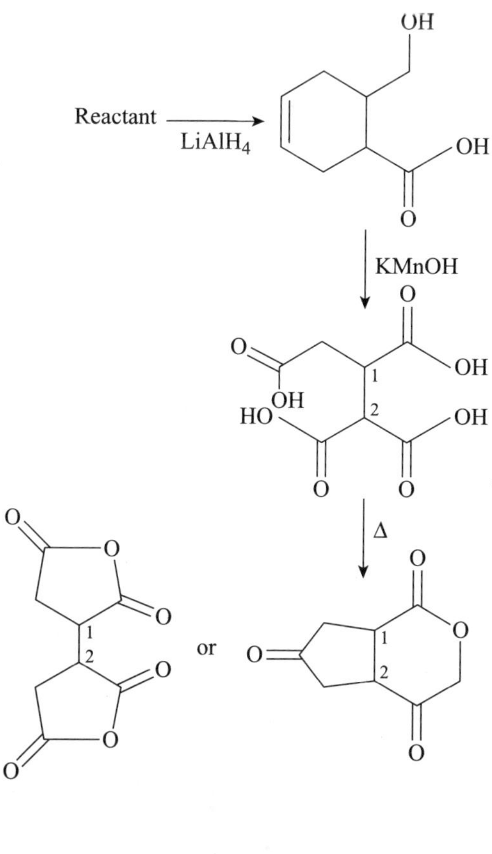

Ans: (a) and (b)

78.

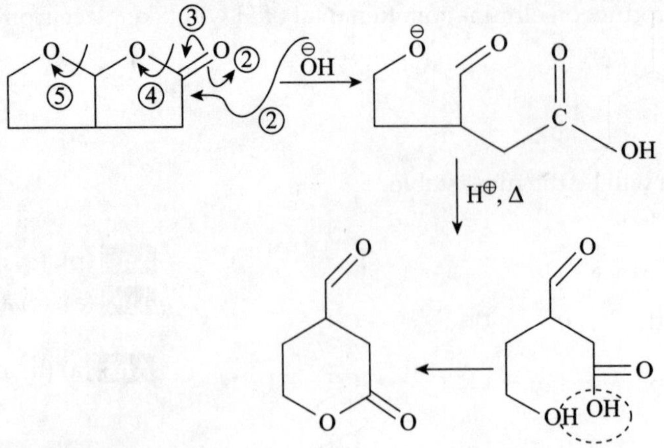

Ans: (a)

Comprehension Type

Passage 1

79. Enantiomeric excess $= \dfrac{70-10}{70+10} \times 100 = 75\%$ **Ans:** (c)

Passage 2

(c) Trace amount of H_2O, helps only to create carbocation.

Passage 3

85. (a) C — F bond is too strong to give SN_2. **Ans:** (c)

86. (c) Using above method on 3° alkyl halide gives elimination product. **Ans:** (c)

Matrix type

103. (a)

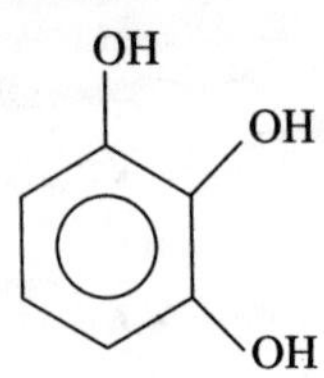

+ $3CH_3I$

Ans: (q) and (s)

(b)

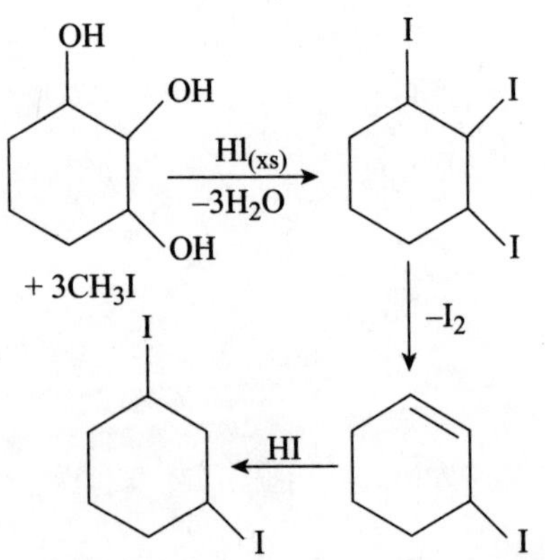

Ans: (p), (q), (r) and (s)

(c)

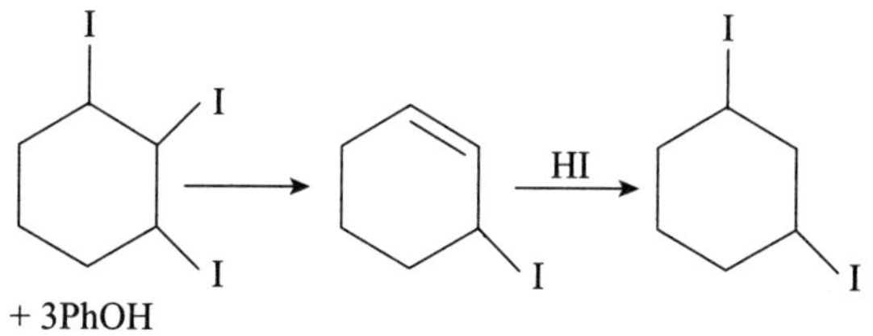

Ans: (p), (q) and (r)

(d)

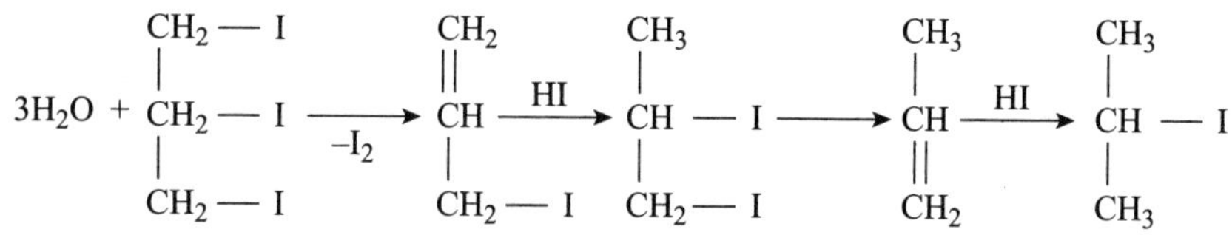

In (b) and (d) H_2O as side product, will react with Na ; while in (c) PhOH.

Ans: (p), (q) and (r)

109. (b) Solvation of TS (as S+ charge on TS), decrease activation energy.

Ans: (p), (r) and (t)

(c) Solvation of reactant > solvation of TS in polar protic solvent, so increase in activation energy.

Ans: (q), (r) and (t)

(d) TS also form in SN_1 between reactant and intermediate.

Ans: (p), (s) and (t)

112. (d)

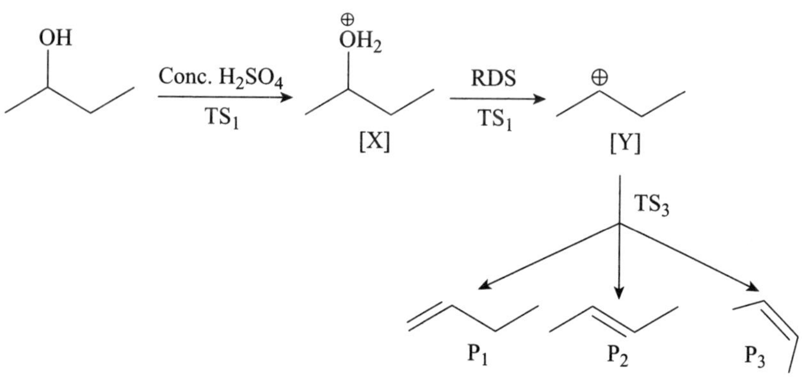

Transition state.

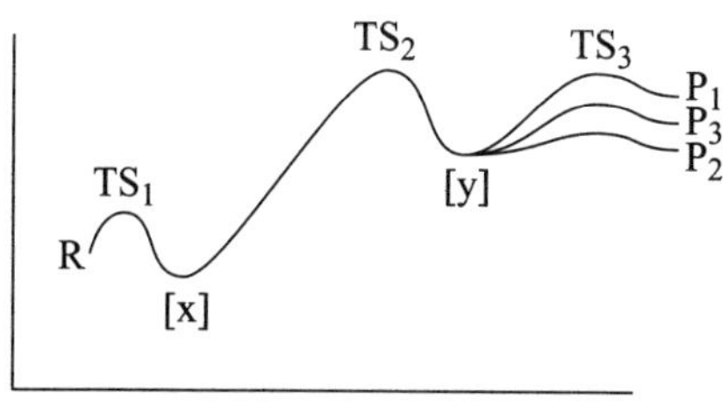

Ans: (s)

113. SN_1 rate $\propto$ stability of $C^{\oplus}$

114. SN_2 rate $\propto$ stability of $C^{\ominus}$

116. $\% \; SN_1 = \dfrac{5.3\times10^{-6}[\text{substract}]}{2.35\times10^{-5}\times1.5[\text{subs.}]+5.3\times10^{-6}[\text{subs.}]}\times100$

Ans: (8.233)

117. 2, 3, 7, 8, 9 **Ans:** (5)

118. 1, 2, 3, 4, 8, 9 **Ans:** (6)

119.

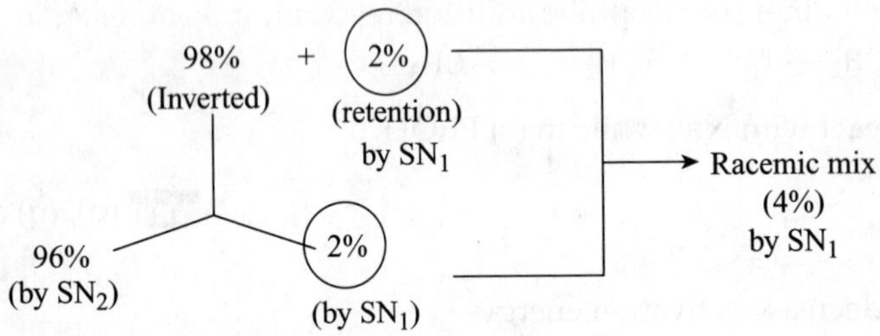

Ans: (4)

120.

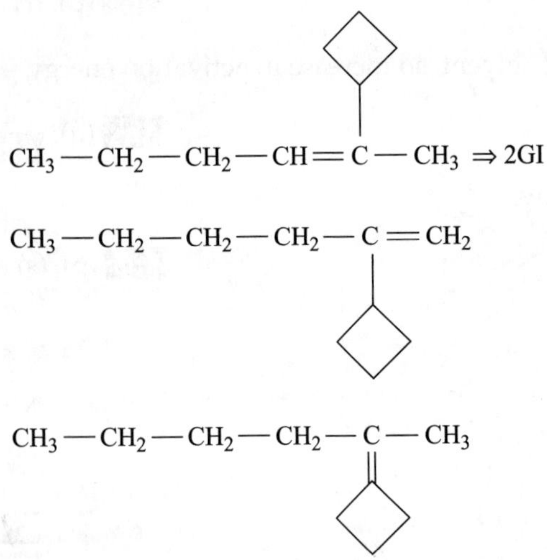

Total 4.

Ans: (4)

121.

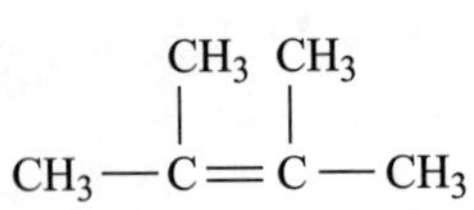

Ans: (12)

CHAPTER

5

Carbonyl Compounds and Acid Derivatives

LEVEL 1

5. Acid base reaction is much faster than any other (nucleophilic addition reaction) type of reaction.

Ans: (d)

8.

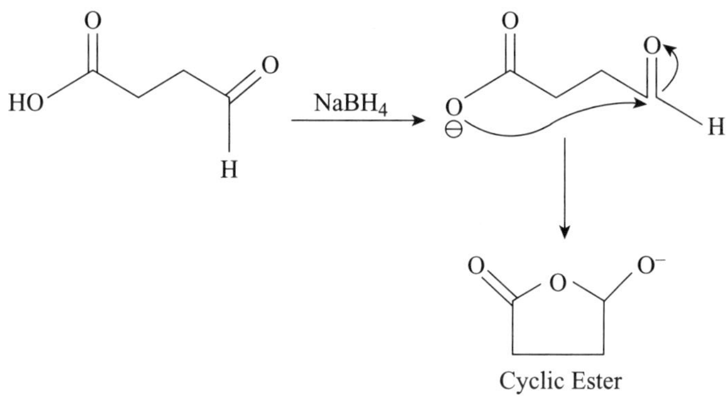

Ans: (d)

9.

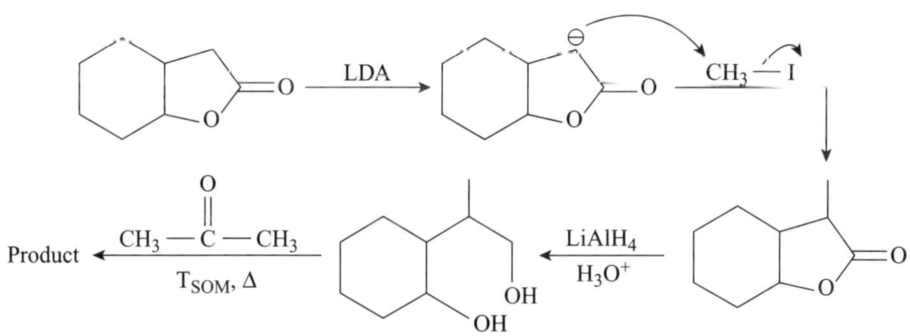

Ans: (c)

10.

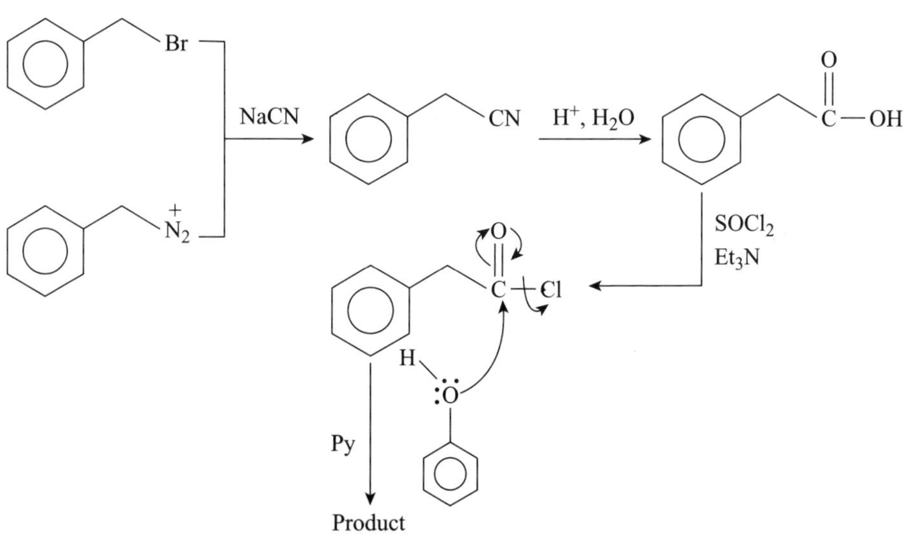

Ans: (b) and (d)

11.

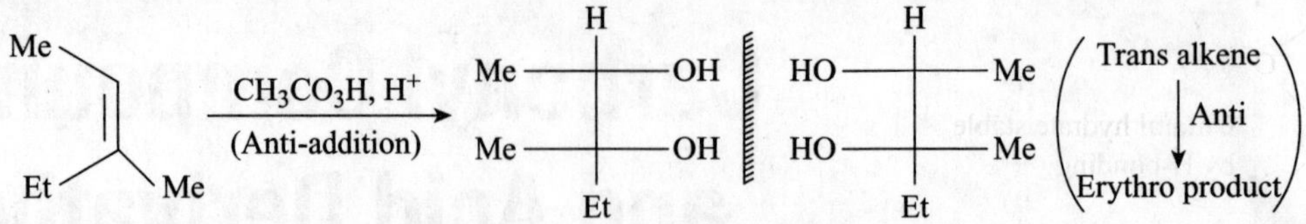

Ans: (b)

12.

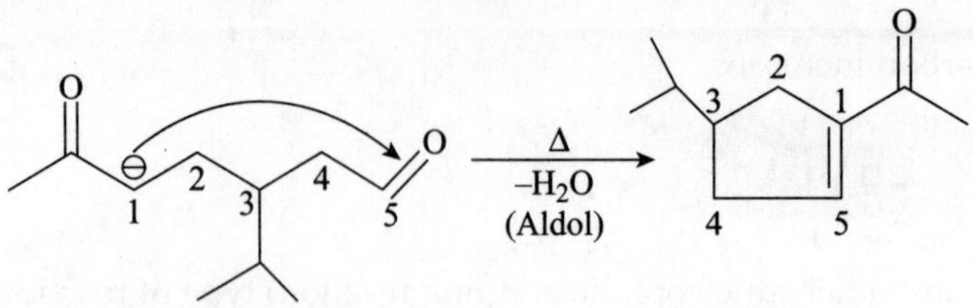

Ans: (a)

13.

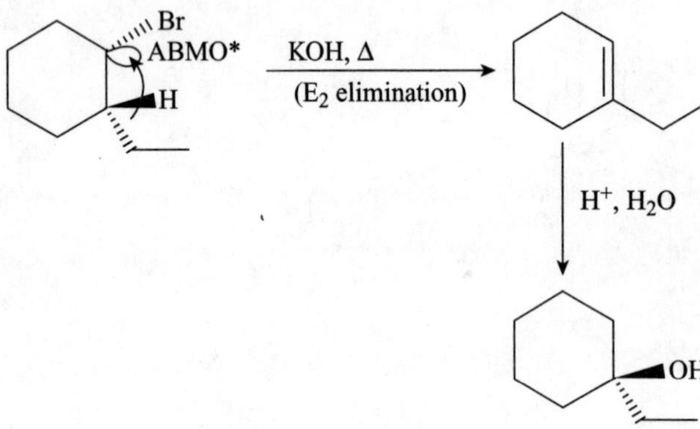

Ans: (a)

14. Trans alkene $\xrightarrow[\text{addition}]{\text{Syn}}$ Threo products

Ans: (c)

15.

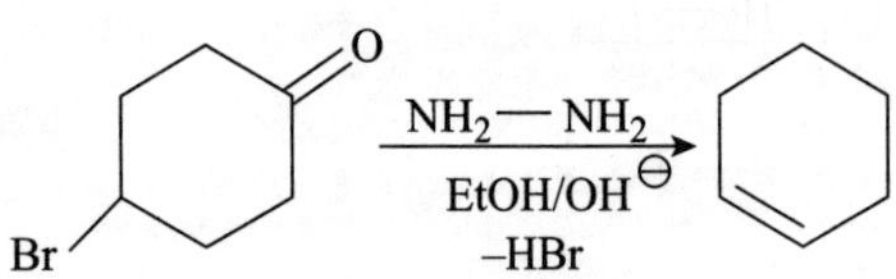

[Wolff-Kishner reduction].

Ans: (c)

16. Intermolecular cannizzaro

Ans: (a) and (b)

17.

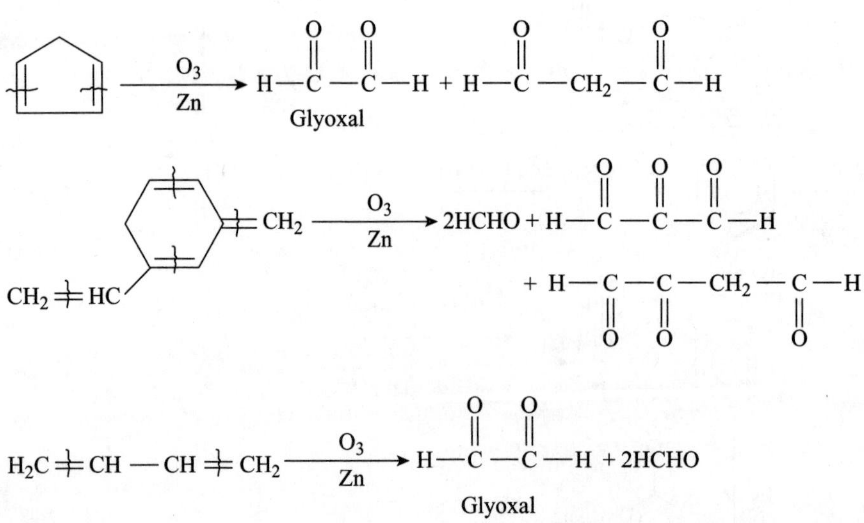

Ans: (d)

18. 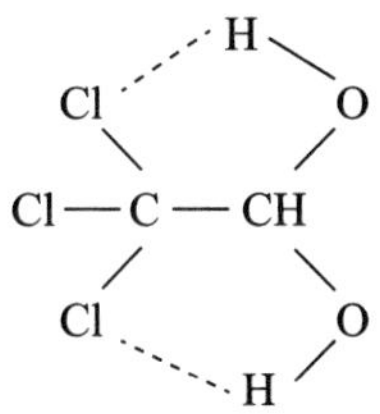

In protonated carbonyl, δ+ charge on carbon increases. **Ans:** (d)

19. 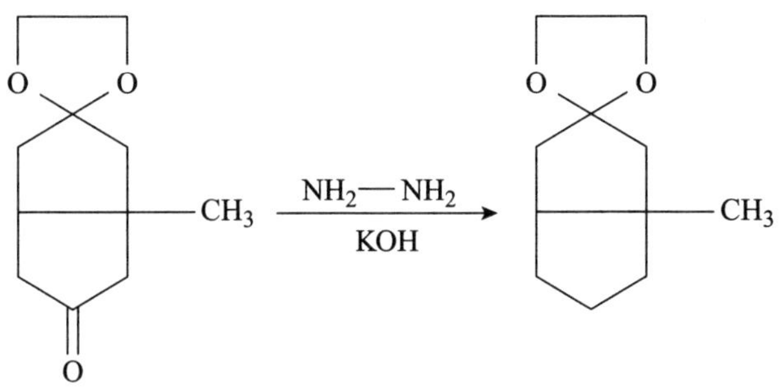

(A)

Ans: (c)

20. Reactivity ∝ δ+ charge on carbon of 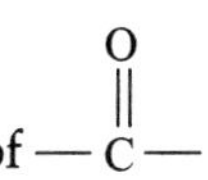 **Ans:** (a)

22. 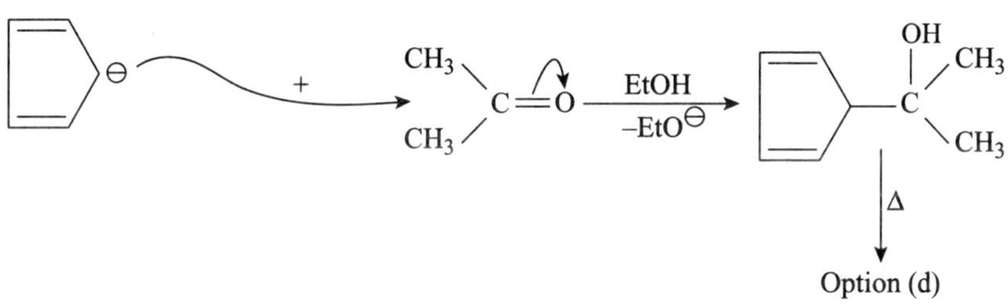

Ans: (d)

23. Wittig reaction

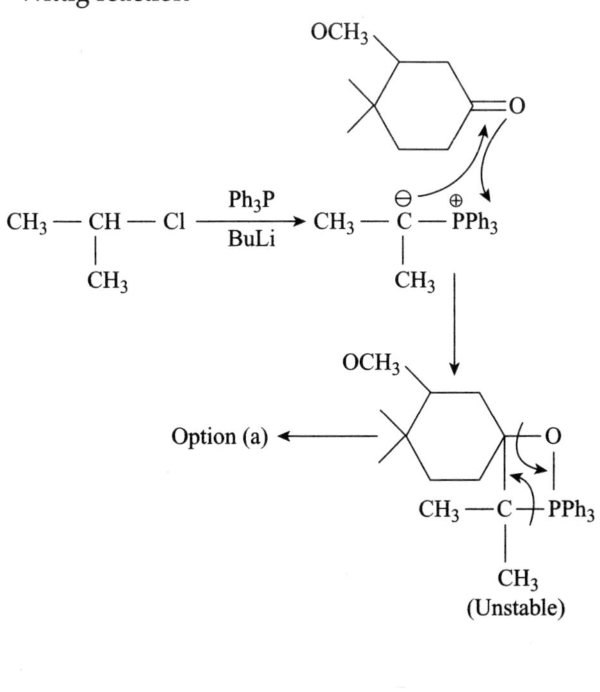

Ans: (a)

25. Intramolecular cannizzaro **Ans:** (b)

26.

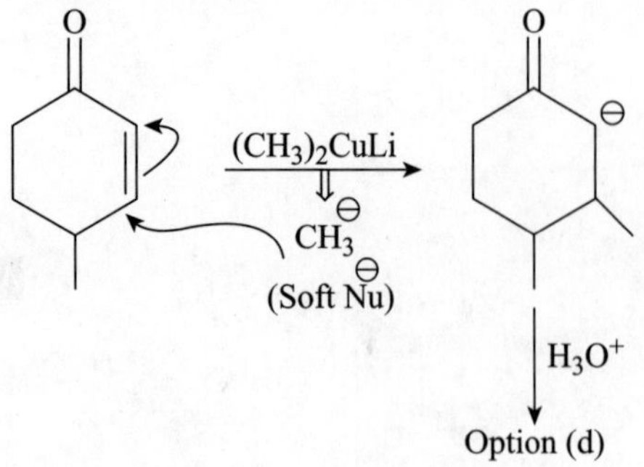

Ans: (d)

27.

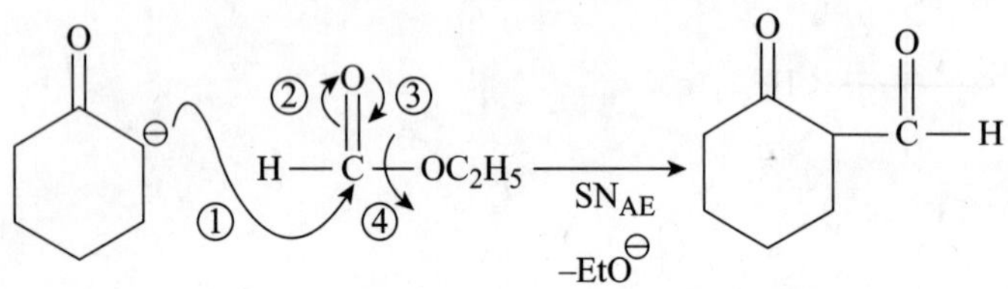

Ans: (c)

28.

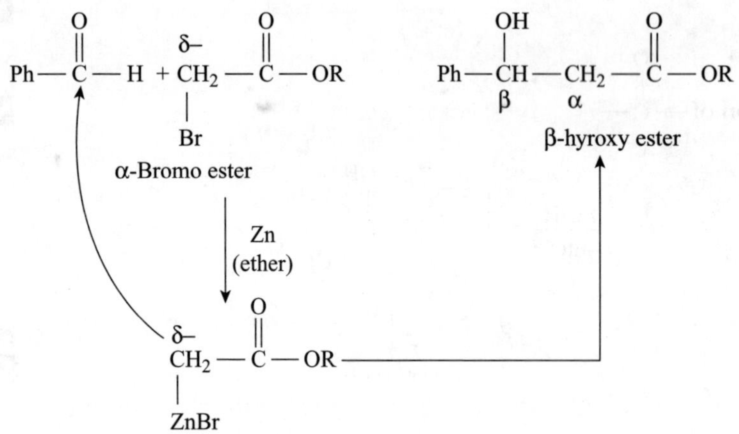

Ans: (b)

29.

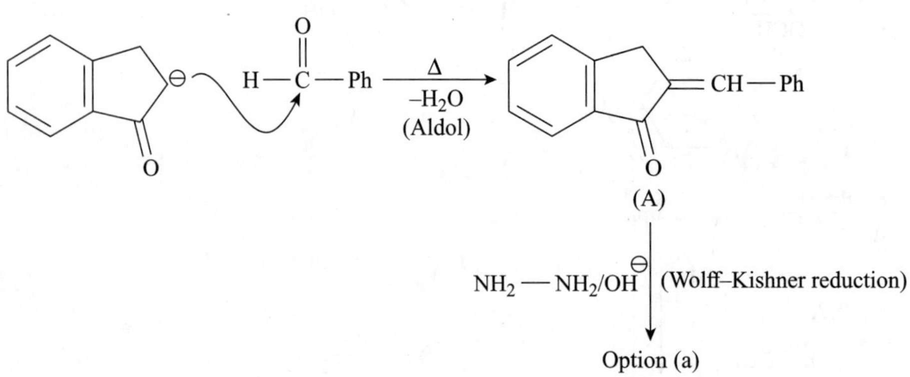

Ans: (a)

30.

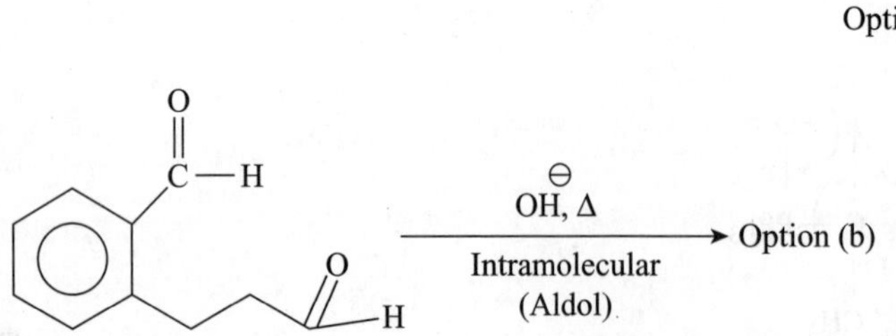

(A)

Ans: (b)

31. $NaHSO_3$ gives white ppt only with aldehyde and methyl ketone. **Ans:** (a) and (b)

33. (a)

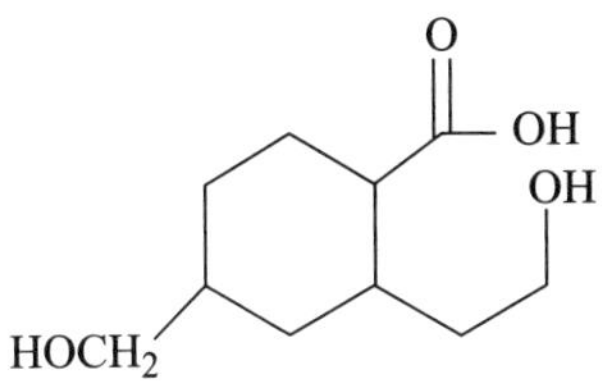

(b)

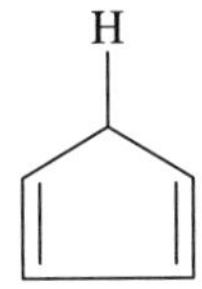

(c)

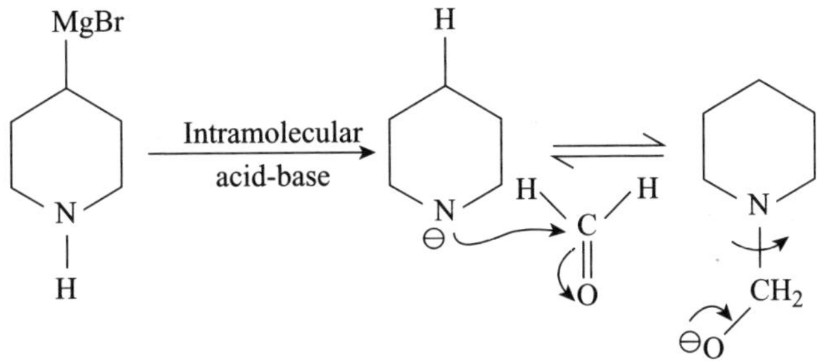

Ans: (d)

34.

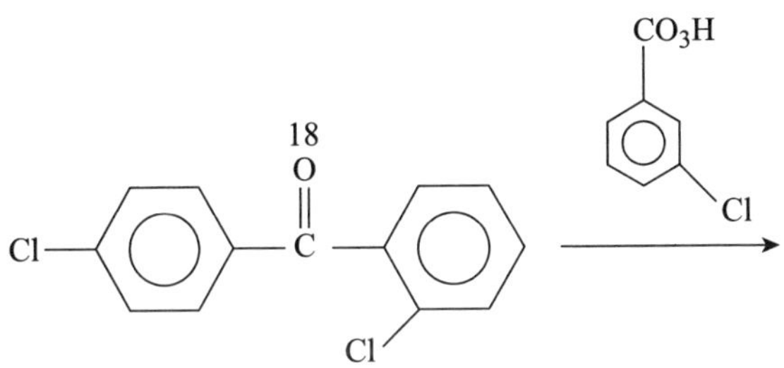

Option (d) (steric factor also play Jmp role).

$-I_{less}$ $\quad$ $-I_{max}$

Better migratory group

Ans: (d)

35.

$$CH_3-\overset{O}{\overset{\|}{C}}-CH_2-\overset{O}{\overset{\|}{C}}-OPh \xrightarrow{NaOH} CH_3-\overset{O}{\overset{\|}{C}}-CH_2-\overset{O}{\overset{\|}{C}}-OH$$

Gives Iodoform test

$$CH_3-\underset{Br}{\overset{Br}{C}}-CH_3 \xrightarrow{NaOH} CH_3-\underset{OH}{\overset{OH}{C}}-CH_3 \longrightarrow CH_3-\overset{O}{\overset{\|}{C}}-CH_3$$

Gives Iodoform test

Ans: (c)

36. Rate of reduction $\alpha \dfrac{1}{\text{steric factor}}$: {aldehyde > ketone} α, δ+ charge on carbon of $-\overset{O}{\overset{\|}{C}}-$

Ans: (a)

37.

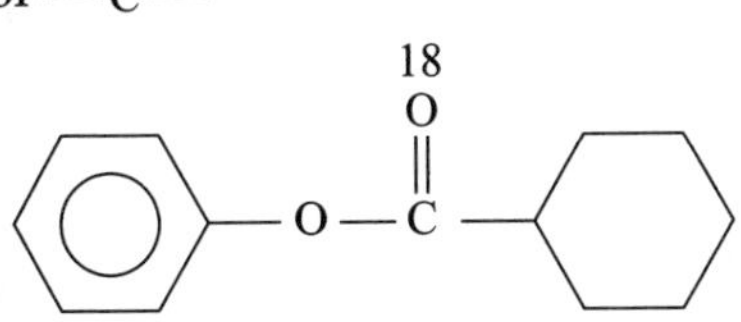

Ans: (d)

38. 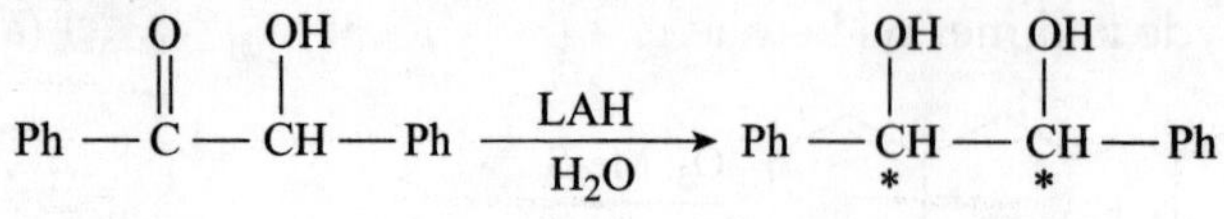

(A)

(after benzoin condensation)

Ans: (c)

39. 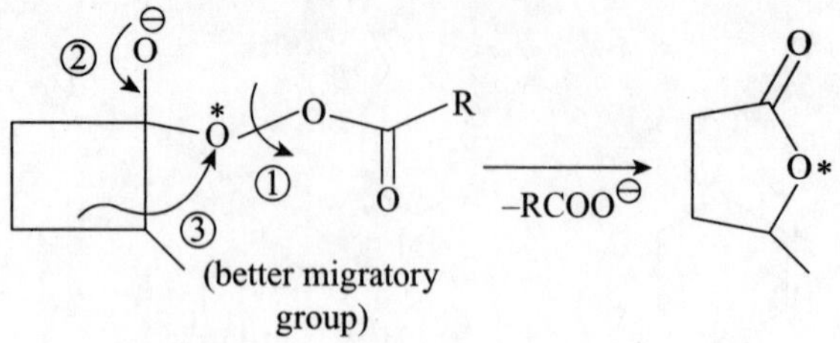

Ans: (b)

41. 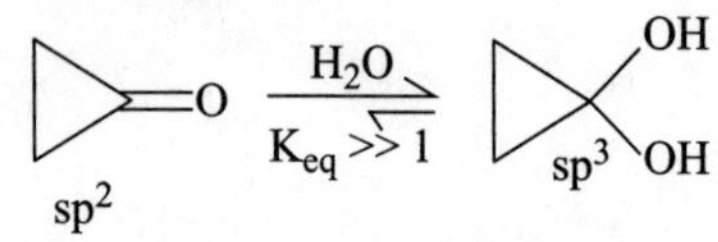

Highly unstable due to angle strain

Ans: (a)

43. 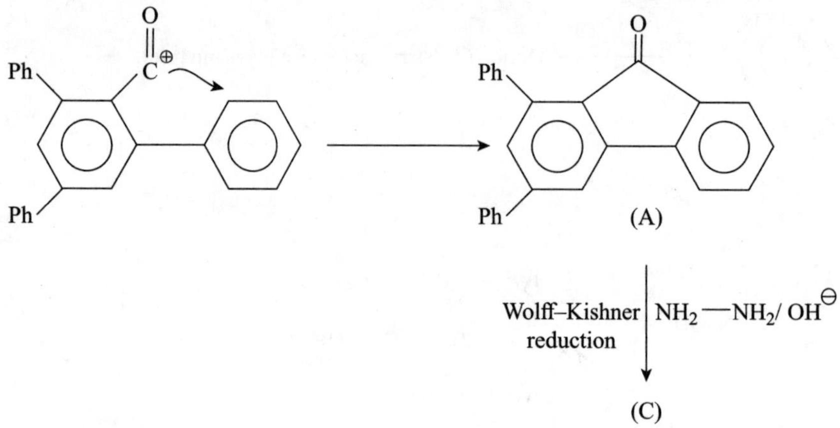

Ans: (c)

51. ROR for nucleophilic addition $\alpha \dfrac{1}{\text{Steric hindrance}}$

α δ+ charge on carbon

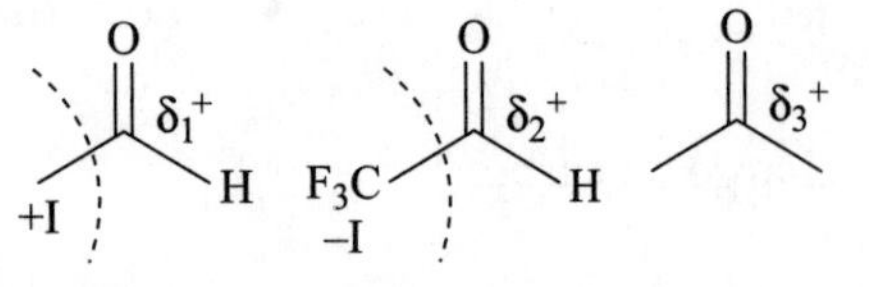

Rate: i, ii > iii (Steric factor)

ii > i > iii (as δ_2+ > δ_1+)

Ans: (d)

52. Hydrate fomation takes place via nucleophilic addition on carbonyl group.

δ_3+ > δ_2+ > δ_1+ > δ_4+ {d+: charge on C of carbonyl group}

Ans: (c)

53.

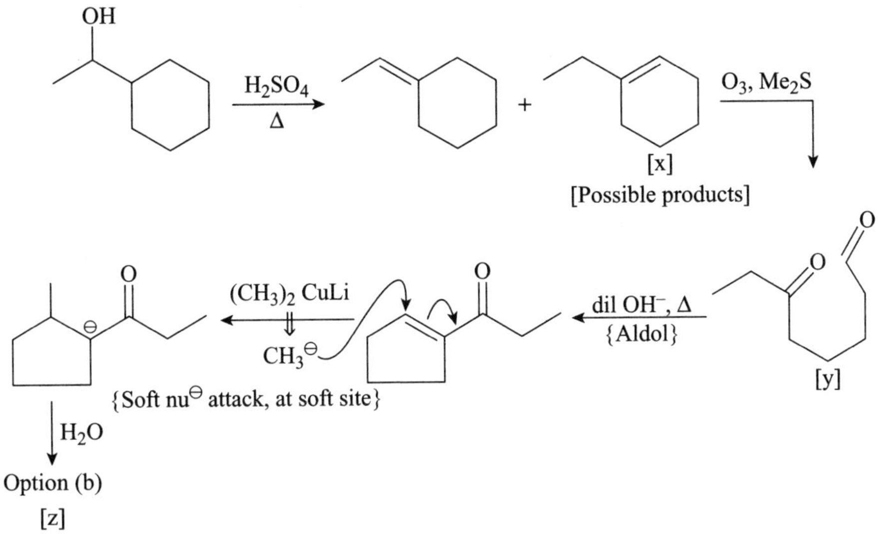

Ans: (b)

54. $P_3 : CHI_3$

P_2 :

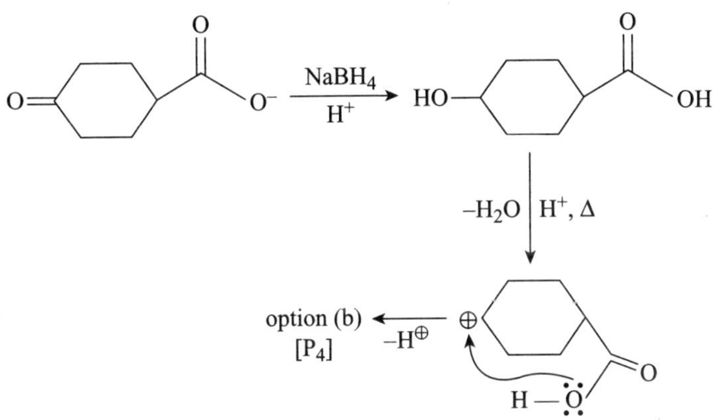

P_1 :

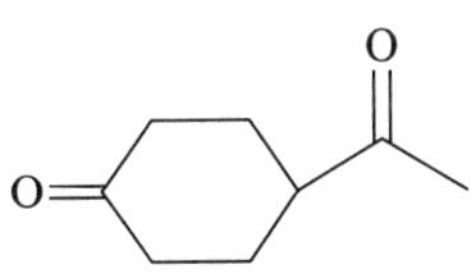

Ans: (b)

55.

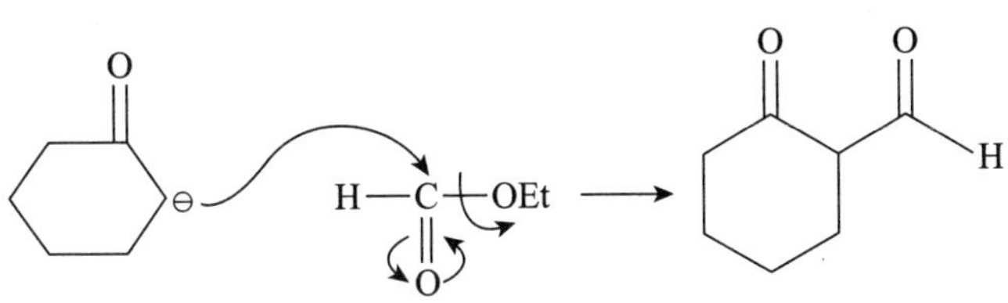

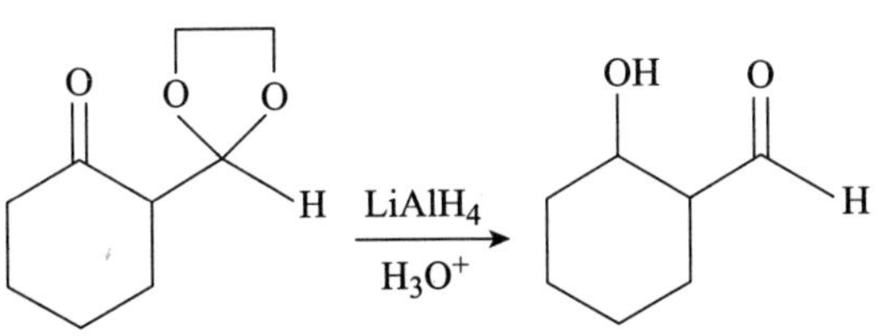

Ans: (c)

56. A: 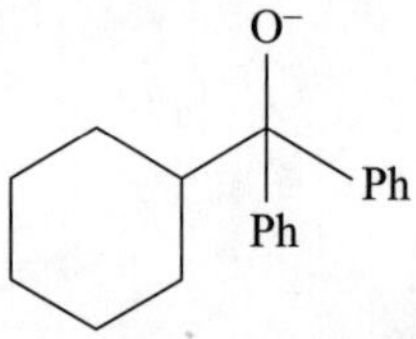 B: 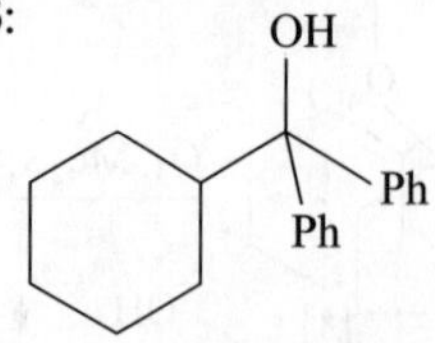

Ans: (a)

57.

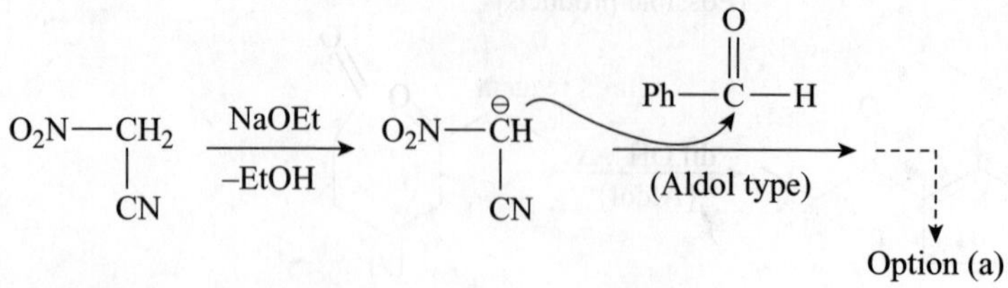

Ans: (a)

59.

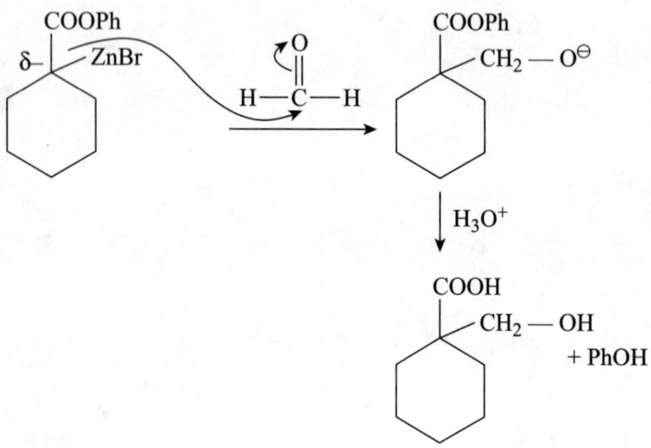

Ans: (c)

60. (b) 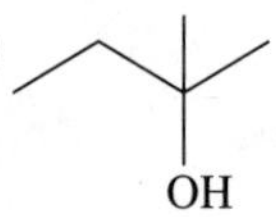 (c) 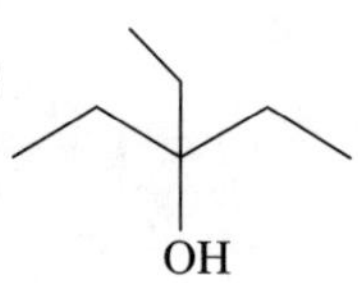

Ans: (b) and (c)

62. A: CHI_3 B: $CH \equiv CH$ C: $CH_3-\overset{O}{\overset{\|}{C}}-H$ **Ans:** (b)

63. Cannizzaro reaction **Ans:** (c)

64. Approach:

$$\underset{\text{(ketone)}}{\underset{[D]}{C_3H_6O}} \xrightarrow[Br_2]{NaOH} C_2H_4O_2^-$$

D: $CH_3-\overset{O}{\overset{\|}{C}}-CH_3$

C: $CH_3-C\equiv CH$ ← Elimination reaction

A: $CH_3CH_2CHCl_2$

B: $CH_3CH_2CHO \xrightarrow{(O)} CH_3CH_2COOH$

Ans: (a)

65. P_1 : 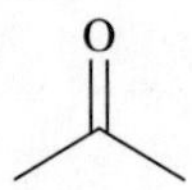

Ans: (c)

67.

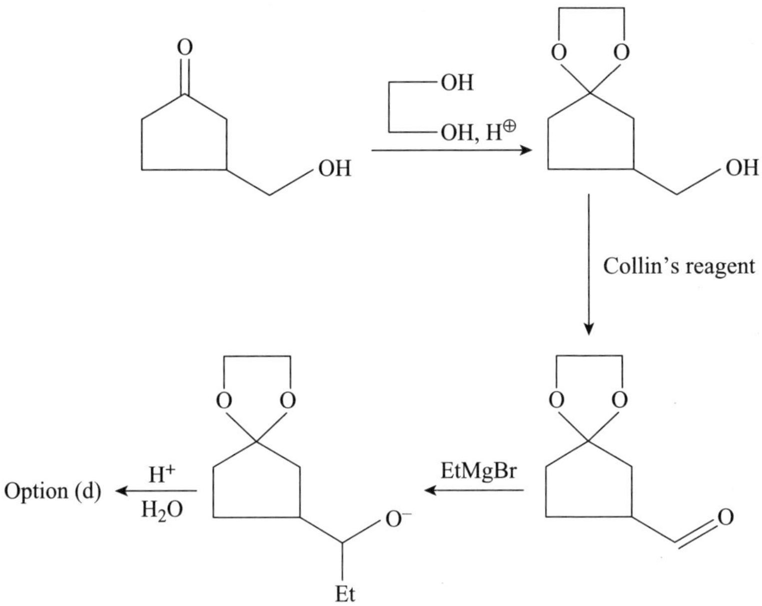

Ans: (d)

69.

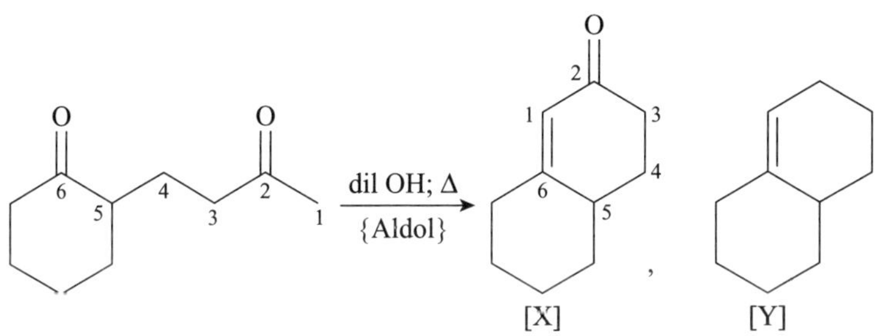

Ans: (c)

70.

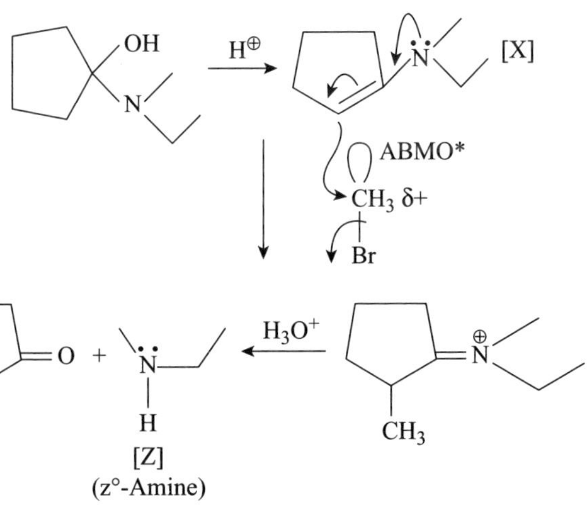

Ans: (b)

73. A:

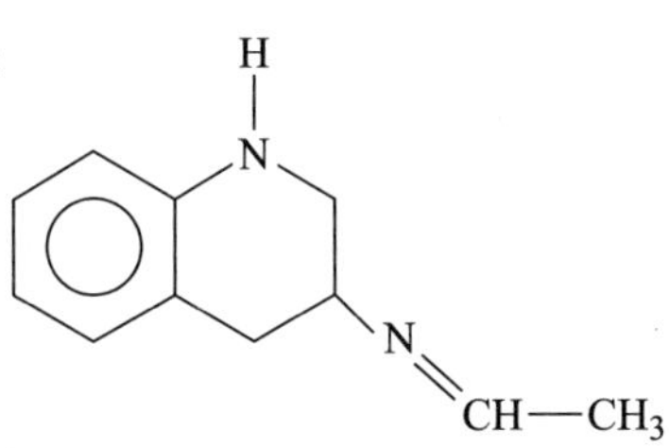

Ans: (c)

74. A: In acidic $K_2Cr_2O_7$ (non-oxidisable) and gives alkene with H_2SO_4, Δ

$\Downarrow$

means A is 3° Alcohol

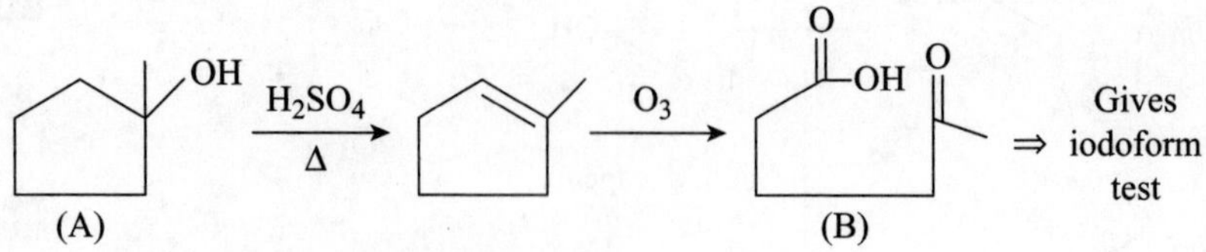

Ans: (b)

75.

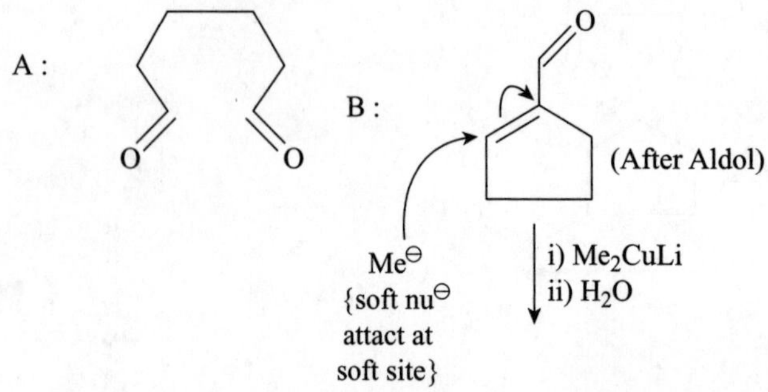

Ans: (d)

76.

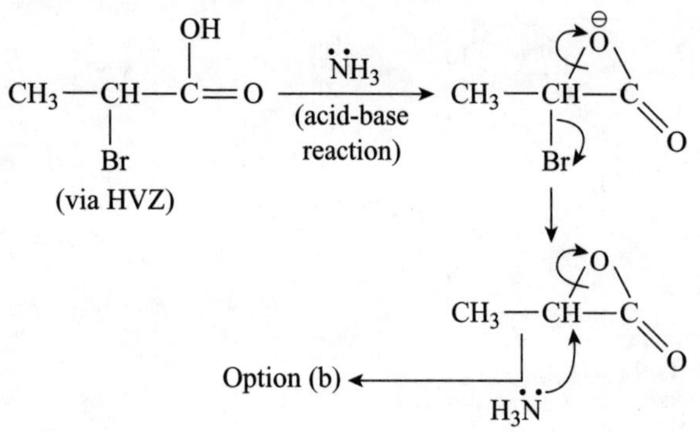

Ans: (b)

78.

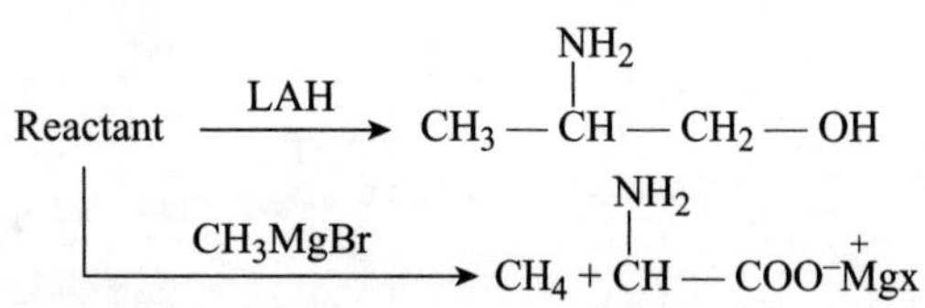

Ans: (b)

79. Reactivity α $\delta+$ charge on carbon of $-\overset{\overset{O}{\|}}{C}-$ α L.G. tendency

Ans: (b)

84. and 92. B.P. α dipole–dipole interaction (H-bonding)
(as all compound have comparable molecular weight)

Ans: (b)

91.

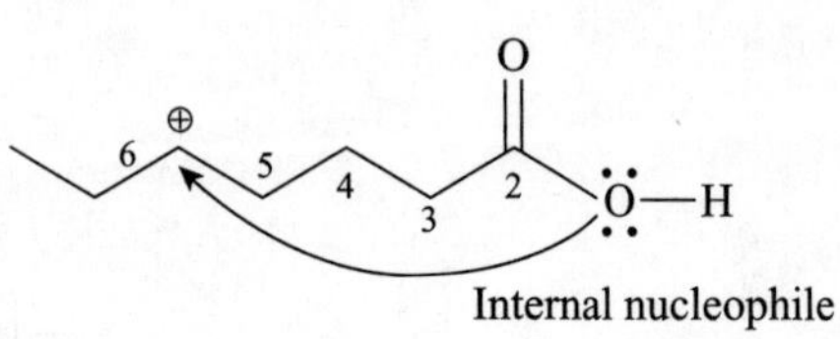

Ans: (d)

93.

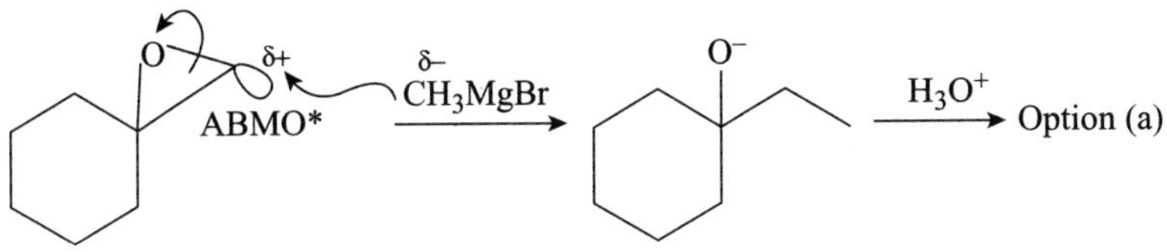

Ans: (a)

94.

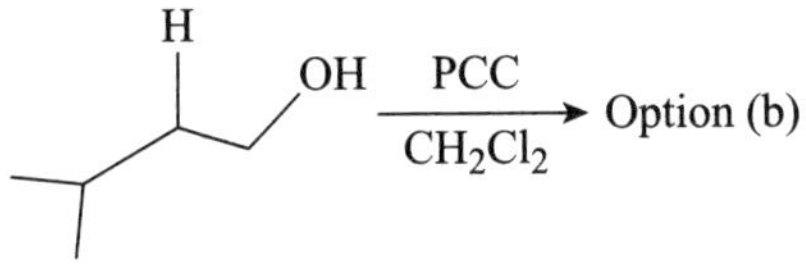

Ans: (b)

95.

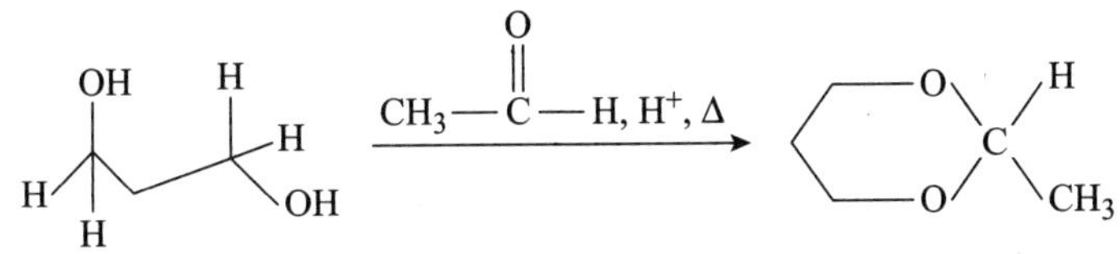

Ans: (a)

96.

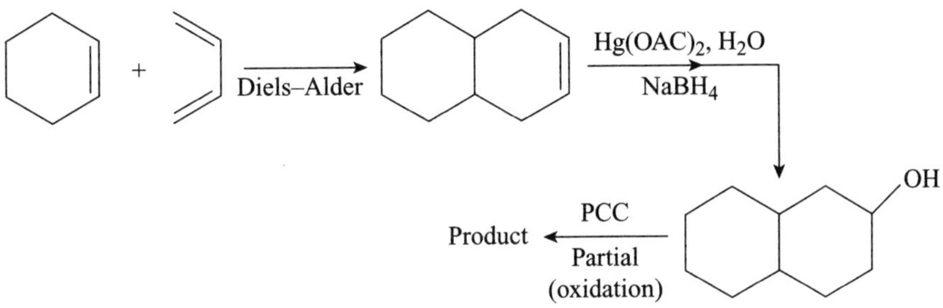

Ans: (a)

97.

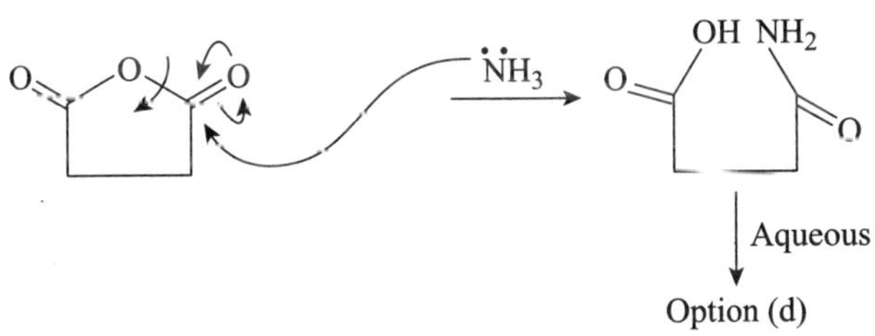

Ans: (d)

99.

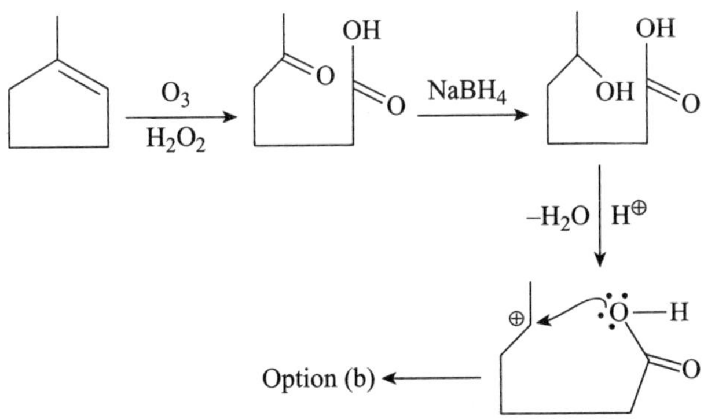

Ans: (b)

100.

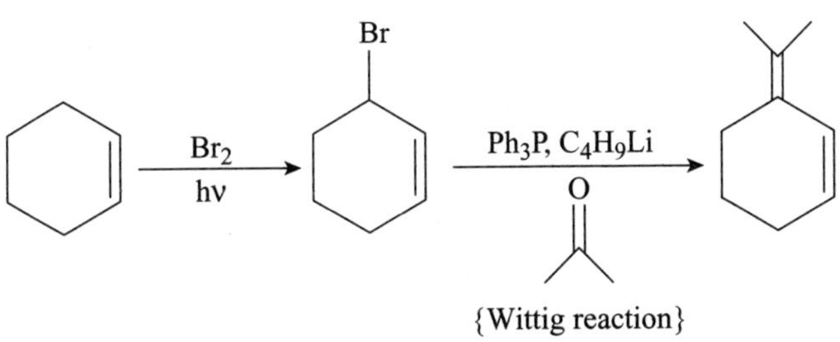

Ans: (a)

101.

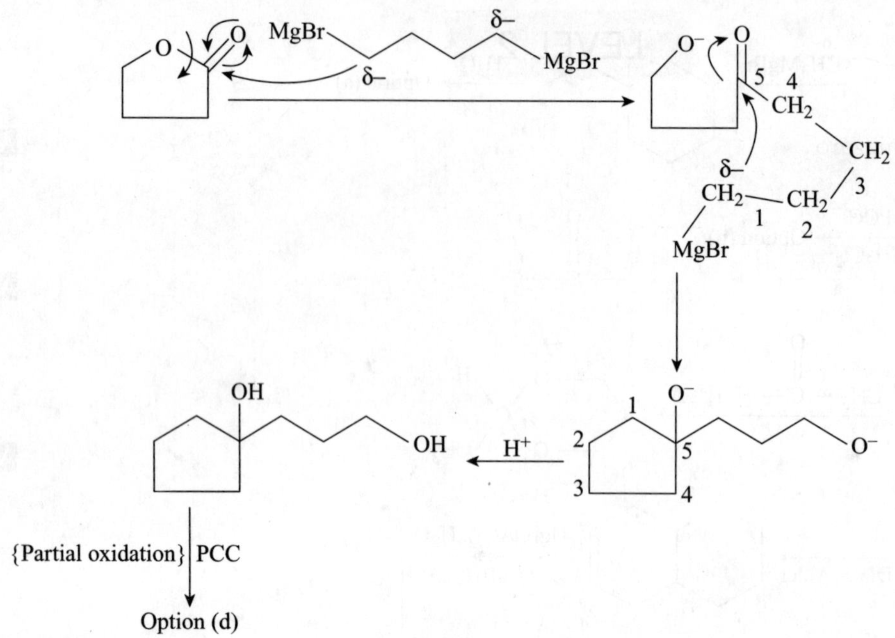

Ans: (d)

102. 1: Birch reduction.

2: Electrophilic addition reaction.

Ans: (c)

103.

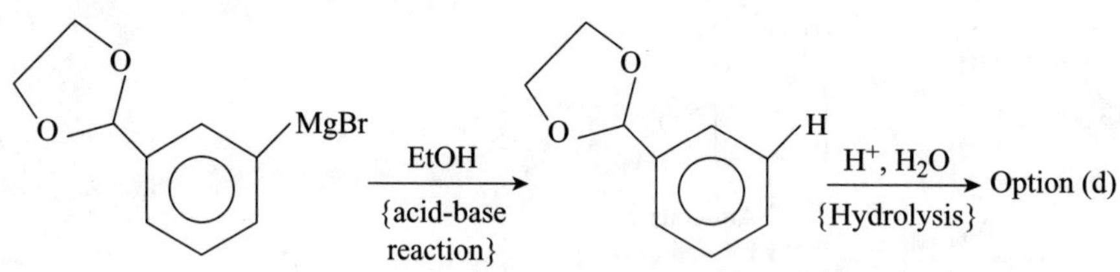

Ans: (d)

104.

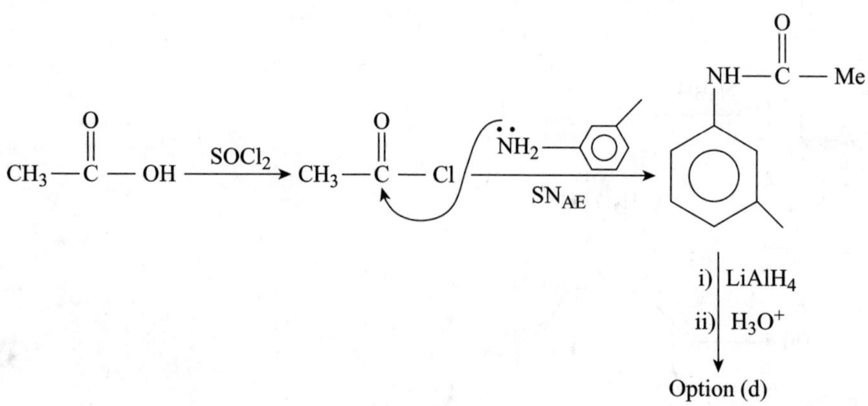

Ans: (d)

LEVEL 2

1. Intramolecular cannizzaro

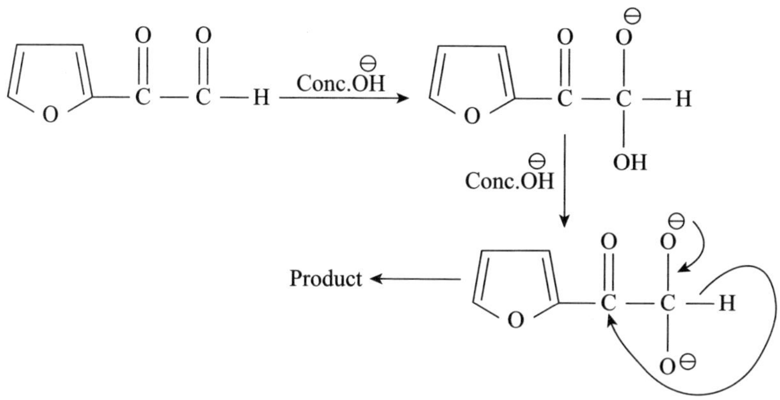

Ans: (c)

2. A:

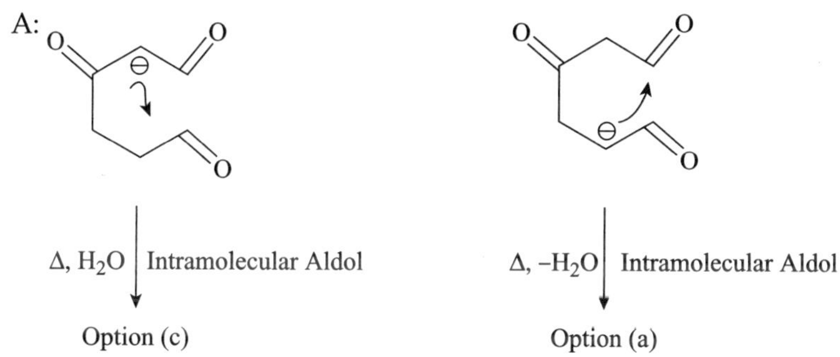

Ans: (a) and (c)

3.

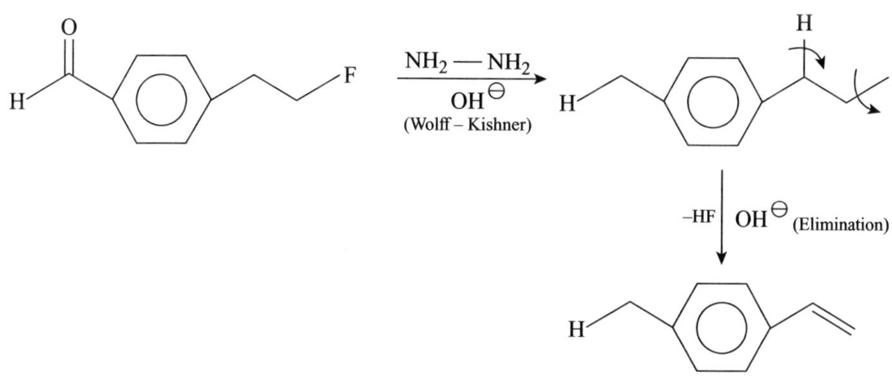

Ans: (b) and (c)

4. Wittig reaction

Ans: (a) and (b)

5. Glucose/Fructose/Mannose

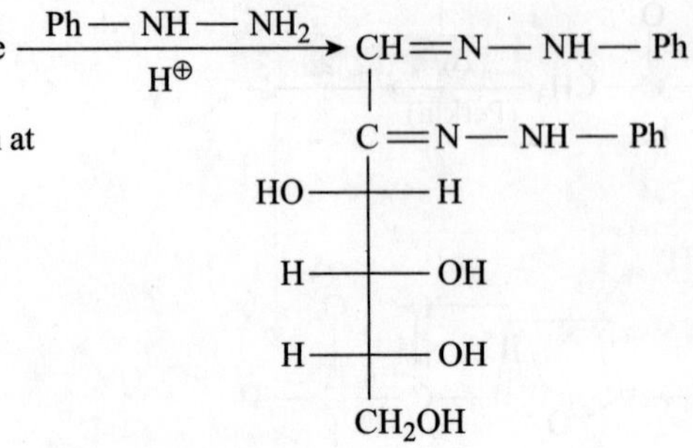

(all have same configuration at C-3, C-4, C-5)

Osazone

Ans: (a), (b) and (c)

7.

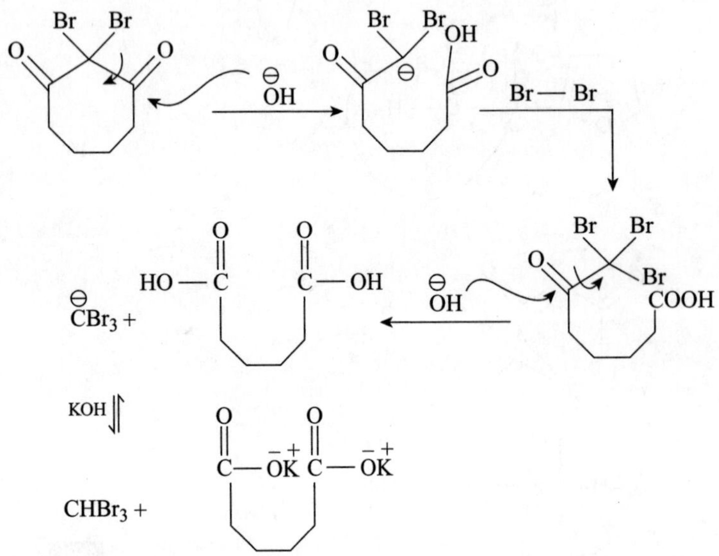

(A) (B)

Ans: (a) and (b)

8. Favorskii reaction

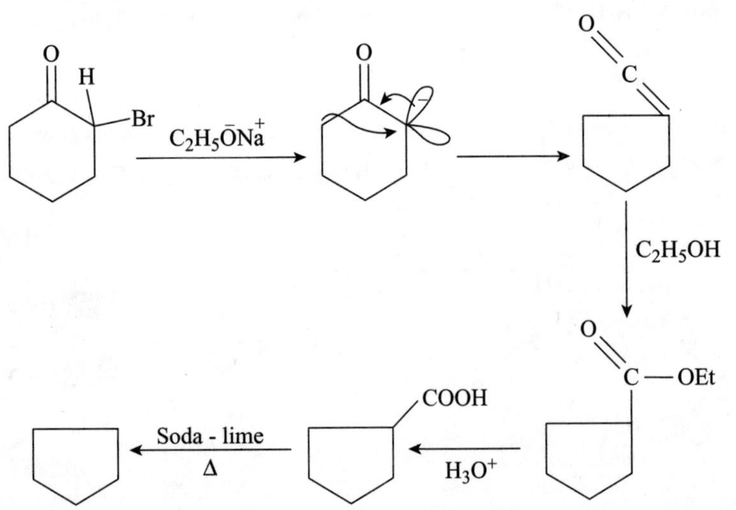

(B) (A)

Ans: (b) and (d)

9. 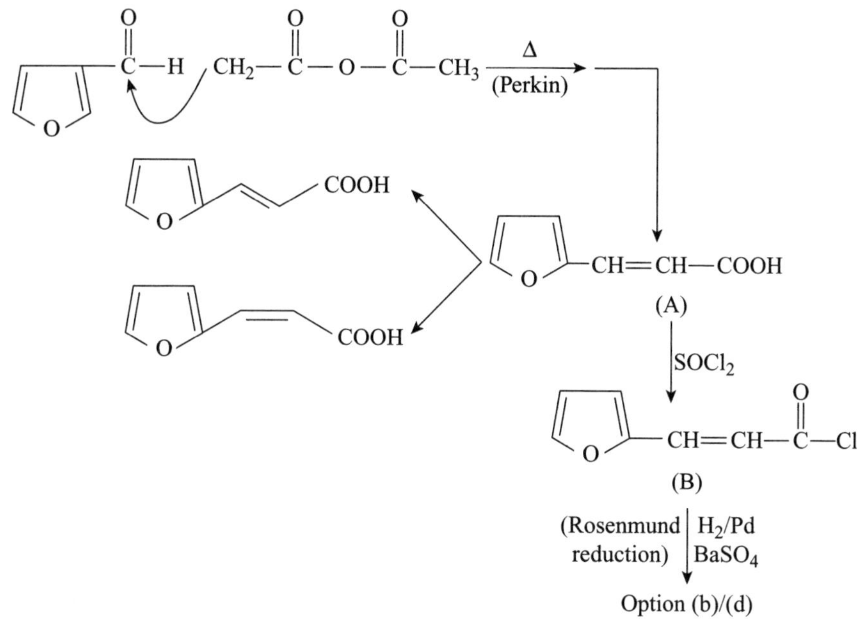

Ans: (b) and (d)

10. A: 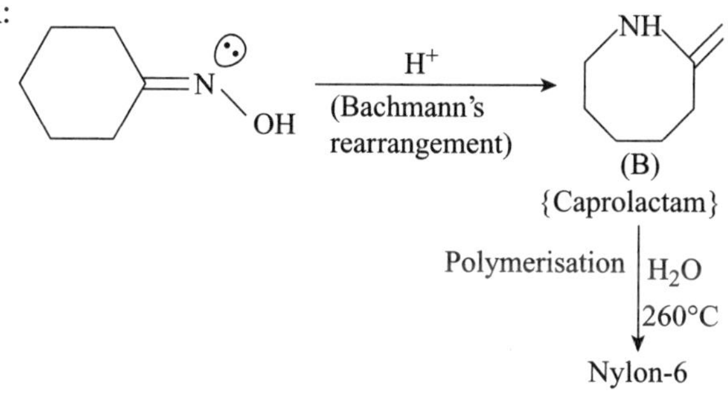

Ans: (b), (c) and (d)

11. 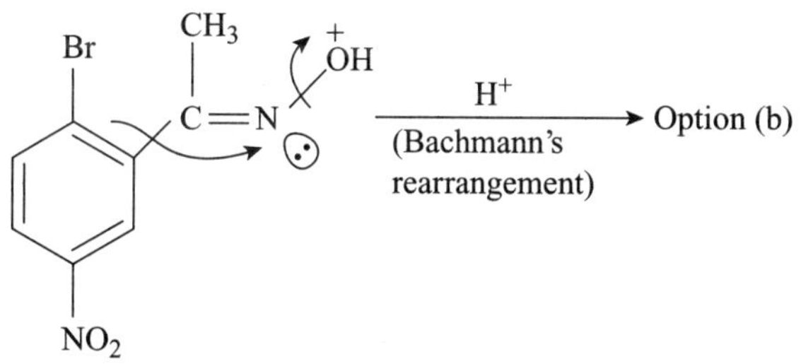

Ans: (b)

12. (a) Claisen Condensation 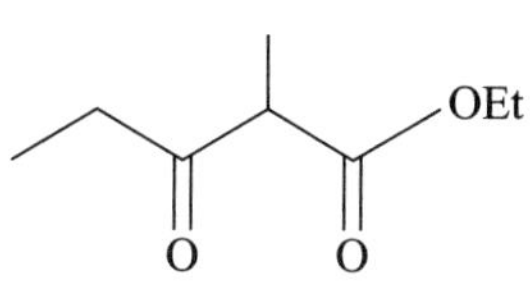

(b) Aldol Condensation

(c)

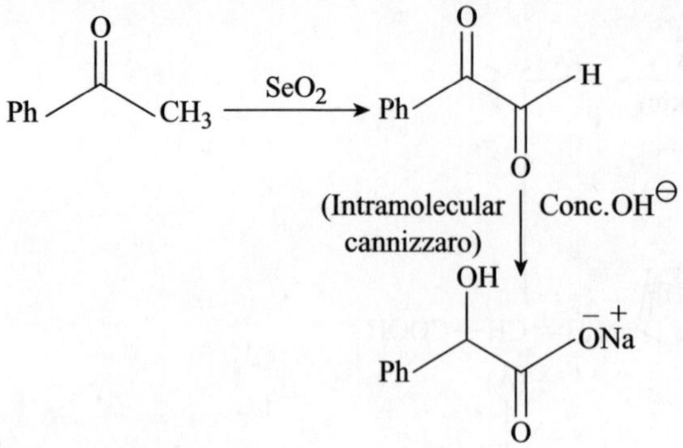

(d) Perkin reaction **Ans:** (b), (c) and (d)

15. P_1: $2H-\overset{\overset{O}{\|}}{C}-H$

P_2:

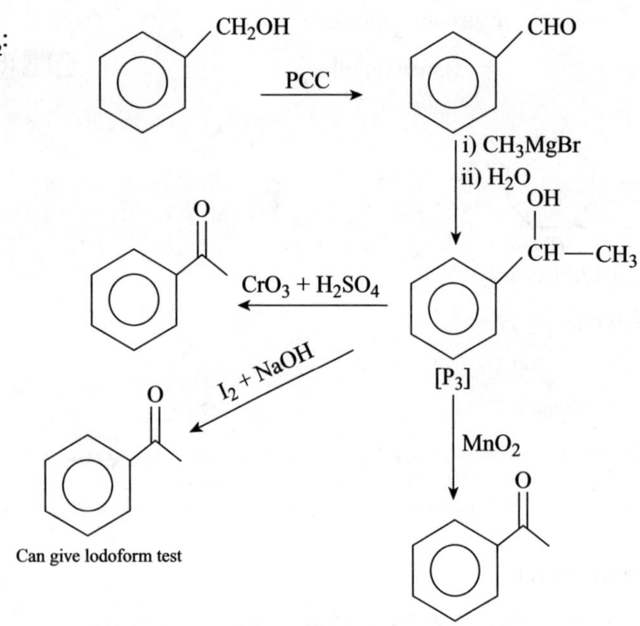

Ans: (a), (b) and (c)

16. (c) $Me-\overset{\overset{OH}{|}}{C}=NH \rightleftharpoons Me-\overset{\overset{O}{\|}}{C}-NH_2$ **Ans:** (a), (b) and (d)

18. (a) AgCl : White ppt

(c) $Me-C\equiv C^{-}Ag^{+}$: White ppt

(b) and (d) itself doesn't give any ppt, rather they give silver mirror ppt of Ag. **Ans:** (a) and (c)

19. $P_1 : H-C\equiv C^{-}Na^{+}$

P_2: 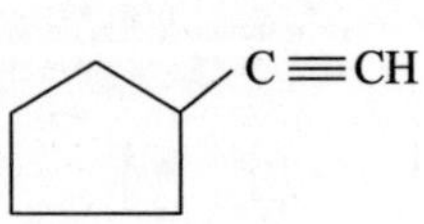 P_3: 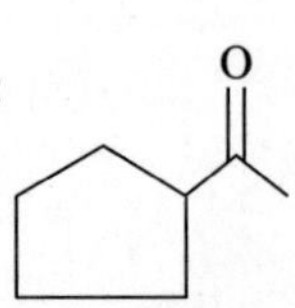

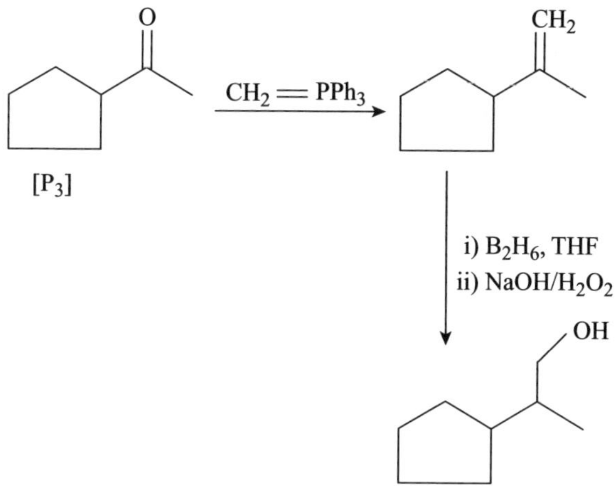

Ans: (a), (b) and (c)

20. Path-1:

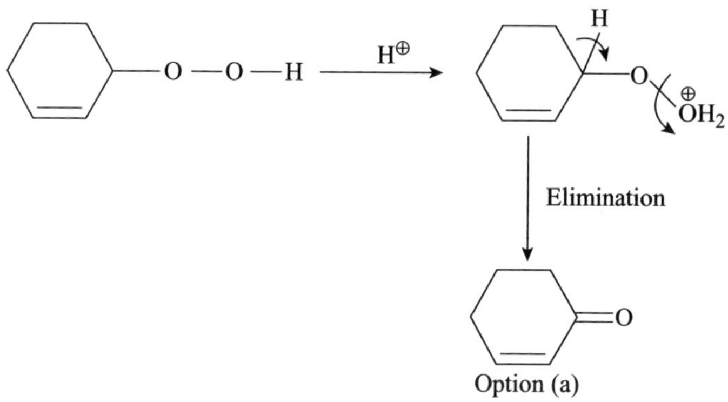

Path-2:

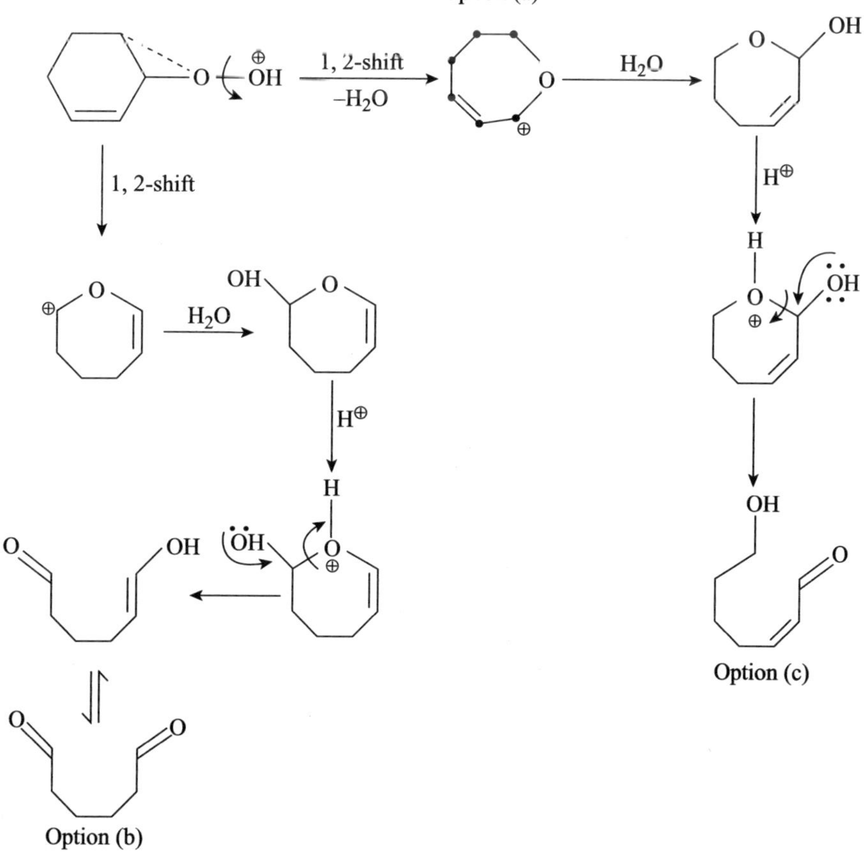

Ans: (a), (b) and (c)

21.

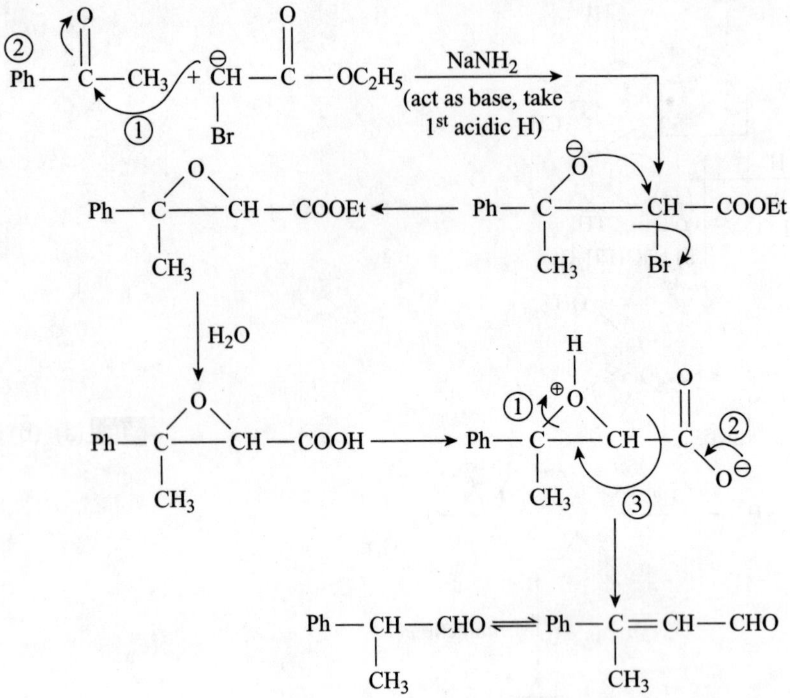

Ans: (b) and (d)

22.

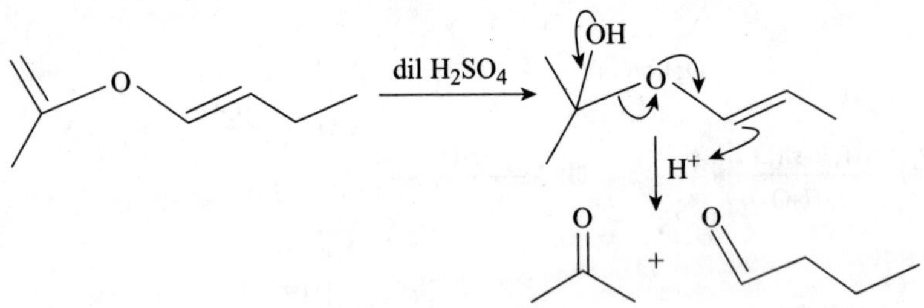

Ans: (a) and (b)

23.

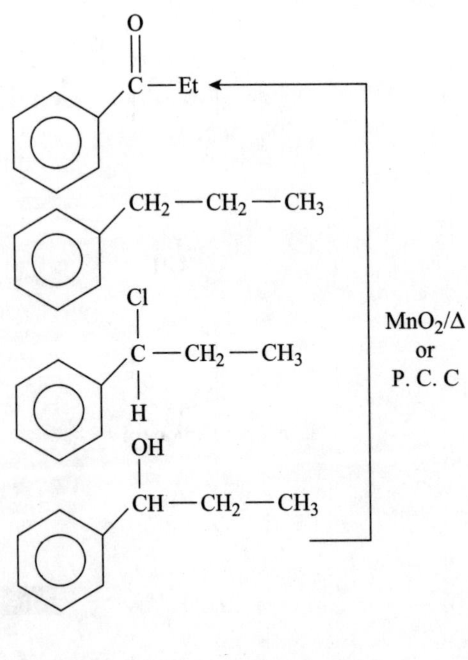

Ans: (a) and (b)

24. (a) 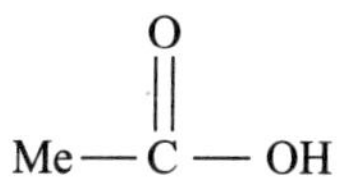

(b) 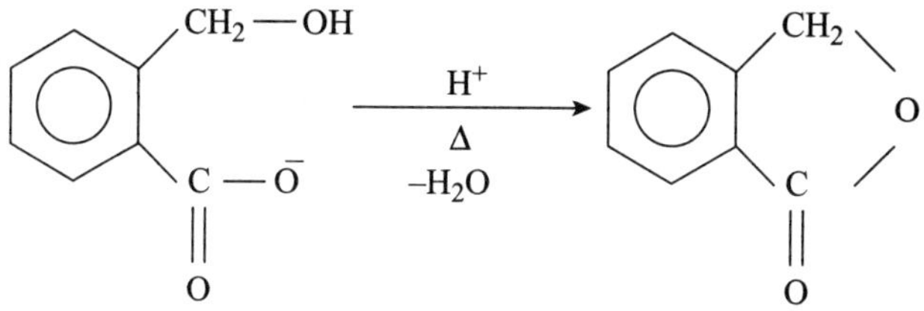

Lactone
(Cyclic ester)

(c) 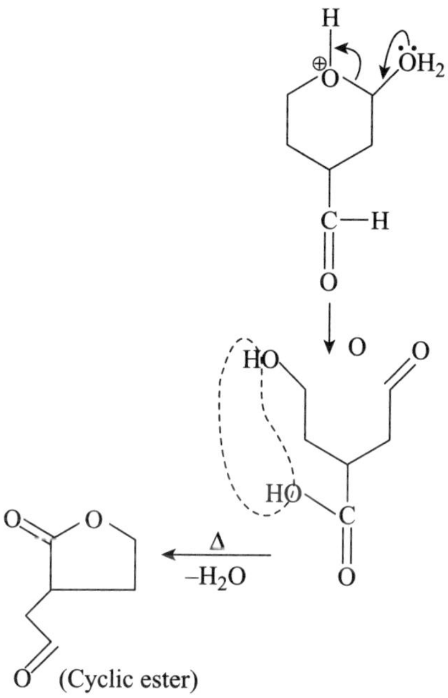

(Cyclic ester)

(d) 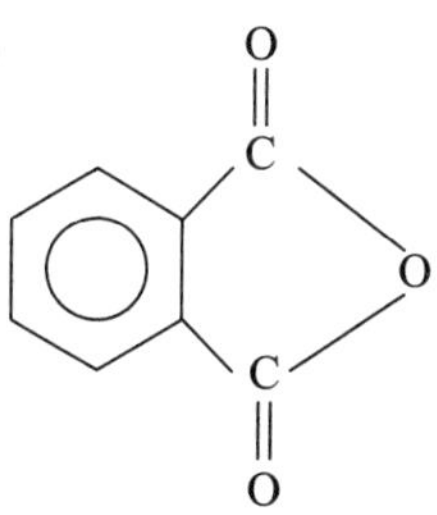

Anhydride

Ans: (b) and (c)

25. (a) 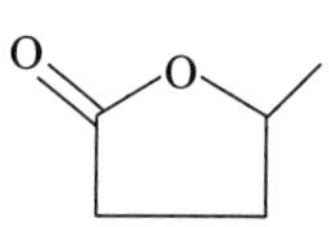 (b) 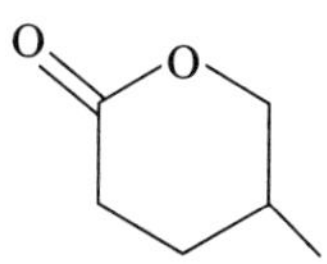

(c) 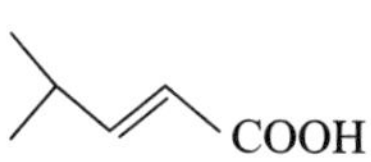 (d) 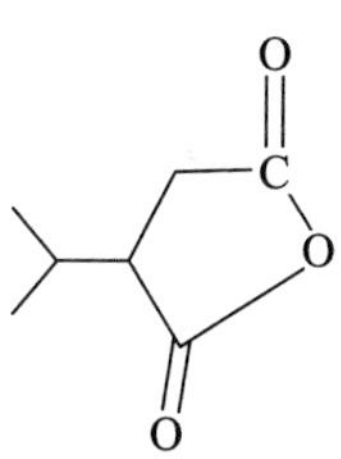

Ans: (a) and (b)

26.

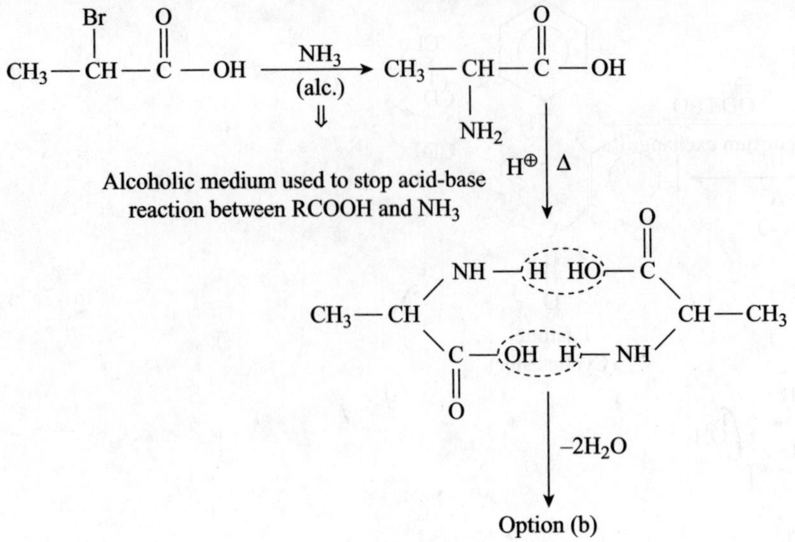

Ans: (b) and (d)

28.

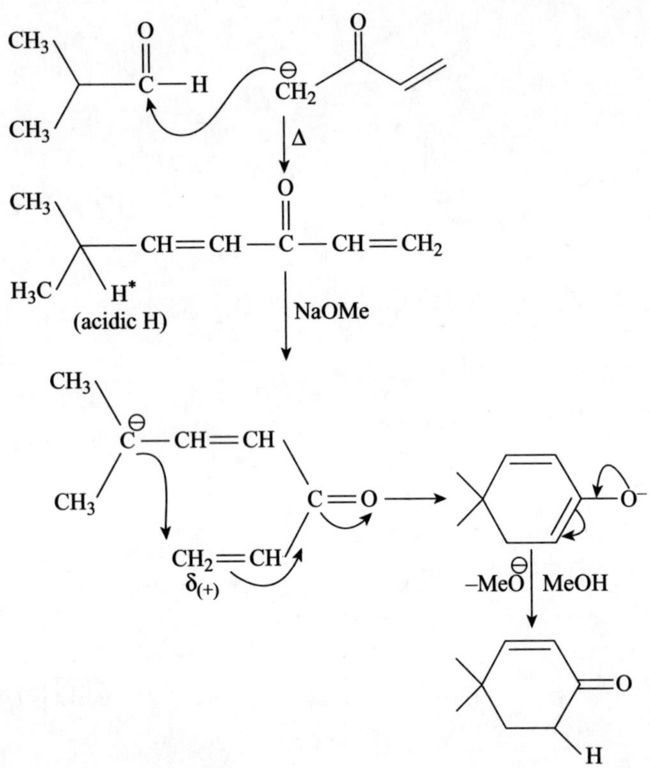

Ans: (c)

29.

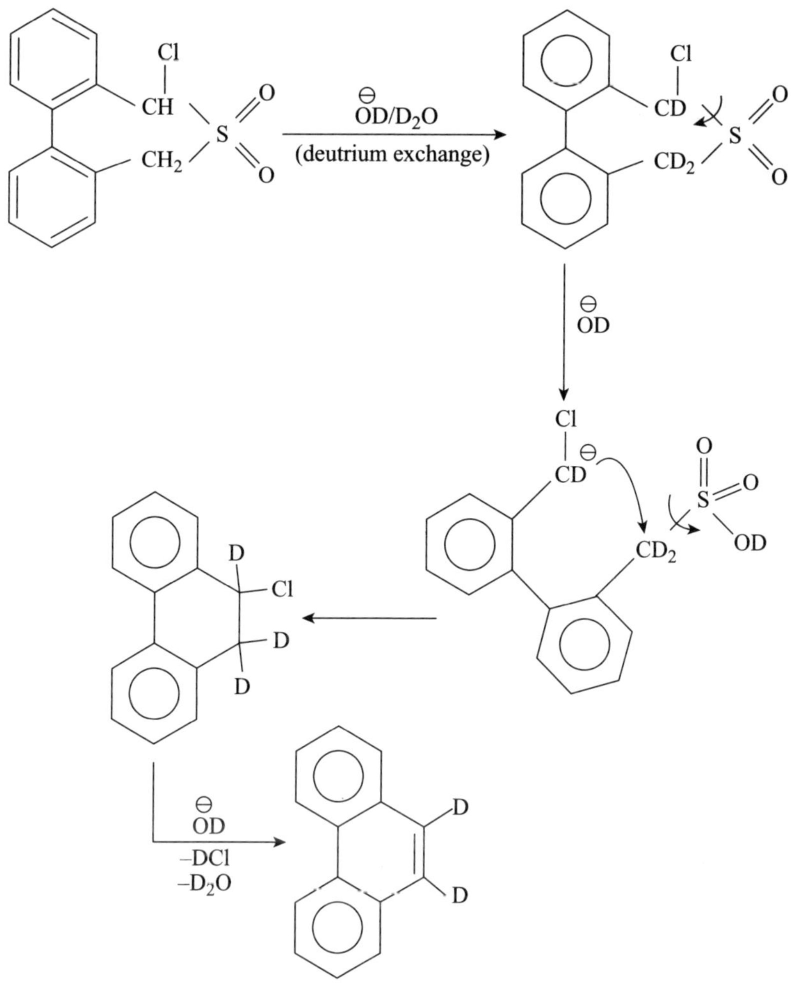

Ans: (d)

30.

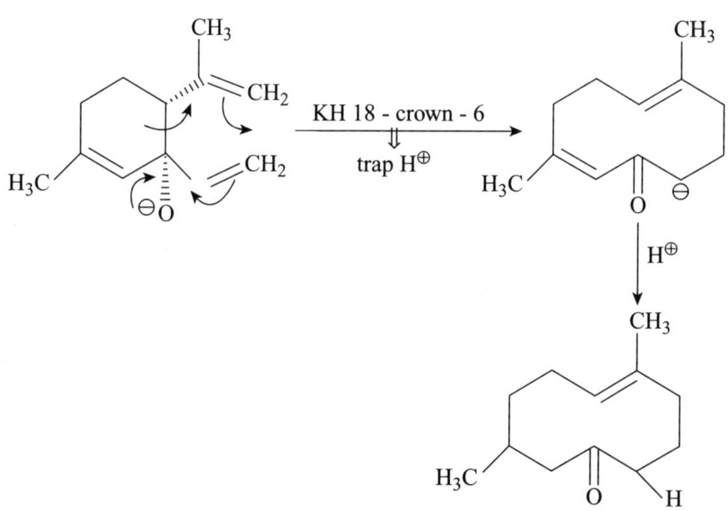

Ans: (b)

31.

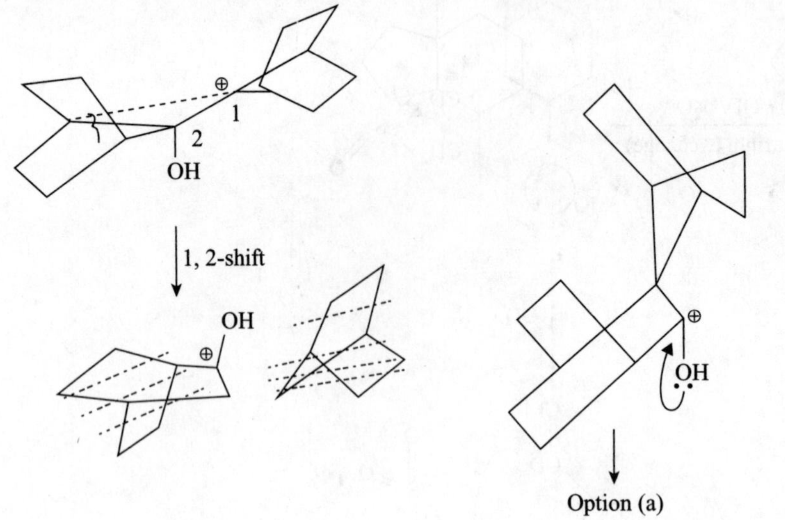

Ans: (a)

32.

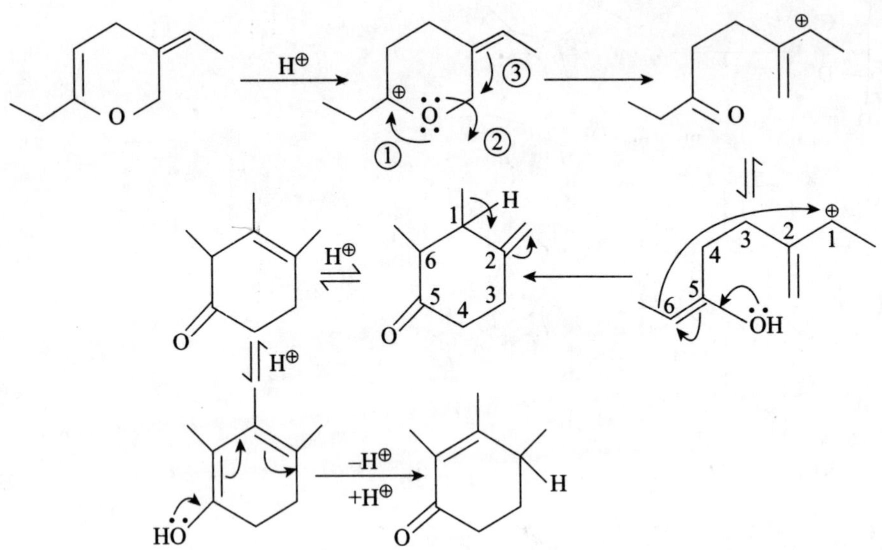

Ans: (a)

33. $3\ CH_3-CH_2-\overset{\overset{O}{||}}{C}-CH_3 \xrightarrow[\text{(Aldol in acidic condition)}]{\text{Conc. } H_2SO_4}$

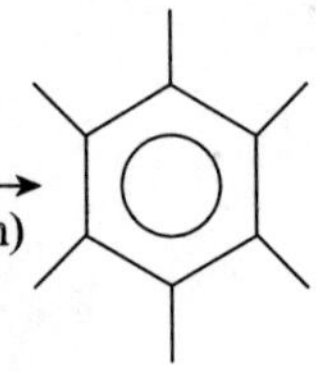

Ans: (b)

34.

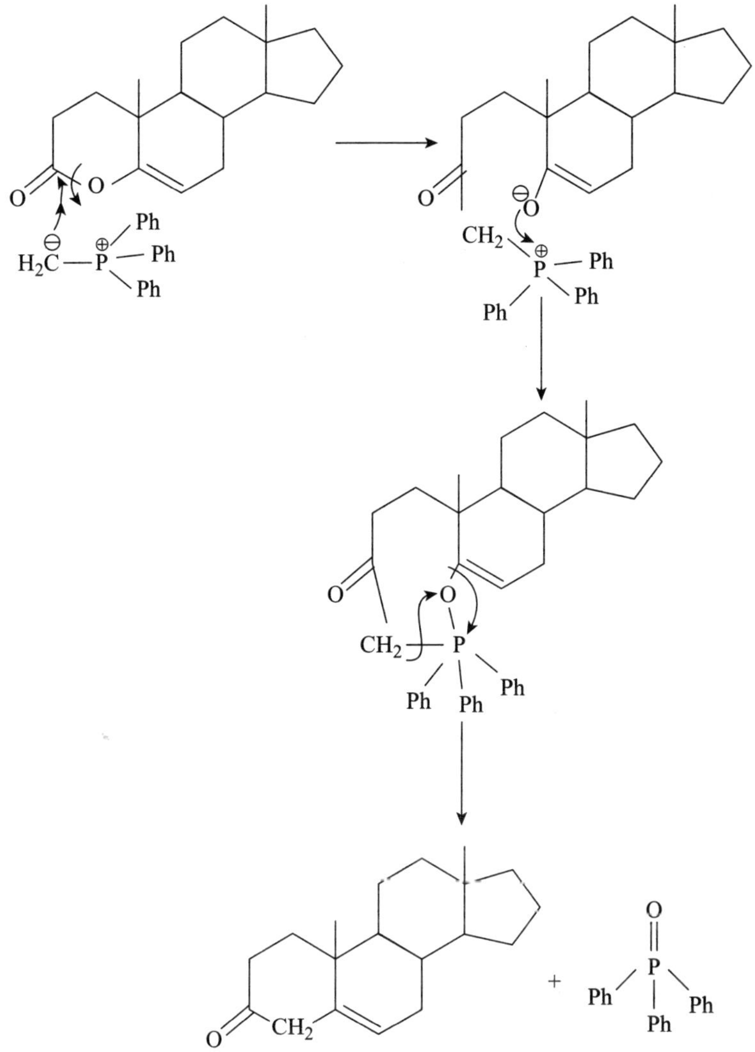

Ans: (b)

35.

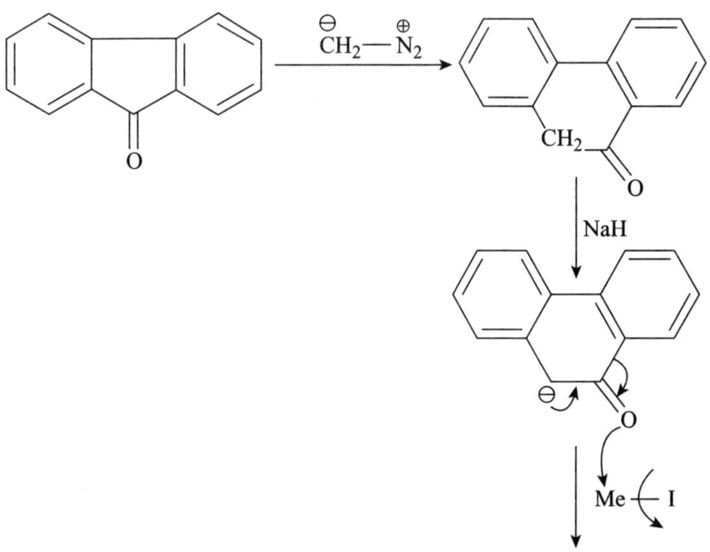

Option (b)
{Aromatic compound}

Ans: (b)

36.

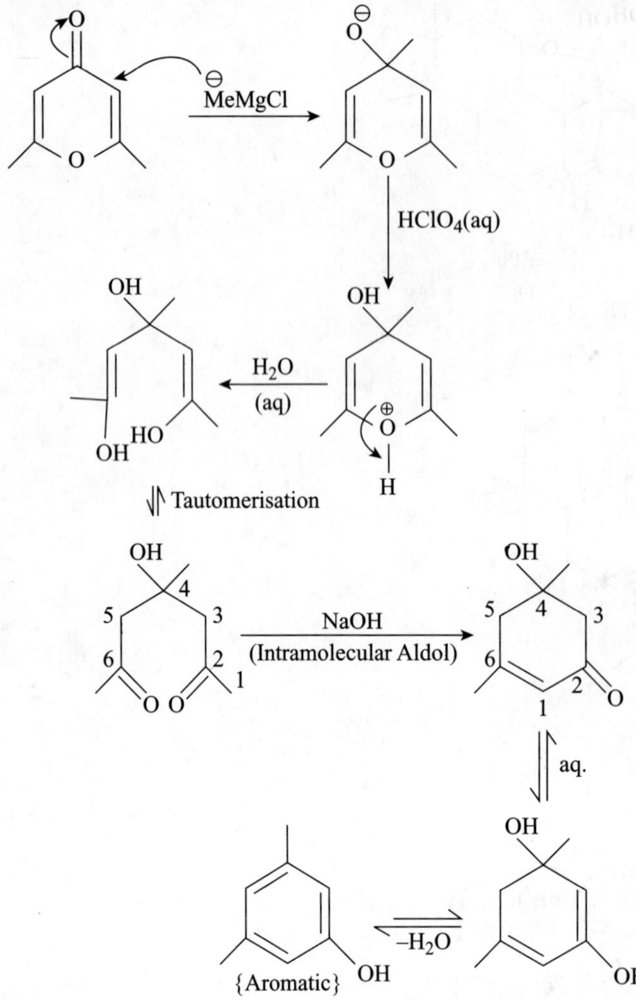

Stable compound form, therefore no need of temperature for elimination of H_2O. **Ans:** (d)

37.

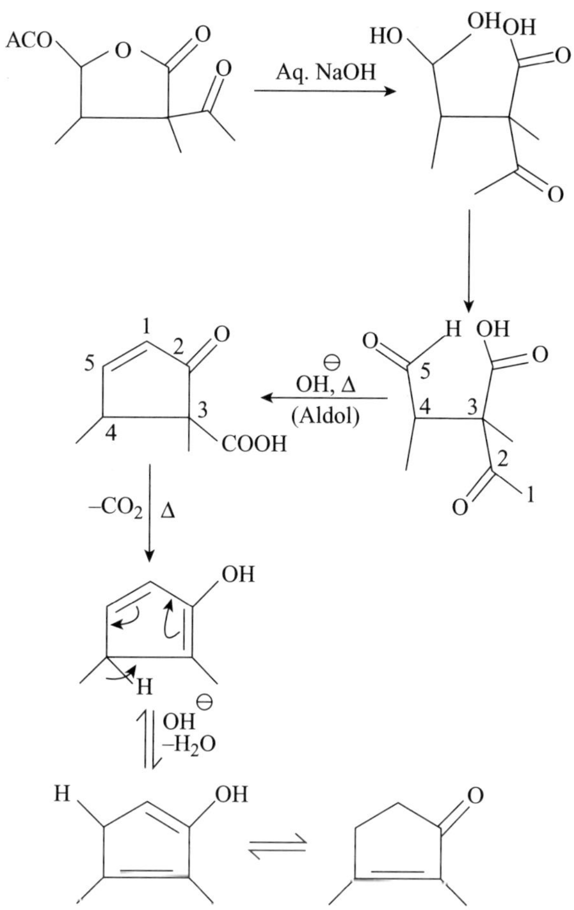

Ans: (a)

38.

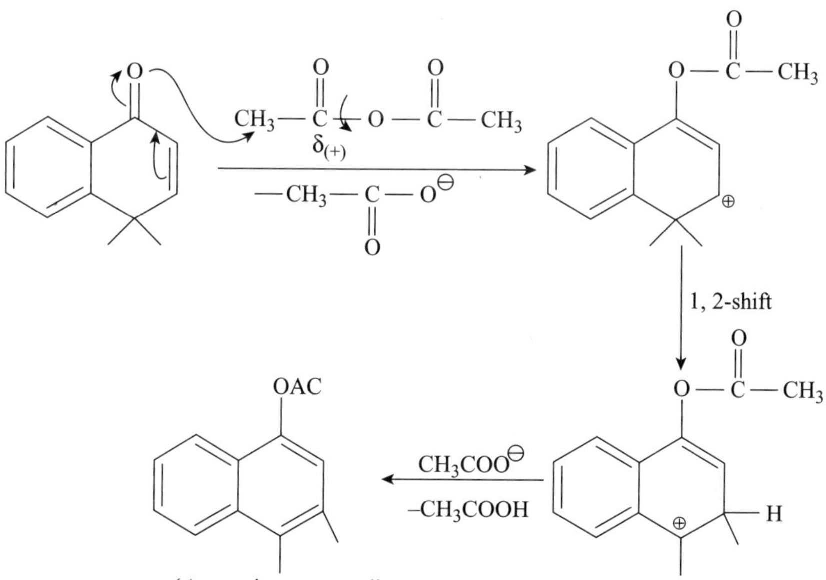

Ans: (c)

39.

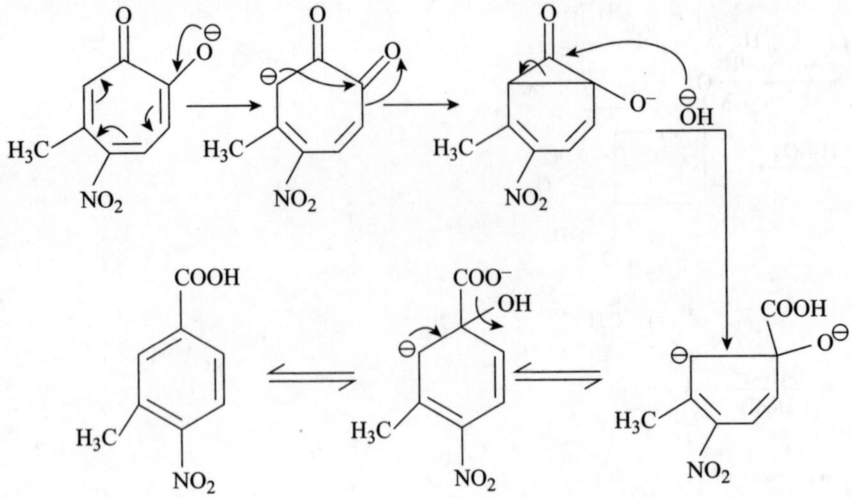

Ans: (b)

40.

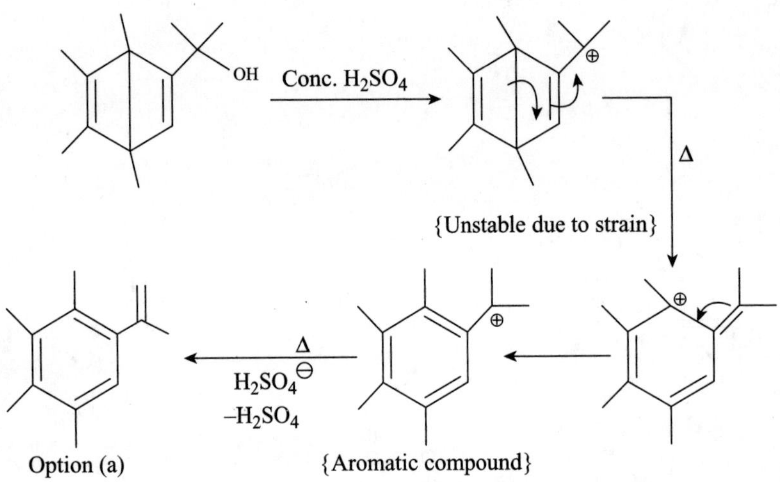

Ans: (a)

41.

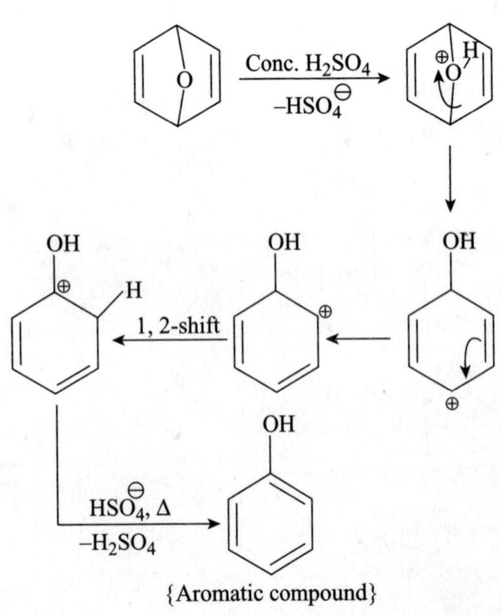

Ans: (a)

42. Path-1:

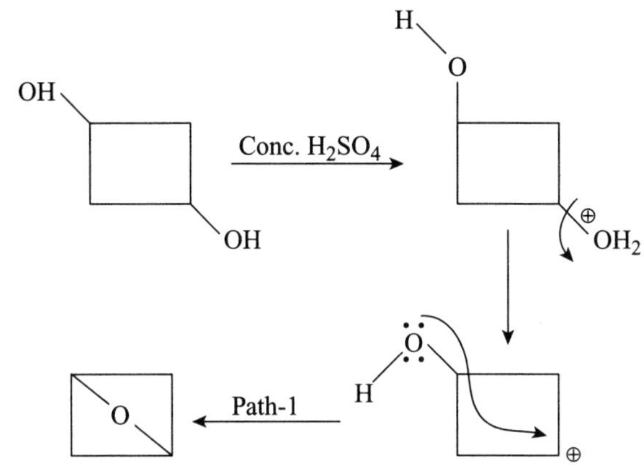

Minor
(Unstable due to strain)

Path-2:

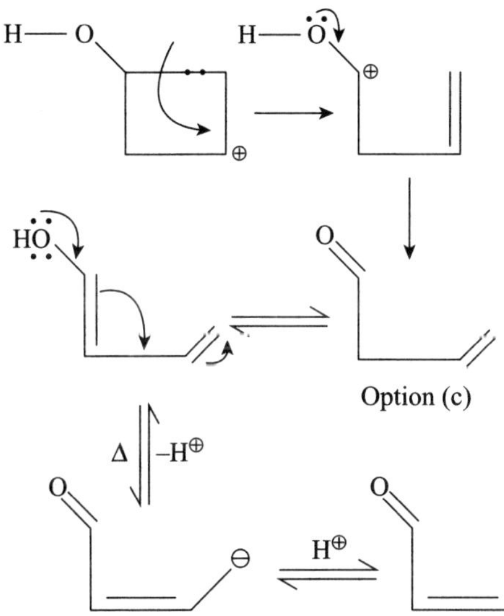

Major product
{Stable alkene having more αH}

Note: At high temperature TCP forms. **Ans:** (b)

43.

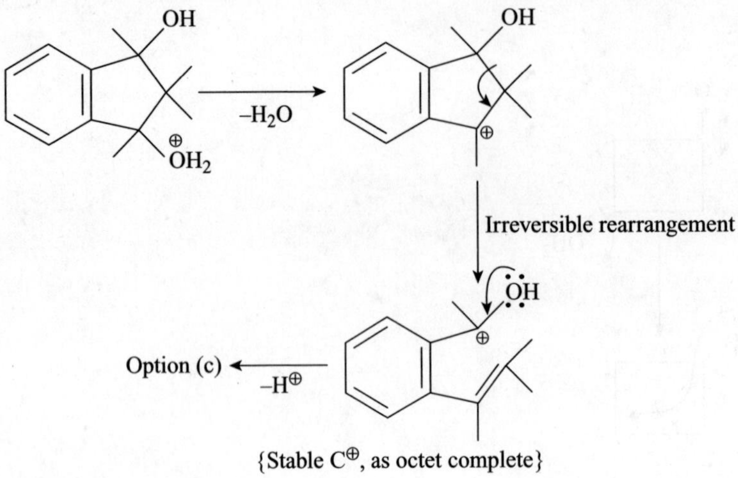

Ans: (c)

44.

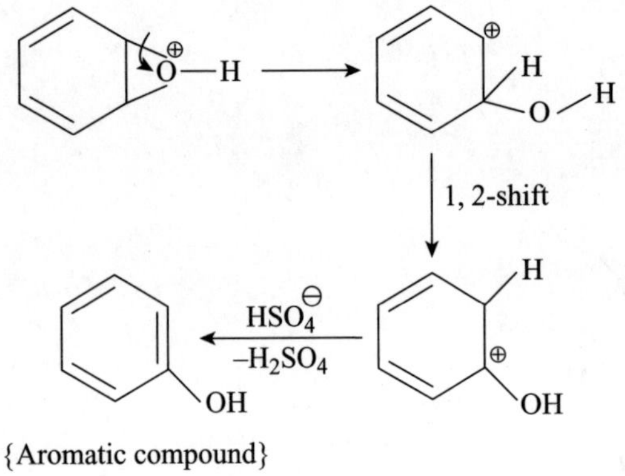

Ans: (a)

45.

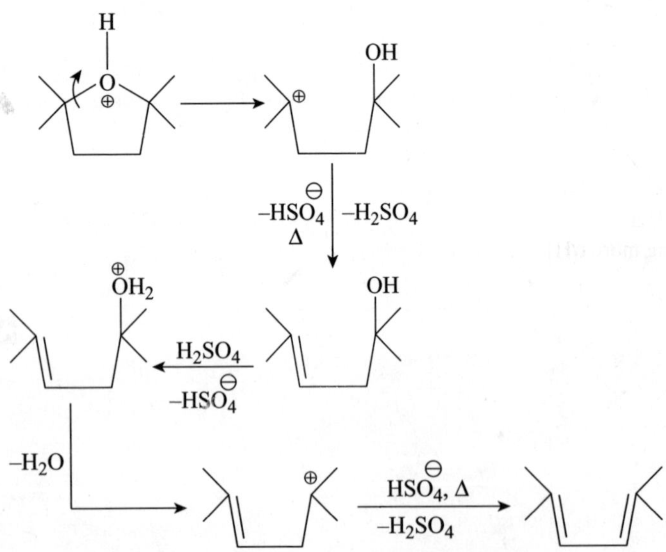

Ans: (b)

46.

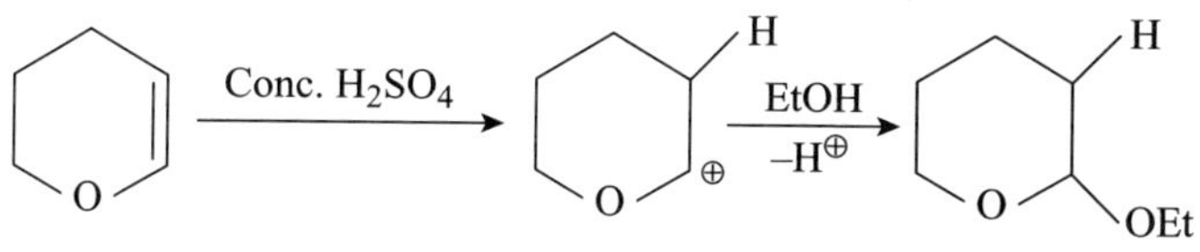

Ans: (a)

47.

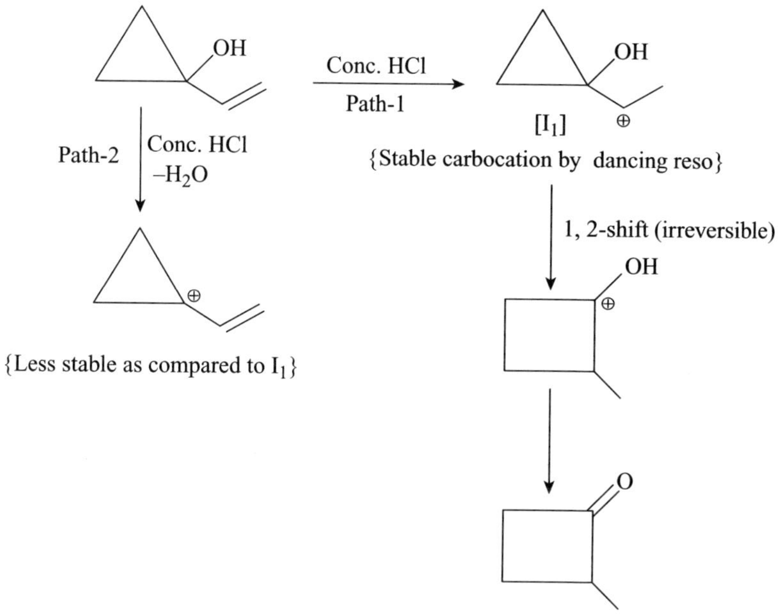

Note: Intramolecular carbocation leading to formation of —C(=O)— bond will be faster compared to attack of external $Cl^{\ominus}$ on $[I_1]$.

Ans: (b)

48.

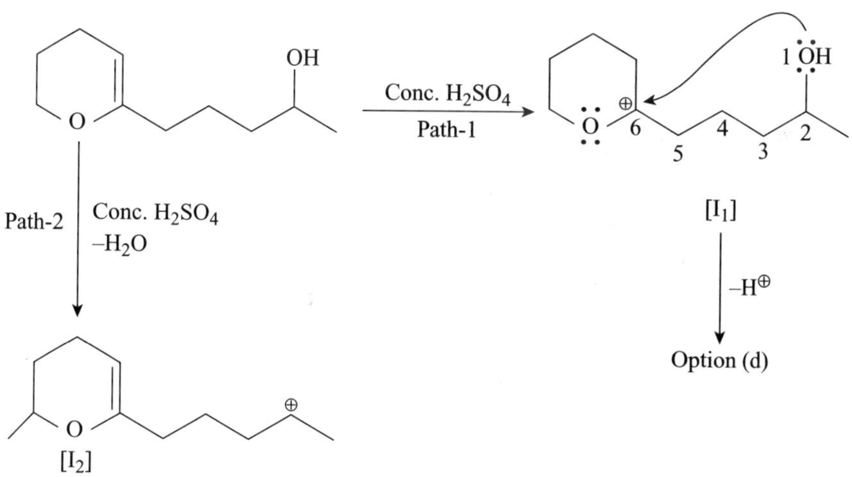

Stable : $[I_1] > [I_2]$: as octet is complete of carbocation in $[I_1]$.

Ans: (d)

49.

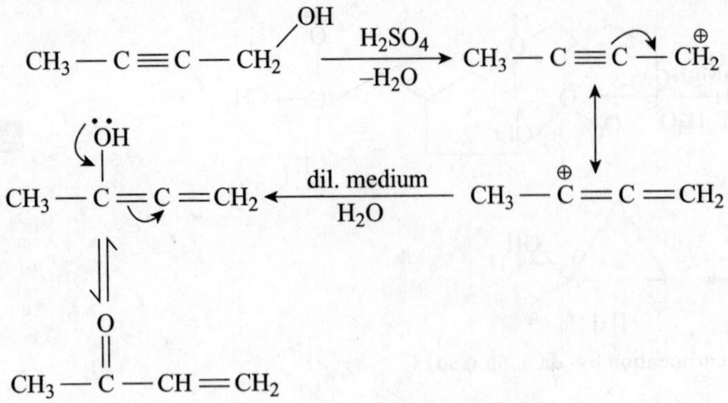

Ans: (b)

50.

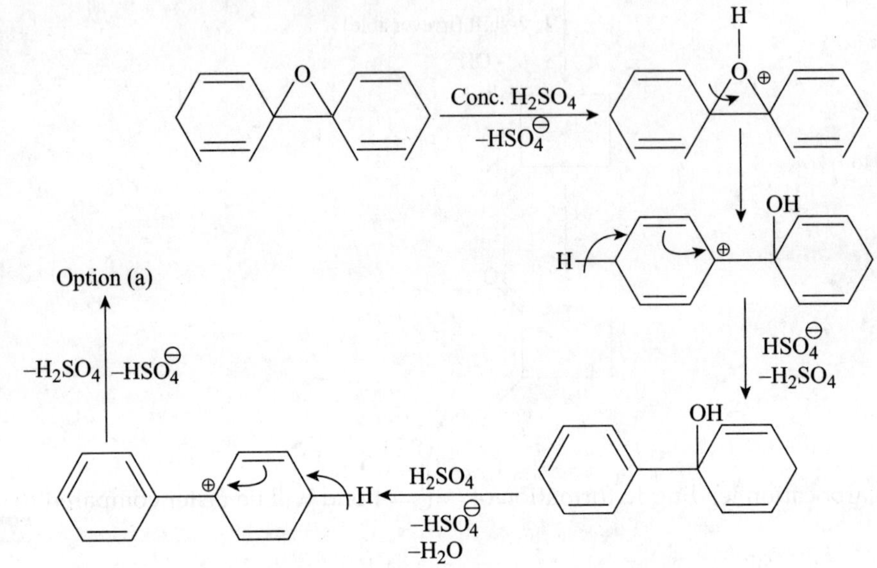

Ans: (a)

51.

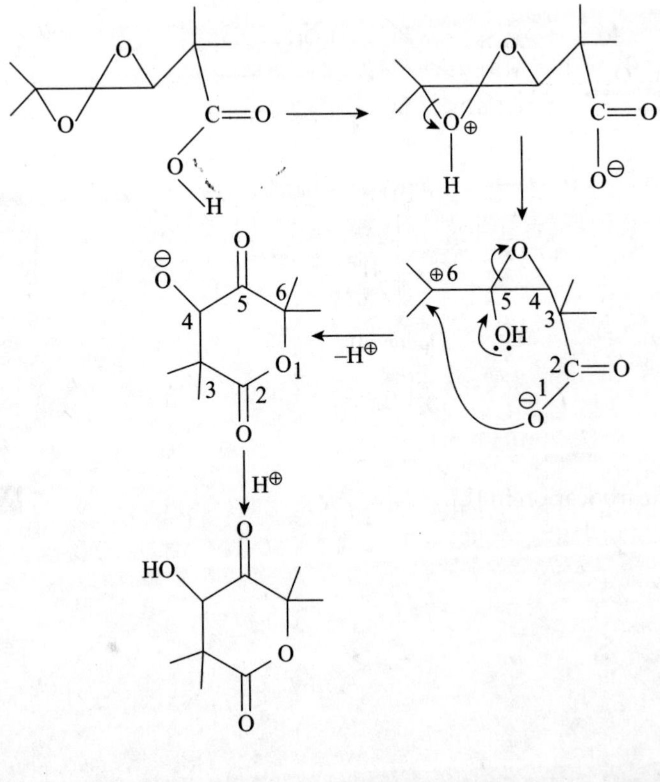

Ans: (a)

52.

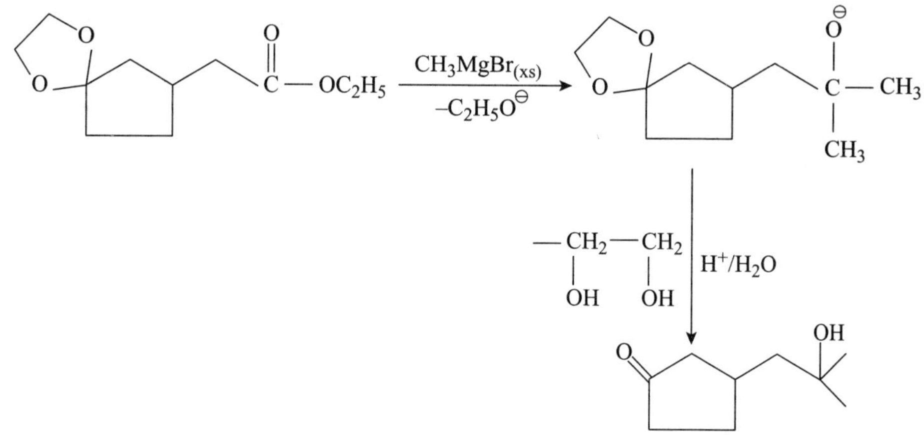

Ans: (c)

53.

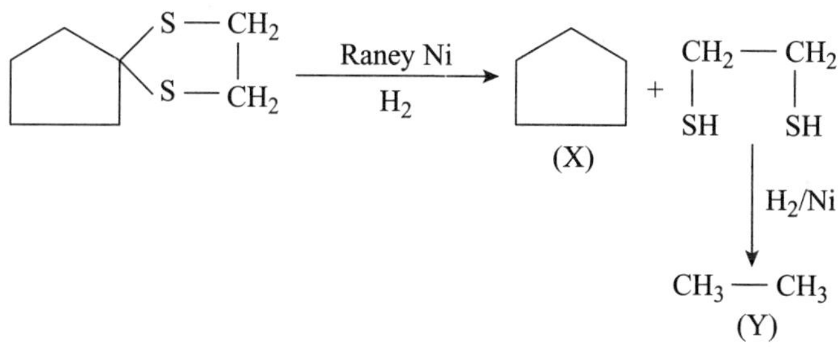

Ans: (b)

54.

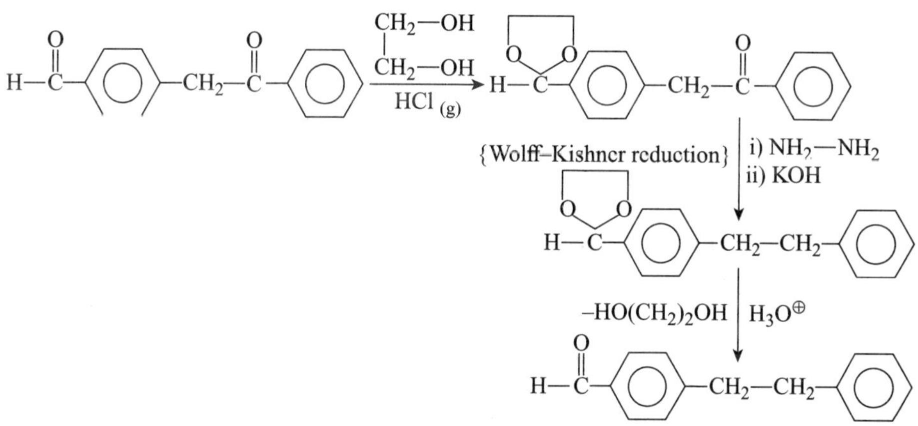

Ans: (b)

55.

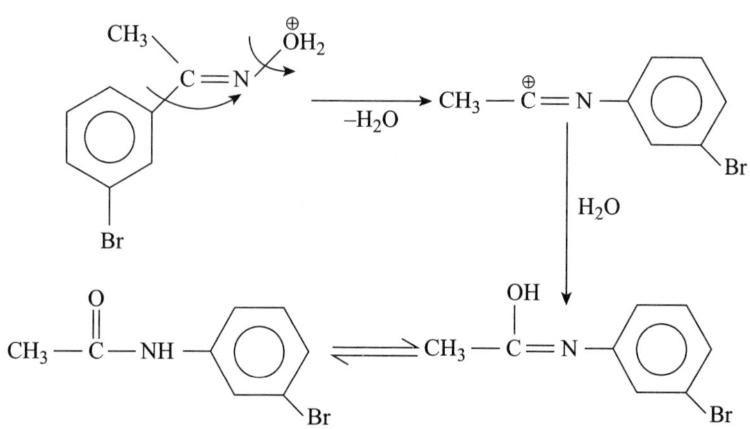

Ans: (b)

56.

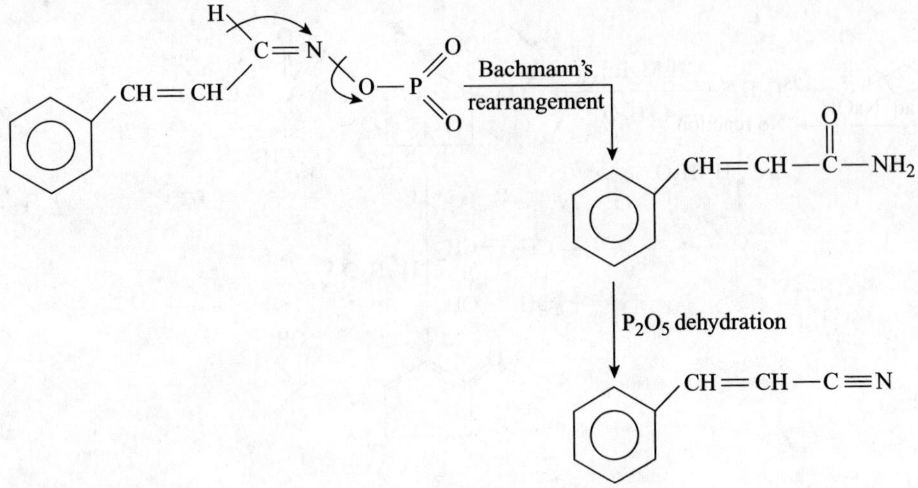

Option (a)

Ans: (a)

57.

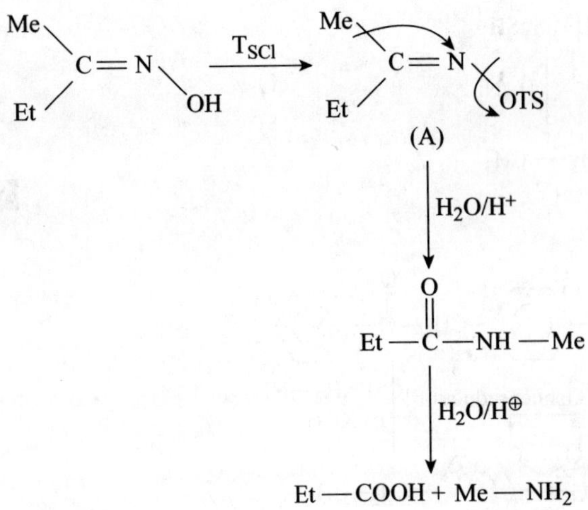

Ans: (a)

58. A: 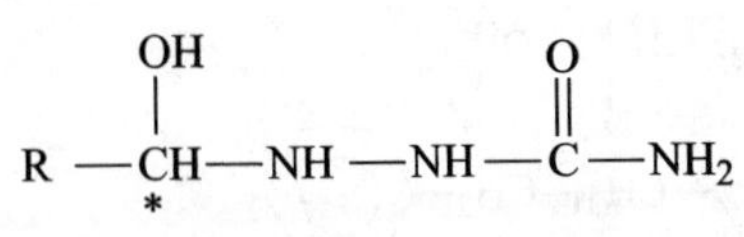

B: 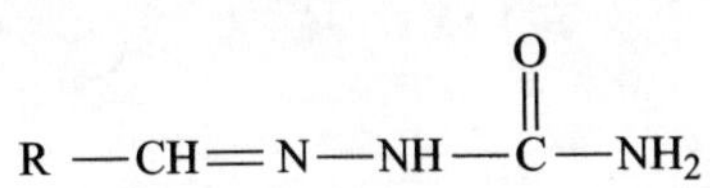

Ans: (b)

61. to 63.

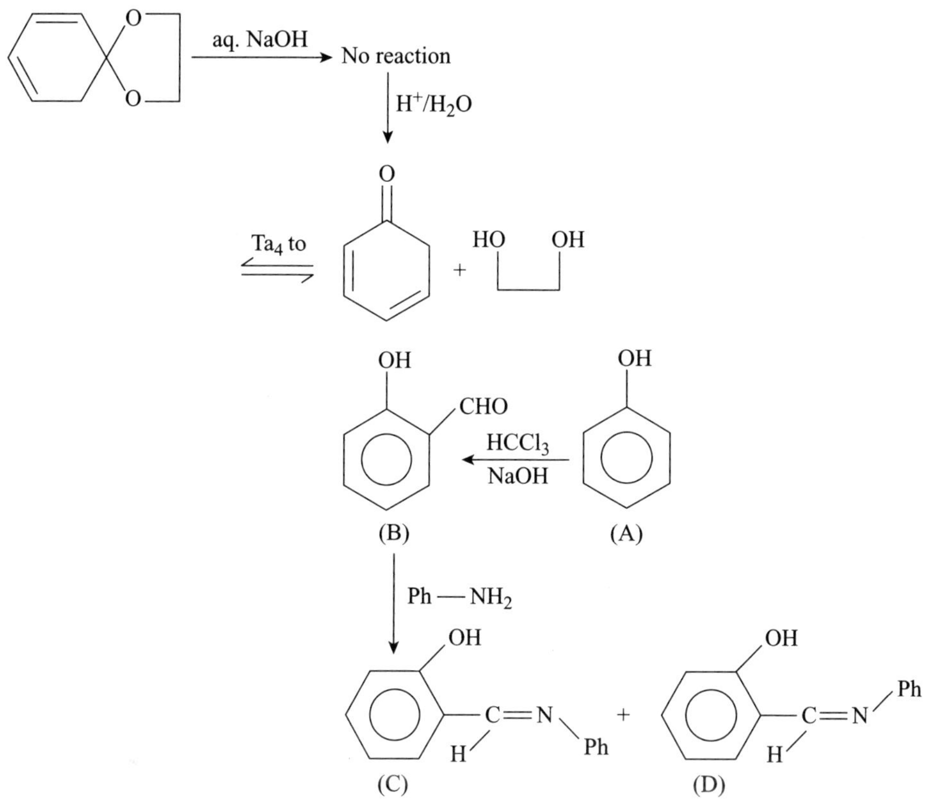

65.

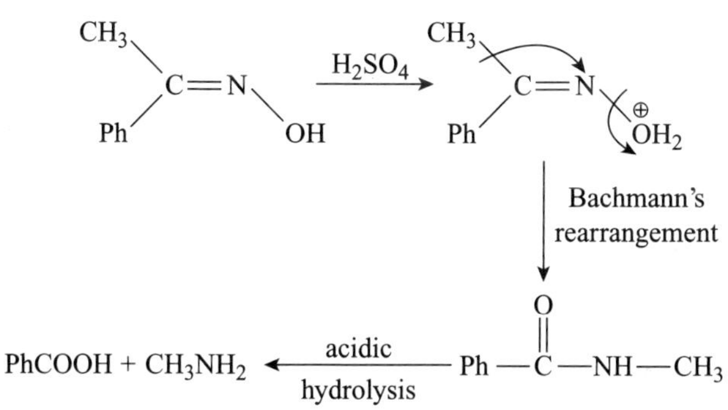

Ans: (d)

66.

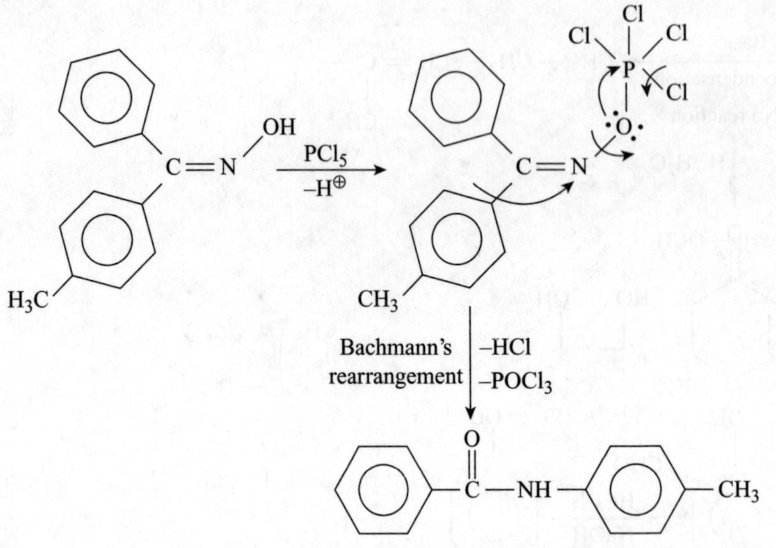

Ans: (b)

67. Test tube 1: $2CH_3-\overset{O}{\overset{\|}{C}}-H \xrightarrow[\{\text{Aldol condensation}\}]{\text{Base}} \underset{\beta}{CH_3-CH}=\underset{\alpha}{CH}-\overset{O}{\overset{\|}{C}}-H$

(A)

α, β unsaturated compound

$\downarrow$ Tollen's reagent

[B]

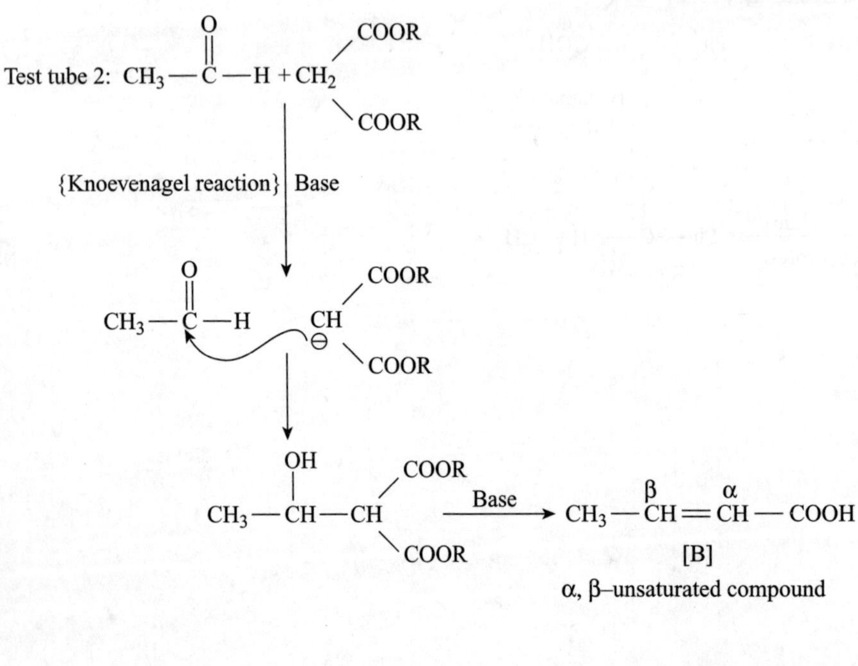

Ans: (c)

69. $CH_3-CH_2-\overset{\overset{O}{\|}}{C}-H \xrightarrow[\{\text{Aldol condensation}\}]{\text{Base}} CH_3-CH_2-CH=\underset{\underset{CH_3}{|}}{C}-$

Ans: (b)

70. A: 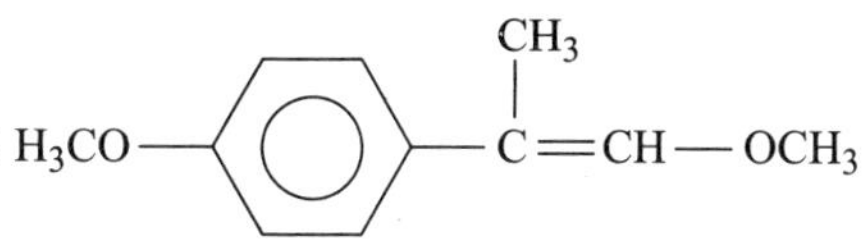

B: 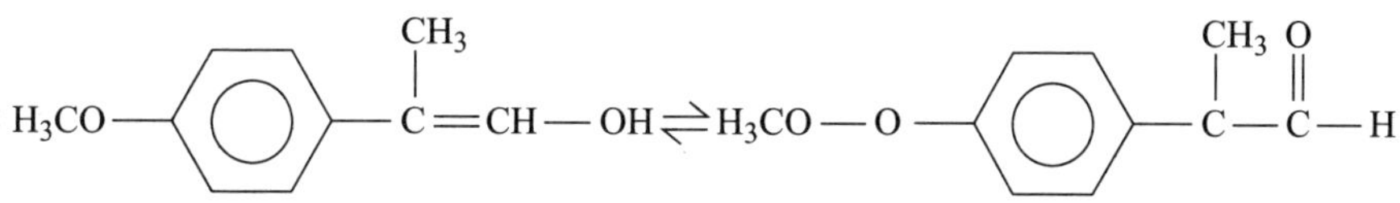

Ans: (d)

71. Atleast 1 H should be present on carbon having halogen, for formation of ylide. **Ans:** (b)

72. X: 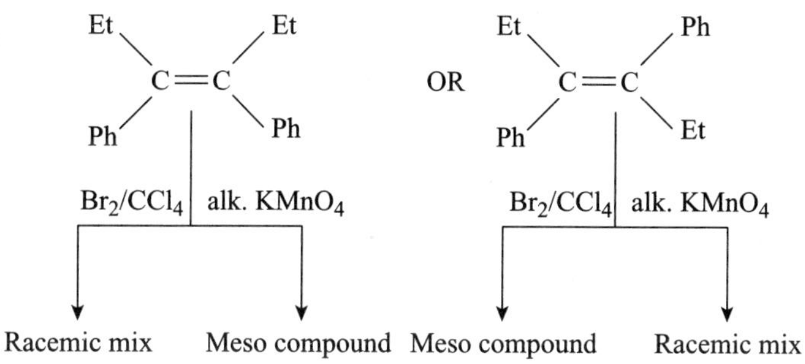

Ans: (d)

73. A: 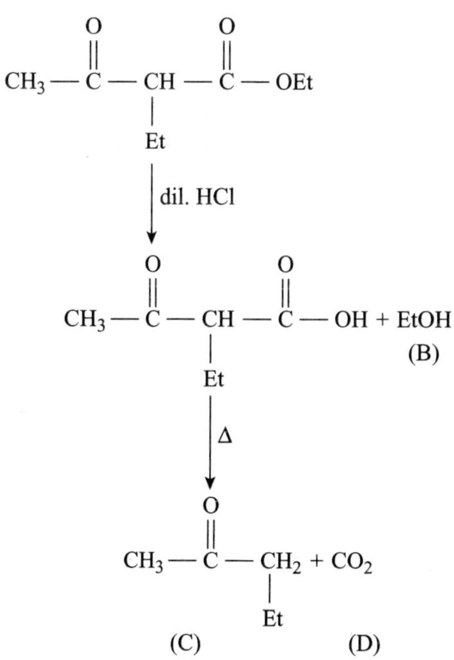

Ans: (d)

74.

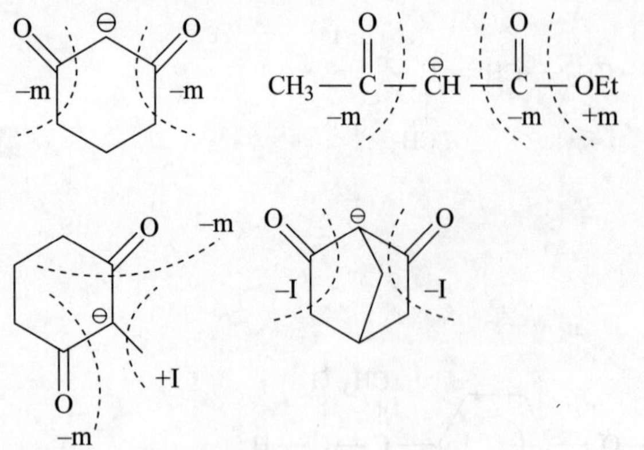

Ans: (a)

75. A :

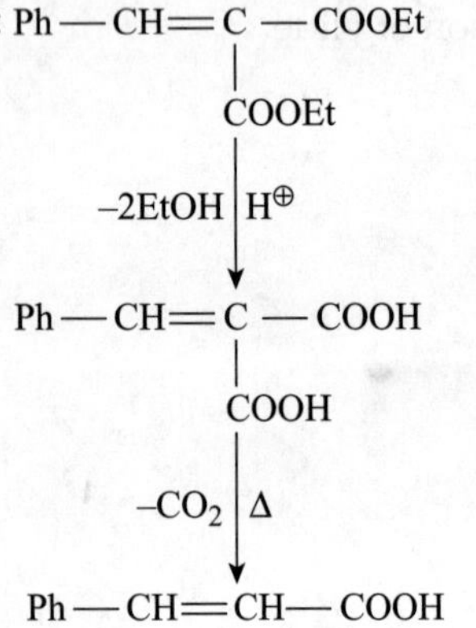

Ans: (c)

76. to 78.

Approach :

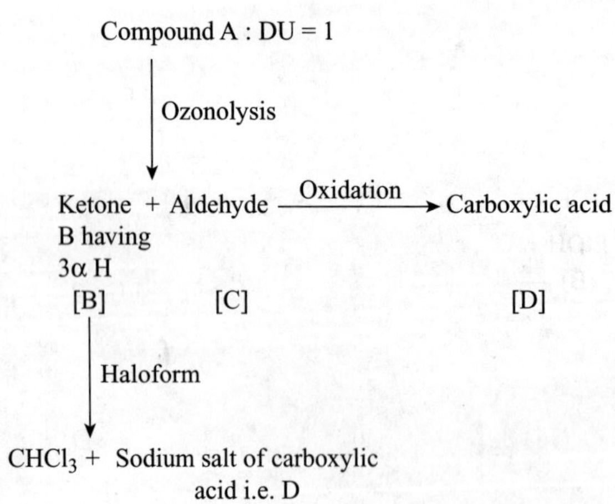

A:

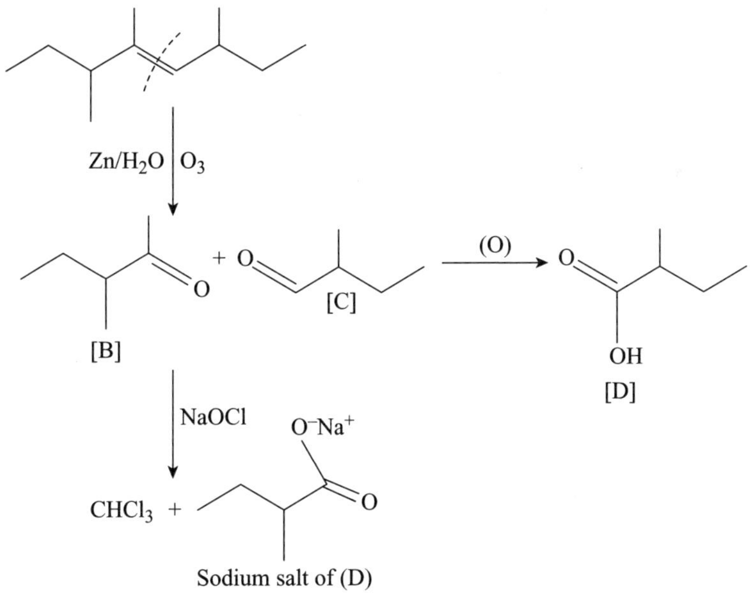
Zn/H2O O3
[B]
[C]
(O)
OH
[D]
NaOCl
O⁻Na⁺
CHCl3 +
Sodium salt of (D)

E:

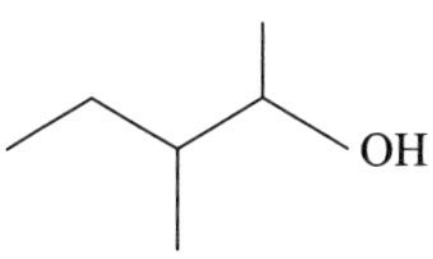
OH

F:

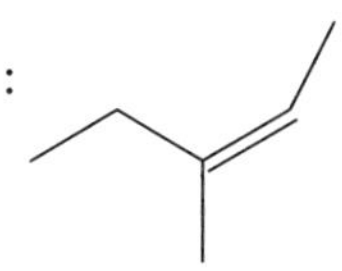

G:

H:

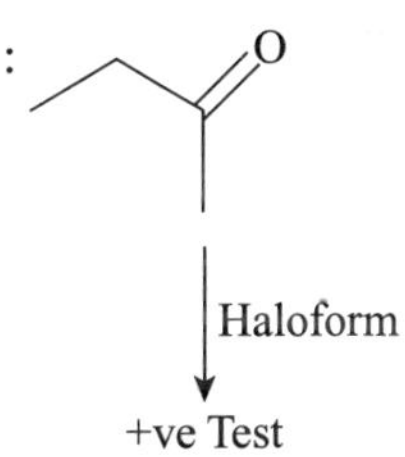
O
Haloform
+ve Test

X:

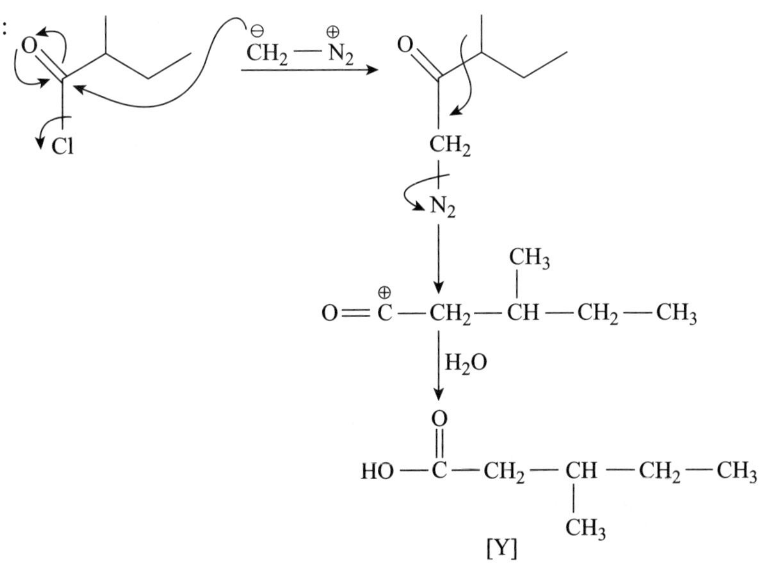
Cl
CH2 — N2
CH2
N2
CH3
O = C — CH2 — CH — CH2 — CH3
H2O
HO — C — CH2 — CH — CH2 — CH3
CH3
[Y]

W : 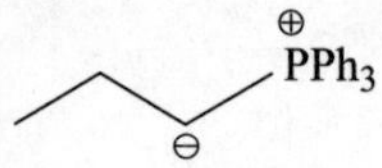

T : 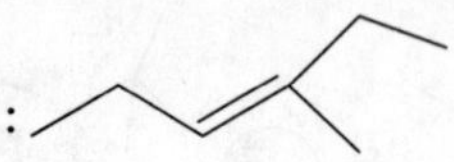

79. to 81.

Approach : E : 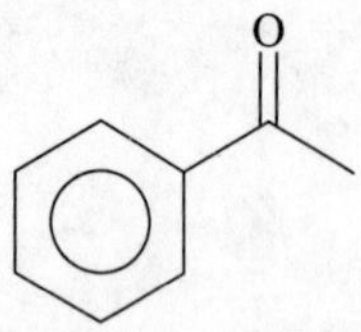

F : Can give haloform test

B : Can't show GI (alkene)

C : Optically active

A : DU = 4 and has acidic H

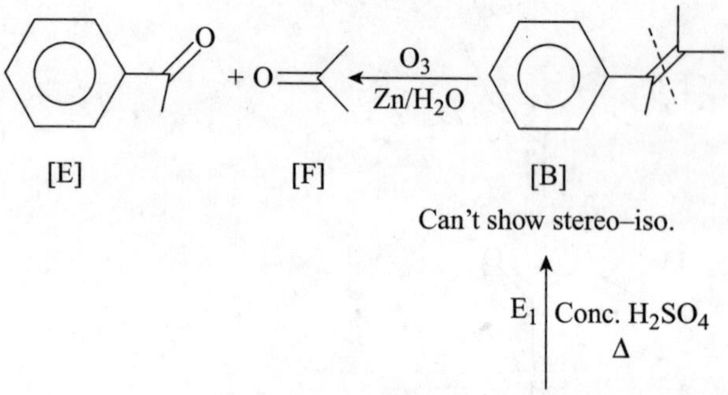

[A]

Alcohol (having acidic H)

C : 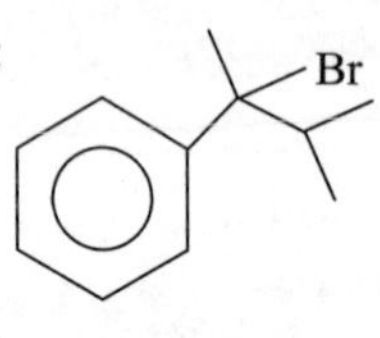

D : 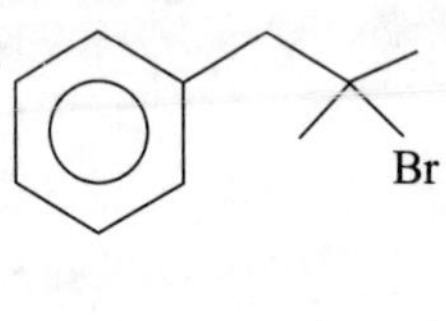

X : 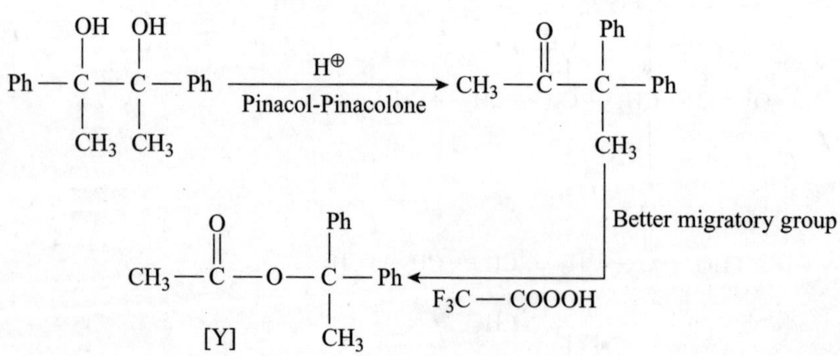

A :

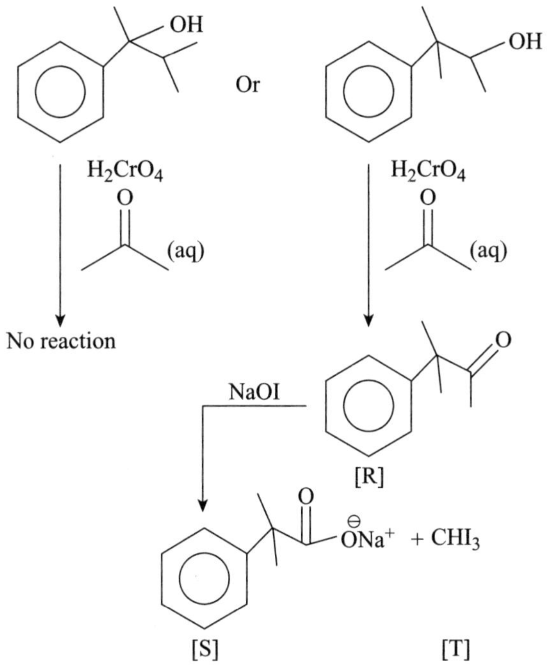

82.

$$\underset{\text{(Conjugated system)}}{CH_3-\overset{\delta+}{C}H=CH-\overset{\overset{O}{\|}}{\underset{\delta+}{C}}-H} \xrightarrow[\text{(soft nu}^{\ominus}\text{)}]{NaBH_4} CH_3-CH_2-CH_2-\overset{\overset{O}{\|}}{C}-H \xrightarrow{NaBH_4} CH_3-CH_2-CH_2-CH_2-OH$$

$$CH_3-CH=CH-\overset{\overset{O}{\|}}{C}-H \xrightarrow[\text{(Hard nu}^{\ominus}\text{)}]{LiAlH_4} CH_3-CH=CH-CH_2-OH$$

$$\underset{\text{(Conjugated system)}}{Ph-CH=CH-\overset{\overset{O}{\|}}{C}-H} \xrightarrow{Pd-C/H_2} \text{Reduce only } C=C$$

$$CH_2=CH-CH_2-\overset{\overset{O}{\|}}{C}-H \xrightarrow{Pd/H_2} CH_3-CH_2-CH_2-CH_2-OH$$

83. (a)

A : 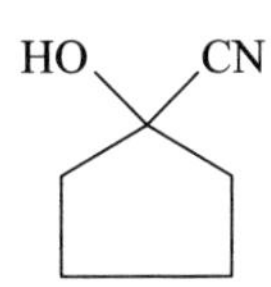 B : 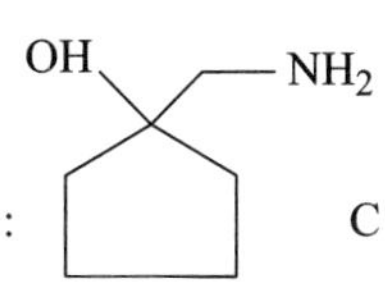 C : 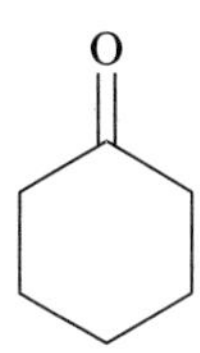

Ans: (p), (q) and (s)

(b)

A : 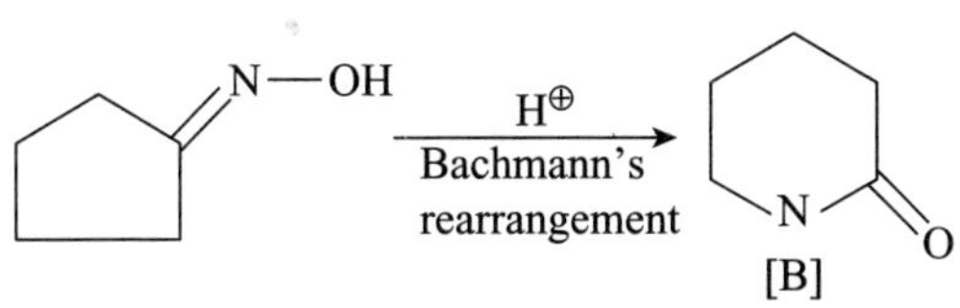

Ans: (p)

(c)

A : 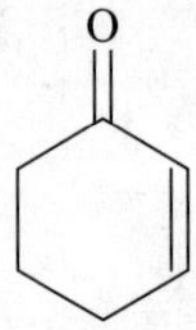 (After intramolecular Aldol)

Ans: (p), (q) and (s)

(d)

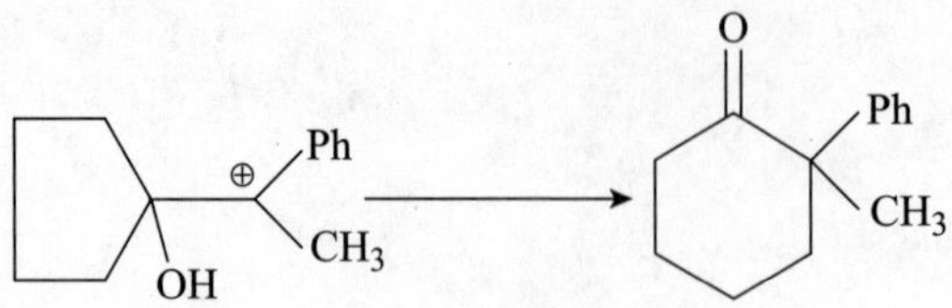

[A]

Ans: (p), (q) and (s)

84. (a) $CH_3—CH_2—OH \xrightarrow{I_2 + NaOH} CH_3—\overset{\overset{O}{\|}}{C}—H$ (Can give Iodoform test further)

$\text{Glucose} \xrightarrow[\text{(Oxidising agent)}]{\text{Tollen/Fehling}}$ both give +ve Test

$H—\overset{\overset{O}{\|}}{C}—\overset{\overset{O}{\|}}{C}—H \xrightarrow{\text{Tollen/Fehling/Schiff reagent}}$ Give all three +ve test

Glyoxal

Ans: (s)

85. $R—\overset{\overset{O}{\|}}{C}—R' \xrightarrow[\text{(Clemmensen red)}]{Zn—Hg/HCl} R—CH_2—R'$

$\xrightarrow{\text{Red P + HI}}$, $\xrightarrow{Ni/H_2}$, $\xrightarrow{LiAlH_4/NaBH_4}$ $R—\underset{}{\overset{\overset{OH}{|}}{CH}}—R'$

$R—X \xrightarrow{LiAlH_4/NaBH_4} R—H$

$R—\overset{\overset{O}{\|}}{C}—OR'$

$\xrightarrow{NaBH_4} X$

$\xrightarrow{LiAlH_4} R—CH_2—OH + R'OH$

$\xrightarrow{\text{Red P + HI}} R—CH_3 + R'H$

86.

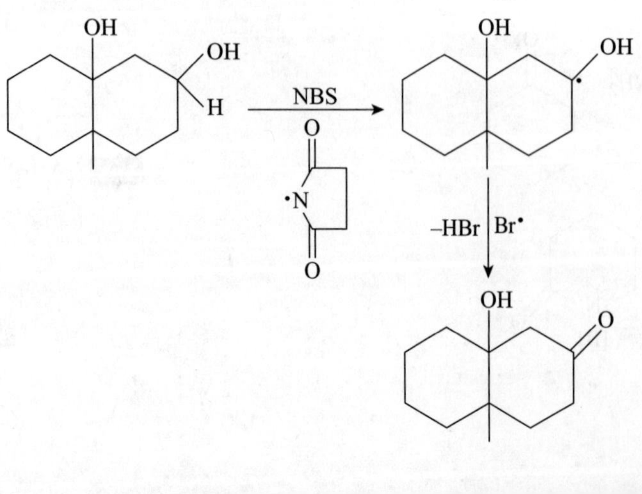

NBS : Oxidising agent

$$\underset{(1^\circ/2^\circ)}{\text{Alcohol}} \xrightarrow[\text{(Partial oxidation)}]{\text{PCC}} \text{aldehyde/ketone}$$

$AlPO_4$: Dehydrating agent

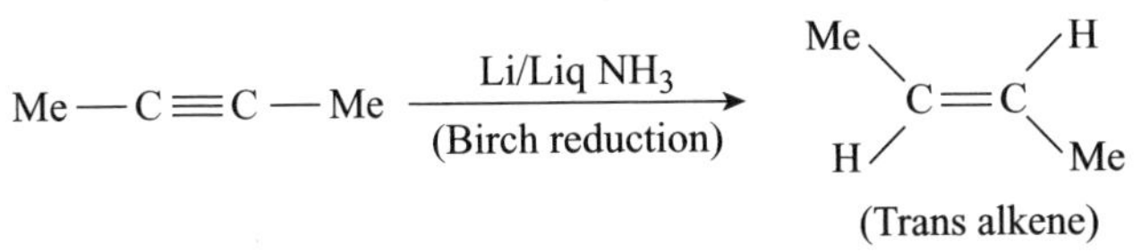

87. 2, 4-DNP test : Carbonyl compound (aldehyde and ketone)

Tollen's Fehling : Aldehyde, Formic acid

Haloform test : $R—\underset{\underset{O}{\|}}{C}—CH_3$

88.

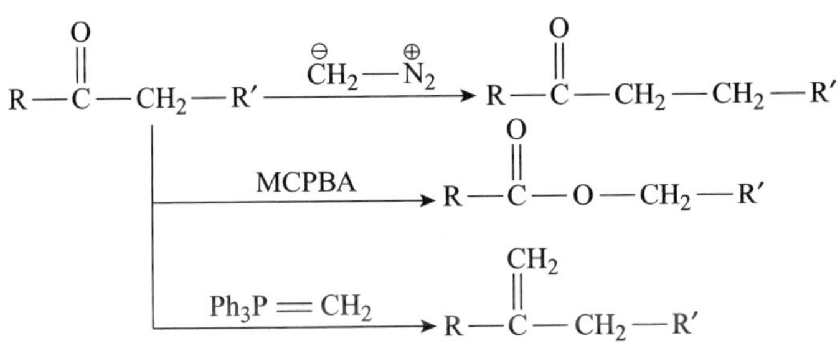

90. (a)

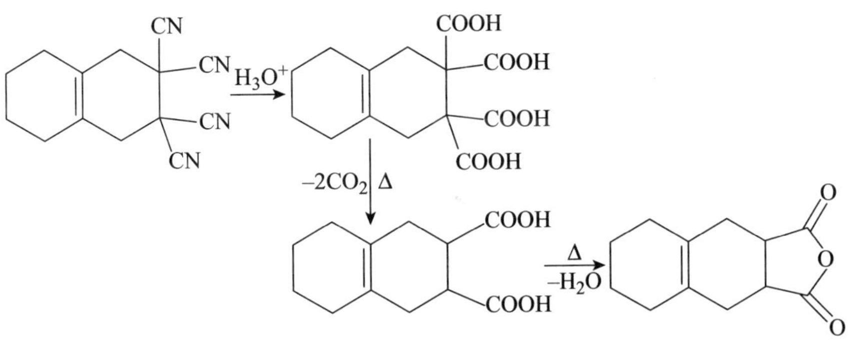

Ans: (p) and (r)

(b)

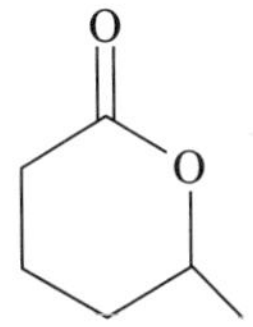

Ans: (s) and (t)

(c)

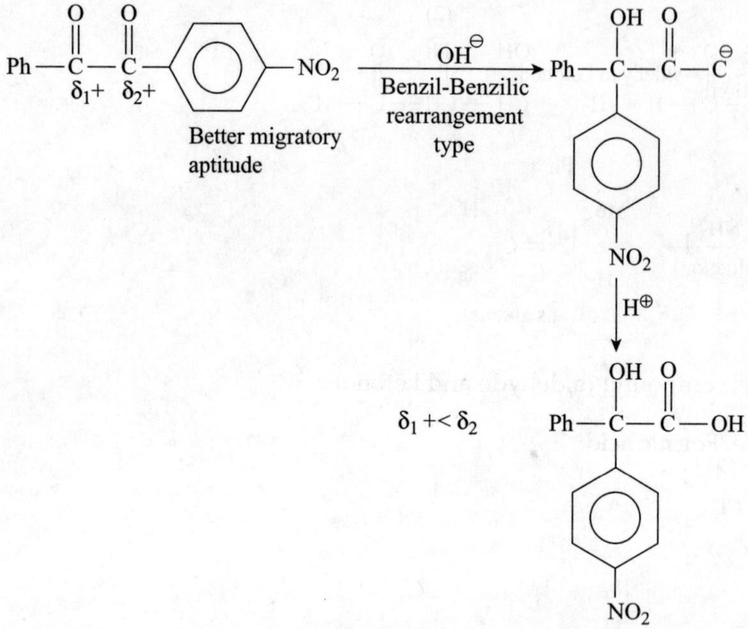

Ans: (q) and (s)

91.

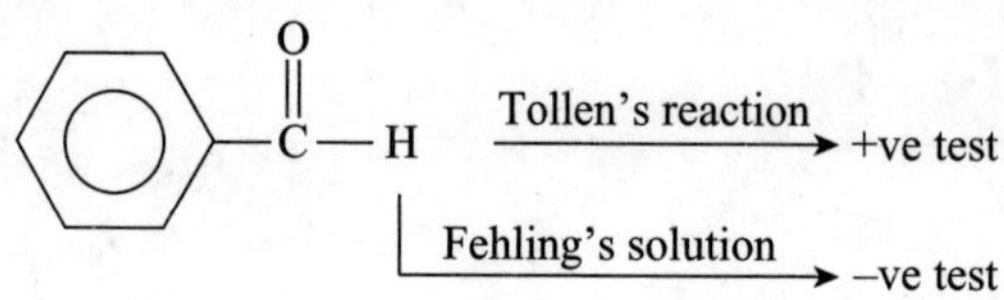

92.

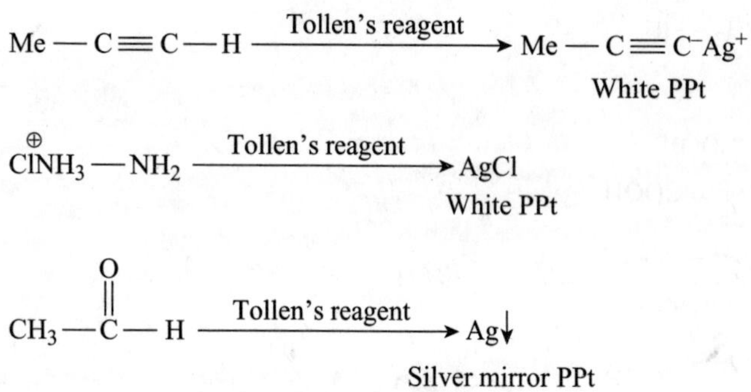

93. (a) Aldol condensation **Ans:** (p) and (q)

(b), (c) Cannizzaro reaction **Ans:** (s)

(d) Reimer–Tiemann **Ans:** (t)

95.

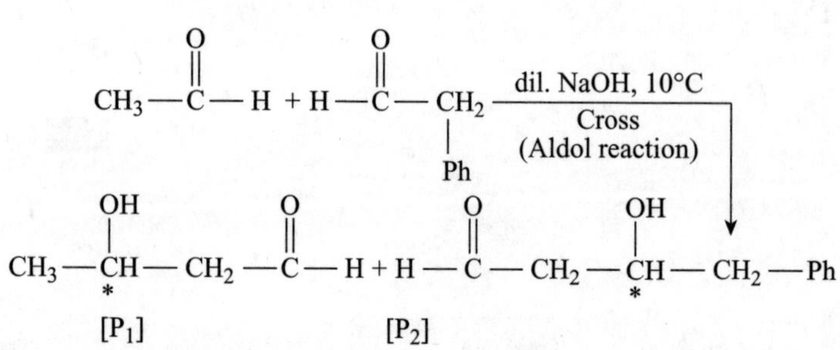

Stereo–iso: (2) (2)

$CH_3-\overset{*}{C}H(OH)-\overset{*}{C}H(Ph)-C(=O)-H + H_2C(Ph)-\overset{*}{C}H(OH)-\overset{*}{C}H(Ph)-C(=O)-H$

[P_3] [P_4]

Stereo–iso: (4) (4) **Ans:** (12)

96. Gas: CH_3-CH_3

X : $CH_3-C\equiv\overset{\ominus}{C}\,\overset{\oplus}{Mg}Br$

Y : $CH_3-C\equiv C-CH_2-CH_3$

Z : $CH_3-C(=O)-CH_2-CH_2-CH_3$

P : $CH_3-CH(OH)-CH_2-CH_2-CH_3$

q : $CH_2{=}CH-CH_2-CH_2-CH_3 + CH_3-CH{=}CH-CH_2-CH_3$

$CH_2{=}CH-CH_2-CH_2-CH_3 \xrightarrow{\text{Ozonolysis}} \underset{(a)}{H-C(=O)-H} + \underset{(b)}{CH_3-CH_2-CH_2-C(=O)-H}$

$CH_3-CH{=}CH-CH_2-CH_3 \xrightarrow{\text{Ozonolysis}} \underset{(c)}{CH_3-C(=O)-H} + \underset{(d)}{CH_3-CH_2-C(=O)-H}$

No. of product (including stereo)

$a + b \xrightarrow[\Delta\ \text{(Aldol condensation)}]{\text{dil. OH}} CH_2{=}C(Et)-C(=O)-H$ — 1

$b + b \xrightarrow[\Delta\ \text{(Aldol condensation)}]{\text{dil. OH}} C_3H_7-CH{=}C(Et)-C(=O)-H$ — 2 (GI)

$c + d \xrightarrow[\Delta\ \text{(Aldol condensation)}]{\text{dil. OH}} H-C(=O)-CH{=}CH-Et$ — 2 (GI)

$d + c \xrightarrow[\Delta\ \text{(Aldol condensation)}]{\text{dil. OH}} CH_3-CH{=}C(Me)-CHO$ — 2 (GI)

$c + c \xrightarrow[\Delta\ \text{(Aldol condensation)}]{\text{dil. OH}} CH_3-CH{=}CH-CHO$ — 2 (GI)

$d + d \xrightarrow[\Delta\ \text{(Aldol condensation)}]{\text{dil. OH}} CH_3-CH_2-CH{=}C(CH_3)-CHO$ — 2 (GI)

Total : 11

Ans: (11)

97. i, ii, vi, vii, viii, ix **Ans:** (6)

98. A, B, C, D, J, N **Ans:** (6)

99. 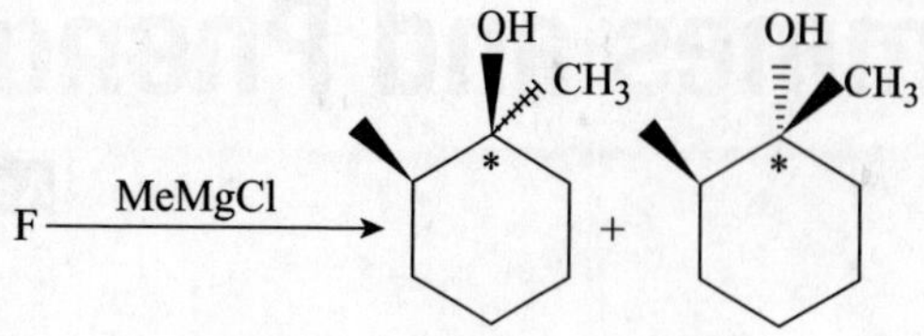

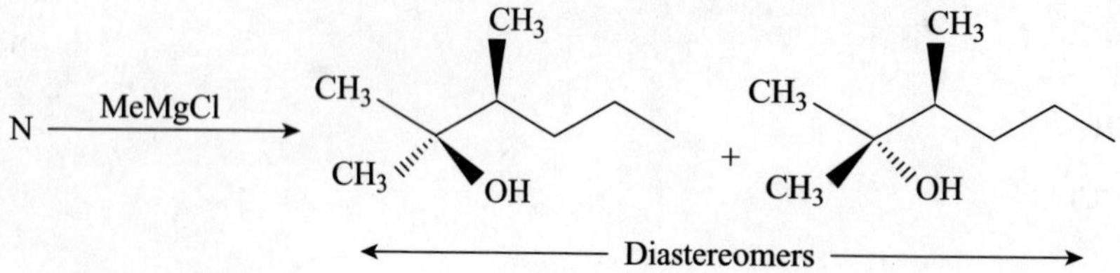

Ans: (2)

100. A, B, E, H, I, L **Ans:** (6)

CHAPTER 6

ESR Amines and Phenols

LEVEL 1

1.

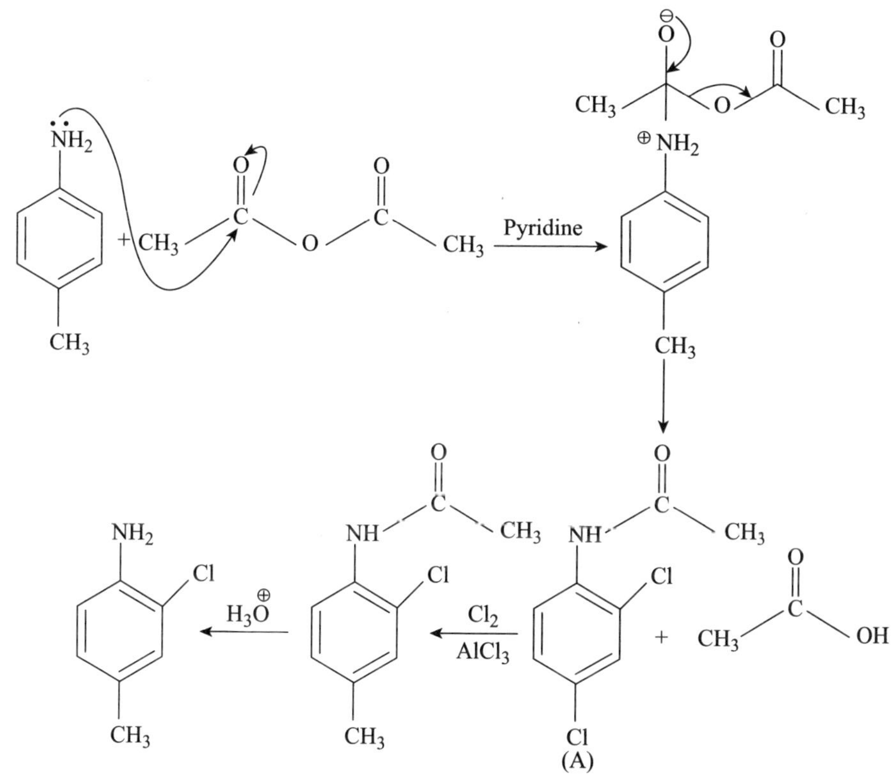

Ans: (b)

2. Groups that increases electron density on benzene ring will make it more reactive towards nitration

$-\ddot{\underset{\cdot\cdot}{O}}-CH_3$ is strong activating group.

$-\ddot{N}H-\underset{\underset{O}{\|}}{C}-CH_3$ and $-\ddot{N}(CH_3)-N{=}O$ Moderate activating groups

$-\underset{\underset{O}{\|}}{C}-CH_3 > Br$ De-activating order.

Groups that shows –I and –R effect on benzene ring are deactivating in nature. Ans: (b)

3. If there are two groups on benzene ring and both are activating in nature then substitution takes place according to strong activating group.

If both have common directing position then substitution takes place there.

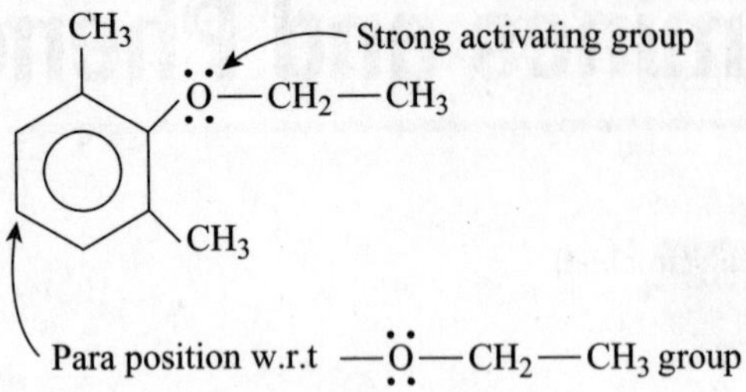

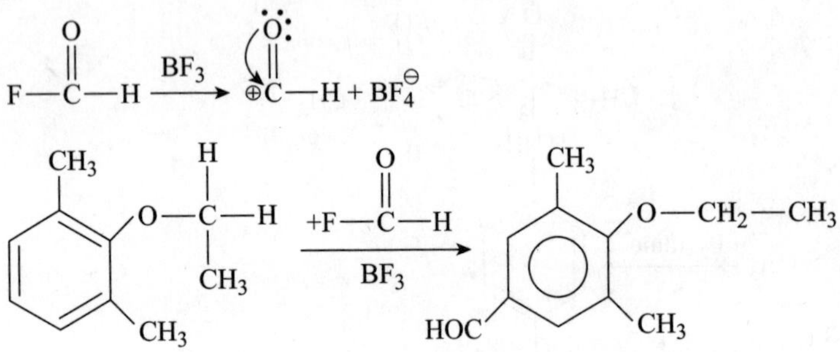

Ans: (c)

4. 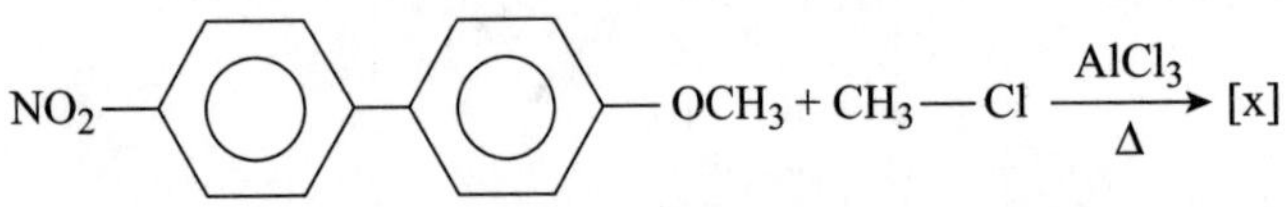

In biphenyl compounds or fused rings, if one ring is having activating group and other have deactivating group then substitution takes place on the activated ring.

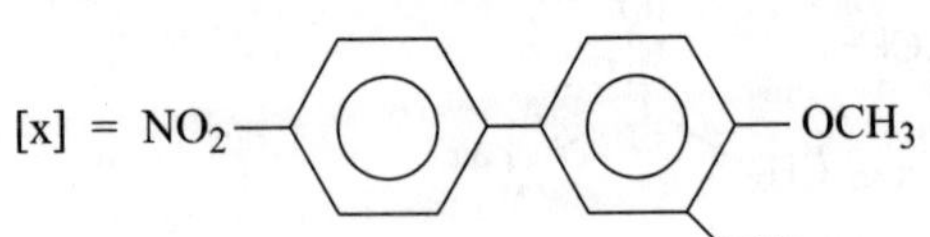

Ans: (c)

6. 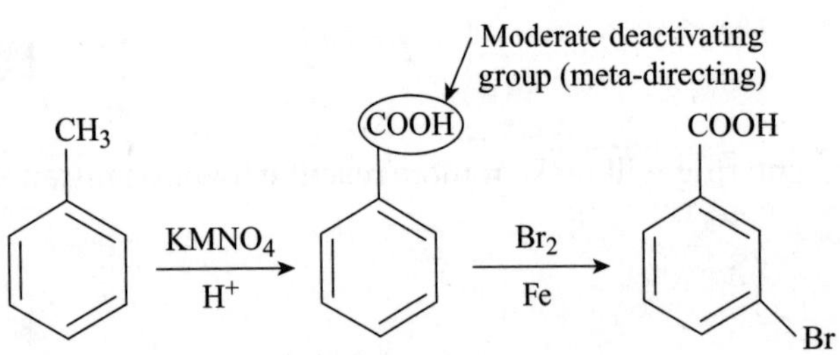

$2\,Fe + 3\,Br_2 \longrightarrow 2\,FeBr_3$ (Lewis Acid)

$FeBr_3 \xrightarrow{Br—Br} FeBr_4^{\ominus} + Br^{\oplus}$

Ans: (b)

7. 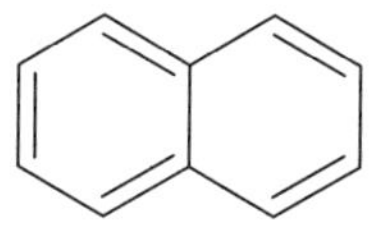 $\xrightarrow[H_2SO_4]{HNO_3}$ (more number of resonating structure when (α) attack takes place).

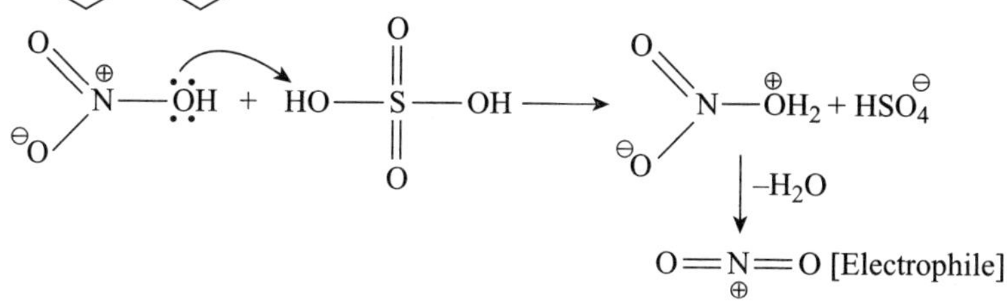

There are two positions where an electrophile can attack.

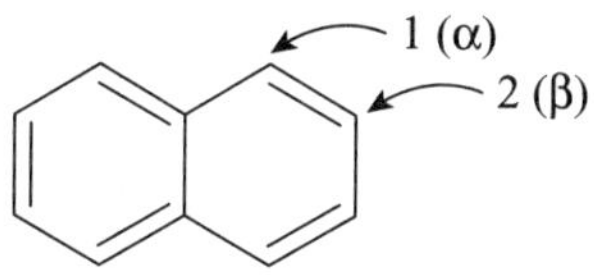

If electrophile attacks on position 1 (α), then,

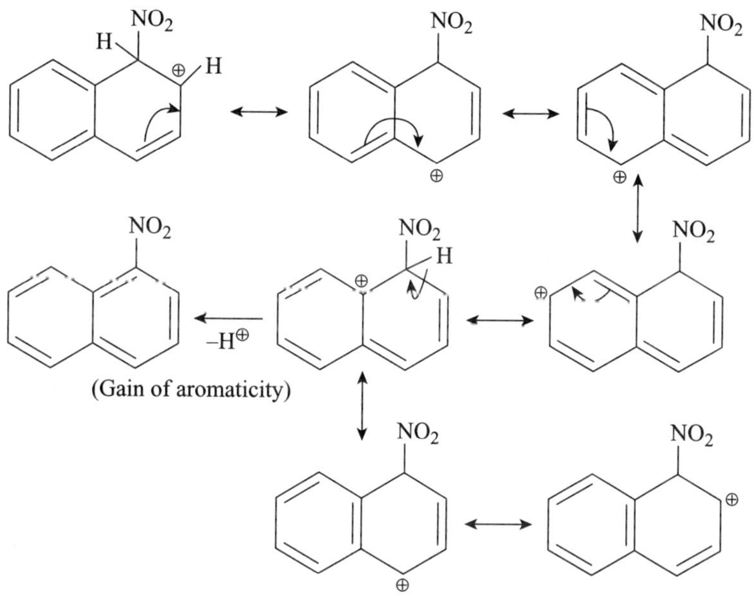

If electrophile attacks on position 2(β), then

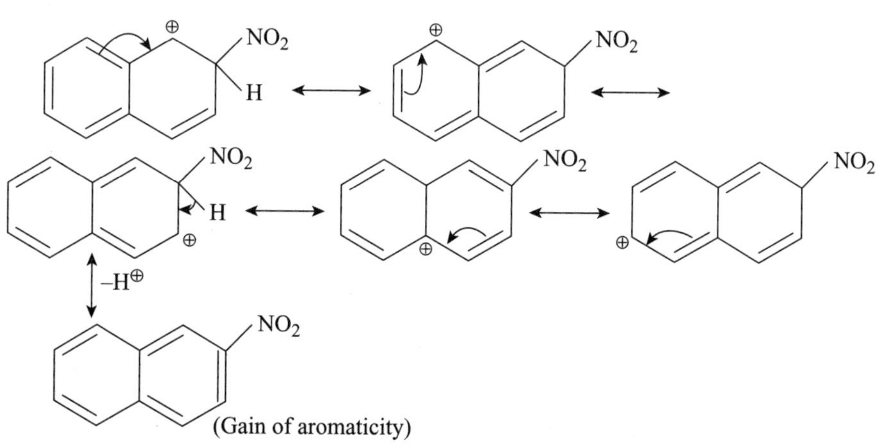

When electrophile attacks on position 2(β) then the aromaticity of ring is lost in second step but when attacks on position 1(α) the aromaticity is lost in third step. **Ans:** (a)

8.

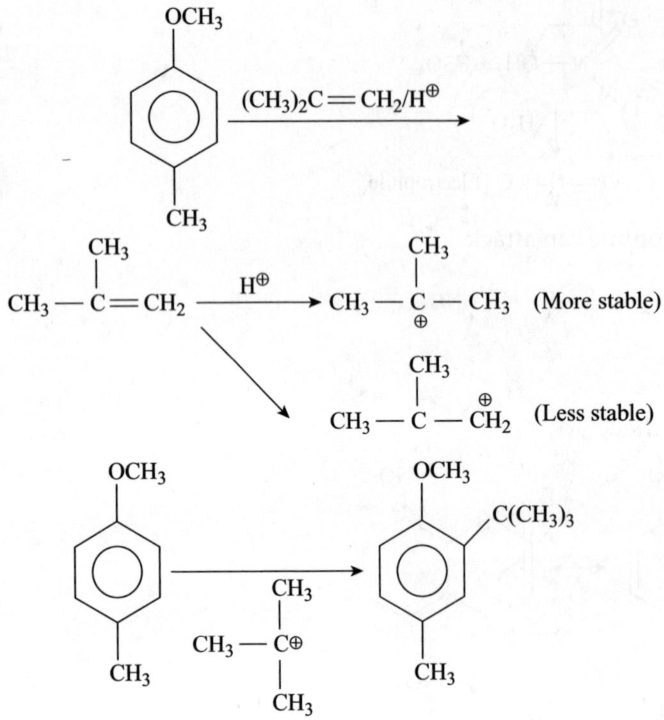

Ans: (a)

9.

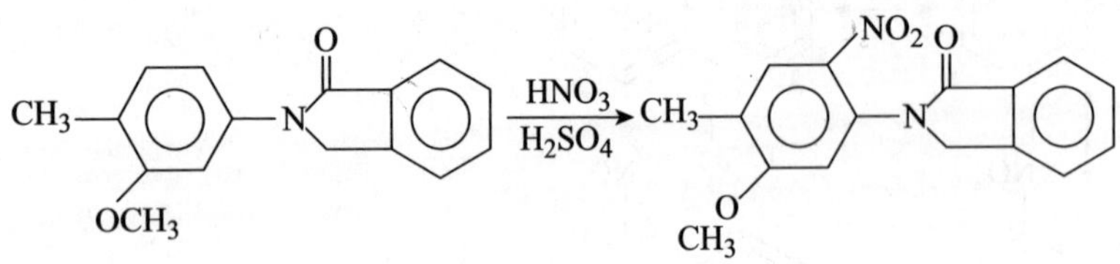

Ans: (c)

Just remember the concept of Question Number 3 ; attack will take place on more activated ring and on ortho and para positions w.r.t strong activating group.

10.

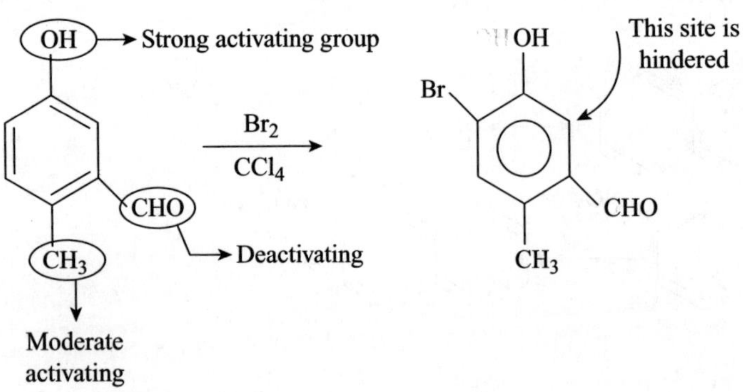

Ans: (b)

12.

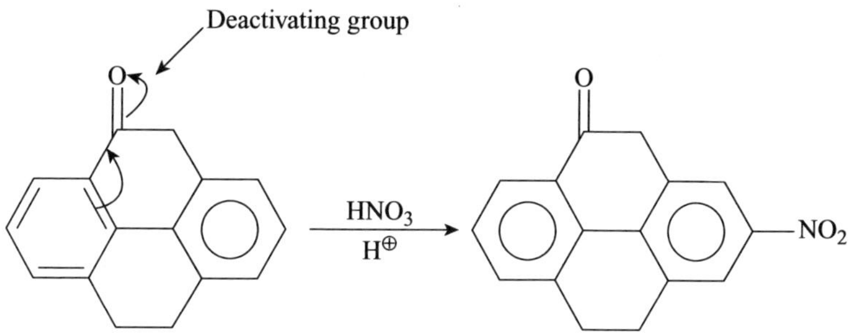

Ans: (c)

13.

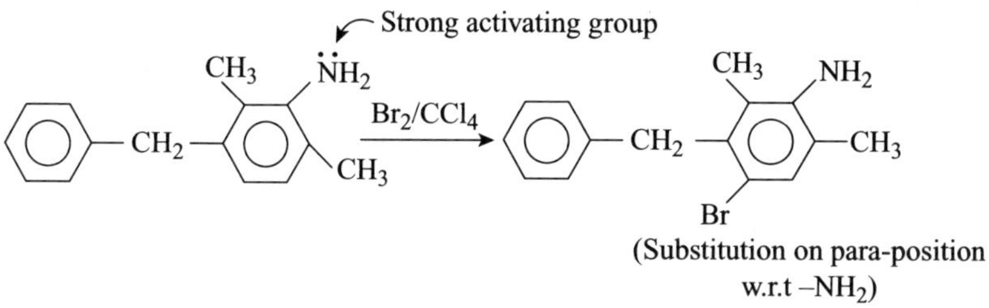

Ans: (c)

17.

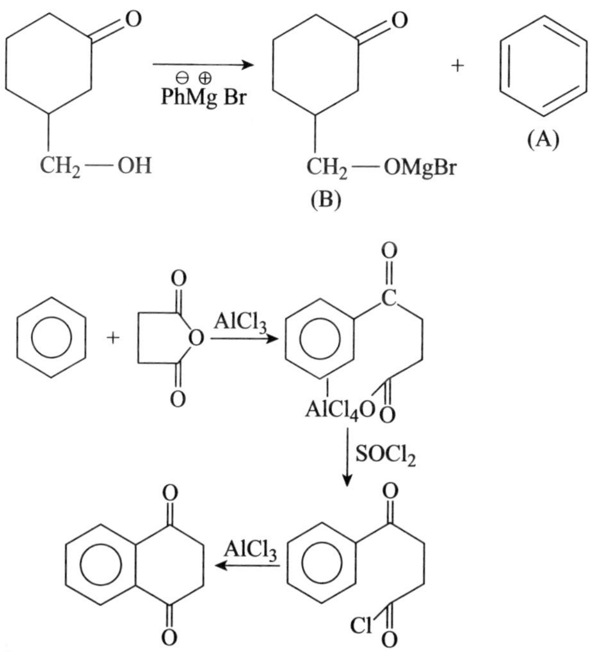

Ans: (a)

18.

⟶ Electrons with drawing groups will increase the reactivity

Ans: (b)

20. (a) $CH_2{=}CH{-}CH_3 \xrightarrow{HF} CH_3{-}\overset{\oplus}{C}H{-}CH_3$

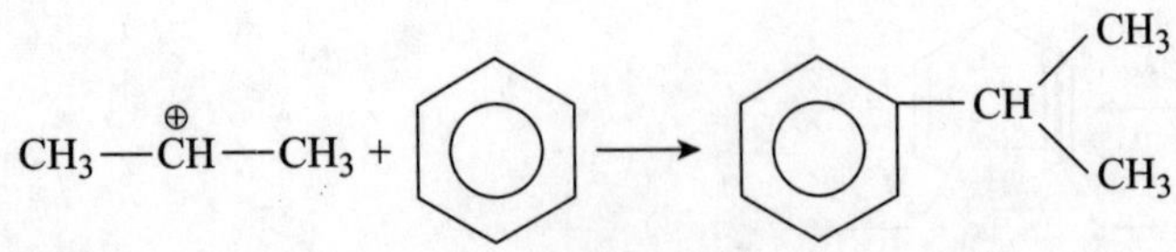

(b) $CH_3{-}CH_2{-}CH_2{-}Cl + AlCl_3 \longrightarrow CH_3{-}CH_2{-}\overset{\oplus}{C}H_2{-}\overset{\ominus}{A}lCl_4$

$\updownarrow$

$CH_3{-}\overset{\oplus}{C}H{-}CH_3$

(c) $CH_3{-}CH(Cl){-}CH_3 + AlCl_3 \longrightarrow CH_3{-}\underset{\oplus}{C}H{-}CH_3{-}\overset{\ominus}{A}lCl_4$

Ans: (d)

22.

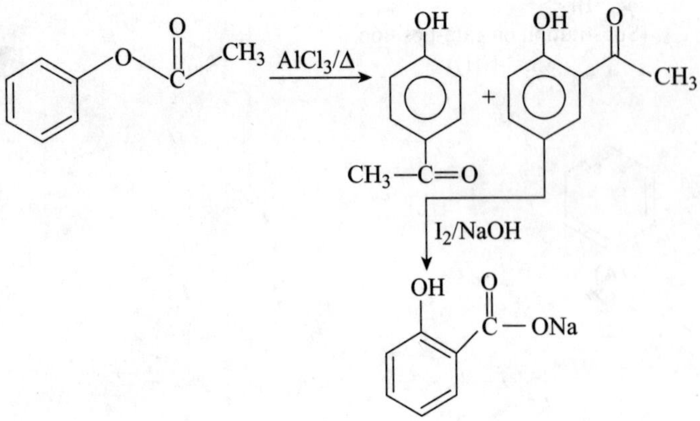

Ans: (d)

23.

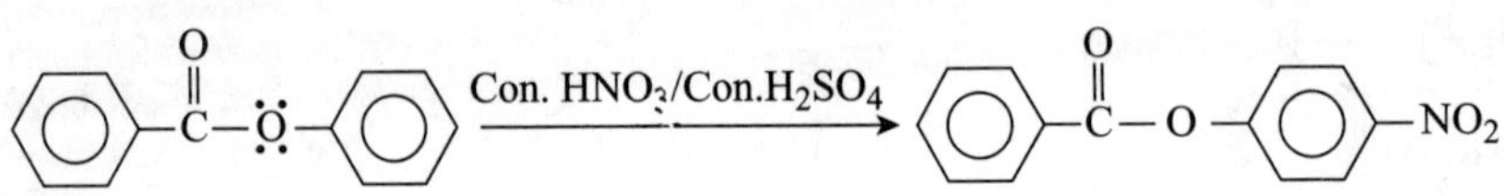

Ans: (d)

Note:

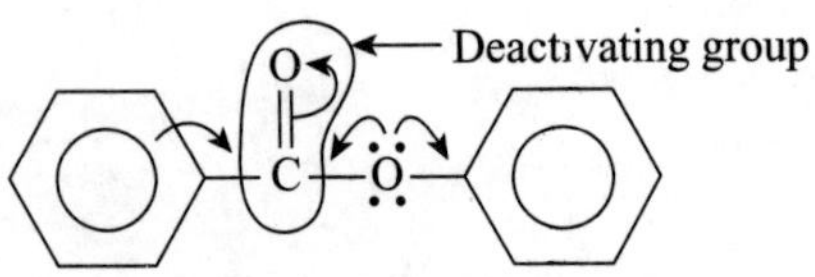

$-C(=O)-\ddot{O}-H$ Deactivating $\qquad H-C(=O)-\ddot{O}-$ Activating

Ans: (d)

24.

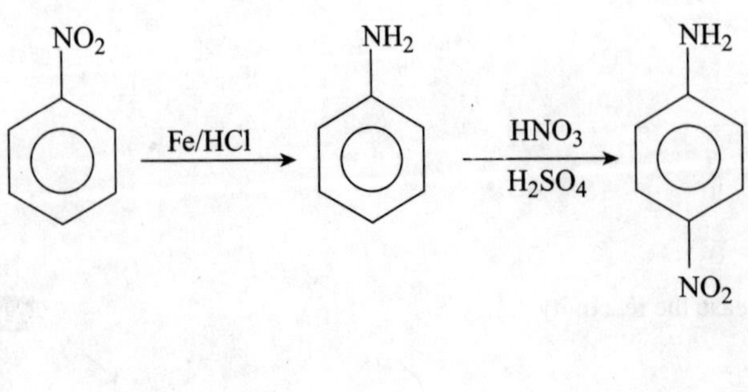

Ans: (b)

25.

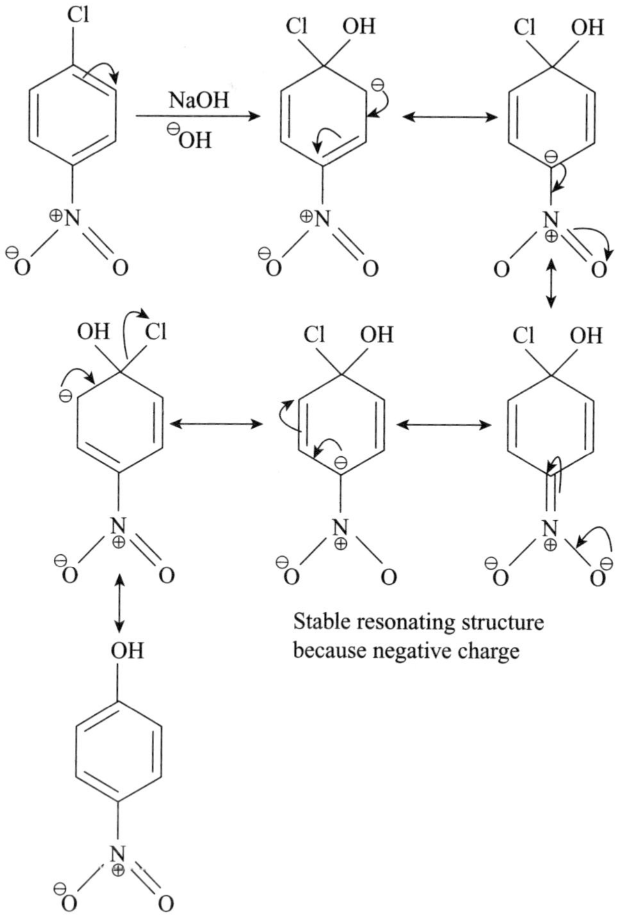

Groups will stabilise the intermediates will make it more reactive towards nucleophilic substitution if leaving group is present on ortho or para position w.r.t electrons withdrawing group through resonance will be more reactive than the compound in which leaving group is on meta position w.r.t electron withdrawing group.

Ans: (d)

26.

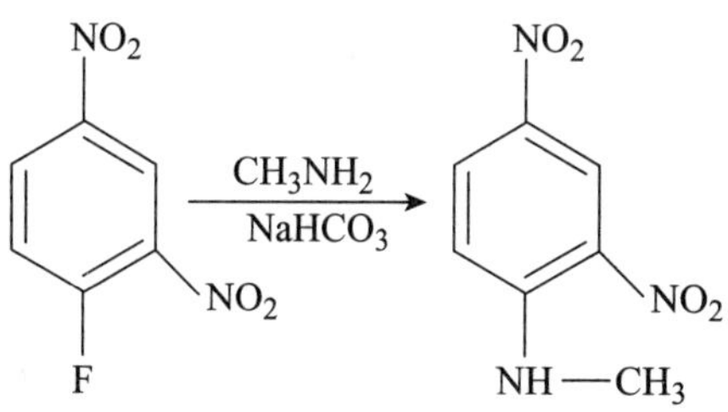

Hint:

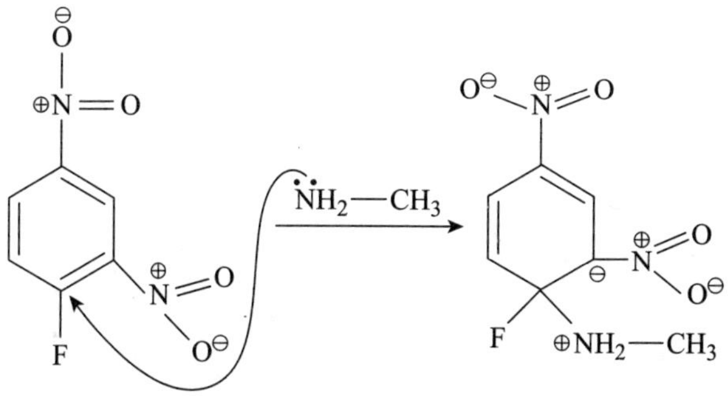

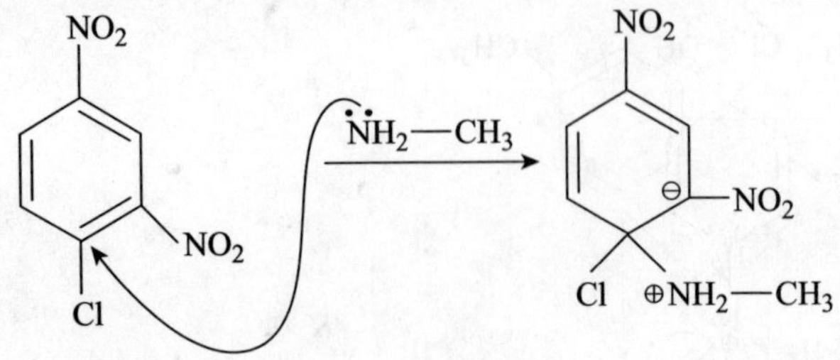

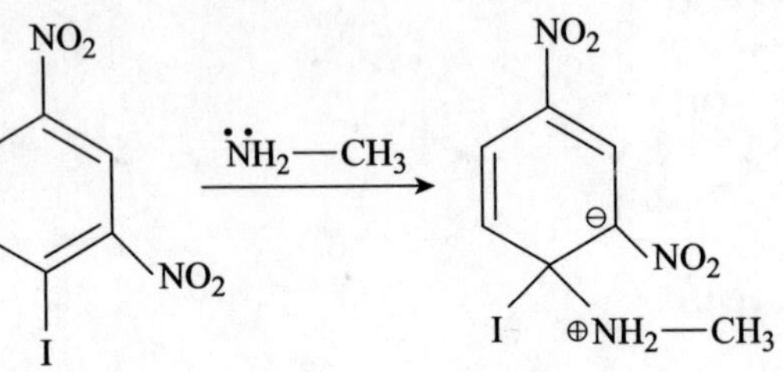

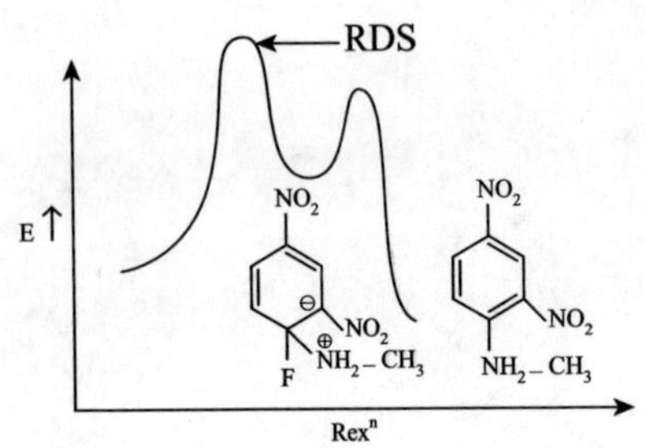

F has value of electronegativity as compared to Cl and I stabilises the intermediate carbanion formed.

Secondly, 1st step is RDS not the second one in which F, Cl and I leave with –ve charge, hence rate does not depend upon leaving ability. **Ans:** (b)

28.

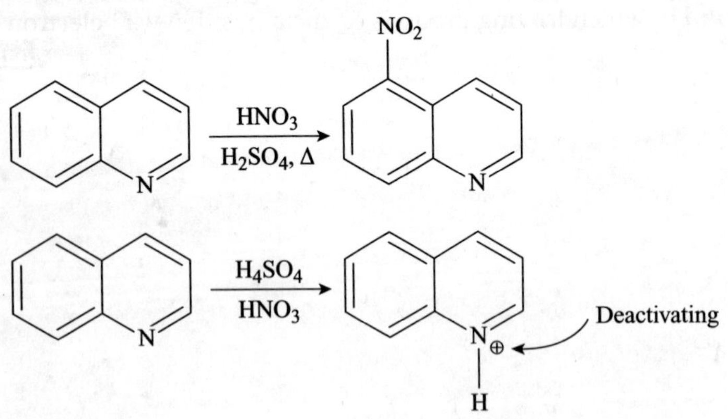

Ans: (b)

33.

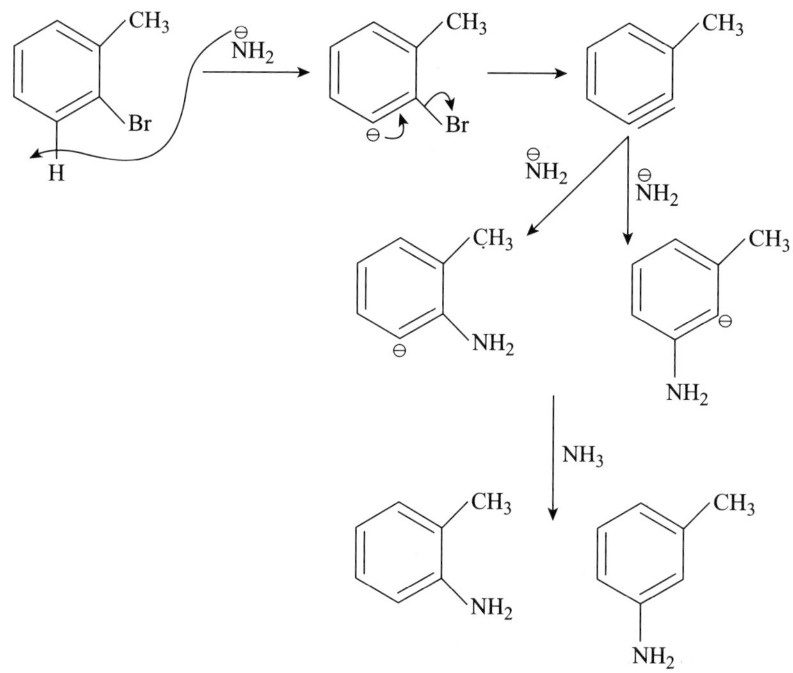

Benzyne is formed as an intermediate and it is electrophile in nature. $-NH_2$ can attack on either side of the triple bond.

Ans: (d)

35.

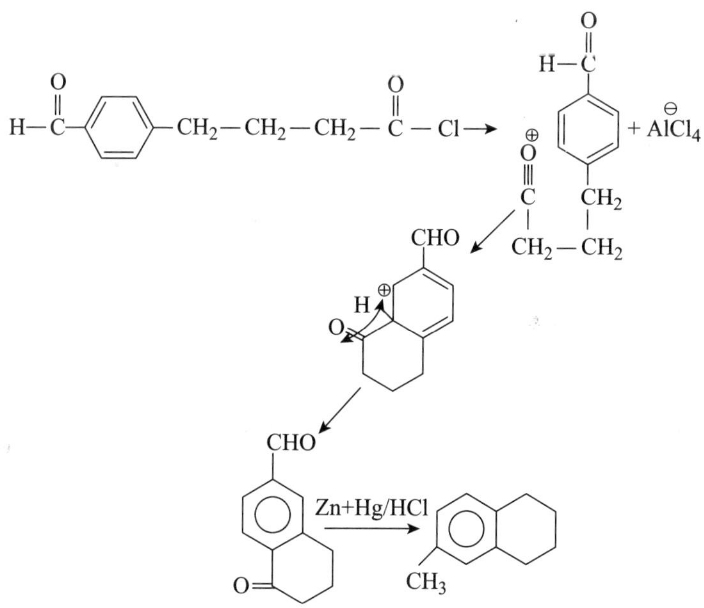

Hint: —C(=O)—H group is meta-directing.

Ans: (b)

36.

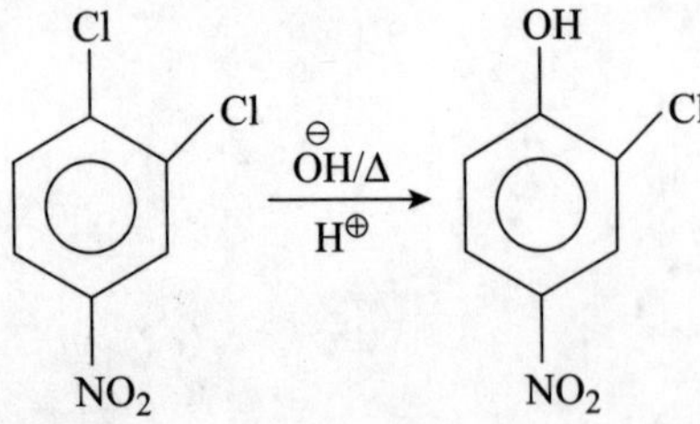

Chlorine which on para-position w.r.t $-NO_2$ group will be substituted (see explanation of Question Number 25 of this excercise).

Ans: (b)

37.

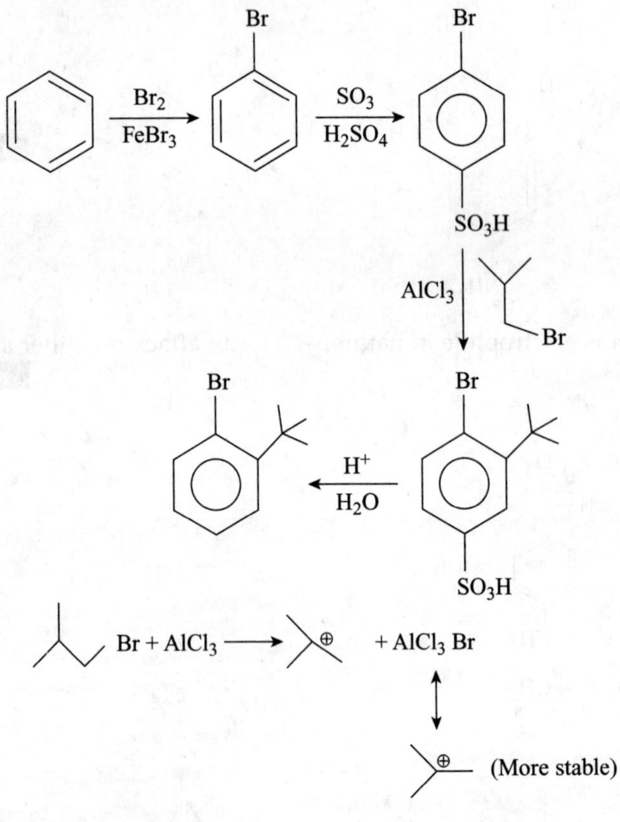

Ans: (c)

38. If leaving group is present on ortho/para positions of electrons with drawing group then it will be more reactive towards nucleophilic substitution reaction.

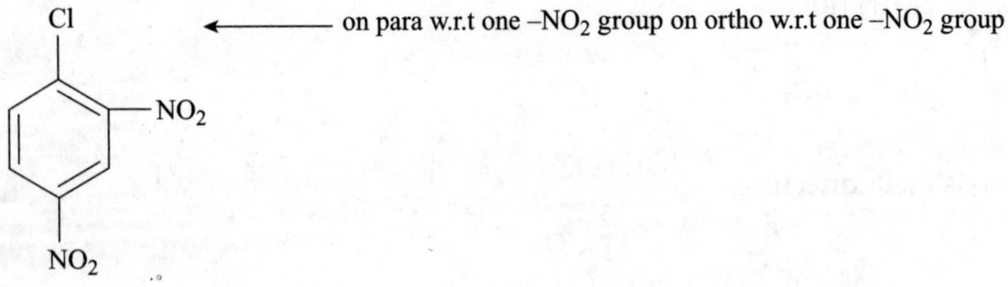

Ans: (d)

41.

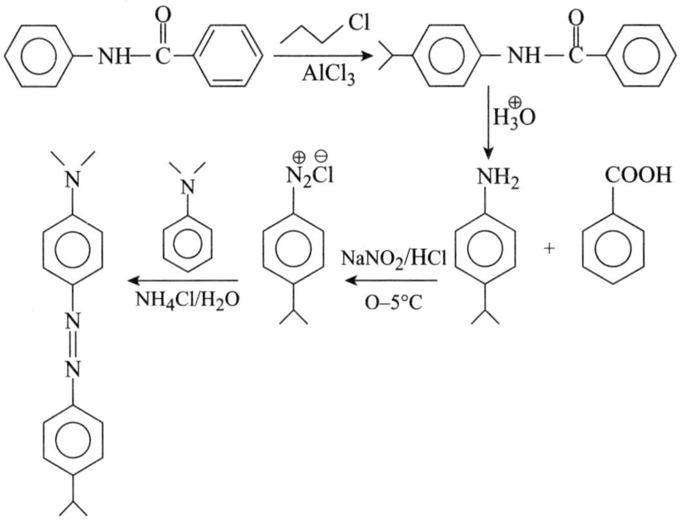

Ans: (b)

43.

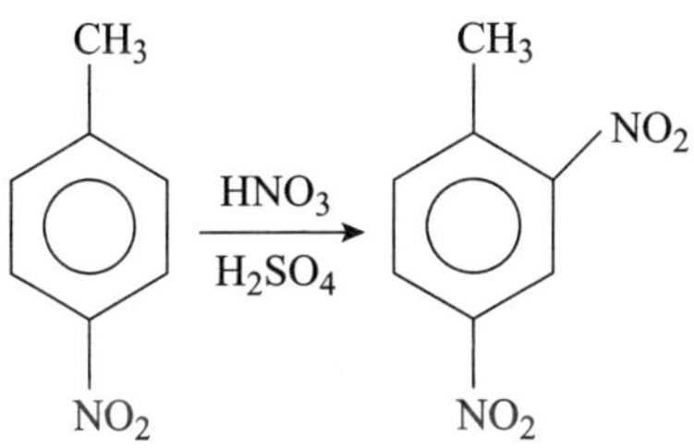

Here $-CH_3$ is moderate activating group and $-NO_2$ is strong deactivating group, substitution takes place at ortho of $-CH_3$ and meta of $-NO_2$ where both group supports. **Ans:** (a)

44.

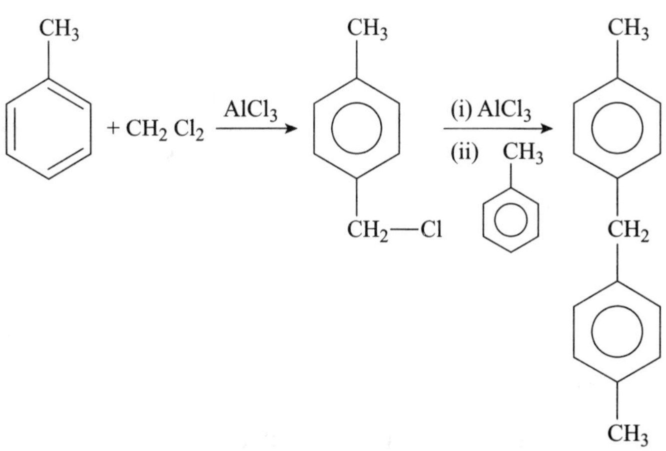

Ans: (d)

46.

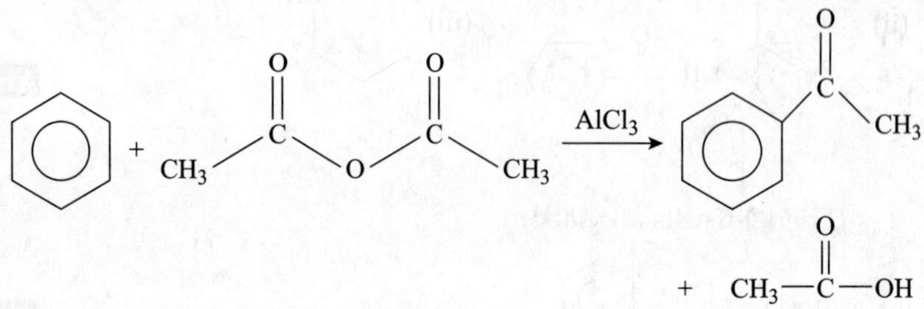

Ans: (c)

45

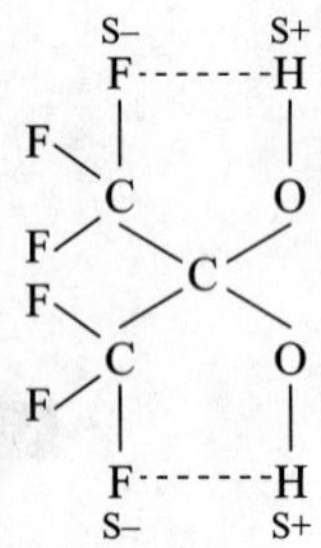

Ans: (b)

46.

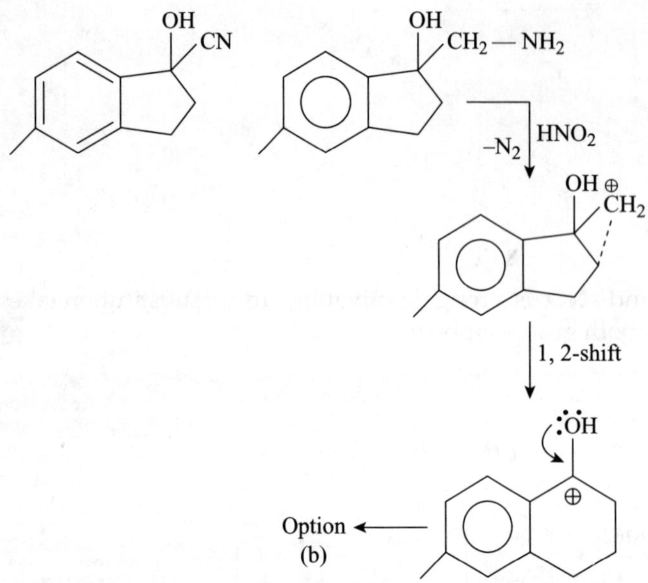

Ans: (c)

47. $PH = Pk_a + \log_{10}\frac{[CH_3COO]}{[CH_3COOH]}$

$= 4.76 + \log_{10} 10$

$= 5.76.$

Ans: (b)

49. Acidic strength : $NH3 < R-C\equiv CH < H_2O$

Acid-base equilibrium shifts from strong acid and strong base towards weak conjugate acid or weak conjugate base.

$$\underset{SA}{CH_3-C\equiv CH} + \underset{SB}{\overset{\ominus}{N}H_2} \xrightleftharpoons{Keq > 1} \underset{\text{Weak conjugate base}}{CH_3-C\equiv \overset{\ominus}{C}} + \underset{\text{Weak conjugate acid}}{NH_3}$$

Ans: (b)

50. (i) 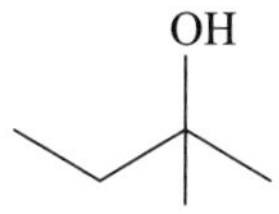 (ii) 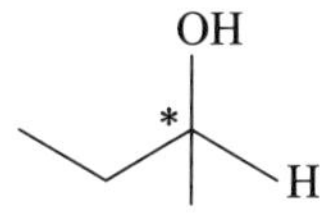 (iii) 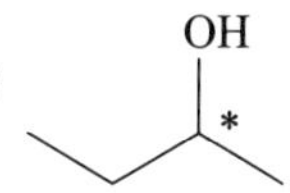

Ans: (b)

51. 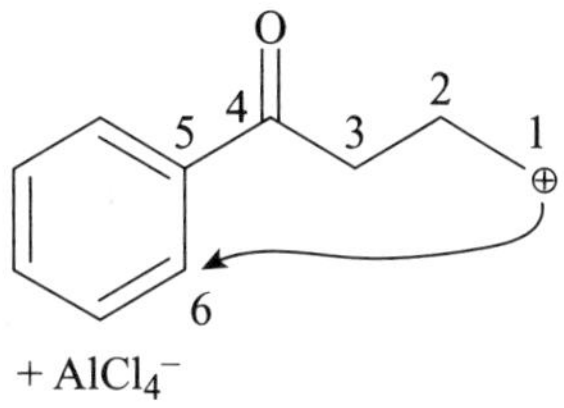

(Friedel-Crafts alkylation)

Ans: (b)

52. 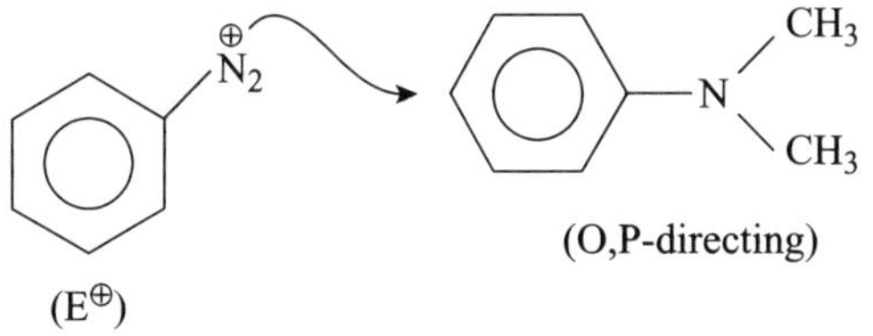

Ans: (b)

54. 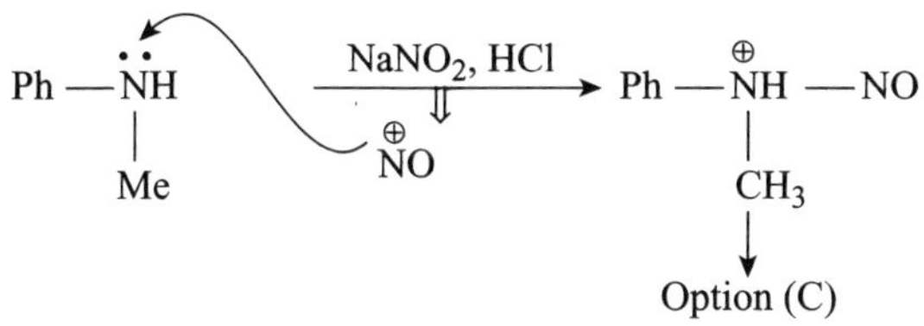

Ans: (c)

55. (1) Friedel-Crafts acylation

(2) Wolff-Kishner reduction

Ans: (b)

56. 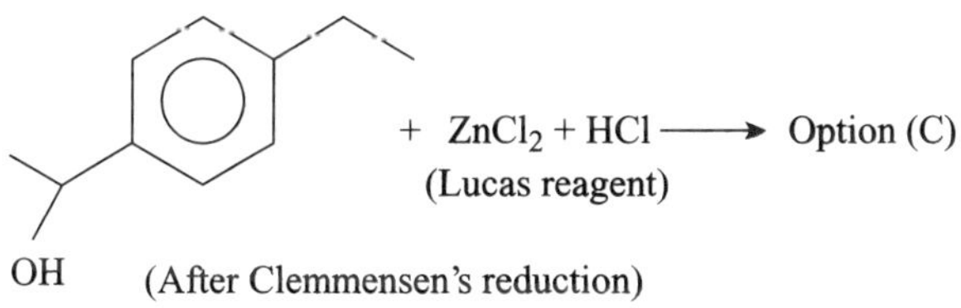

Ans: (c)

57. 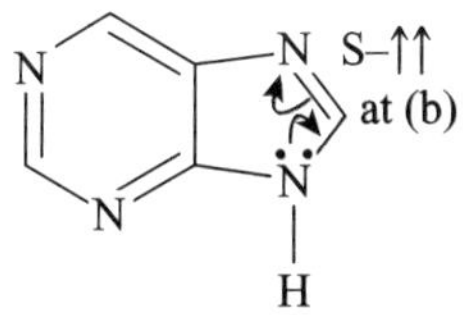

so will donate LP easily.

Ans: (b)

58.

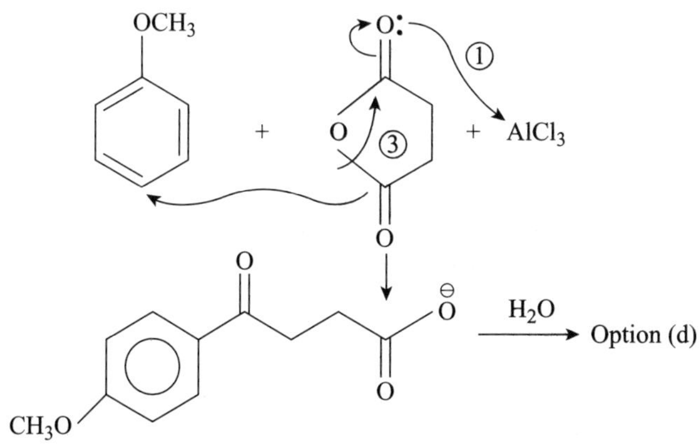

Ans: (d)

61. Refer Chapter-4, Level 2: Question Number 21. **Ans:** (d)

62. $NaNO_2$, HCl ⇒ $NO^{\oplus}$(as $E^{\oplus}$). **Ans:** (c)

63. EDG increase stability of a-complex while EWG decrease stability. **Ans:** (c)

64. 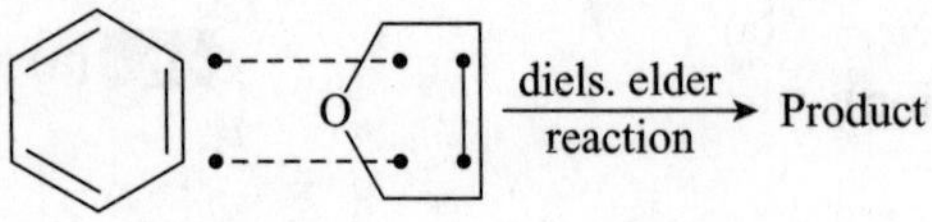

Ans: (d)

65. Nucleophilic aromatic substitution. **Ans:** (a)

66. $-CF_3$: Meta directing **Ans:** (c)

67. 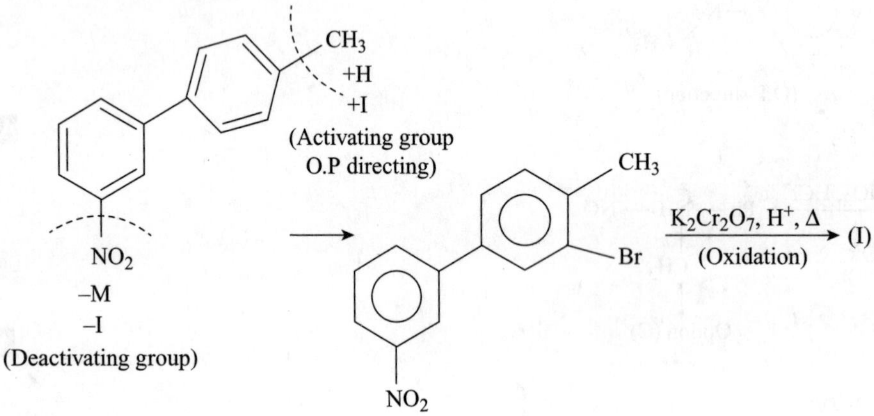

Ans: (a)

68. Path-1:

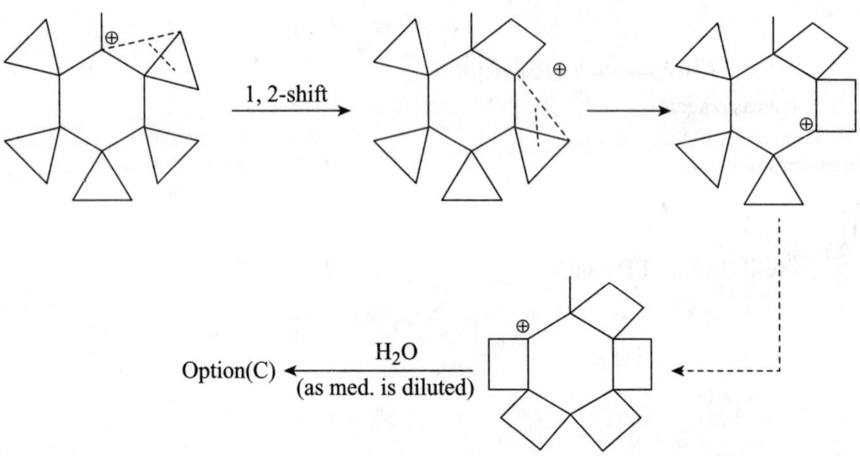

Path-2:

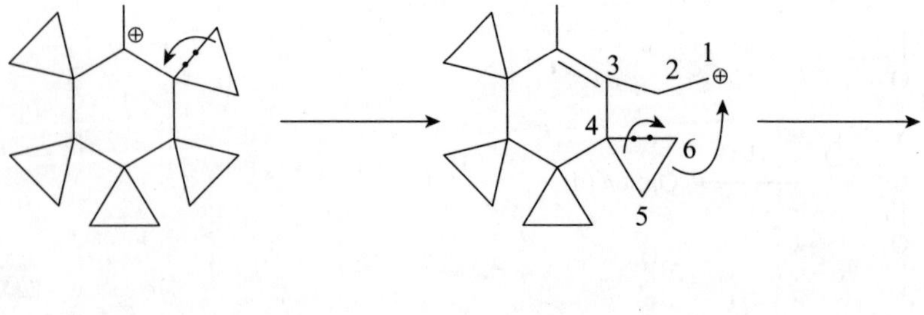

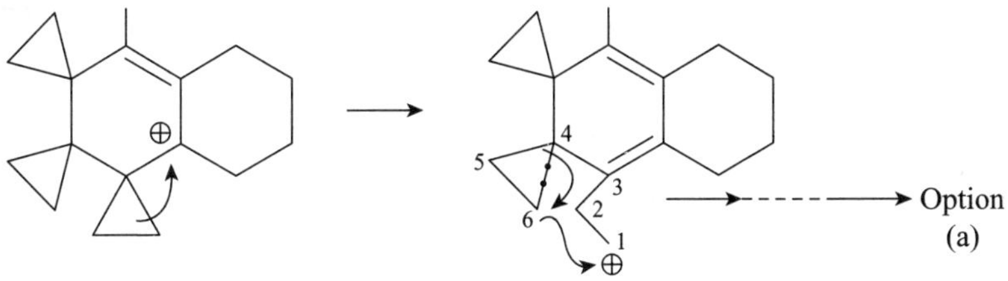

Ans: (a) and (c)

72. 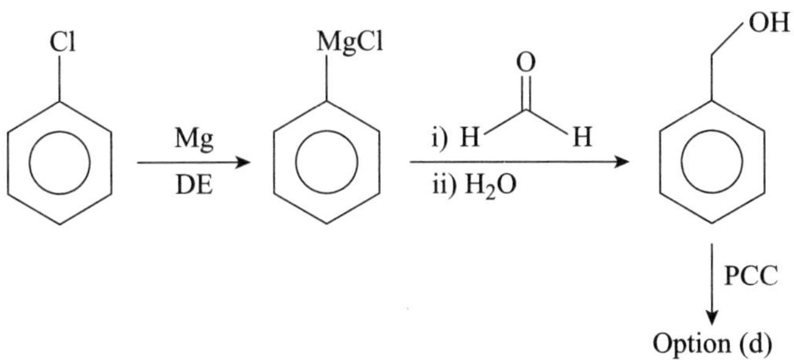

PCC: used for partial oxidation (alcohol to aldehyde)

Ans: (c)

76. P_2O_5 : dehydrating agent

Ans: (b)

85. (a), (c), (d): 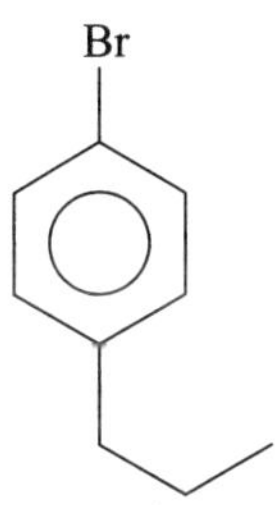

Ans: (b)

88. 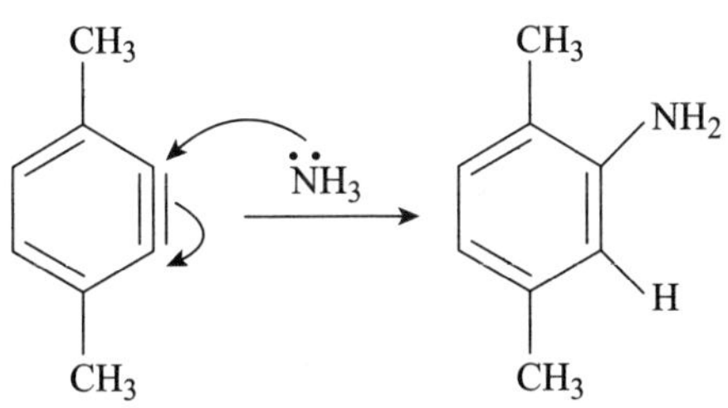

Ans: (b)

LEVEL 2

2.

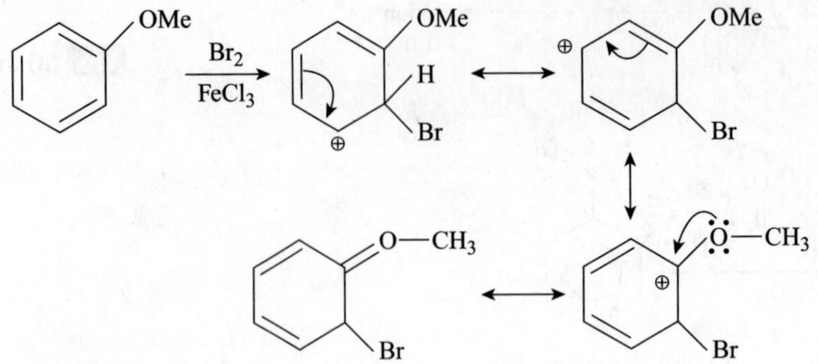

Ans: (a)

4.

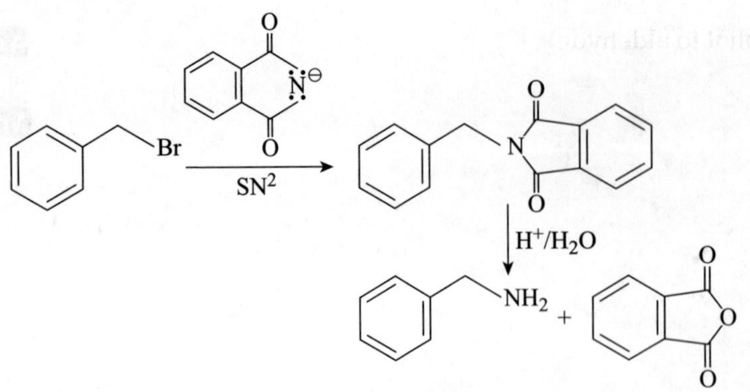

Ans: (b)

9.

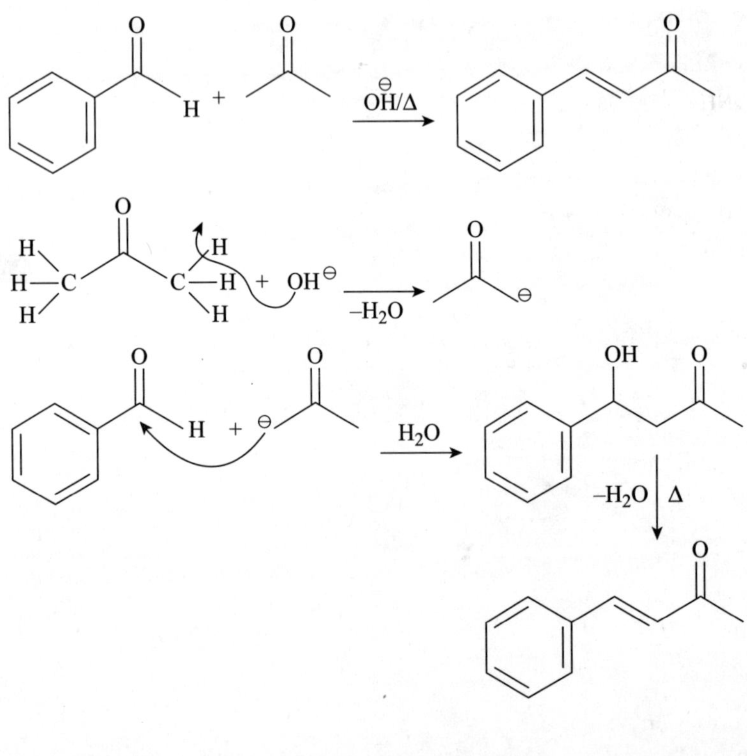

Ans: (c)

10.

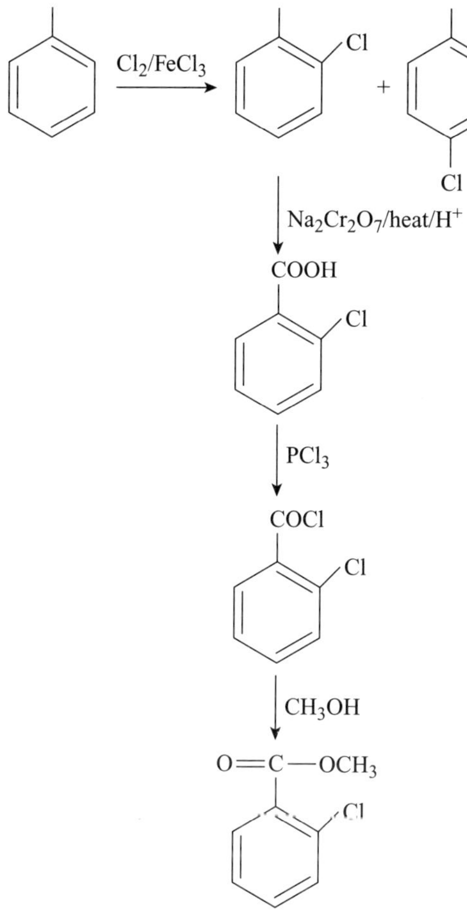

Ans: (a)

11.

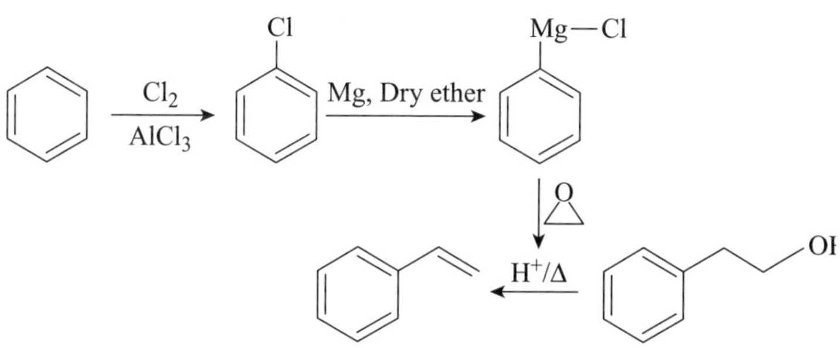

Ans: (b)

12.

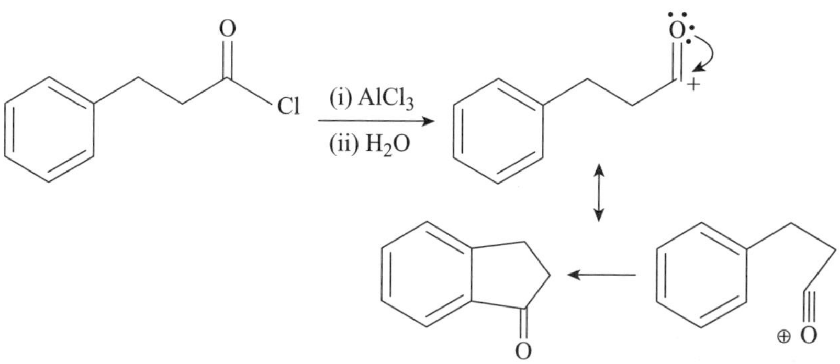

Ans: (b)

13.

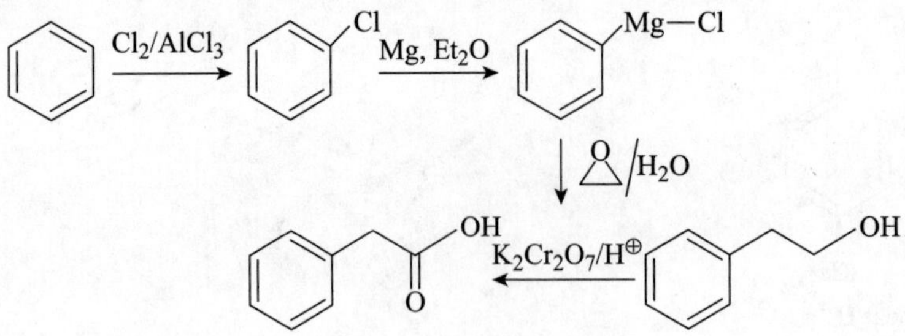

Ans: (b)

15.

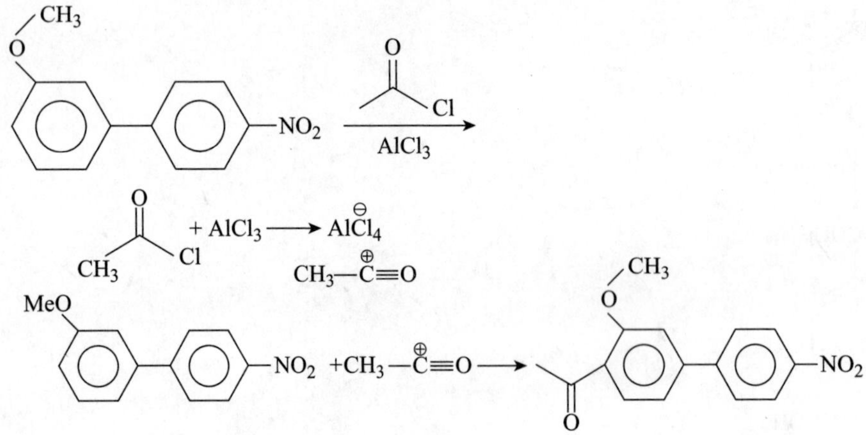

Do you remember ? We have learnt that in case of biphenyl compounds ESR takes place on that ring which is more activated.

— O — CH_3 → Activating group

— NO_2 → Strong deactivating group

Ans: (c)

17.

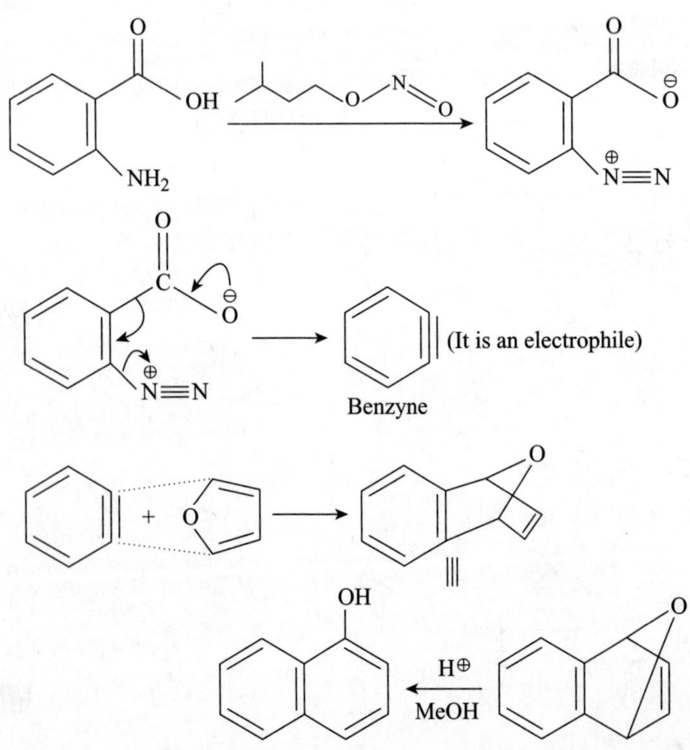

Ans: (a)

19.

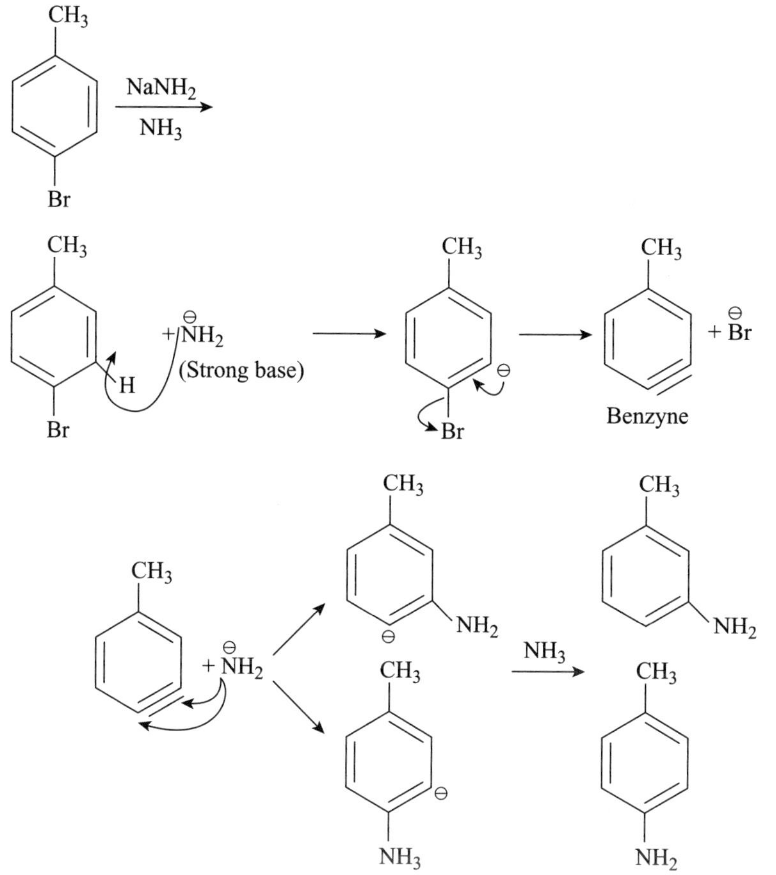

Ans: (c)

24.

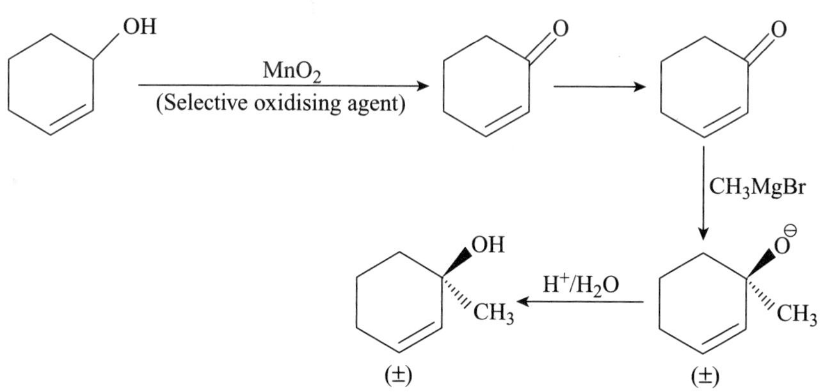

Ans: (a)

25.

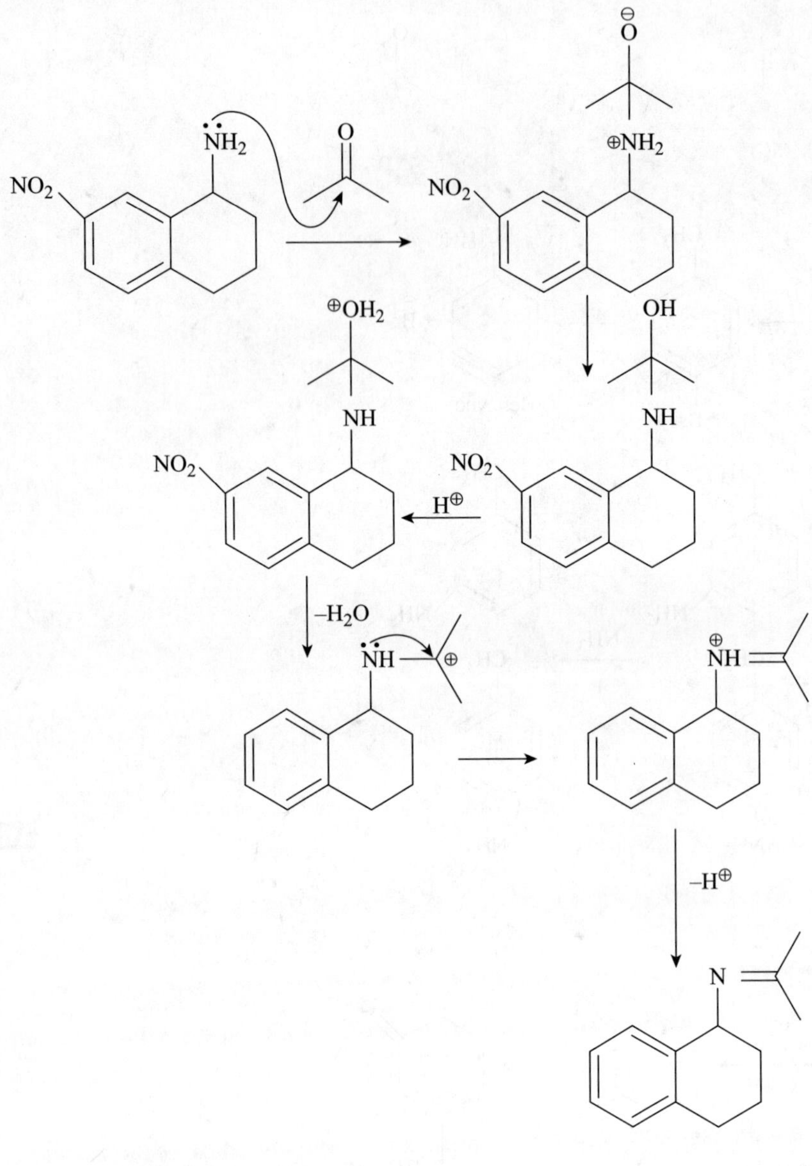

Ans: (b)

29.

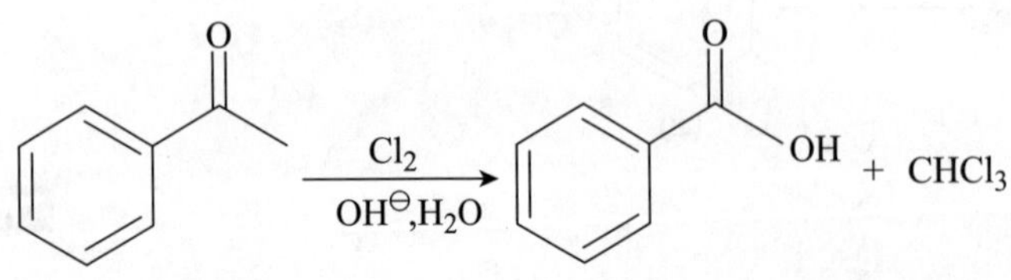

Ans: (d)

30.

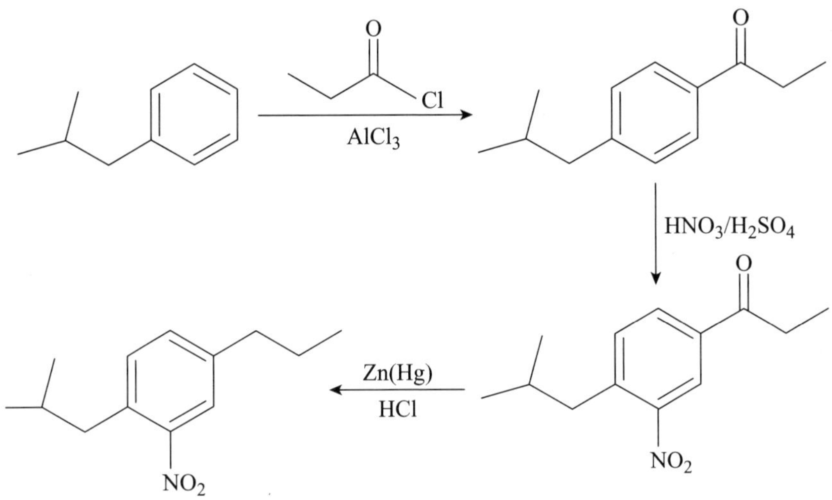

Ans: (b)

31.

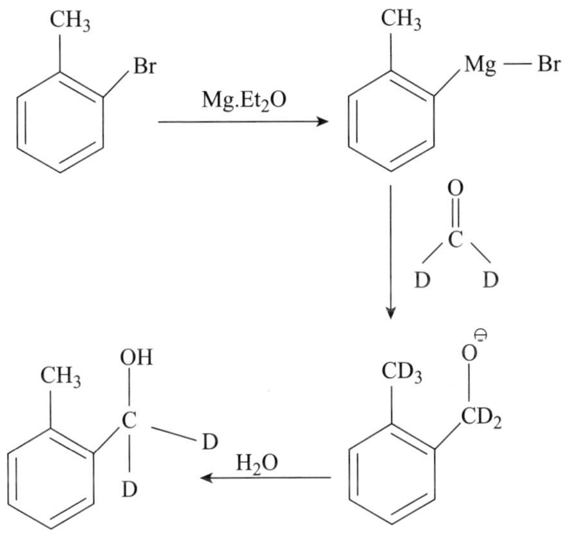

Ans: (d)

32. (a)

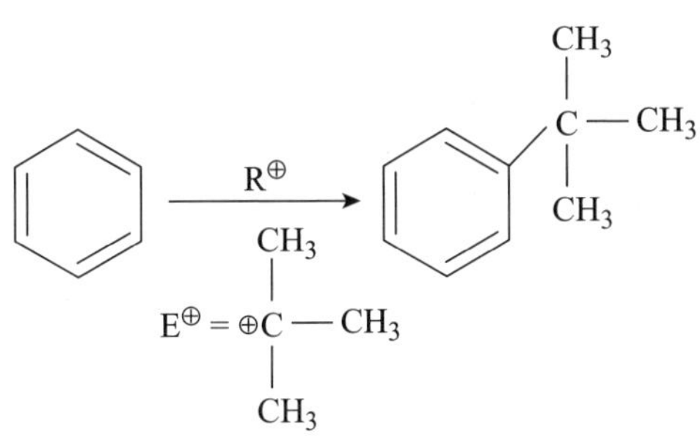

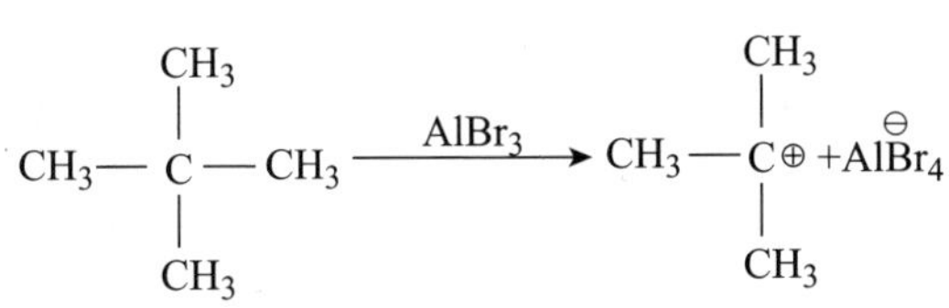

(b)

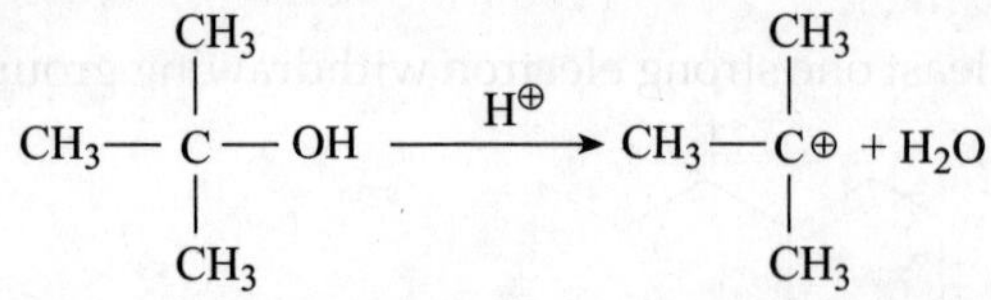

(c)

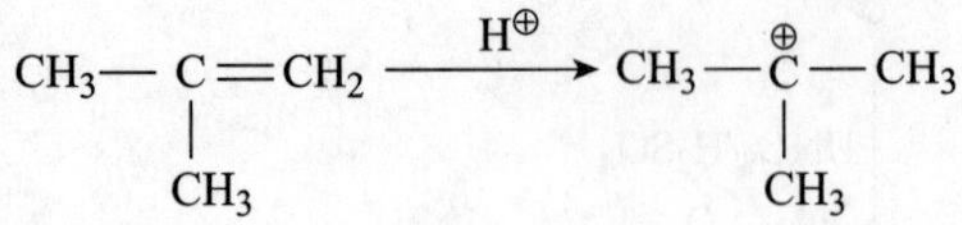

Ans: (a) (b) and (c)

33. If any deactivating group is present on the ring the Friedel-Crafts reaction do not occur on the ring because electrophile formed during the reaction are soft electrophile.

(a)

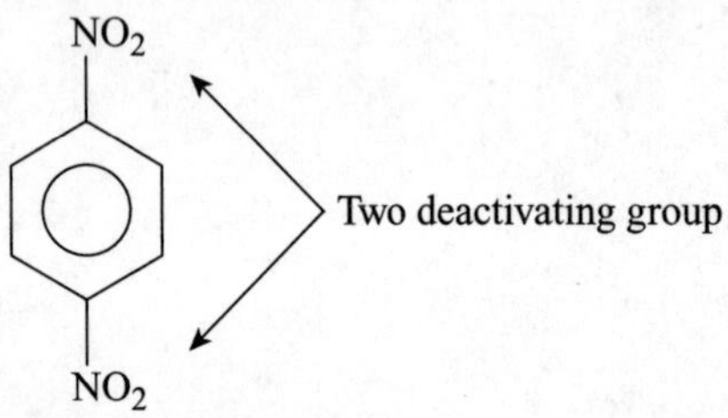

(b)

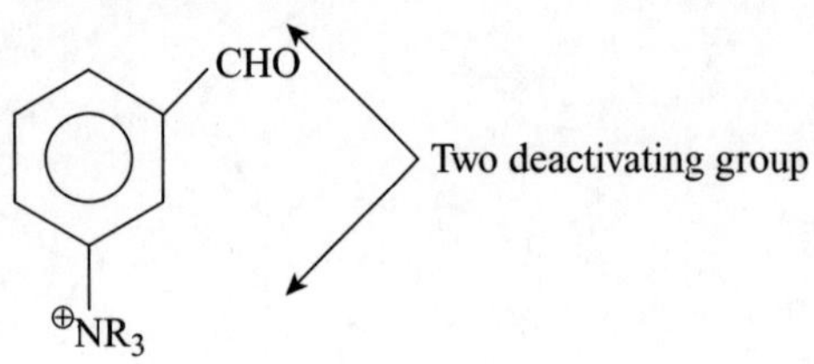

(c)

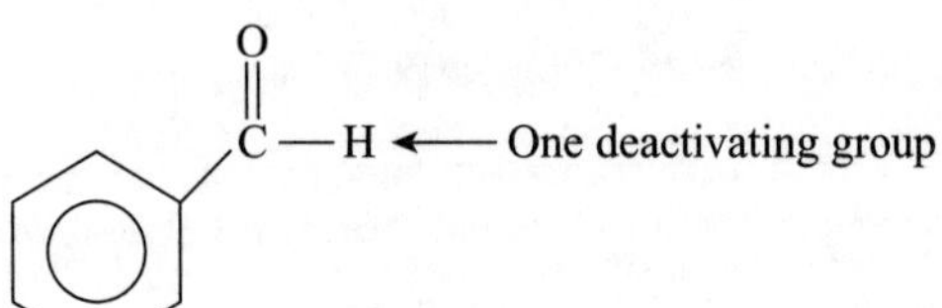

(d)

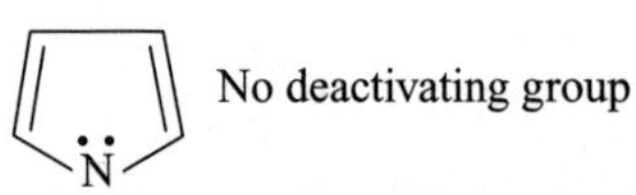

Ans: (a) (b) and (c)

34. We have already discussed that in case of biphenyl compounds ESR takes place at that ring which more activated at that site which is less hindered. If ortho and para positions w.r.t activating group are equally hindered then reaction takes place on both positions.

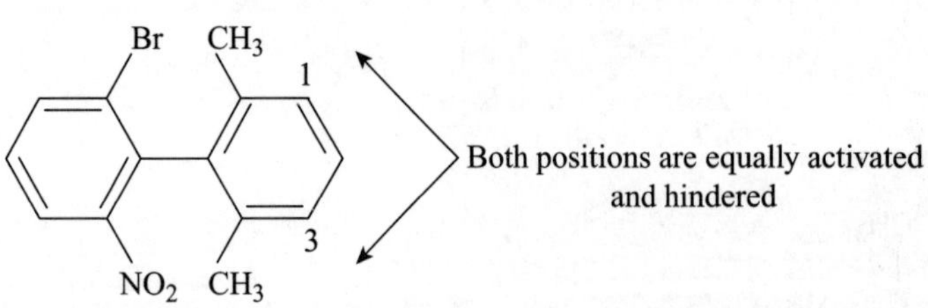

Ans: (b) and (c)

37. For nucleophilic substitution reaction on benzene atleast one strong electron withdrawing group must be present on the ring.

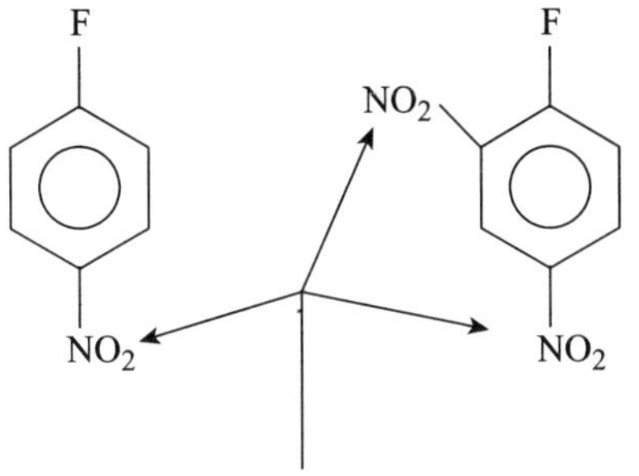

Electron drawing with groups.

Ans: (b) and (c)

40. Anti-aromatic compounds are unstable at room temperature. **Ans:** (a), (c) and (d)

42. (a)

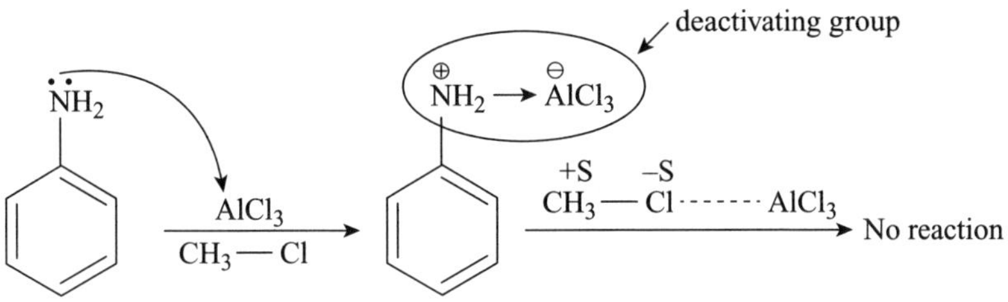

(b)

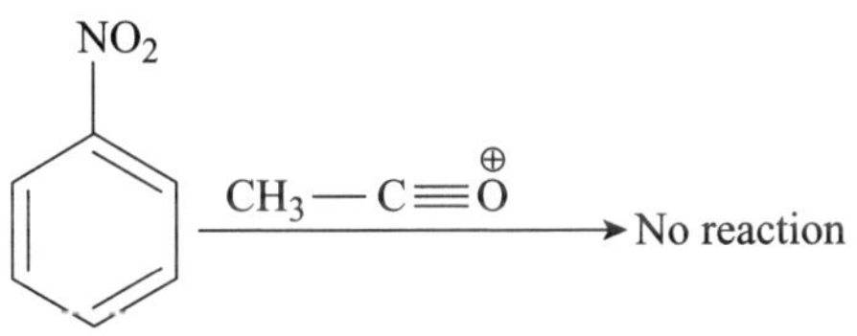

If a single deactivating group is present on the ring Friedel-Crafts reaction does not occur. (See solution of Question Number 33). **Ans:** (a) and (b)

43. In non-polar solvent tribromo substituted product is not formed.

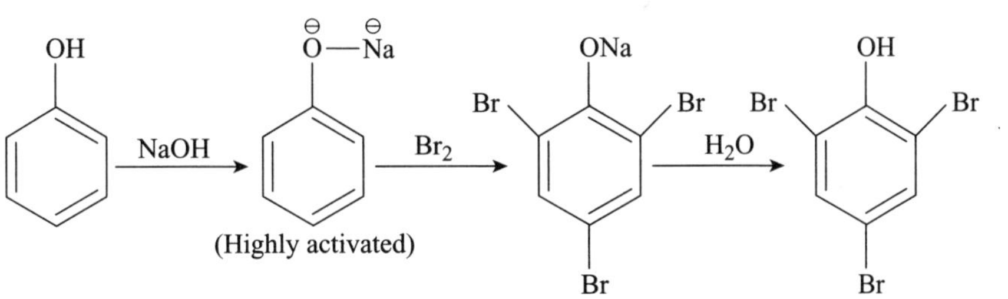

(d)

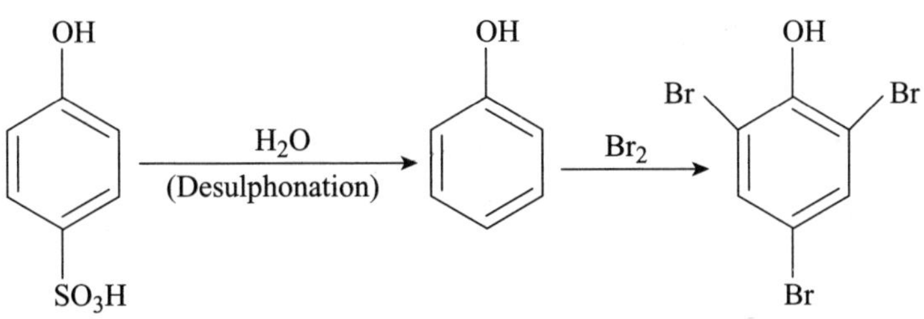

Ans: (a), (b) and (d)

51. (a) $(NH_4)_2CO_3 \xrightarrow{\Delta} 2NH_3 + H_2CO_3$

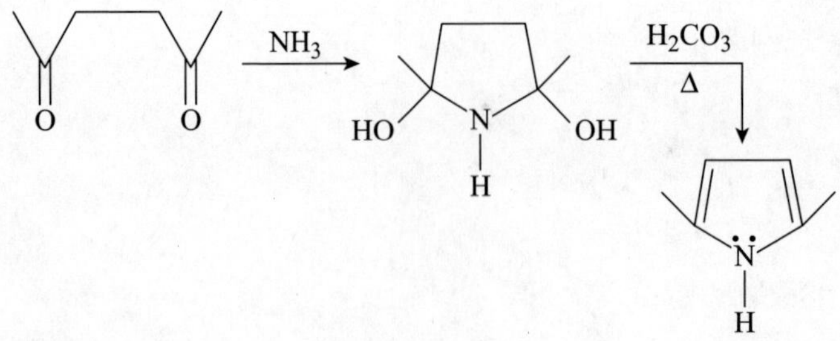

(Aromatic)

(b) 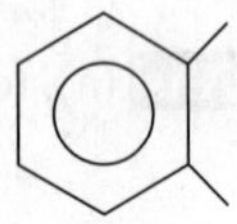 (by Wolff–Kishner reduction)

(c) 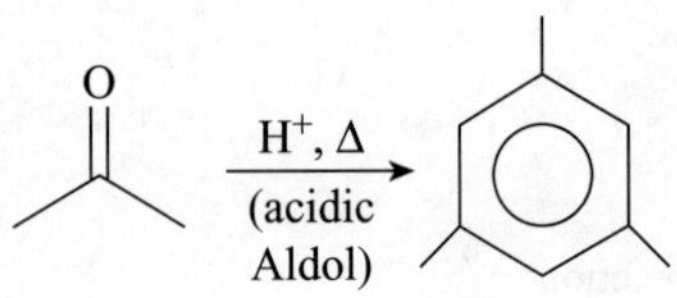

Ans: (a), (b) and (c)

52.

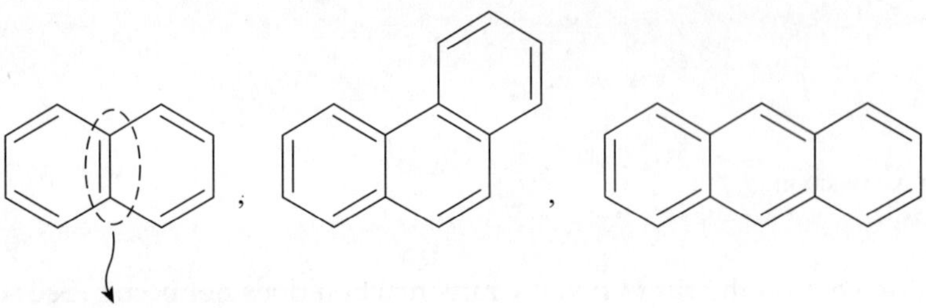

This π doing cross conjugation making both ring aromatic simultaneously; called annealation effect.

Ans: (b), (c) and (d)

54. 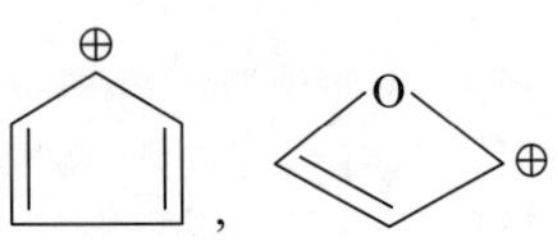 are anti-aromatic.

Ans: (a) and (c)

55. During migration "R" does not change its configuration.

Ans: (a)

57. ROR α migration aptitude of "R"

Migratory aptitude : 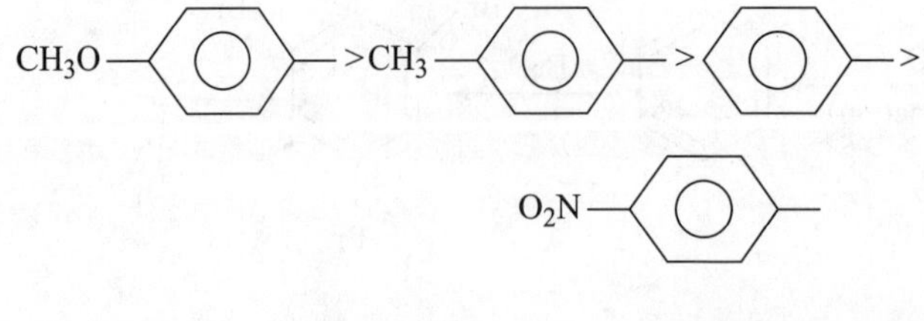

Ans: (c)

60. 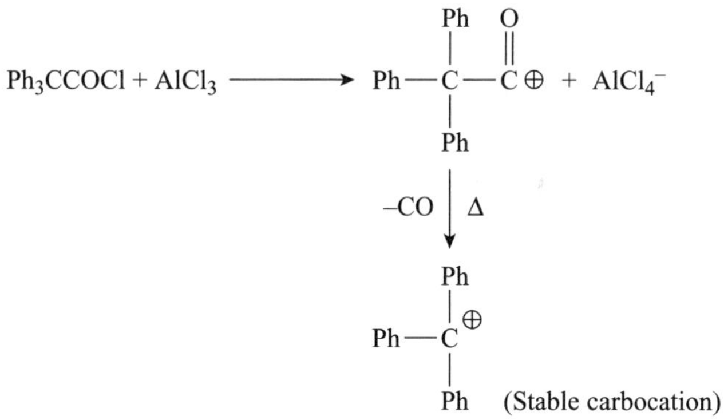

{act as $E^{\oplus}$ NOIN} **Ans:** (b)

63. 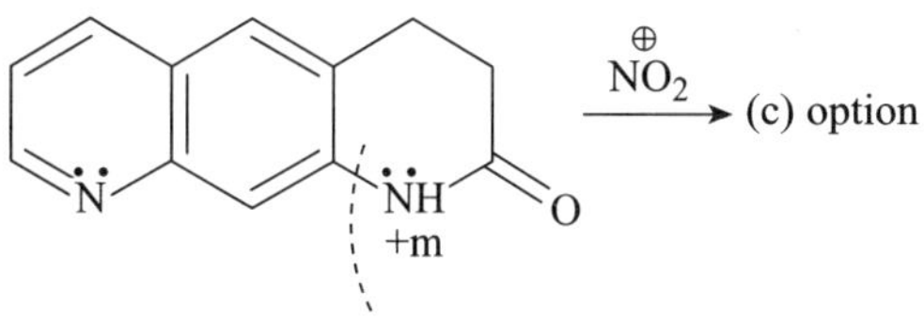

Ans: (c)

64. In case of G as EWG or EWG attach to ortho or para position with respect to group G, s-complex (arrehenium ion) will be destabilised leading to decrease of ROR for ipso substitution. **Ans:** (c)

65. 2nd NO_2^+ will not attach to benzene ring, as after 1st attachment of $-NO_2$ to ring, e–density in ring decreases very much. **Ans:** (d)

66. 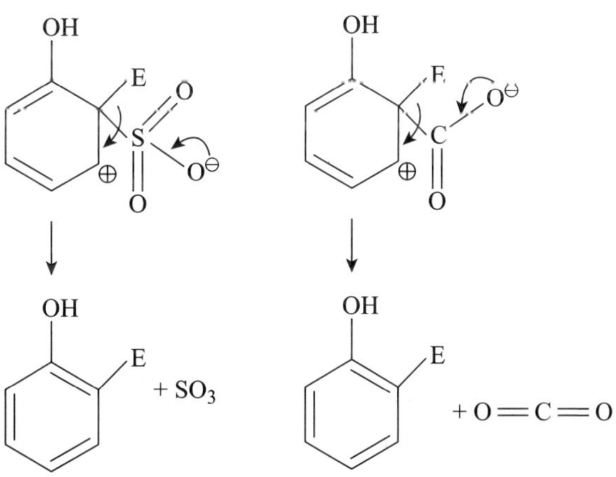

$-SO_3$ and CO_2 are good LG as formation of C = C and S = O bond provides thermodynamic stability.

Ans: (c)

69. 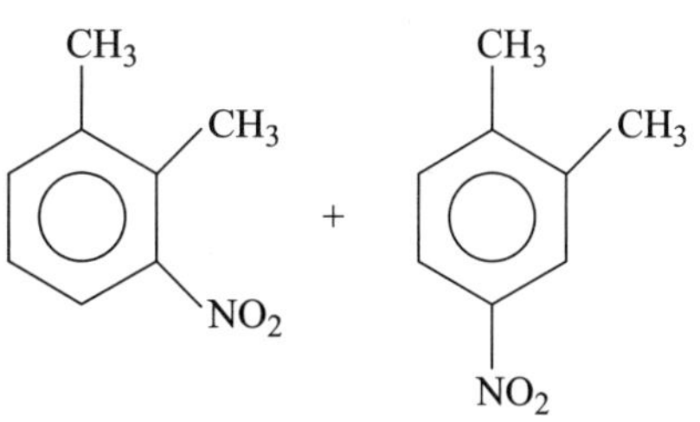

Ans: (a)

70.

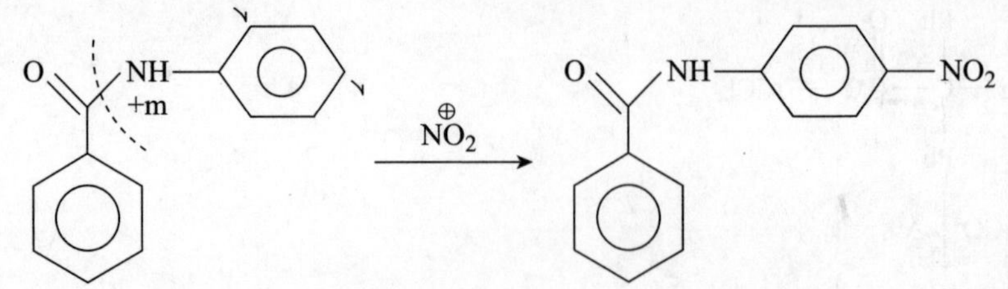

Ans: (d)

71.

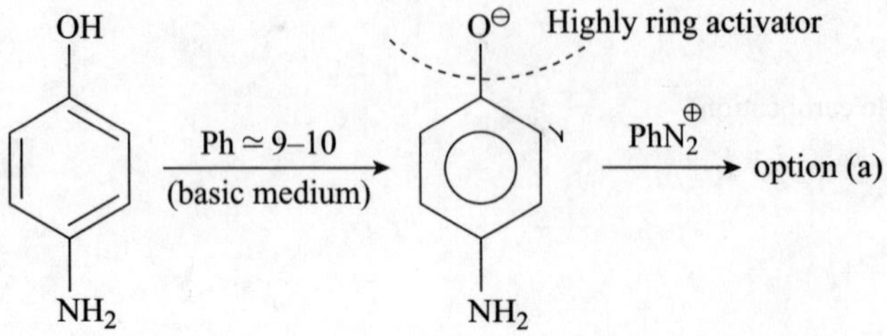

Ans: (a)

Passage 7

X : 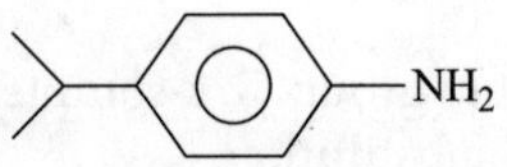

Y : 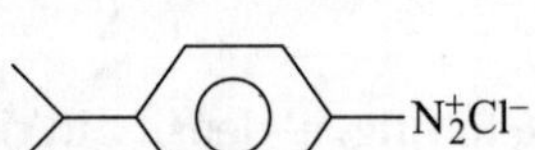

P : 

Q : 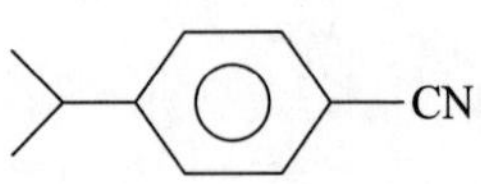

R : 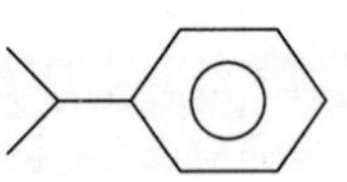

S : 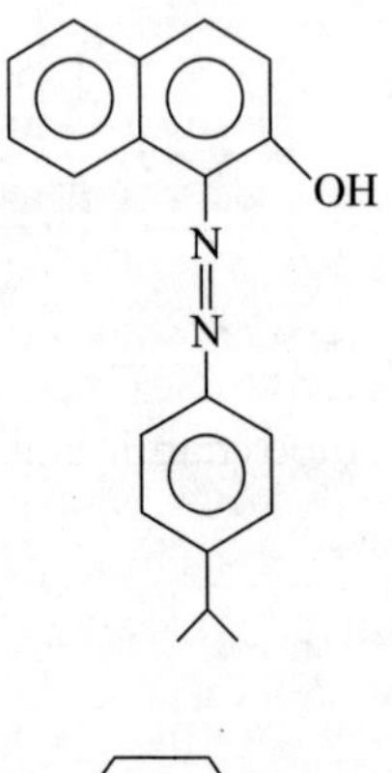

73.

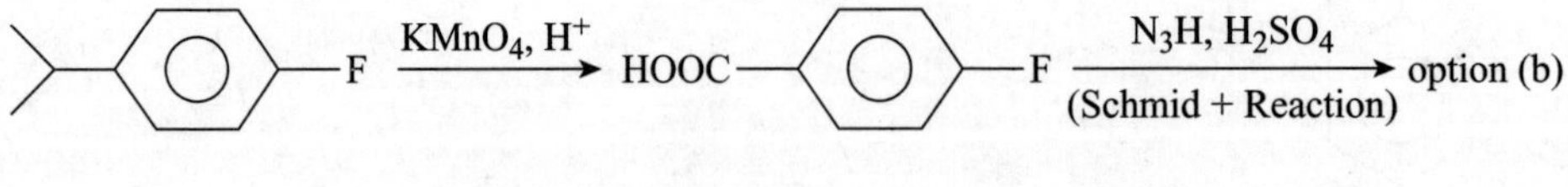

Ans: (b)

74.

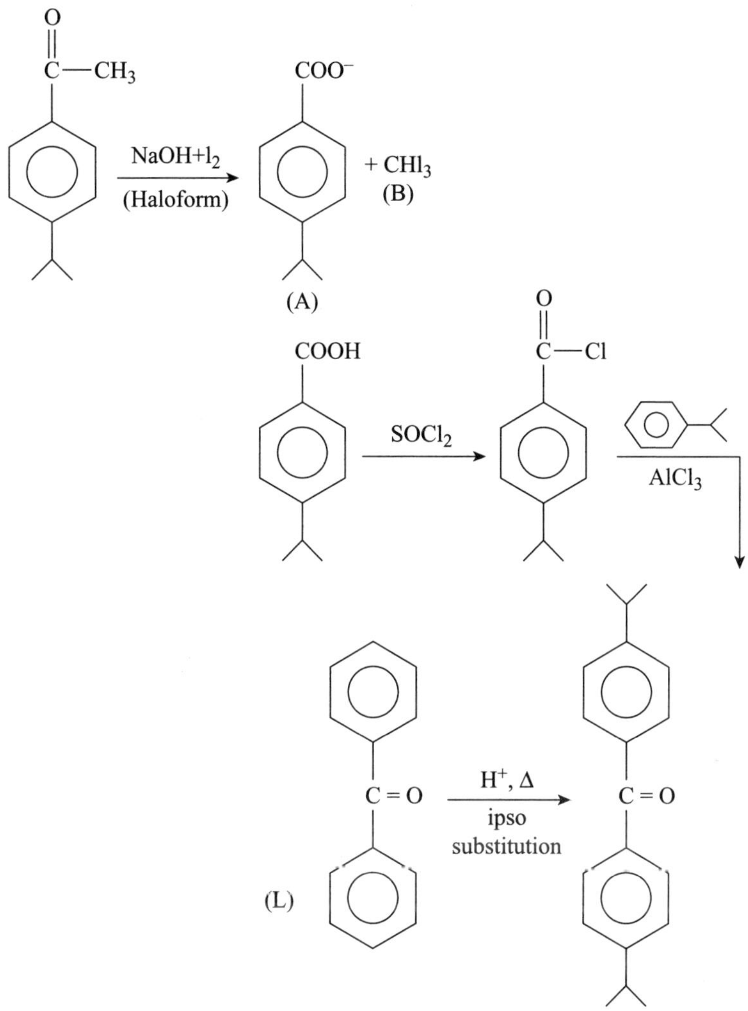

Ans: (b)

76.

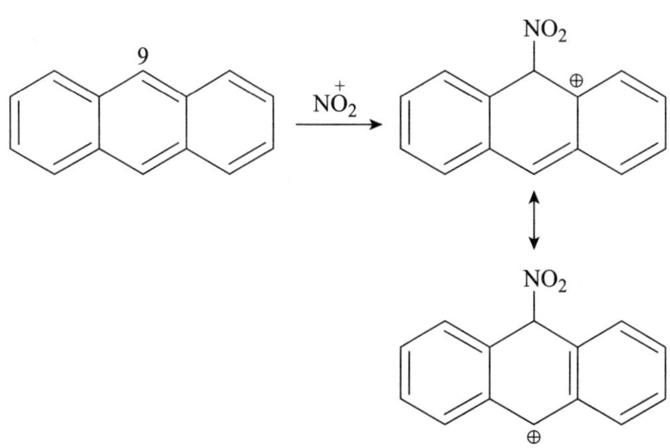

Stable intermediate
(as two rings are aromatic
simultaneously)

Ans: (b)

77. and 78.

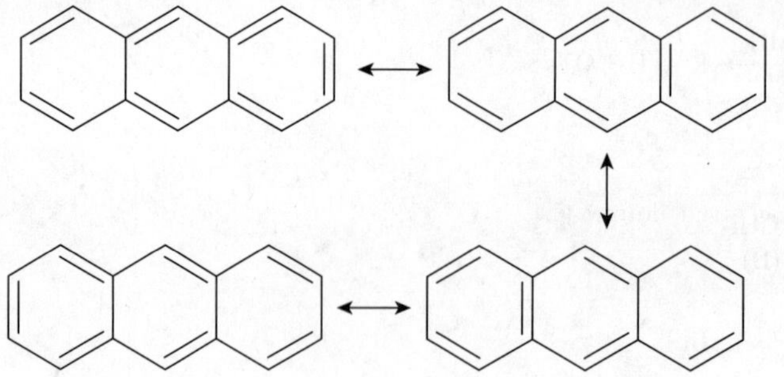

Ans: (b) and (b)

81.

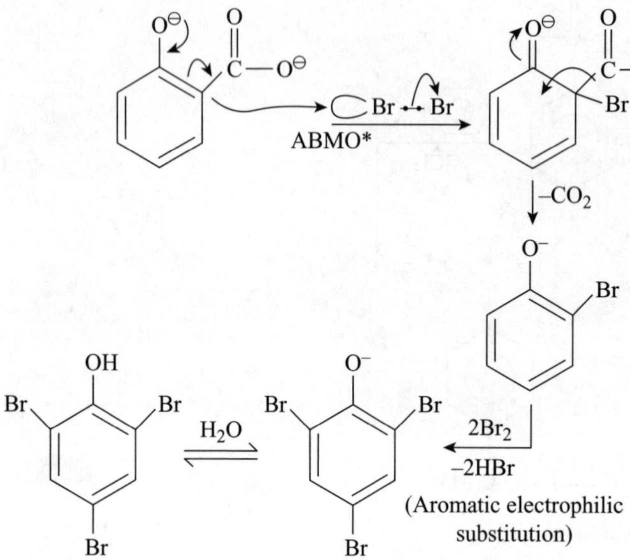

Ans: (c)

85. (a) in 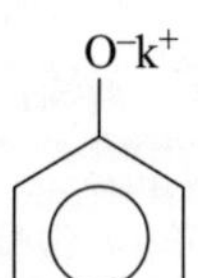 para product will be major due to steric factor (as large size of k+) **Ans:** (s)

(b) While in 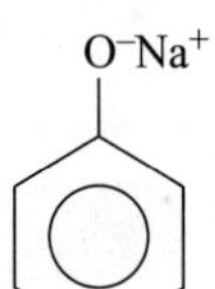 Steric factor is not dominating (as small size of Na+), thus ortho product will be major as stabilised by chelation. **Ans:** (qr)

(d) **Ans:** (p)

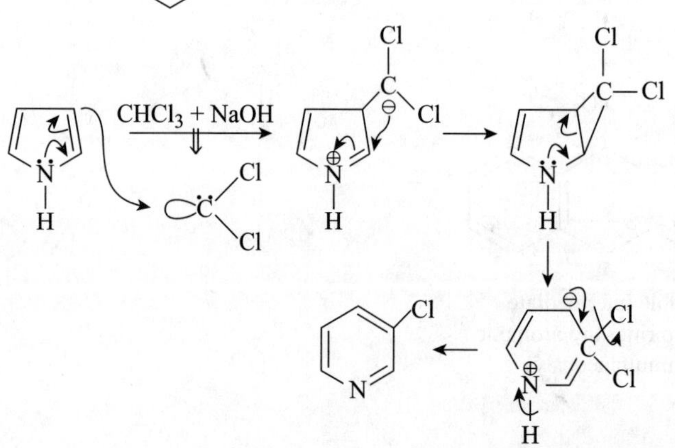

86. (b) $R—OH \xrightarrow{TSCl} R—OTs \xrightarrow[\Downarrow\ H^-]{LiAlH_4} R—H + \overset{\ominus}{O}Ts$ **Ans:** (q)

88. (a) 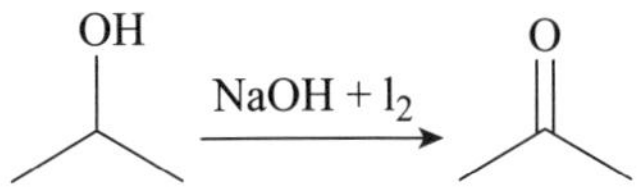

can give Iodoform test **Ans:** (r) and (s)

(b) 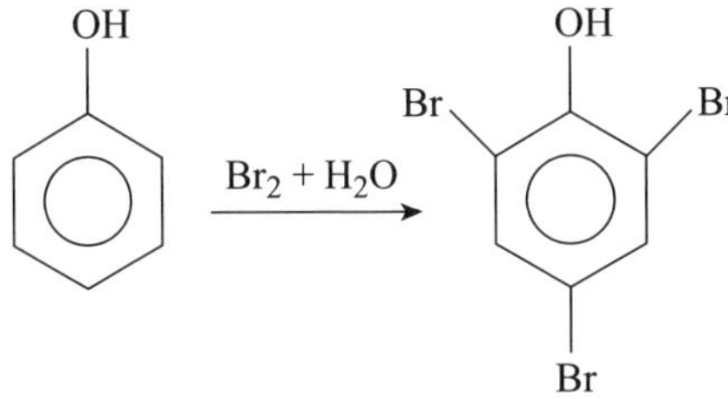

{Brown colour of bromine water decolourises} **Ans:** (p) and (q)

(c) 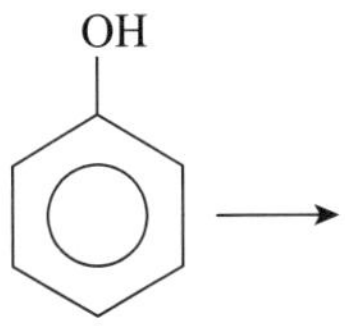

⟶ Positive Tollen's test **Ans:** (q) and (r)

(d) $Me—CH_2—C \equiv CH \xrightarrow{Tollen} Me—CH_2—C \equiv C^-Ag^+$ White ppt.

$\xrightarrow{[(CuCNH_3)_2]OH} Me—CH_2—C \equiv C^-Cu^+$ Red ppt. **Ans:** (q) and (t)

89. 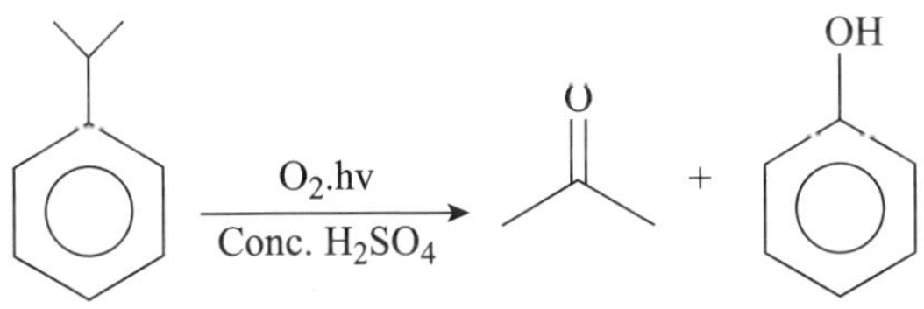

; Industrial method for preparation of phenol **Ans:** (p)

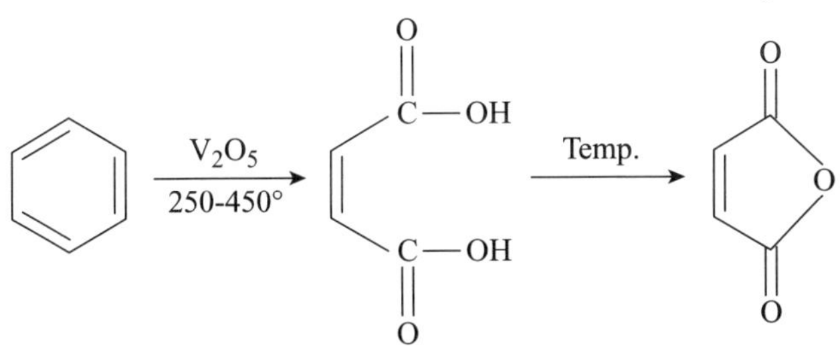

90. (a) 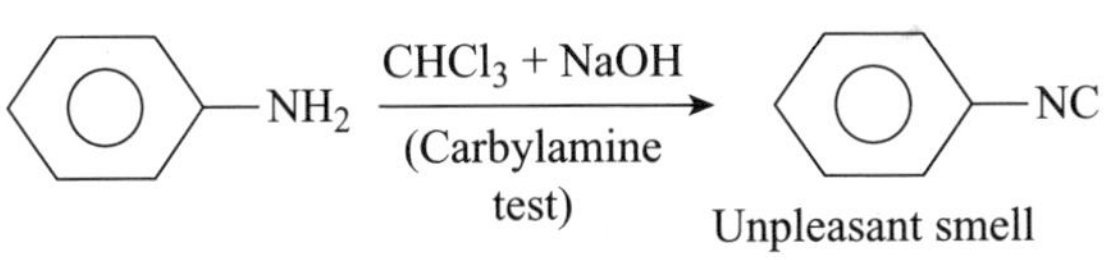

Ans: (q)

(d) 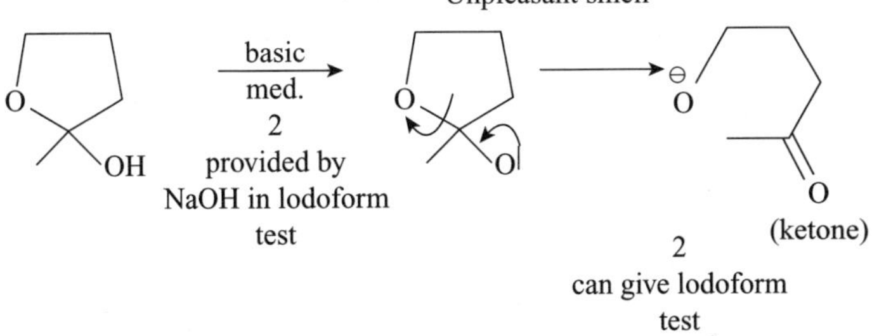

Ans: (r)

91. (d)

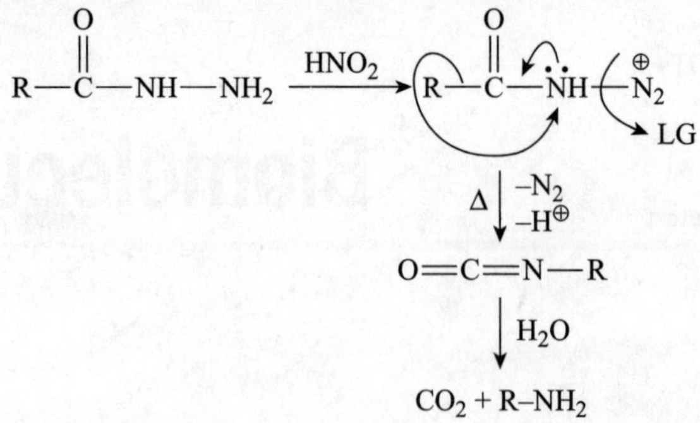

Ans: (s)

93. (a) : Hofmann rearrangement (Hofmann bromamide degradation) **Ans:** (p), (q) and (s)

(b) : Schmidt rearrangement **Ans:** (p) and (s)

(c) : $CH_3-\overset{O}{\overset{\|}{C}}-O^{-}\overset{+}{N}H_4 \xrightarrow[-H_2O]{\Delta} CH_3-\overset{O}{\overset{\|}{C}}-NH_2$ **Ans:** (r)

(d) : Curtius rearrangement **Ans:** (p) and (s)

94. Beilstein test : Cu wire $\xrightarrow[\text{ii) dip wire in solution of halogen; iii) } \Delta \text{ in flame}]{\text{i) } \Delta \text{ in presence of } O_2}$ Cux_2 green flame

x : Cl, Br, I

Baker-Mulliken's test :

$R-NO_2 \xrightarrow[NH_4Cl,\ H^+]{Zn} R-NH-OH \xrightarrow{\text{Filter}} \text{Filterate}$

$\text{Filterate} \xrightarrow{\text{Tollen's reagent}} R-NO + Ag$ (black ppt.)

96. Wurtz : follow both ionic and free-radical mechanism

Diazotisation : $R-N_2^{\oplus} \xrightarrow{-N_2} R^{\oplus}$

–Carbocation

–Electrophile

97. i, iv, v, viii, ix. **Ans:** (5)

98. vii, x, xi, xii. **Ans:** (4)

99. i, ii, vii. **Ans:** (3)

100. i, ii, viii, ix, x, xi. **Ans:** (6)

101. iii, iv, v, vi, vii, viii, xii. **Ans:** (7)

CHAPTER 7

Biomolecules

LEVEL 1

1.

$$\begin{matrix}CHO\\ |\\ (CH-OH)_n\\ |\\ CH_2-OH\end{matrix} \xrightarrow{\text{Kilani Fischer}} \begin{matrix}CHO\\ |\\ CH-OH\\ |\\ (CH-OH)_n\\ |\\ CH_2OH\end{matrix} + \begin{matrix}CHO\\ |\\ HO-CH\\ |\\ (CH-OH)_n\\ |\\ CH_2OH\end{matrix}$$

$$\xrightarrow[\text{Method}]{\text{Ruff's degradation}} \begin{matrix}CHO\\ |\\ (CH-OH)_{n-1}\\ |\\ CH_2OH\end{matrix} + CO_2$$

Ans: (d)

2. Sucrose $\xrightarrow[\text{hydrolysis}]{\text{hydrolysis}}$ D-glucose + D-fructose

Lactose $\longrightarrow$ D-glucose + D-galactose

Maltose $\xrightarrow{\text{hydrolysis}}$ 2(D-glucose)

Ans: (b)

3.

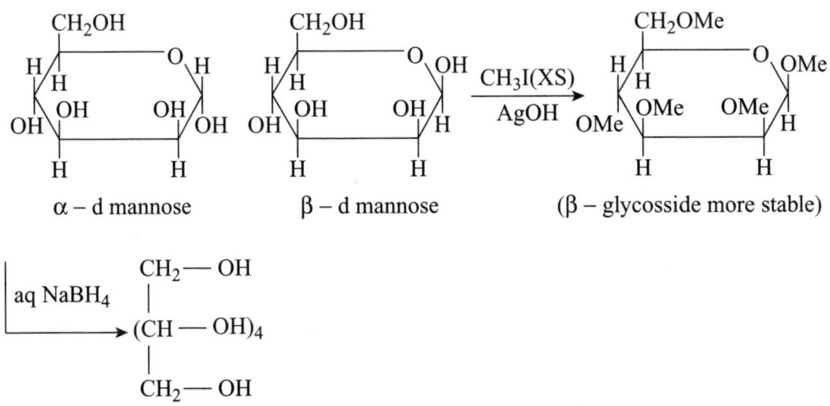

(ring opens 1st in aq medium)

Ans: (d)

4. X and Y give same osazone which means they have same constitution and configuration at C-3, C-4 and C-5.

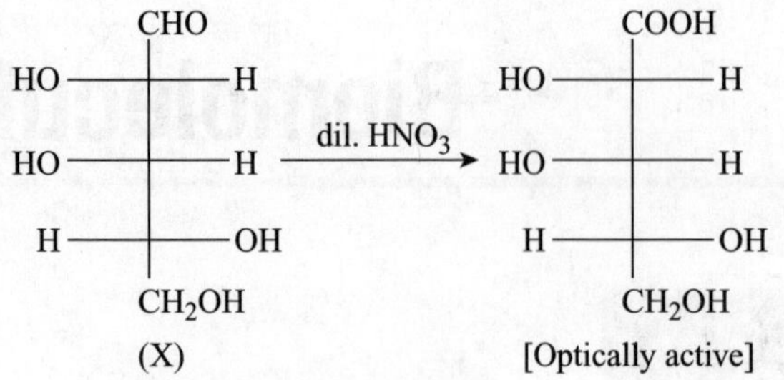

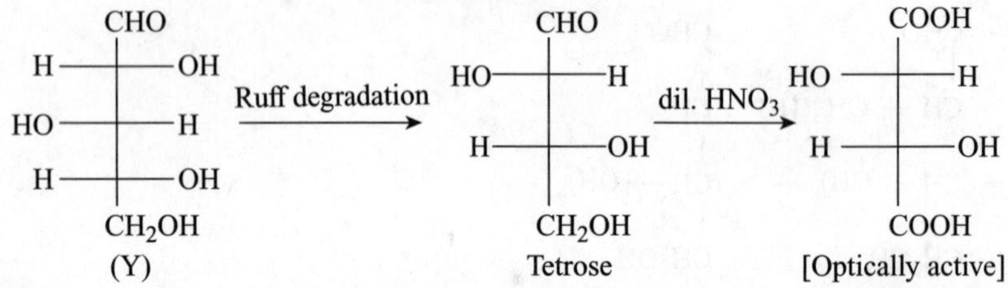

Aldaric acid : $COOH-(CH(OH))_n-COOH$

Ans: (c)

5.

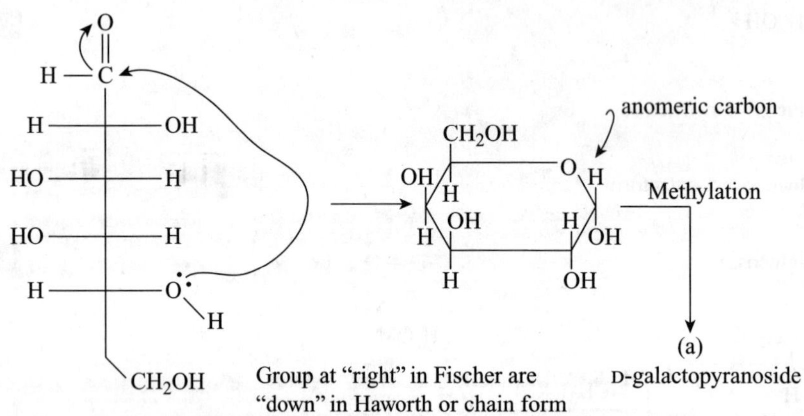

Ans: (a)

7. In Haworth projection of furanose the anomeric carbon (C-1) is presened at right hand side and primary alcohol is drawn up from back left hand corner. β-indicates –OH at anomeric carbon will be up.

Ans: (d)

8.

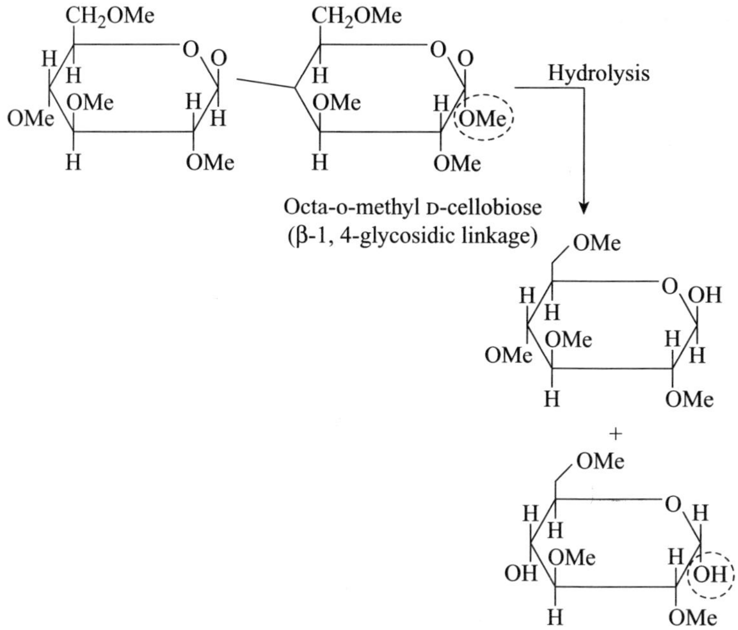
Hydrolysis
Octa-o-methyl D-cellobiose
(β-1, 4-glycosidic linkage)

Ans: (c)

9.

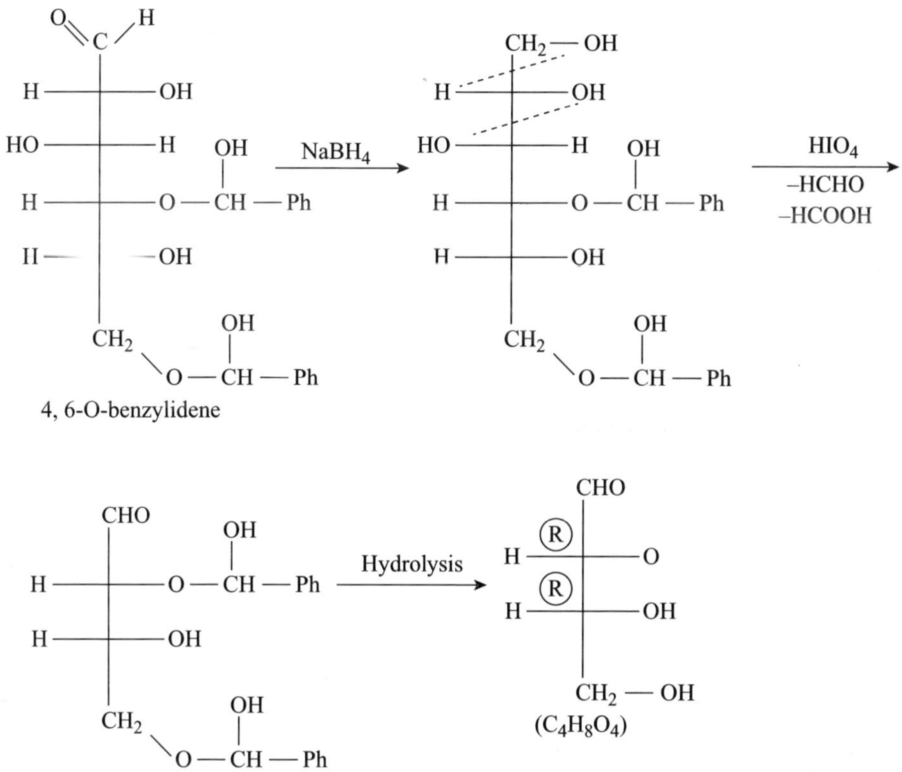
NaBH$_4$
HIO$_4$
–HCHO
–HCOOH
4, 6-O-benzylidene
Hydrolysis
(C$_4$H$_8$O$_4$)

Ans: (c)

10. By (i) and (ii): β-Glycosidic linkage is present between 2 D-glucose molecule.

By (iii) and (iv):

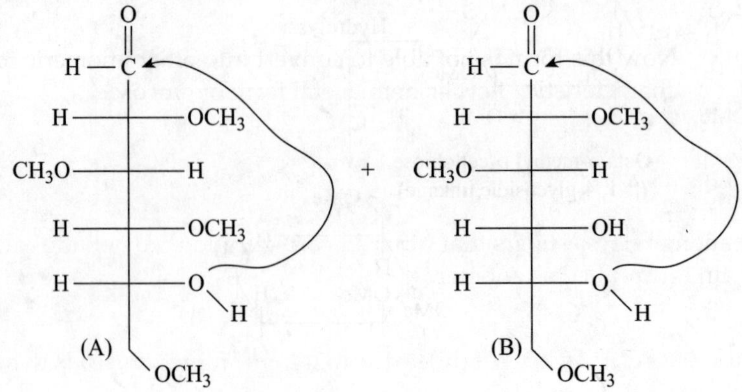

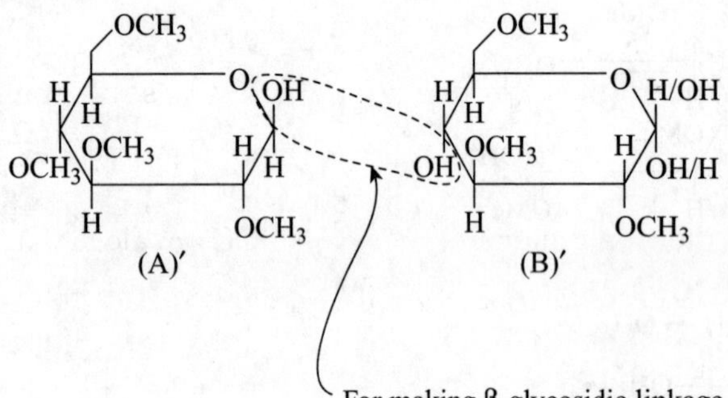

For making β-glycosidic linkage it must 1, 4-β-glycosidic linkage.

So

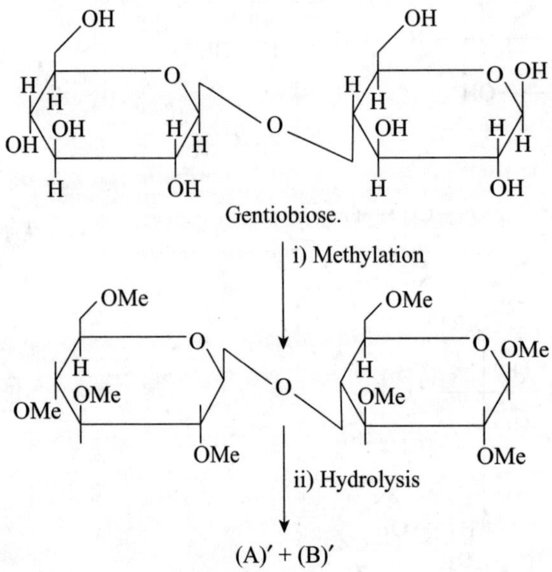

(A)′ + (B)′

Ans: (d)

11. $$PH = Pk_a + \log\frac{[\text{Conjugate base}]}{[\text{Acid}]}$$

If pH > $pK_a \Rightarrow$ [Conjugate base] > [Acid]

If pH > $pK_a \Rightarrow$ [Conjugate base] < [Acid]

$pK_{a1}(4.2) < pH\ (7.2) < pK_{a2}(9.8)$

Ans: (b)

12. Sucrose. **Ans:** (d)

13.

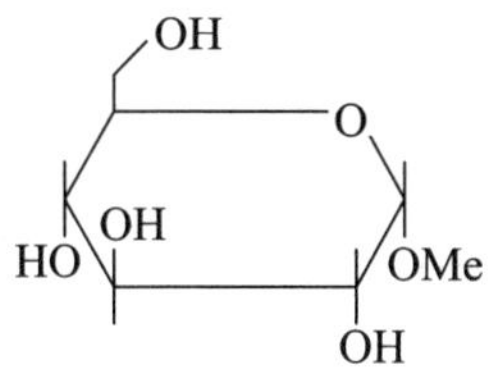

Now this form is not able to convert into other anomeric form as mutarotation is characteristic of cyclic hemiacetal form of glucose. **Ans:** (a)

14. C1 - Epimers and anomers; two sugars that differ in configuration only on the carbon that was carbonyl carbon in open chain form called as anomers. **Ans:** (a)

15. Zwitter-ionic form at isoelectric point solidifies due to ion-ion attraction; leads to maximum viscosity and minimum solubility. **Ans:** (c)

16. Monosaccharides may be optically inactive. Fructose gives positive test with Fehling's solution by first converting to glucose in basic medium. **Ans:** (a) and (b)

17. Surfactant are classified as cationic, anionic, zwitter-ionic and non-ionic (ex-long chain alcohols).

Ans: (b)

19. Nylon 6.6: Copolymer and condensation polymers. Nylon 6: Homopolymer and self-condensation polymer.

Ans: (c)

20. (a) 21 (c) 22 (d) 23 **Ans:** (a)

24. to 29. Nucleoside: attach of base to i position of sugar.

Nucleotide: linkage of phosphoric acid at S-position of sugar moiety in nucleoside.

Ex(Question 24)

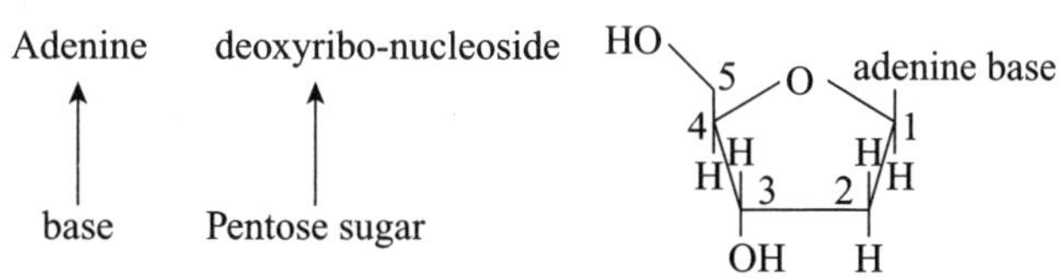

Ans: (a)

Ex(Question 29)

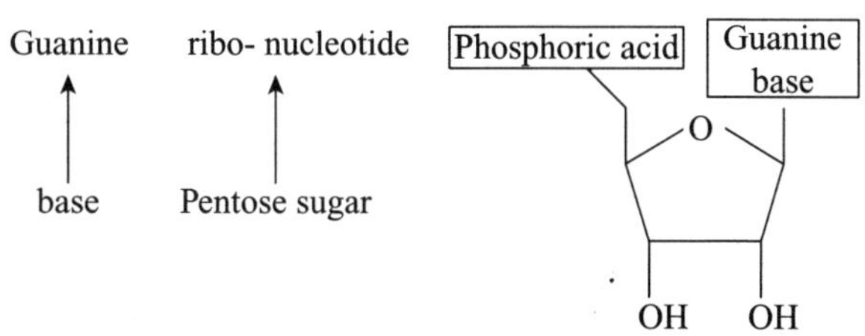

Ans: (a)

30.

$$\underset{\text{diethylene glycol}}{HO-CH_2-CH_2-O-CH_2-CH_2-OH} \xrightarrow{\text{polymerisation}} \left((CH_2)_2 - O - (CH_2)_2 - O \right)_n$$

Ans: (d)

31.

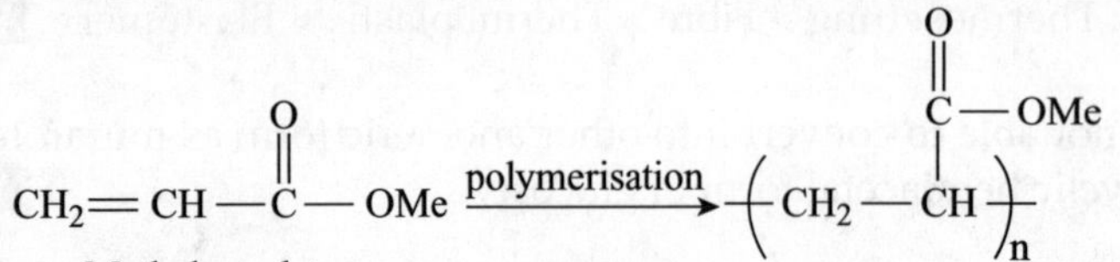

Methyl acrylate

Ans: (d)

32.

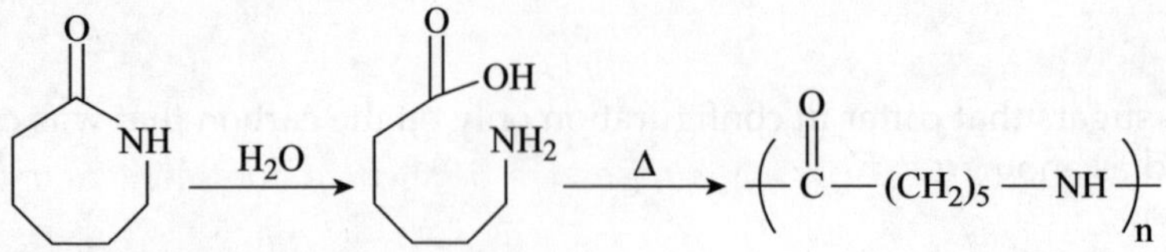

Caprolactum Nylon-6

Ans: (b)

33.

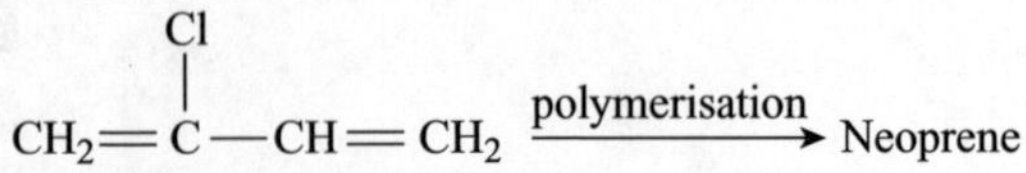

Chloroprene

Ans: (a)

34.

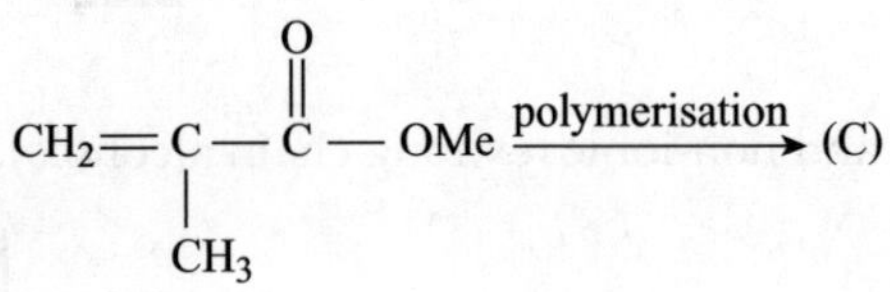

Methyl acrylate meth

Ans: (c)

35. and 36. Homopolymer : formed by only one type of monomer.

Copolymer: formed by more than one type of monomer.

Polymer	Monomer units (S)
Nylon-6	Caprolactum
Buna-N	Butadiene and Acrylonitrile
Buna-S	Butadiene and styrene
Polyethylene	Ethylene
Starch	α - D glucose
PVC	Vinyl chloride
Protein	Different amino acid

37. Paraldehyde used as sleeping pills. **Ans:** (c)

42. Antipyretic. **Ans:** (a)

43. Alizarin dye forms violet coloured complex with Cr^{+3} ion. **Ans:** (a)

44. Polytetrafluoroethylene (PTFE). **Ans:** (c)

46. Copolymer of butadiene and acrylonitrile. **Ans:** (c)

47. Intermolecular force of attraction order : Thermosetting > Fibre > Thermoplastic > Elastomer

Nylon : Fibre : Cellulose : Fibre

PVC : Thermoplastic : Natural rubber : Elastomer **Ans:** (d)

48. Histamine used as antacid. **Ans:** (a)

49. Complimentary DNA sequence : TTAGTCAA

Complementary RNA sequence : UUAGUCAA in DNA bases are bonded in complementary pair as $A \equiv T$ and $G \equiv C$ while in RNA as $A \equiv U$ and $G \equiv C$ **Ans:** (d)

50.

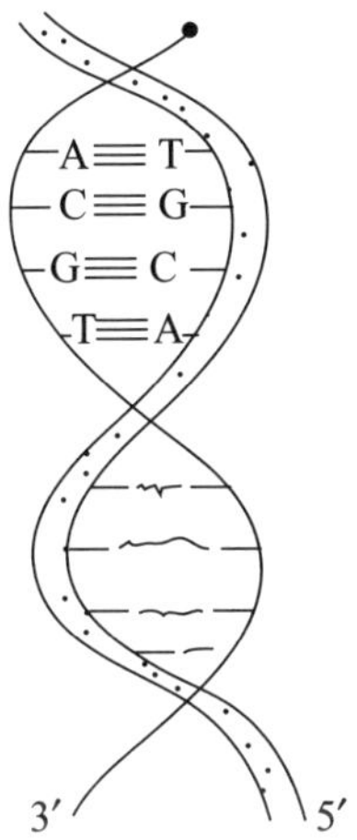

Ans: (b)

51. Mannose. **Ans:** (b)

52. Refer Question Number 11. **Ans:** (c)

53.

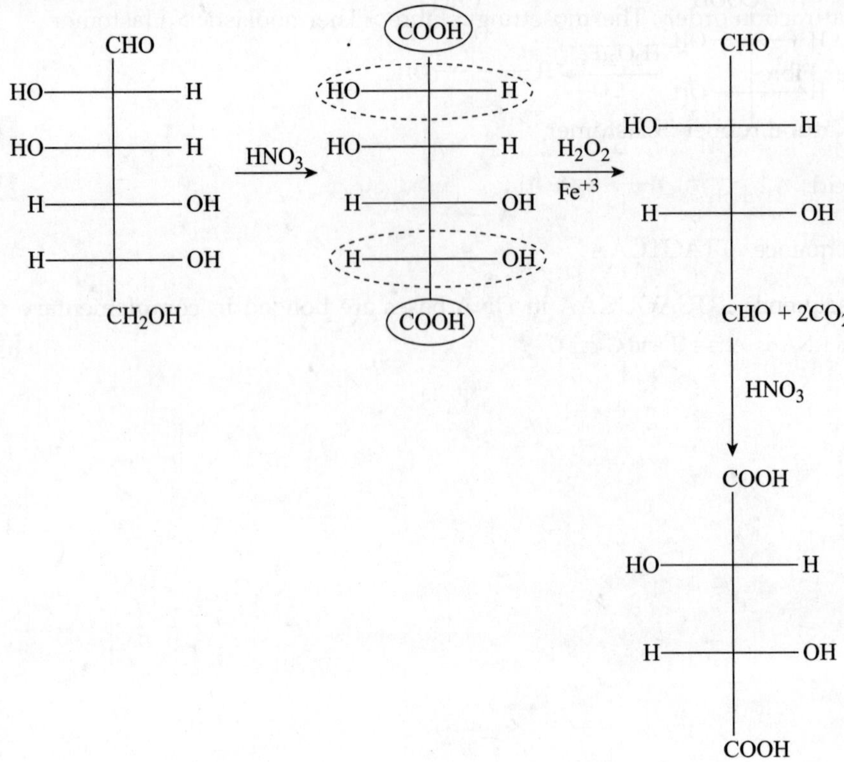

Ans: (a)

54.

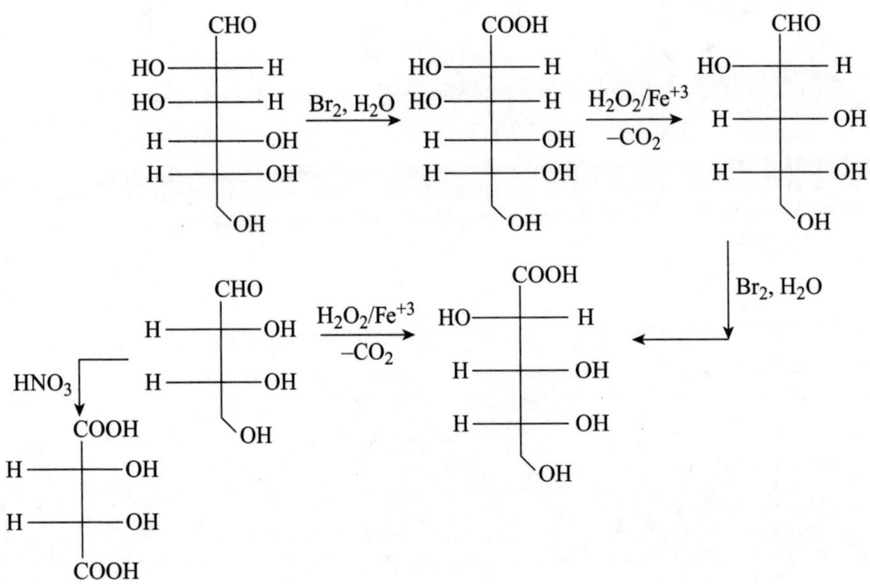

Ans: (b)

55.

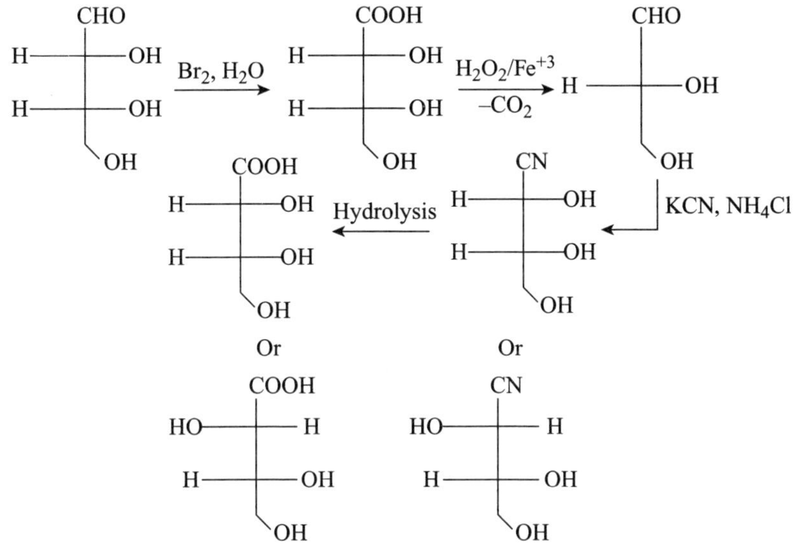

Ans: (a)

56. (a) It indicates –OH at right side at anomeric carbon in Fischer projection.

Furanose : 5-Mem. ring.

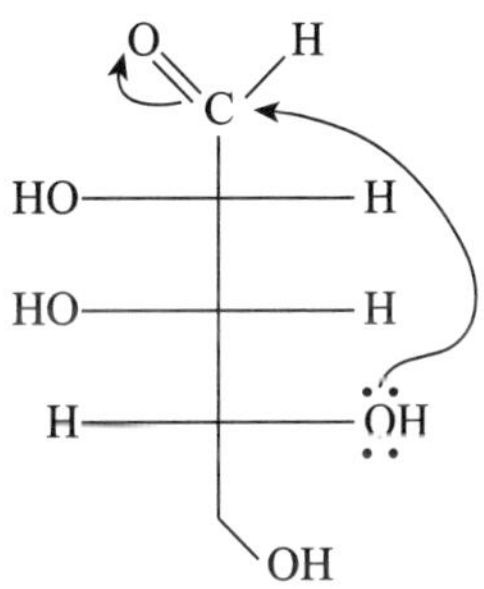

D-Lyxose

Ans: (d)

57. (c)

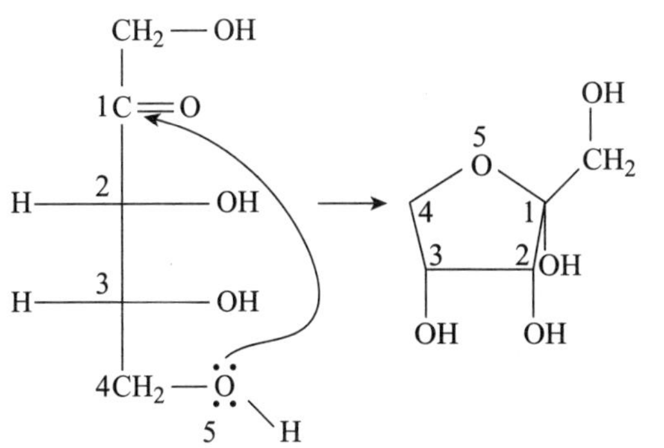

Ans: (c)

58. α indicate –OH at down side at anomeric carbon in Haworth projection.

In Haworth projection of furanose the anomeric carbon (c-1) is presented at right hand side and $-CH_2OH$ (primary alcohol) is drawn up from back left hand corner.

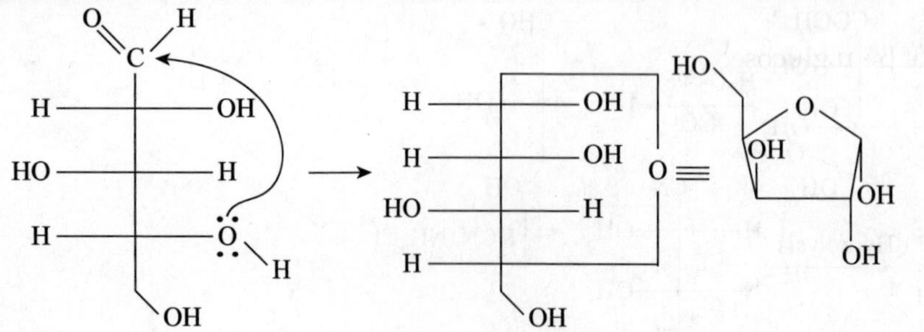

D-xylose

α-D-xylose

(–OH at anomeric carbon will be in right hand side)

(Group in "RHS" in Fiseher are presented "down" in Haworth)

Ans: (a)

60.

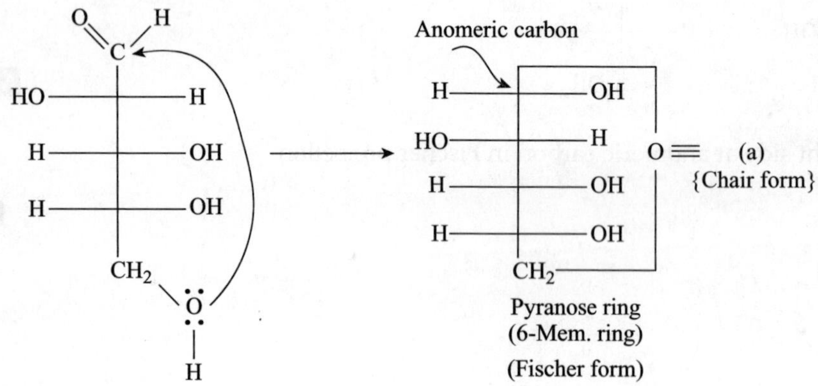

Group at "right" in Fischer are presented "down" in chair form. **Ans:** (a)

64. Atropine. **Ans:** (b)

65. Refer solution of Question Number 11. **Ans:** (a)

66. A. Fleming. **Ans:** (a)

67. Methyl isocyanate (CH_3NCO). **Ans:** (b)

69.

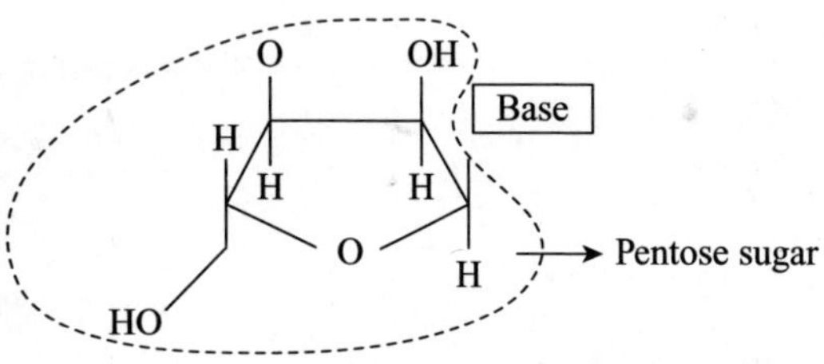

Ans: (b)

70.

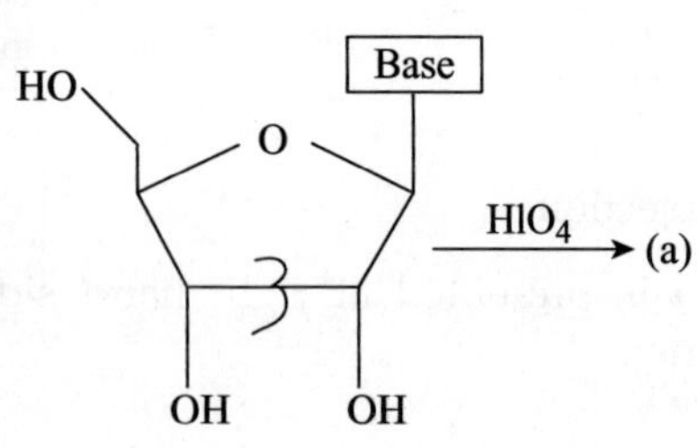

Ans: (a)

71. Cellulose : linear polymer of β - D glucose.

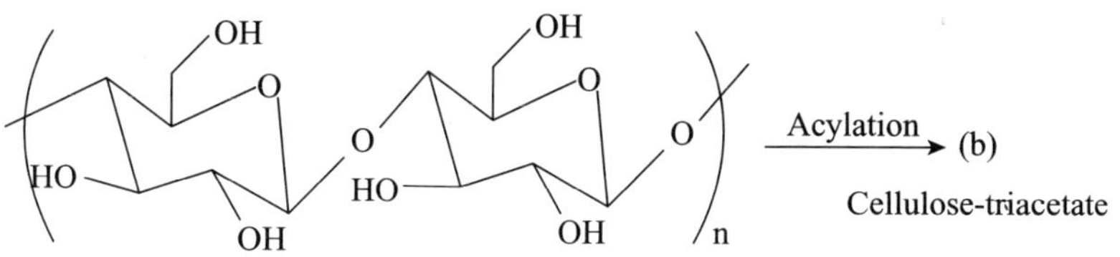

Cellulose

Ans: (b)

72. Groups $-\overset{O}{\overset{\|}{C}}-NH_2$, $-\overset{O}{\overset{\|}{C}}-NH-$, $-CH_2-NH_2$, $-\overset{NH}{\overset{\|}{C}}-NH_2$ and $-\overset{S}{\overset{\|}{C}}-NH_2$

gives biuret test. **Ans:** (b) and (d)

73. –Pyranose : 6-Mem. ring

It has hemi-acetal linkage at anomeric carbon in right hand ring. **Ans:** (d)

78. Natural rubber is polymer of cis 2-methyl-1, 3-butadiene while neoprene is polymer of 2-chloro-1, 3-butadiene.

Ans: (d)

80. Natural polymers that takes part in metabolic process are biopolymers. **Ans:** (b)

81. Psychedelic drug. **Ans:** (c)

82. Polymer of vinyl chloride. **Ans:** (a)

83. Benedict's solution : as urine of patient contains excess glucose. **Ans:** (a)

84. Bakelite, nylon, terylene are condensation polymers. **Ans:** (b)

85. Congo red. **Ans:** (b)

86. Nylon is synthetic polymer. **Ans:** (d)

87. Morphine. **Ans:** (b)

88. Wool. **Ans:** (c)

89. Refer Question Number 1. **Ans:** (c)

90. Polyamide fibre. **Ans:** (b)

91. $CH_2=CH_2$ (Ethylene). **Ans:** (b)

92. Paracetamol. **Ans:** (b)

93. Phenacetin used as analgesic medicine. **Ans:** (a)

94. Transfer of charge between starch and iodide ion in starch-iodide complex results to dark blue solution. **Ans:** (d)

95. Anti-pyretic. **Ans:** (b)

96. Silk. **Ans:** (c)

97. $$NH_2-(CH_2)_6-NH_2 + HO-\overset{O}{\overset{\|}{C}}-(CH_2)_4-\overset{O}{\overset{\|}{C}}-OH \rightarrow \left(NH-(CH_2)_6-NH-\overset{O}{\overset{\|}{C}}-(CH_2)_4-\overset{O}{\overset{\|}{C}} \right)$$

Hexamethylene diammine; Adipic acid

Ans: (d)

98. Chloramphenicol. **Ans:** (c)

99. Nylone-6. **Ans:** (b)

100. Tranquilizer. **Ans:** (d)

101. $-N=N-$ **Ans:** (c)

102. Specific rotation of pure α - D glucose dissolved in water (+112°), reaches ultimately to stable value of (+52.5°) due to mutarotation with β - D glucose form. **Ans:** (a)

103. Antiline yellow. **Ans:** (b)

104. Addition polymer are chain growth polymers. **Ans:** (b)

105. Glucose. **Ans:** (b)

106. Molisch test : Dehydration of carbohydrate by H_2SO_4 in presence of phenol (usually α - naphthol) results to detectable red-coloured compound. **Ans:** (d)

107. Buna-S is addition copolymer of styrene and 1, 3 - butadiene. **Ans:** (a) and (b)

108. Carbohydrate. **Ans:** (b)

110. Refer Question Number 33. **Ans:** (d)

111. Terramycin and chloroquine. **Ans:** (c) and (d)

112. Tuberculosis. **Ans:** (d)

113. Syphilis. **Ans:** (b)

114.

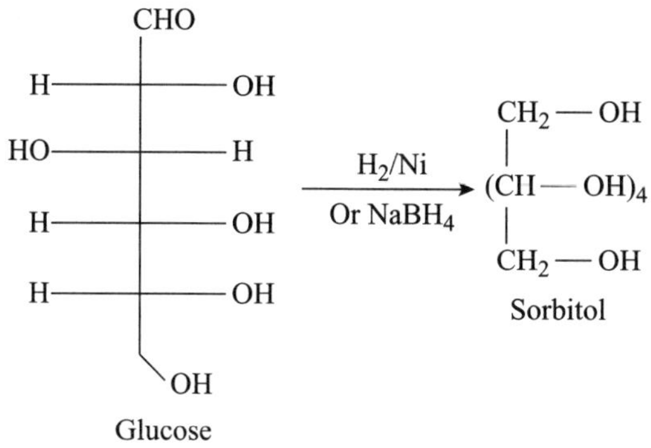

Ans: (b)

115.

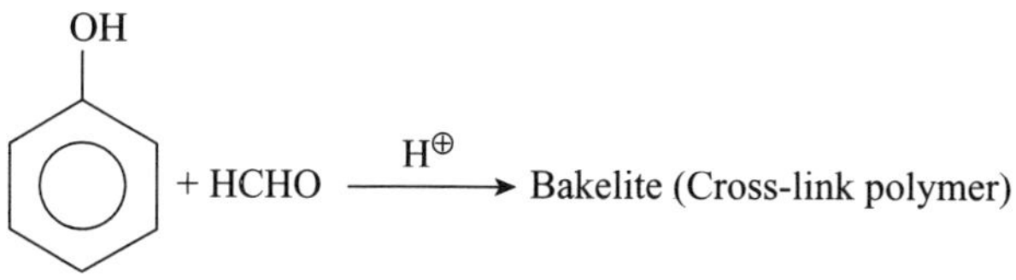

Ans: (c)

116. Cocaine, atropine, nicotine. **Ans:** (d)

117. Adipic acid. **Ans:** (c)

118. Parachute ropes. **Ans:** (b)

119. Quinine. **Ans:** (d)

120. Phenol. **Ans:** (a)

121.

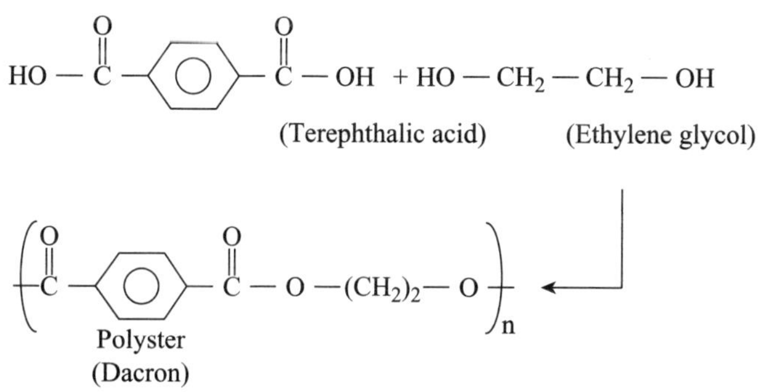

Ans: (d)

122. N_2O **Ans:** (b)

123. Chloroxylenol + terphenyl. **Ans:** (b)

124. Atropine. **Ans:** (a)

125.

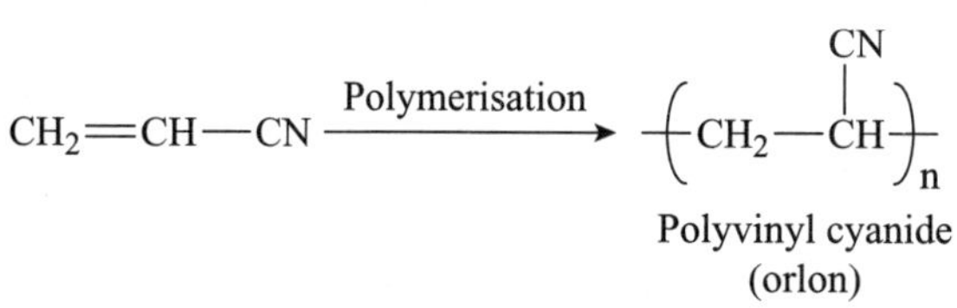

Ans: (a)

126. Alexander Fleming. **Ans:** (d)

127. Tranquilizers. **Ans:** (a)

128. Phthalein dyes. **Ans:** (a)

129. Methyl isocyanate. **Ans:** (b)

130. D-Glucose and L-Fructose **Ans:** (b) and (d)

131. Sucrose. **Ans:** (c)

132. DNA (for detail, Refer Question Number 80). **Ans:** (d)

133. C-2 **Ans:** (c)

134. Bottom most asymmetric carbon. **Ans:** (b)

135. Tetrafluoroethylene. **Ans:** (d)

136.

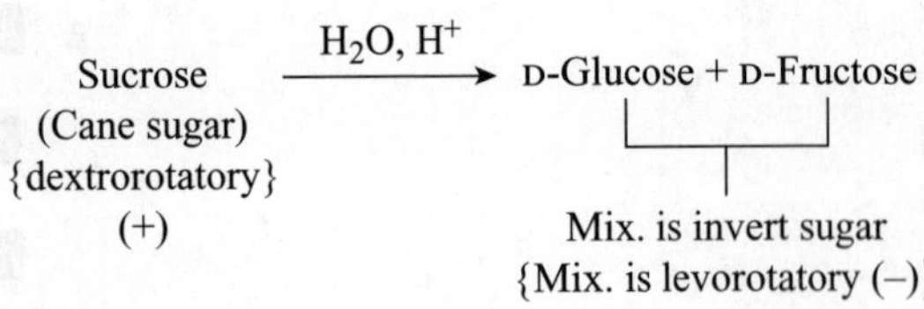

Ans: (b)

137. Sucrose. **Ans:** (d)

138. Buna-S stands for Styrene butadiene rubber. **Ans:** (b)

139. β-D-glucose. **Ans:** (b)

140.

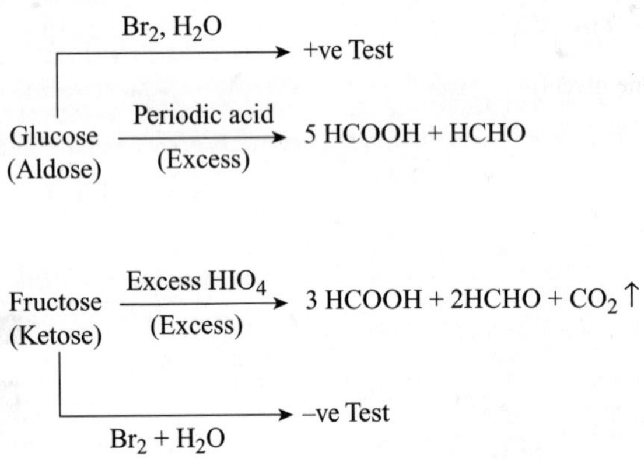

Ans: (d)

141. Glucose, Fructose, Lactose. **Ans:** (d)

142. Vulcanization. **Ans:** (a)

143. D-Mannose. **Ans:** (d)

144. Glucosazone (Refer Question Number 165). **Ans:** (c)

145. Protein. **Ans:** (d)

146. D - (+) - Glucose and D - (-) - Fructose. **Ans:** (b)

147. Refer Question Number 151. **Ans:** (d)

148. Glucose and Galactose. **Ans:** (c)

149. β-D-Glucose. **Ans:** (a)

150. Glucose. **Ans:** (b)

152. Nylon-6, 6. **Ans:** (a)

153. Refer Question Number 151 and 165. **Ans:** (a)

154. 36% and 64%. **Ans:** (c)

155. Cellulose, cellobiose is disaccharide having D-glucose with β-1, 4-glycosidic linkage. **Ans:** (c)

156. Bakelite. **Ans:** (c)

157. Aldehyde group. **Ans:** (a)

158. Aldehyde group and hydroxy group at C - 5. **Ans:** (b)

159. Tollen's reagent gives silver mirror ppt. with glucose. **Ans:** (b)

160. Fructose reduces Tollen's and Fechling reagent because in basic medium fructose isomerises to glucose. **Ans:** (d)

161. Isoprene. **Ans:** (b)

162. C - 2 and C - 5 **Ans:** (c)

163.

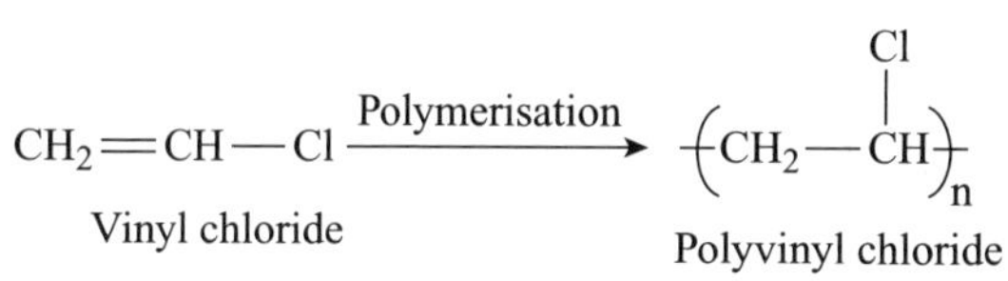

Ans: (d)

164.

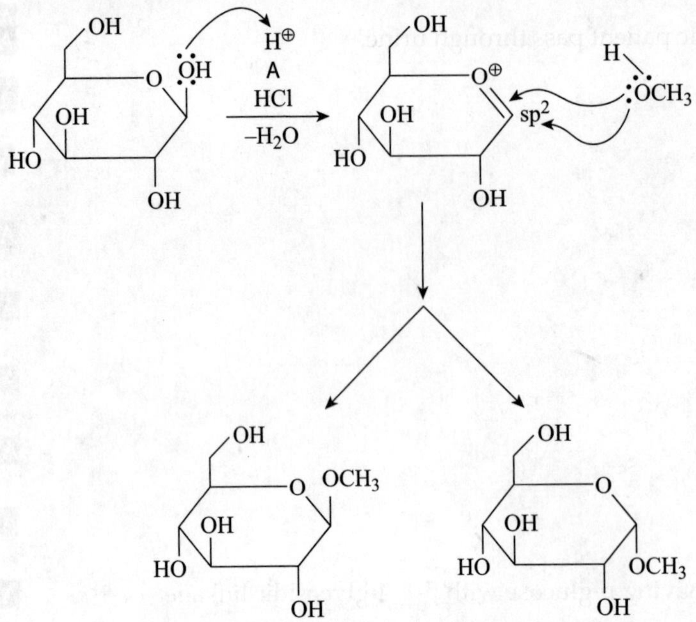

β-Methyl D-Glucopyranoside α-Methyl D-Glucopyranoside

Ans: (d)

165. (A)

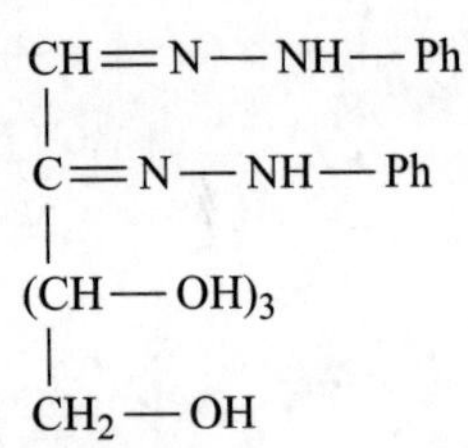

(Osazone)

(B)

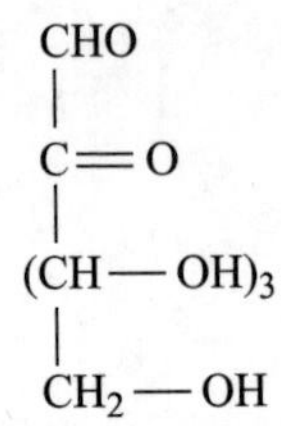

(Osone)

(c)

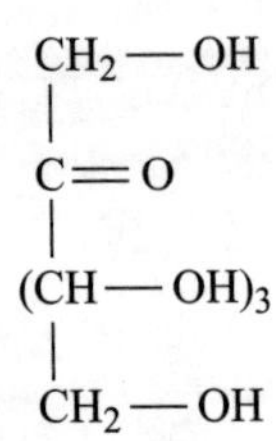

Zn/ACOH is weak reducing agent (it reduces only terminal 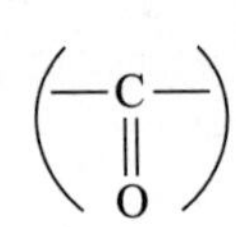

Ans: (b)

166. Nylon. **Ans:** (c)

167. HCN. **Ans:** (a)

168. Antiseptic. **Ans:** (a)

169. and 175.

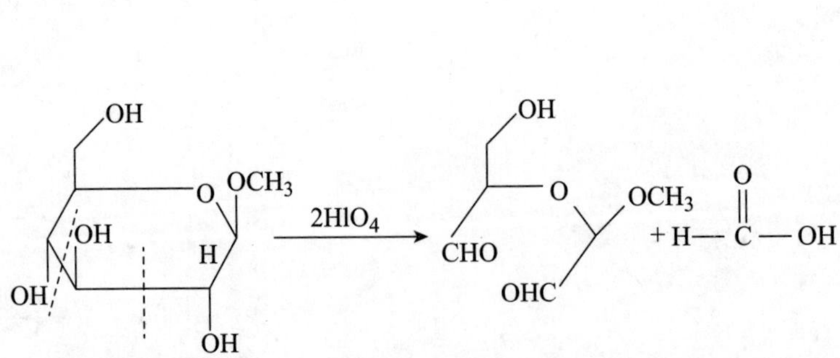

D-Glucopyranoside

Ans: (a)

170. The excess glucose in body of diabetic patient pass through urine. **Ans:** (a)

171. Epimers. **Ans:** (a)

172. Maltose, Sucrose, Lactose. **Ans:** (c)

173. and 193.

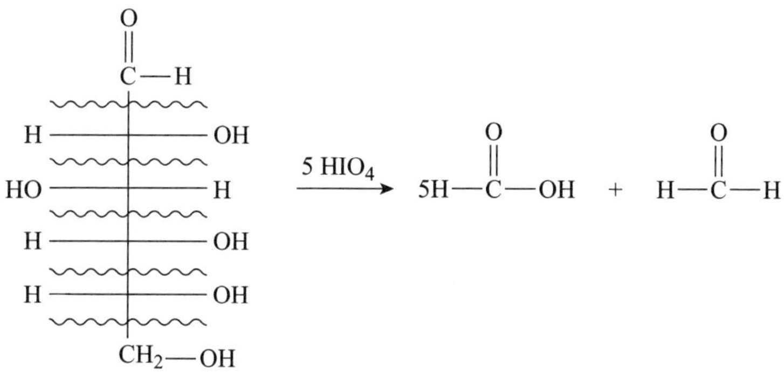

174. Carbohydrate. **Ans:** (a)

176. Starch. **Ans:** (a)

177. Indigo. **Ans:** (a)

178. Amylose and amylopectin. **Ans:** (c)

179. Antibiotic. **Ans:** (a)

180. Dehydration. **Ans:** (c)

181. Natural rubber. **Ans:** (a)

182. Aldehyde. **Ans:** (b)

183. Glucose and Galactose. **Ans:** (a)

184. Antimalarial. **Ans:** (c)

185. Terephthalic acid. **Ans:** (b)

186. $-NO_2$, $-SO_3H$, $-OH$ **Ans:** (a), (b) and (d)

187. Phenol–formaldehyde resin. **Ans:** (d)

188. Orange:

Absorbed colour	VI	B	G	y	O	R
Observed colour	Y	O	R	VI	B	G

Ans: (b)

189. Polythene. **Ans:** (a)

190. Cr^{+3} **Ans:** (a)

191. Its monosaccharide. **Ans:** (d)

192. C-2. **Ans:** (b)

LEVEL 2

1. Acidic strength : $-NH_3 < H_2O < Ph-OH < RCOOH$ ($Pk_a = 7$)

   $pH > pK_a \Rightarrow$ [Conjugate base] > [Acid]

   $pH > pK_a \Rightarrow$ [Conjugate base] < [Acid]

   Ans: (b) and (d)

2. Nylon-6, 6 and bakelite. **Ans:** (a) and (b)

3. Monosaccharides cannot be hydrolysed into smaller molecules. **Ans:** (a) and (b)

4. Nylon 6, 6 , protein and Nylon-6. **Ans:** (a), (b) and (c)

5. The compound having same constitution and configuration at C-3,

   C-4, C-5 and C-6 but different constitution or configuration at C_1 and C-2 which becomes identical by osazone formaton; gives same osazone derivative. **Ans:** (a) and (b)

7. Aldohexose and ketohexose having same constitution and configuration at C-3, C-4, C-5 and C-6 are interconvertible in basic medium. **Ans:** (a), (b) and (c)

8. Buna-N , Buna-S, protein **Ans:** (b), (c) and (d)

9. Terylene, glyptal, protein. **Ans:** (a), (b) and (d)

10. Natural rubber, orlon. **Ans:** (c) and (d)

11. $pH = \dfrac{pKa_1 + pKa_2}{2}$ **Ans:** (b)

12. $pH = \dfrac{pKa_2 + pKa_3}{2}$ **Ans:** (d)

13. Acidic strength: $-COOH > -NH_3$ **Ans:** (a)

14. Polyester is condensation polymer of ethylene glycol and terephthalic acid. **Ans:** (b)

15. Addition polymer of isoprene. **Ans:** (b)

16. Nylon-6 is self-condensation polymer. **Ans:** (a) and (d)

17. Sugar having hemi-acetal/ hemi-ketal linkage at anomeric carbon are reducing sugar as ring opens in basic or aqueous medium.

    Ans: (c)

18. Glucose and Mannose. **Ans:** (b)

19. Lactose gives positive test with Tollen's reagent. **Ans:** (a)

Passage 4

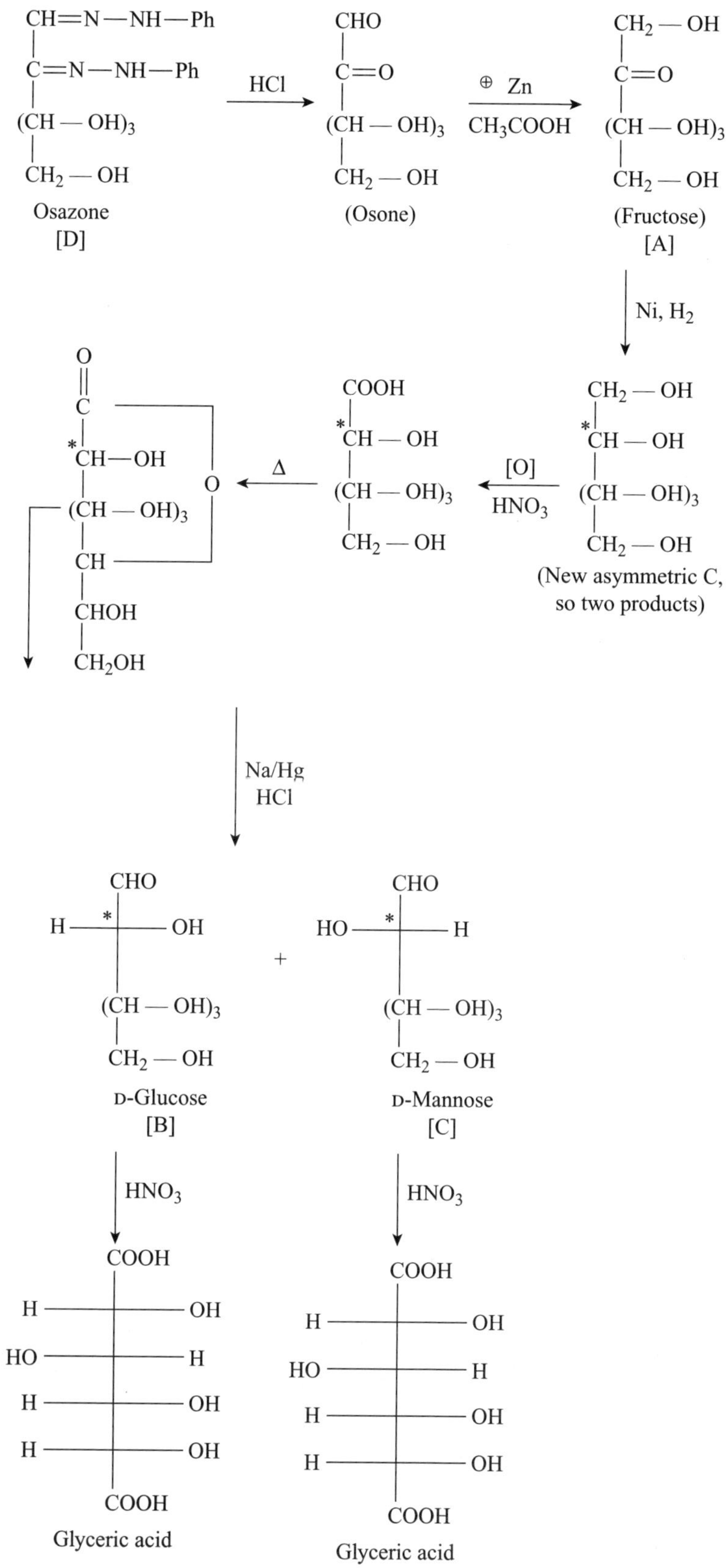
Osazone
[D]
HCl
(Osone)
⊕ Zn
CH3COOH
(Fructose)
[A]
Ni, H2
(New asymmetric C,
so two products)
[O]
HNO3
Δ
Na/Hg
HCl
D-Glucose
[B]
D-Mannose
[C]
HNO3
HNO3
Glyceric acid
Glyceric acid

Glucose, Fructose, Mannose and gives

CHO
|
C=O
|
$(CH-OH)_3$
|
CH_2-OH
(Osone)

Same osazone with excess of phenylhydrazine.

23. After ring opening oxidation of α - D - glucose leads to formation of (A).

(A) → 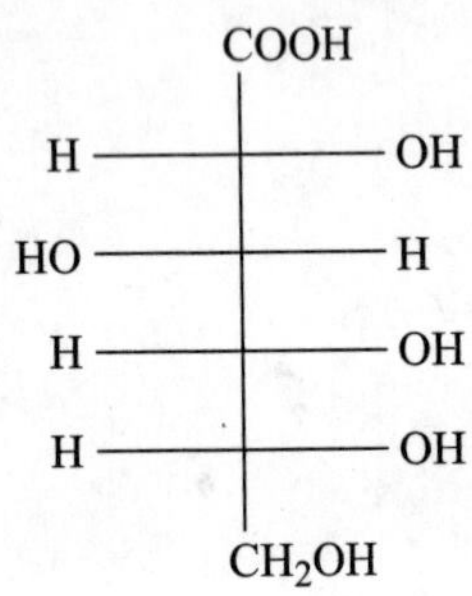

Ans: (4)

24. 5, 6, 7, 8 (4 compounds at bottom line) **Ans:** (4)

25. 2, 3, 5 **Ans:** (3)

26. 2, 3, 4, 8 **Ans:** (4)

27. 4, 5, 6

Explanation for (2)

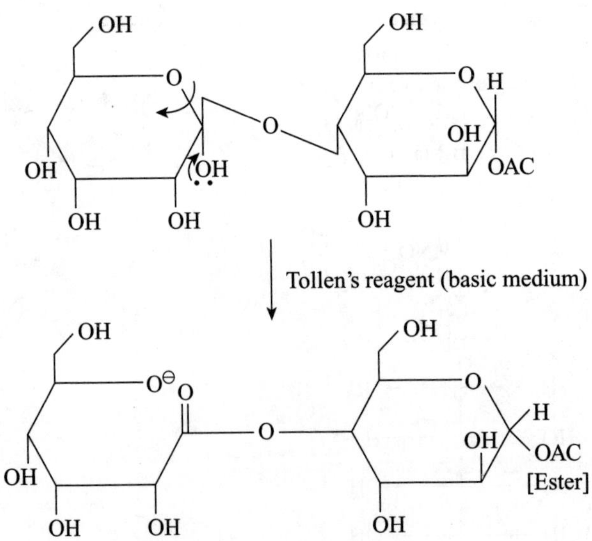

In basic medium rather formation of carbonyl compound (aldehyde), ester forms that gives –ve test further with Tollen's reagent (TR). But for ex. (5) open in basic medium (provided by TR) as,

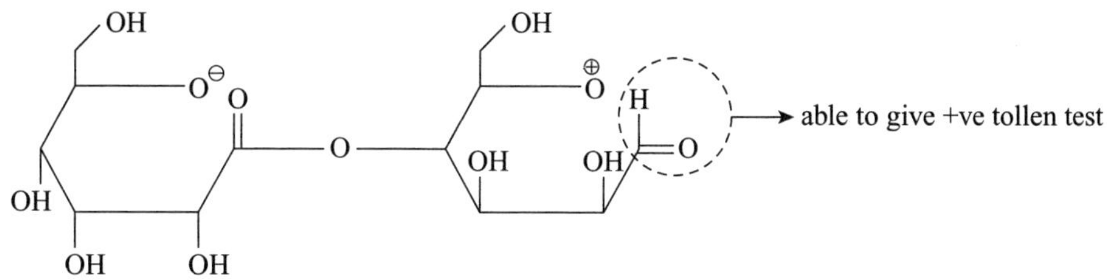

Ans: (3)

CHAPTER 8

Organic Reaction Mechanisms and Reagents

LEVEL 1

4.

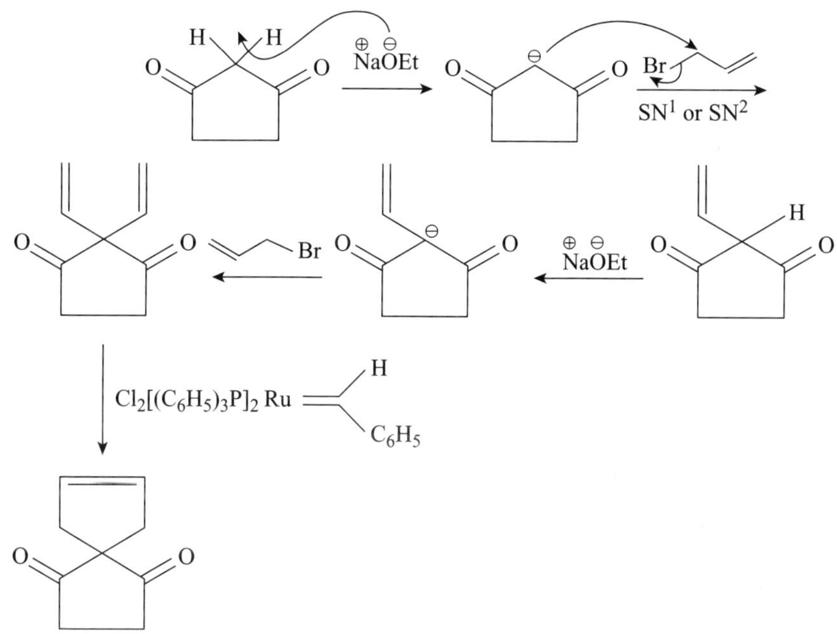

Ans: (d)

6.

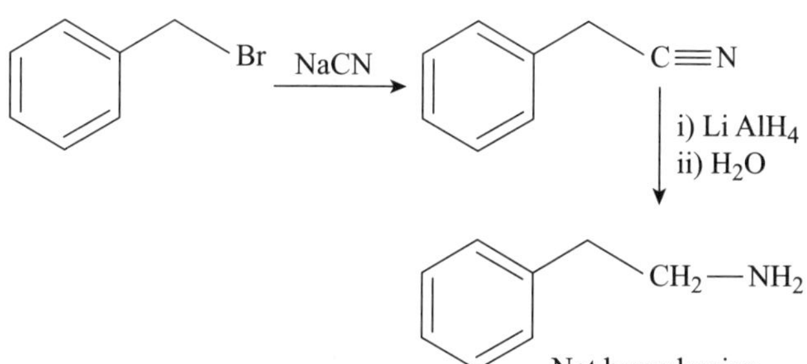

Not benzylamine

Ans: (b)

7.

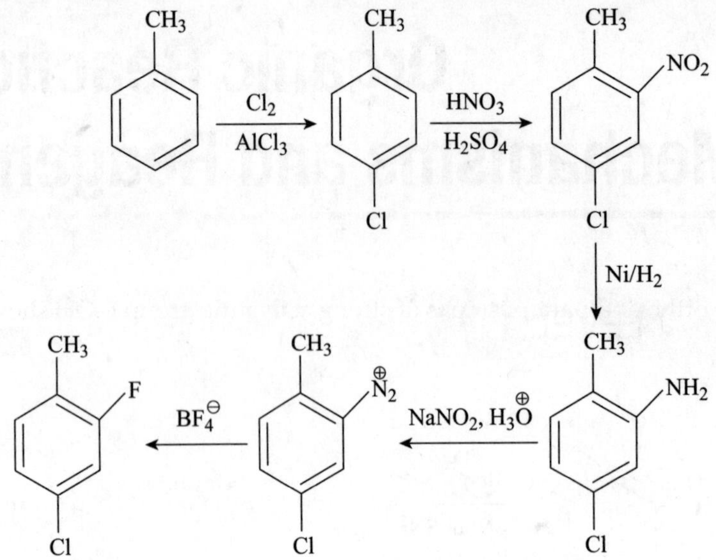

Ans: (a)

8.

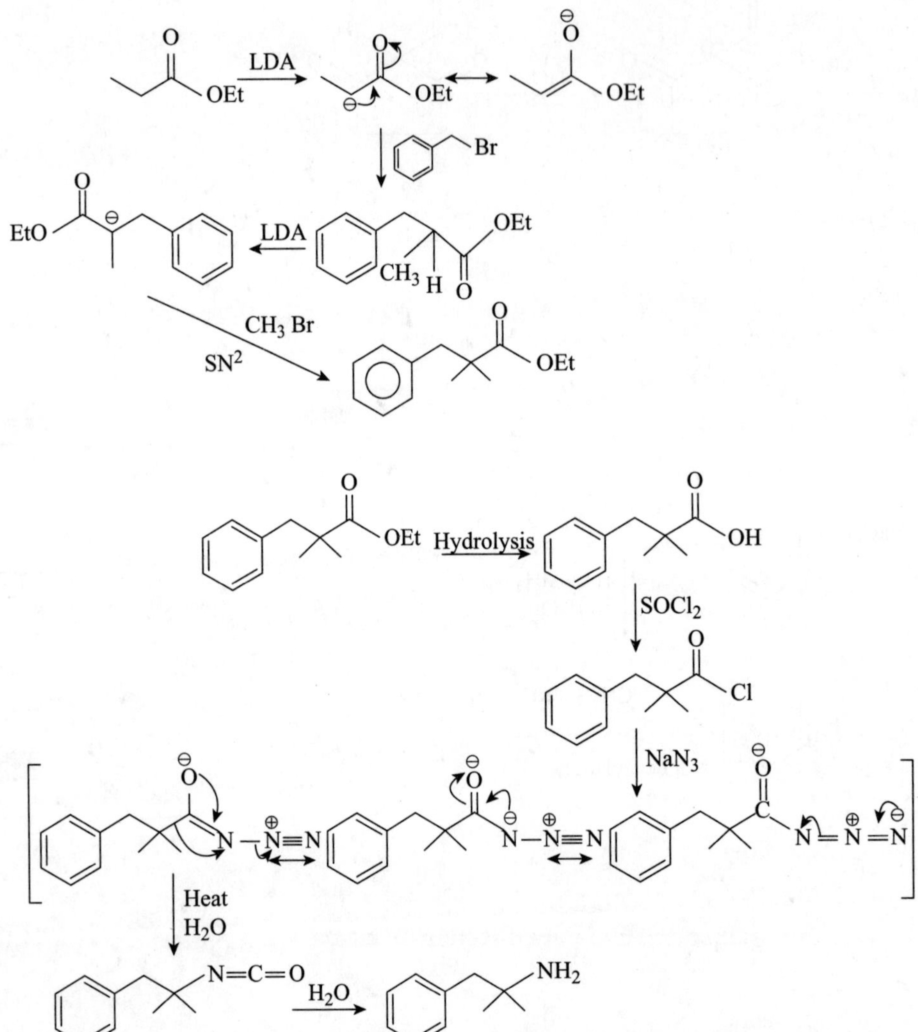

Ans: (d)

9. 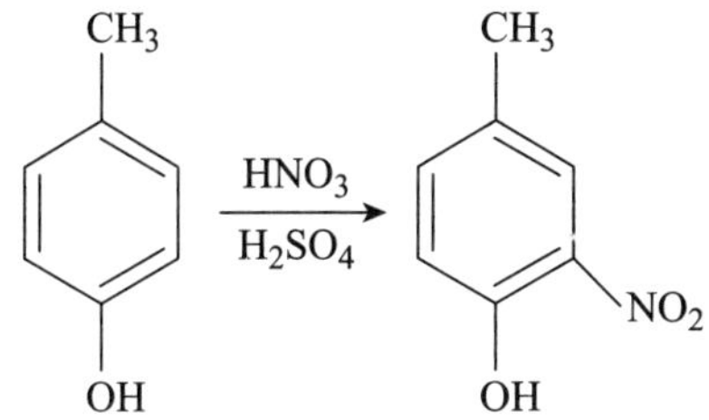

Hint: Substitution takes place at the ortho and para positions of strong activating group (–OH shows +R effect).

Ans: (d)

10. 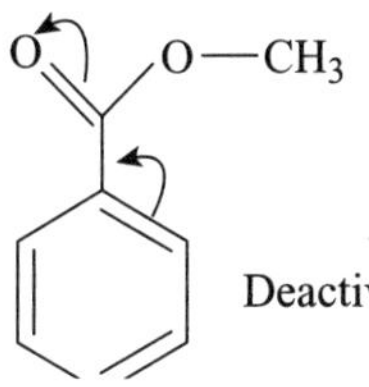

Deactivating group

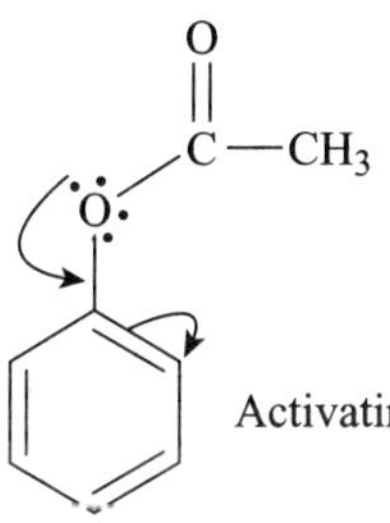

Activating group

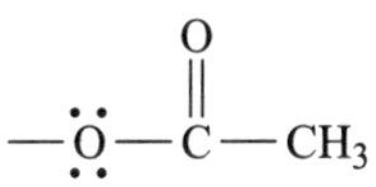

+ R-effect

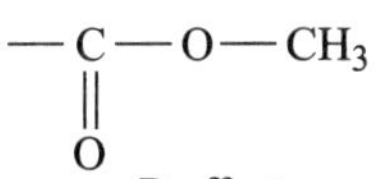

–R effect

Ans: (d)

11. 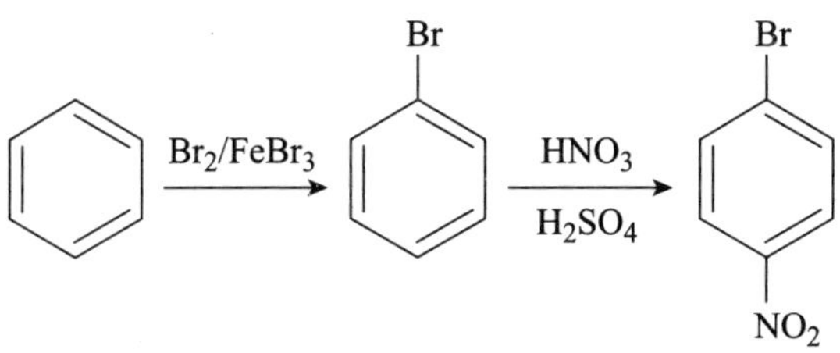

Note: Halogens are deactivating but ortho and para directing in nature.

Ans: (d)

12. 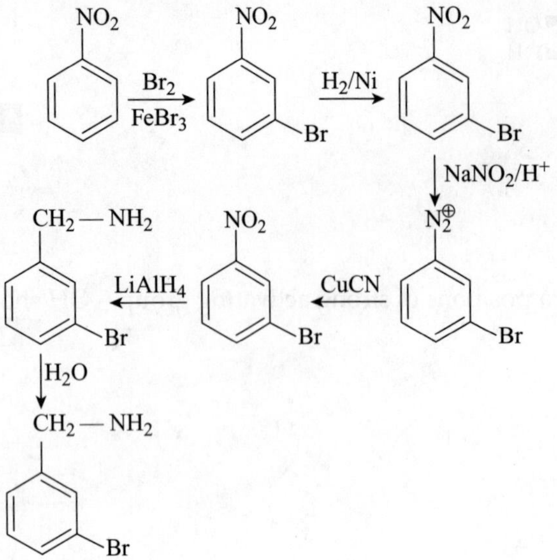

In normal conditions $-NH_2$ is not a good leaving group. **Ans:** (a)

13. 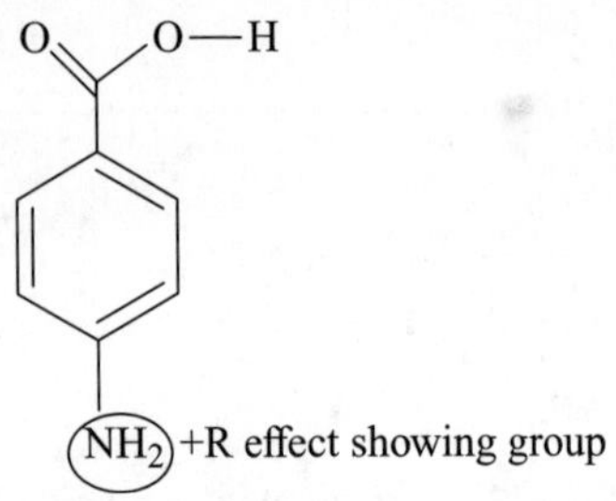

Ans: (c)

14. 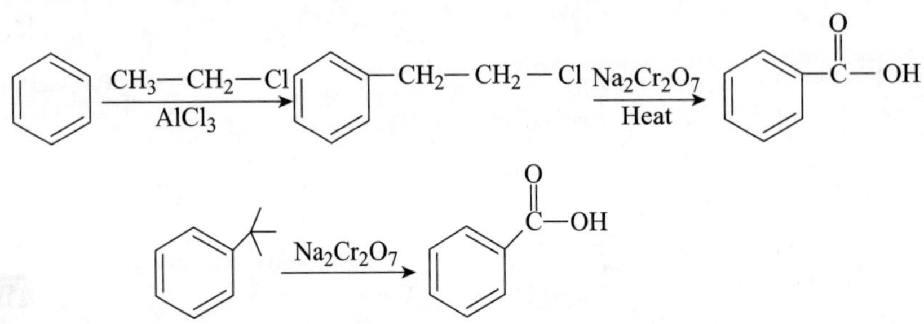

Ans: (a)

15. 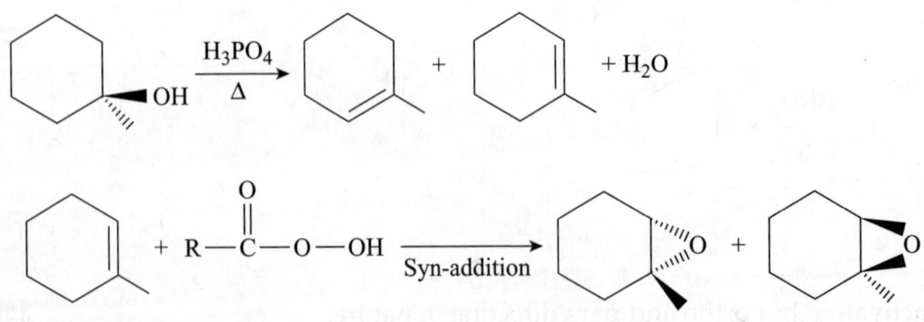

Ans: (a)

16. 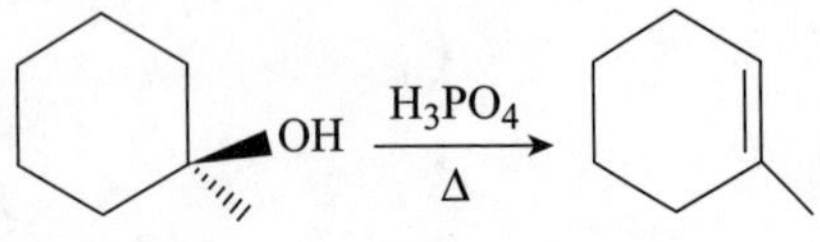

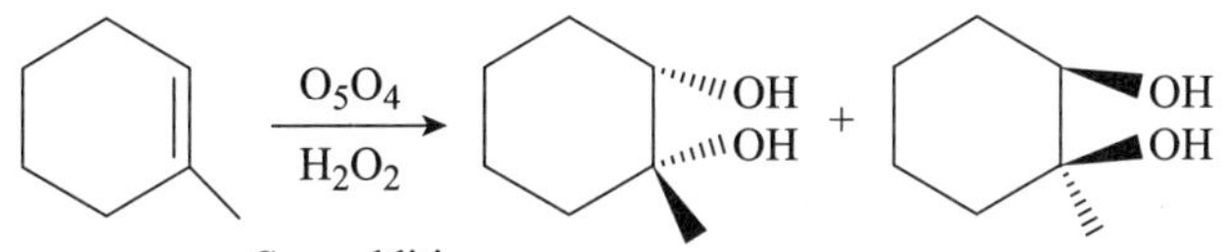

Syn-addition

Ans: (a)

18.

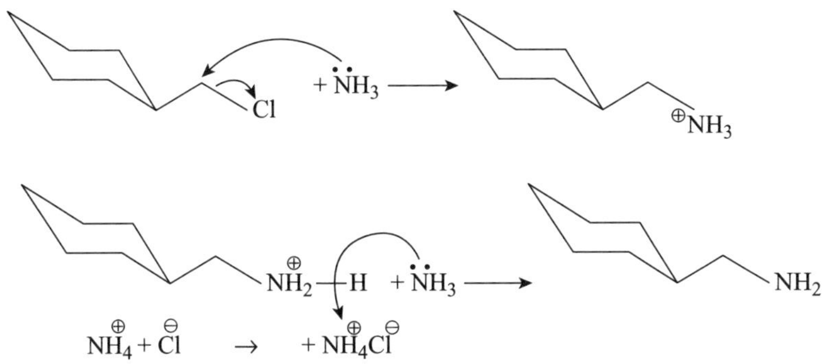

Ans: (b)

19. $LiAlH_4$ shows nucleophilic addition reaction on carbonyl bond 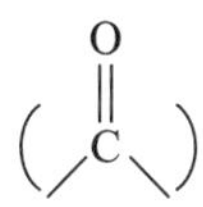

Hence simply check the reactivity of aldehydes and ketones towards nucleophilic addition reaction.

Ans: (d)

20.

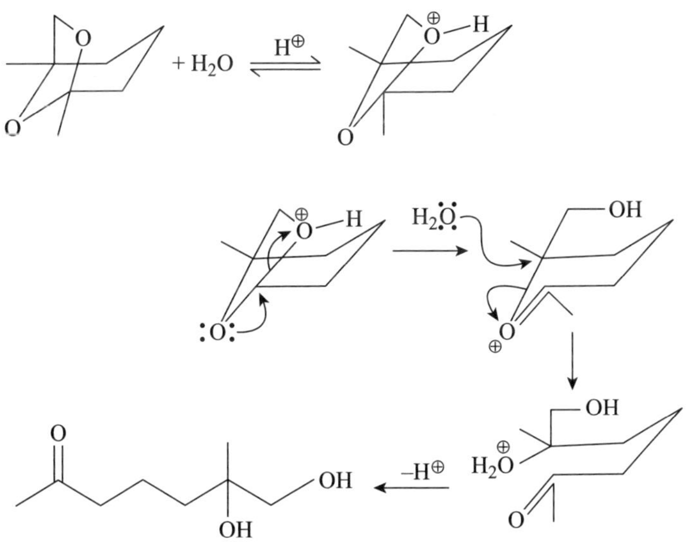

Ans: (a)

22.

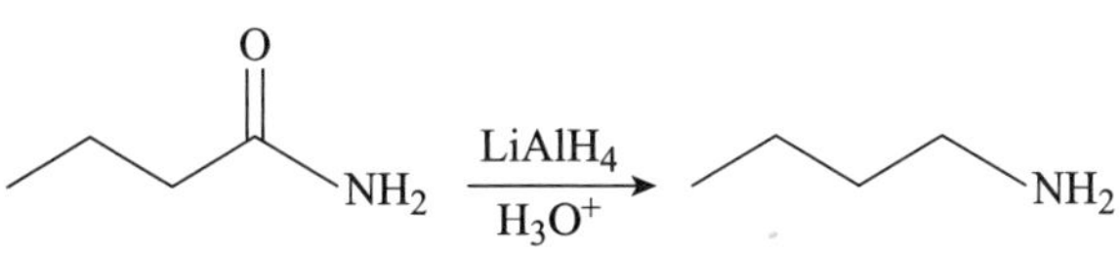

Ans: (b)

23. Halogens are ortho, para directing in nature but deactivating groups.

— Cl deactivating but ortho, para directing.

— OH activating and ortho, para directing.

Ans: (d)

24.

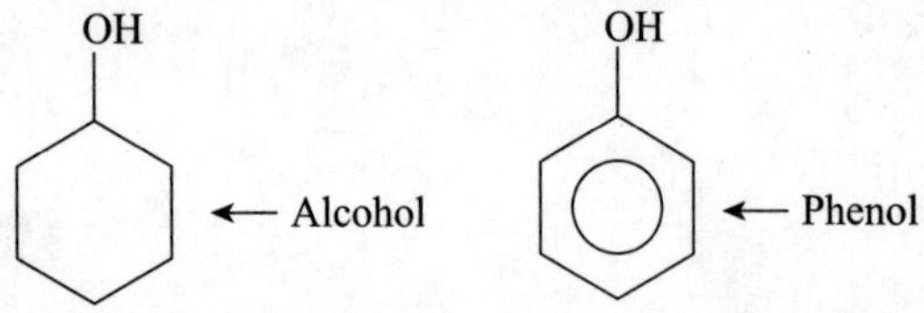

Phenols are more acidic than alcohols. **Ans:** (b)

25.

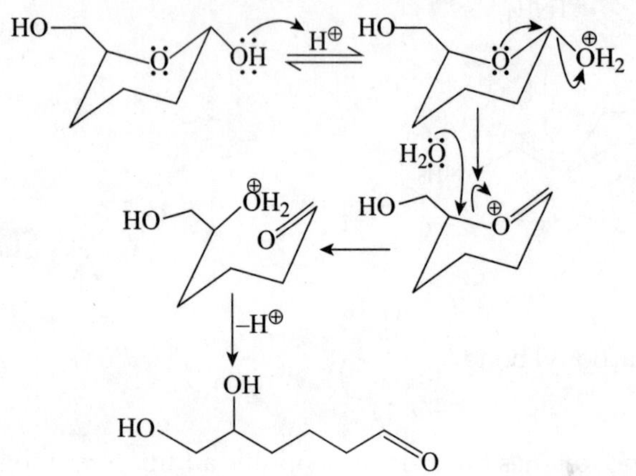

Ans: (a)

26.

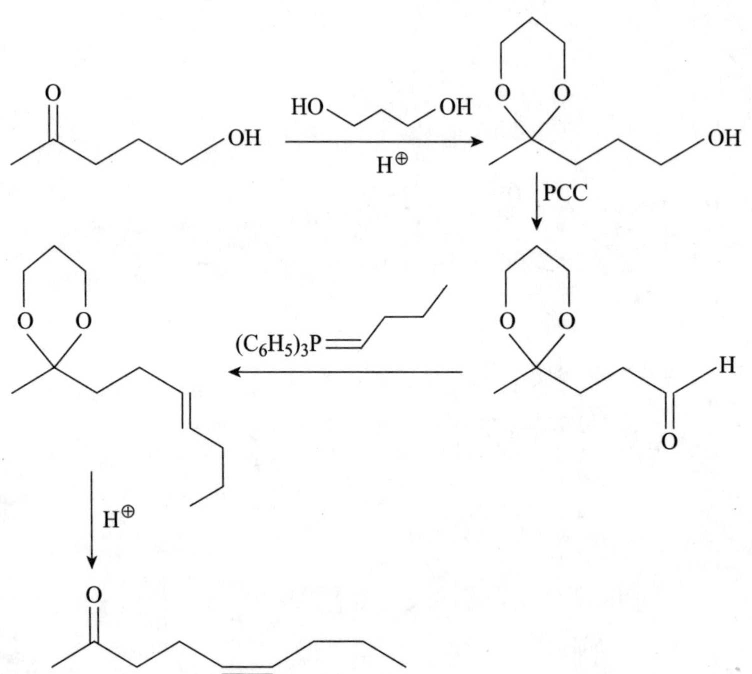

Ans: (d)

27.

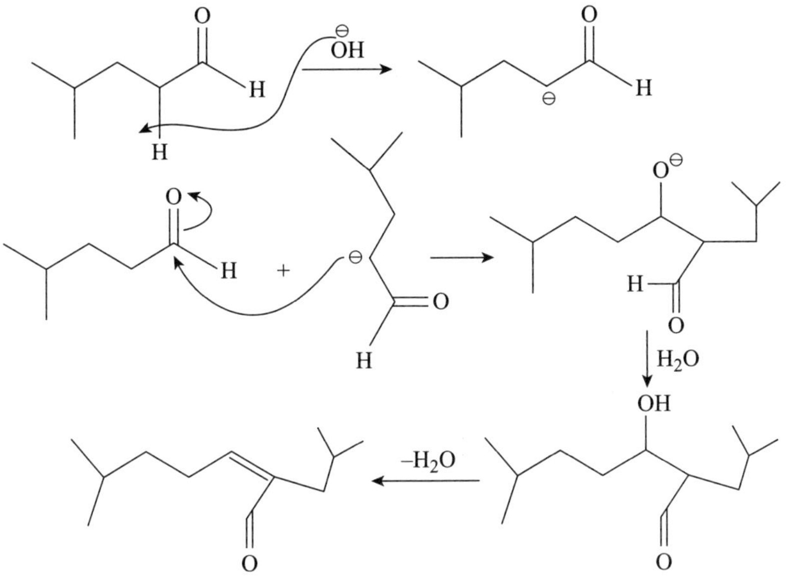

Ans: (c)

28.

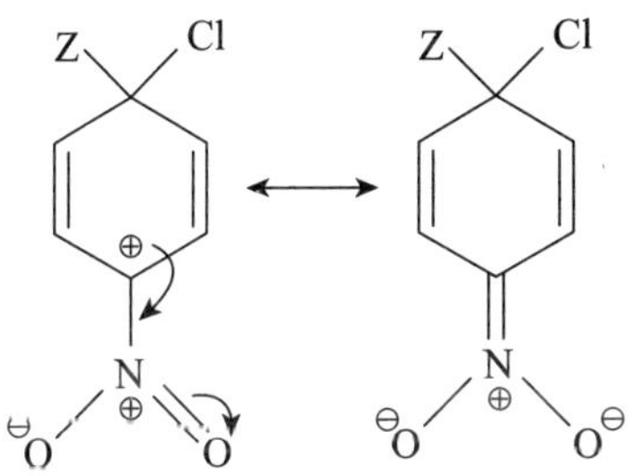

Ans: (b)

29.

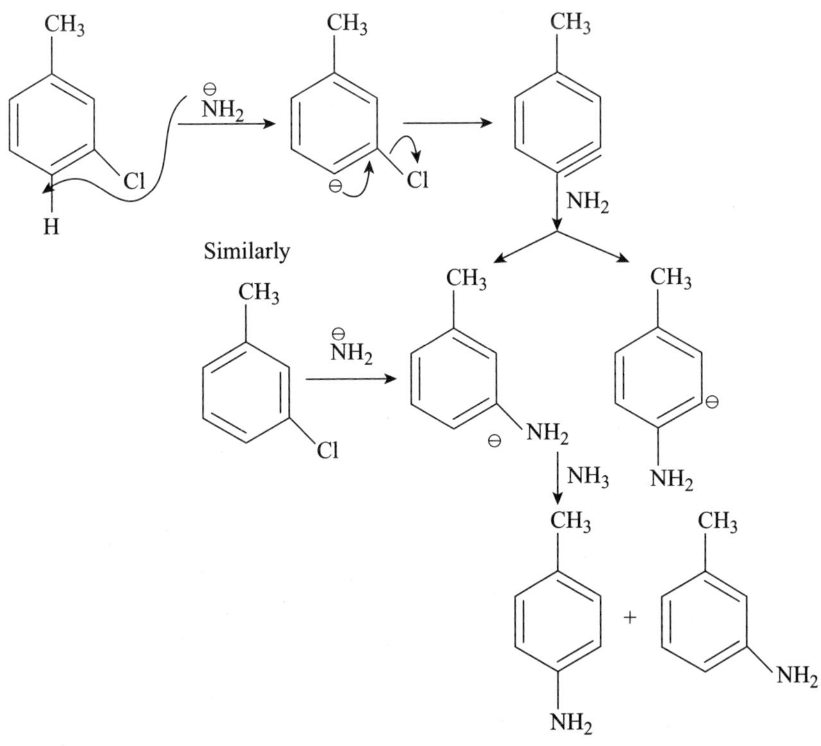

Ans: (d)

30.

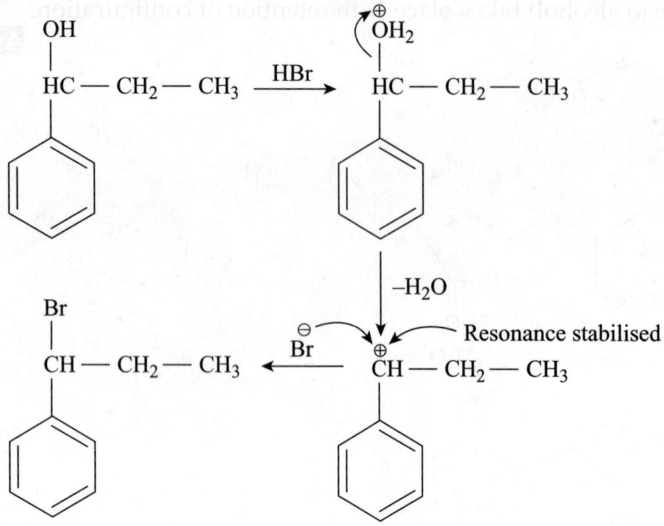

Ans: (b)

31.

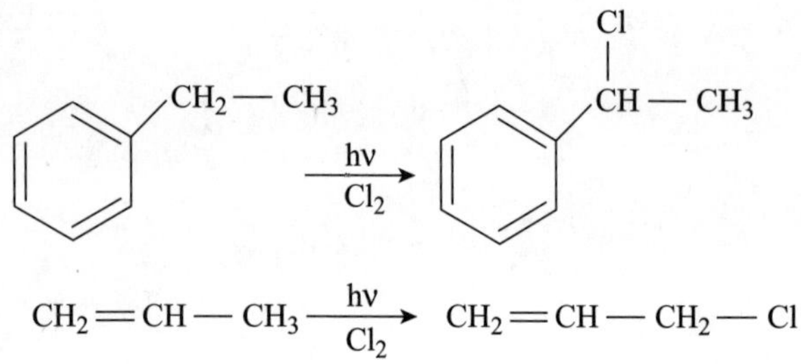

Ans: (b)

32.

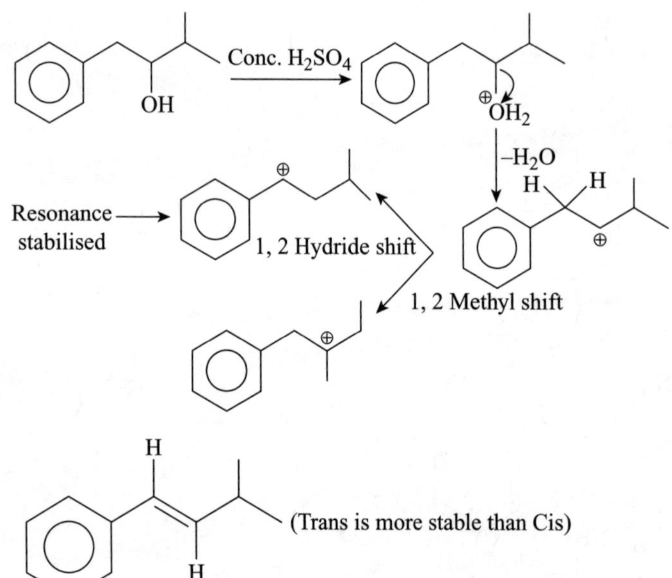

Ans: (b)

33.

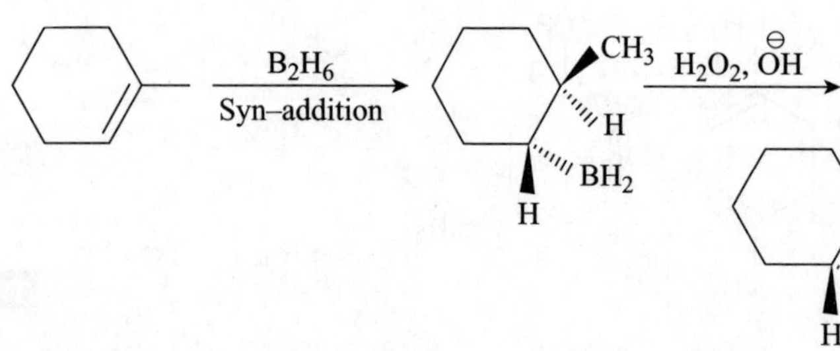

The second step (oxidation of the borane to alcohol) takes place with retention of configuration.

Ans: (b)

34.

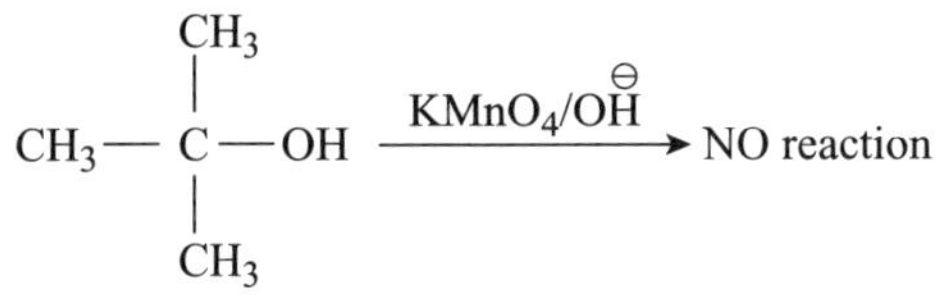

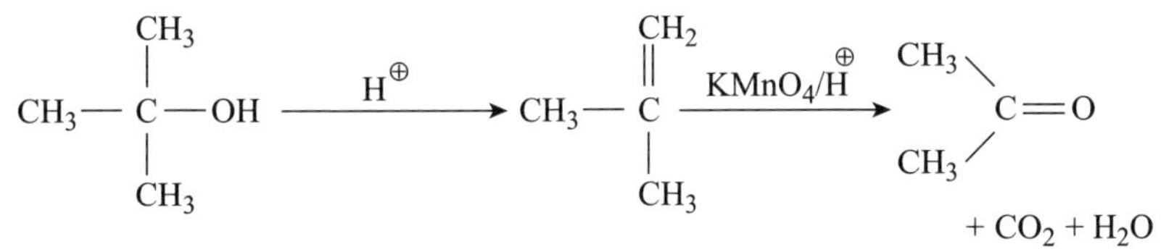

Ans: (c)

41.

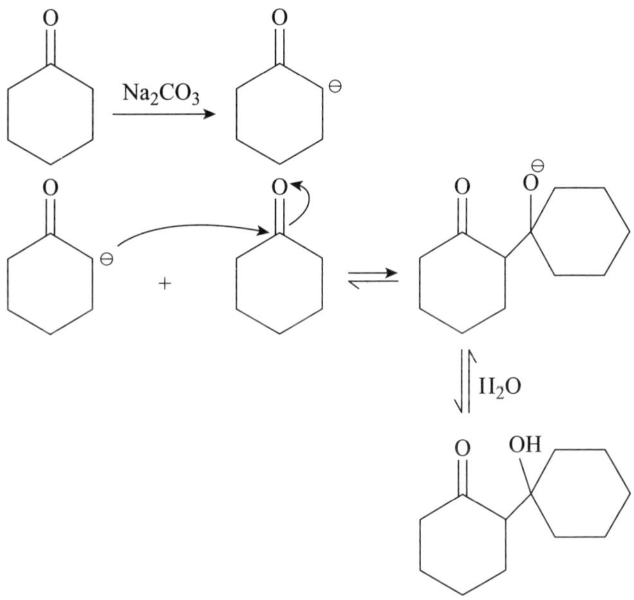

$[NaOH + N_2CO_3 \rightleftharpoons Na_2CO_3 + 2H_2O]$

↑ Basic

Ans: (c)

43.

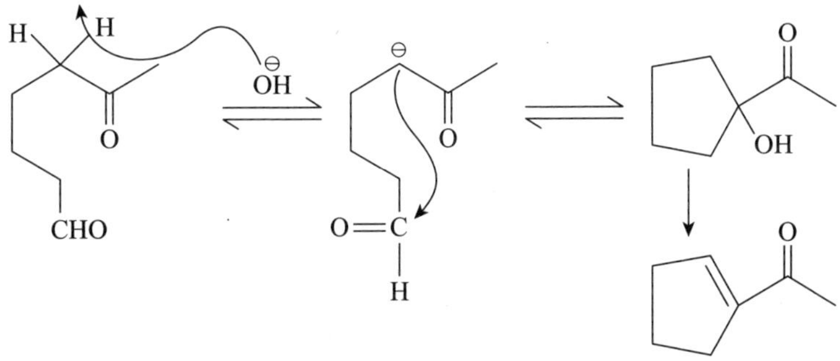

Ans: (d)

44.

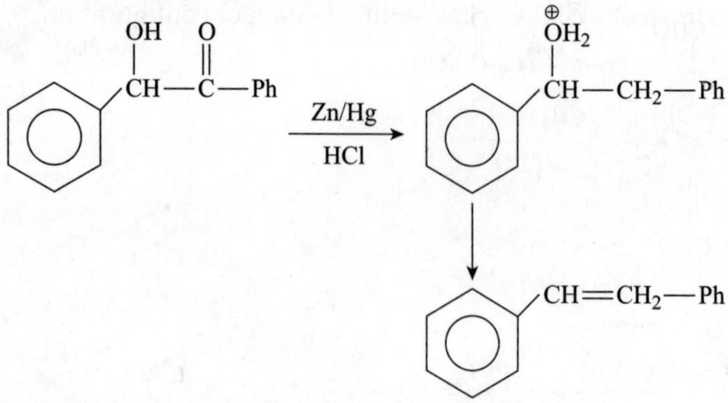

Note: Acid is present in the reaction mixture (HCl).

Ans: (d)

47.

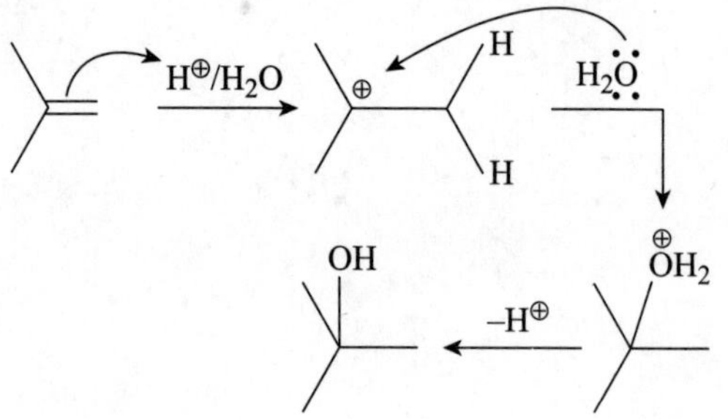

Ans: (c)

51.

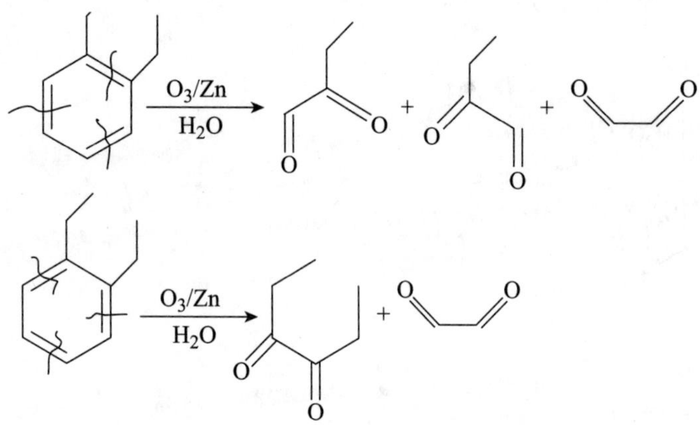

Totally 3 types of products are formed.

Ans: (c)

52.

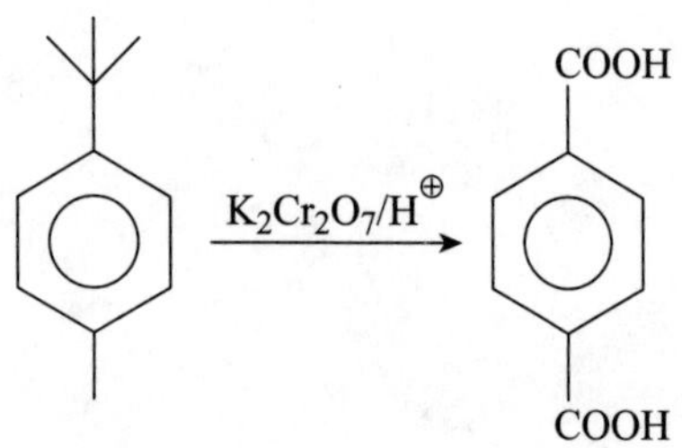

Ans: (c)

53.

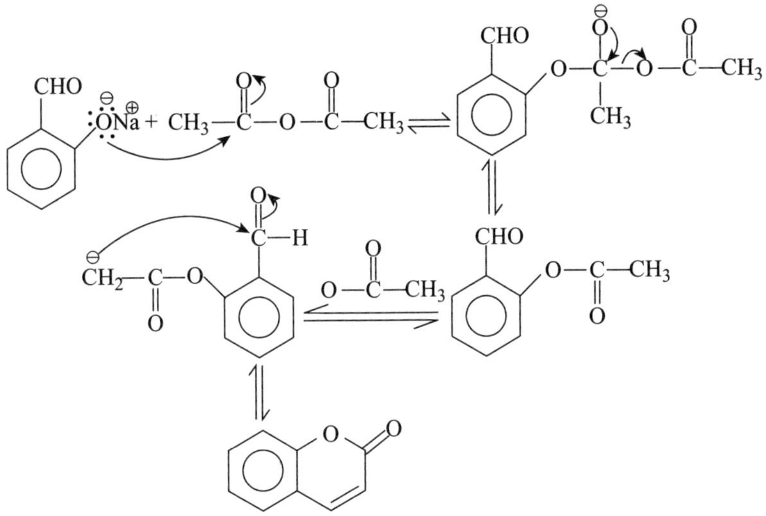

Ans: (d)

55.

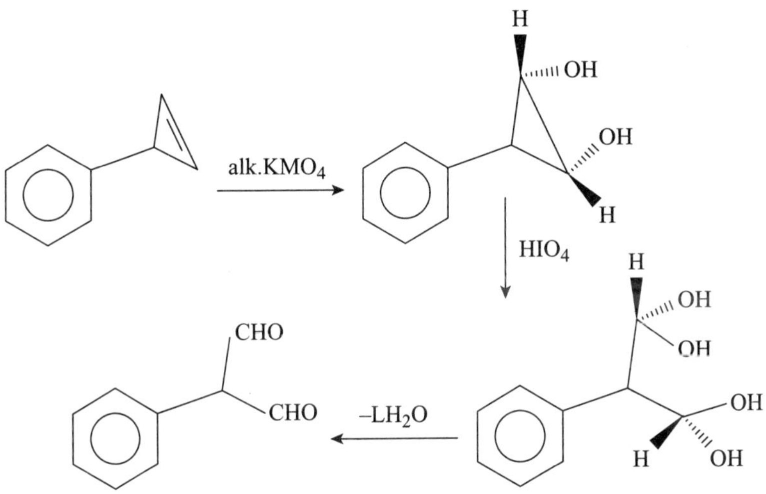

Ans: (c)

56.

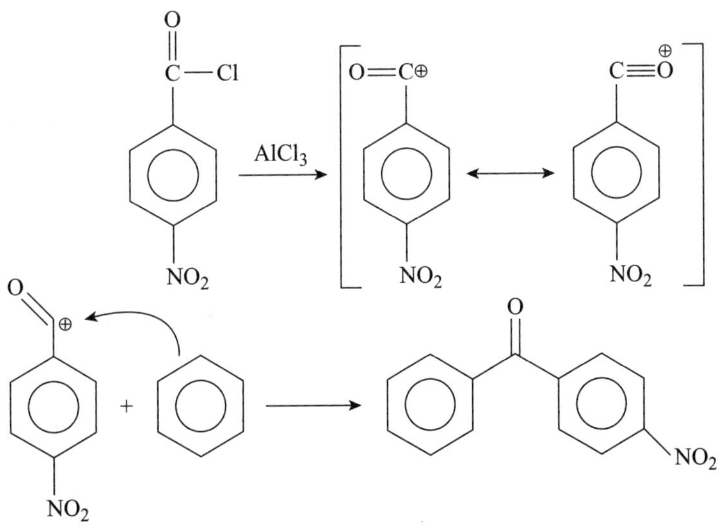

Ans: (c)

58. Benzyne mechanism is followed (similar to Question Number 29). **Ans:** (c)

59.

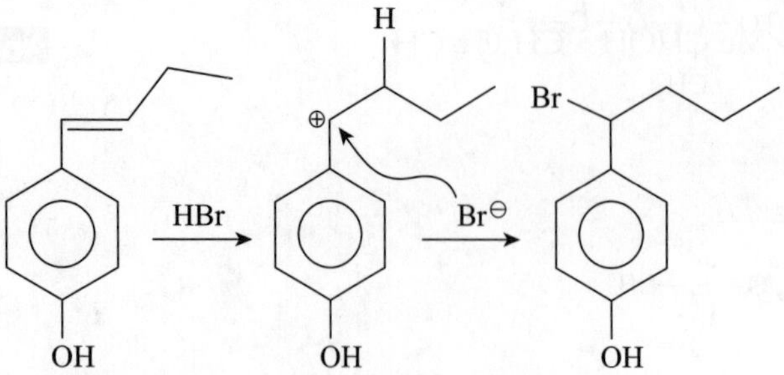

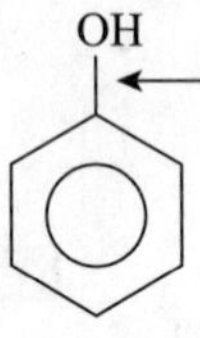

← Partial double bond character does not show substitution reaction like,

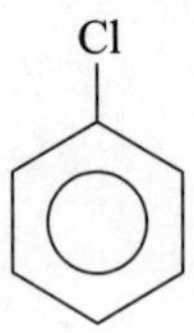

(Chlorobenzene) and $H_2C{=}CH{-}Cl$ (vinylchloride)

Ans: (b)

62. Lone pair present on oxygen will form H-bond with water molecules.

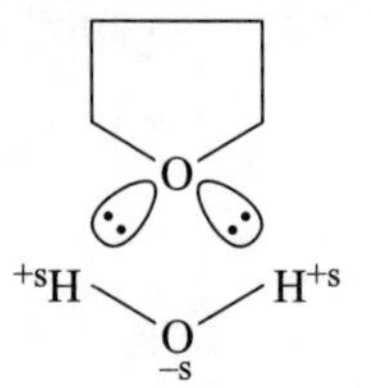

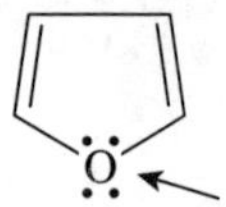

← Lone pair is delocalised (Aromatic)

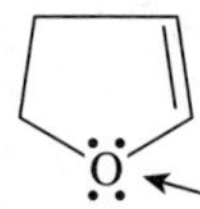

← Delocalised

Ans: (a)

64.

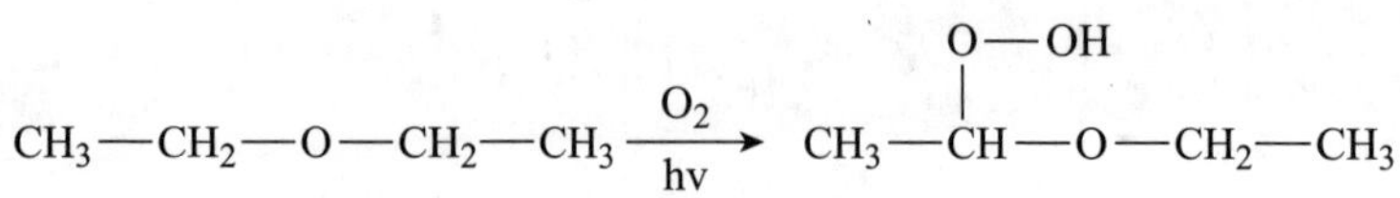

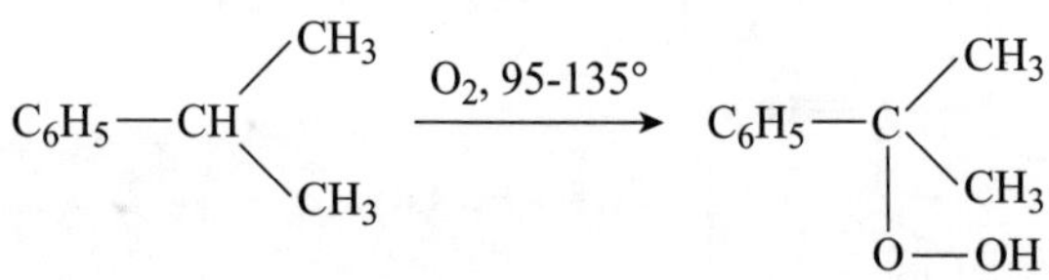

This reaction, called autoxidation, takes place by a free-radical chain mechanism. **Ans:** (c)

66.

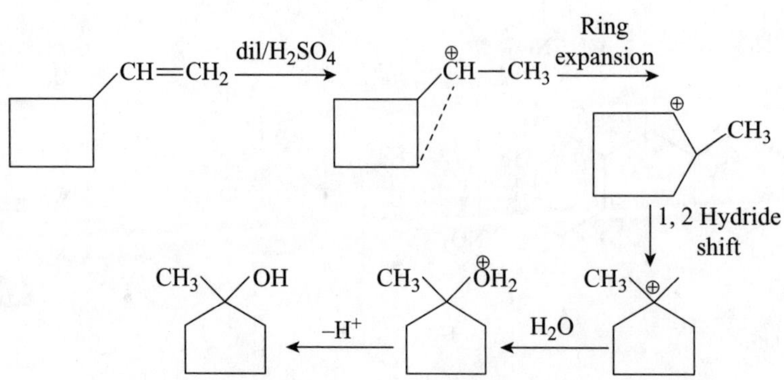

Ans: (c)

67. CH_3OH is more acidic than H_2O pK_a of water is 15.7 with a pK_a of around 16-19 alcohols are generally slightly weaker acid than H_2O.
Hence acidic strength order is $CH_3OH > H_2O > Me_2CHOH > CH_3C \equiv CH$ **Ans:** (d)

68. 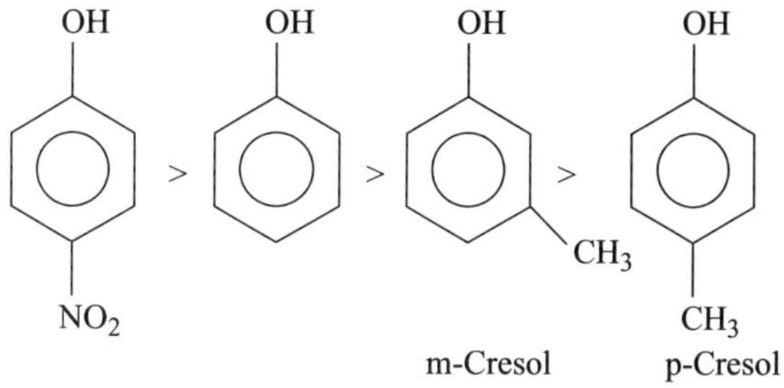

Ans: (b)

69. 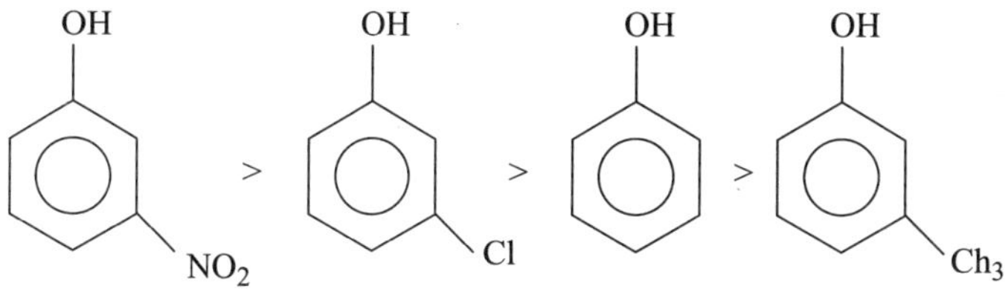

Ans: (c)

72. 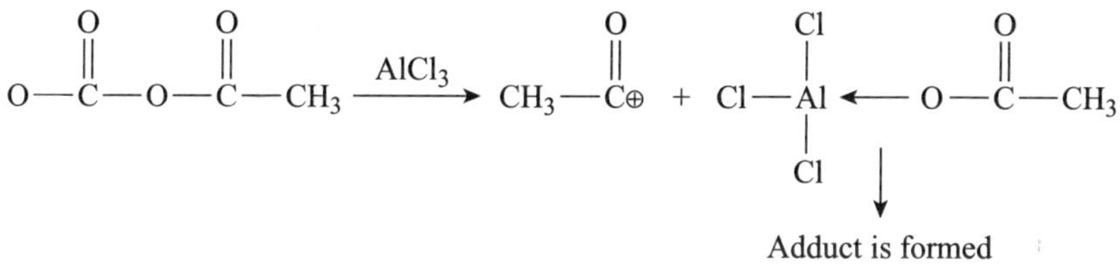

$AlCl_3$ is consumed in this reaction. **Ans:** (c)

73. Acids which are more acidic than 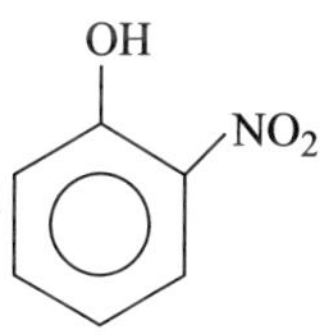 give effervescence with $NaHCO_3$. **Ans:** (c)

76. Cl_3C — CHO is the most reactive and $CH_3-\overset{O}{\overset{\|}{C}}-CH_3$ is the least reactive. **Ans:** (c)

77. $H-\overset{O}{\overset{\|}{C}}-H$ is the most reactive and $CH_3-\overset{O}{\overset{\|}{C}}-CH_3$ is the least reactive.

Hint: Simply consider their relative reactivity. **Ans:** (c)

78. 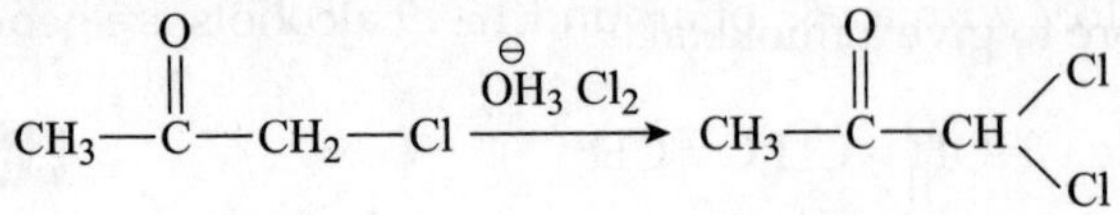

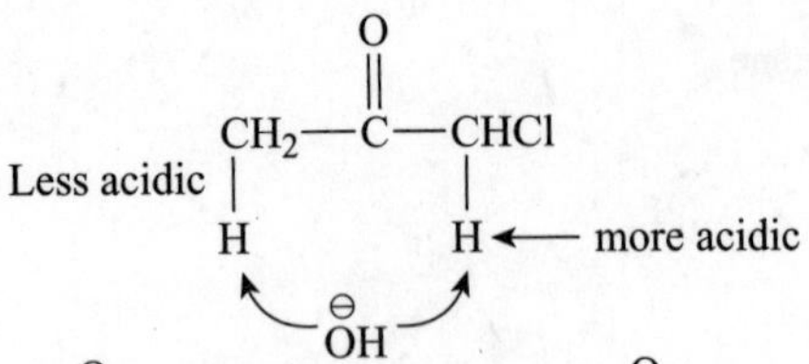

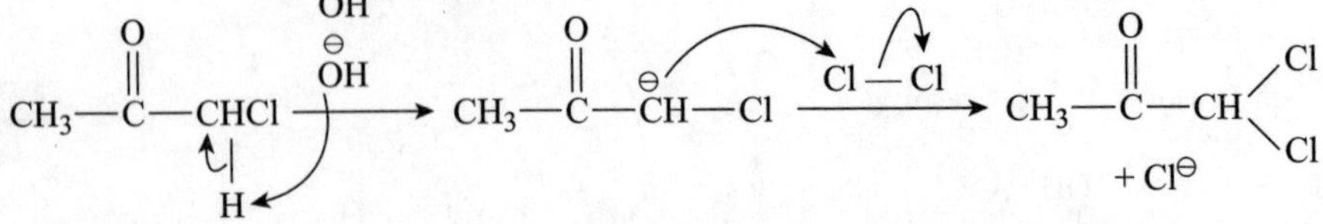

Similarly

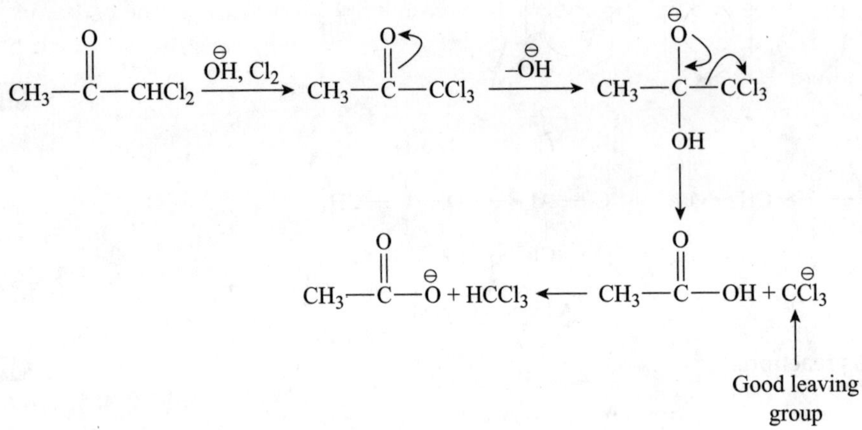

Ans: (b)

80. $CH_3-\overset{O}{\overset{\|}{C}}-CH_3 > CH_3-CH_2-\overset{O}{\overset{\|}{C}}H$

Boiling point order

Ans: (c)

81. 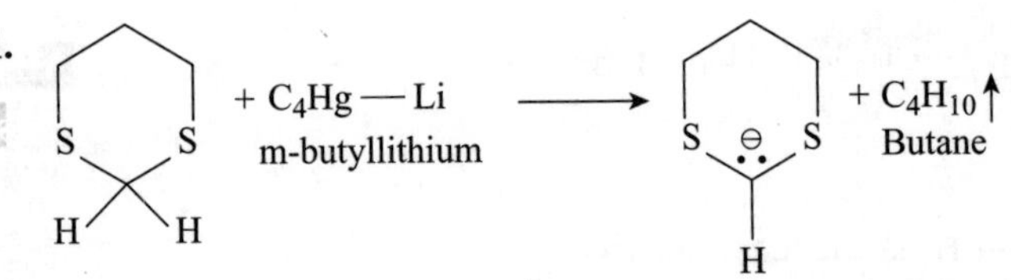

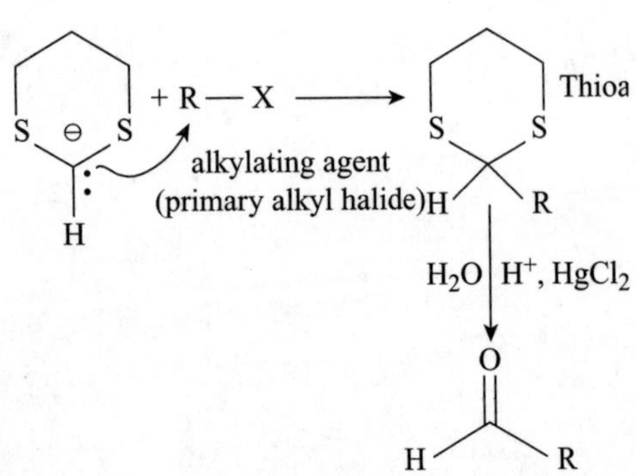

Alternatively, thioacetal can be alkylated once more to give a thioketal.

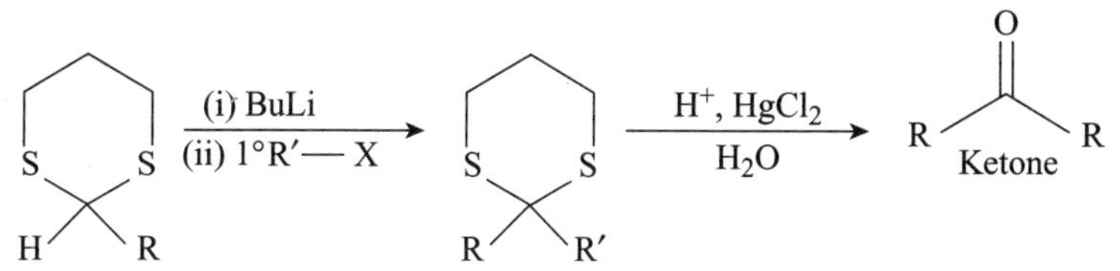

Ans: (c)

82. $$H—\overset{O}{\overset{\|}{C}}—H \underset{}{\overset{H_2O}{\rightleftharpoons}} H—\underset{OH}{\overset{OH}{C}}—H$$

Ans: (d)

83. Certain aldehydes and ketones containing a strongly electron-attracting group form hydrate of sufficient degree of stability to permit their isolation. Chloral is destabilised by strong dipole-2 repulsion between the carbonyl group and the trihalomethyl group. It achieves stability by destroying the carbonyl group by the formation of hydrates. In the hydrate, the carbonyl group dipole is replaced by much weaker dipole of diol. The hydrogen bonding between –OH group and halogen atoms is another factor that stabilises the hydrates.

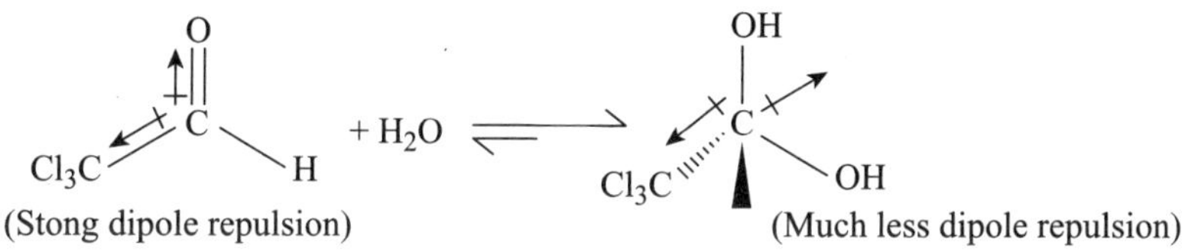

Ans: (b)

84.

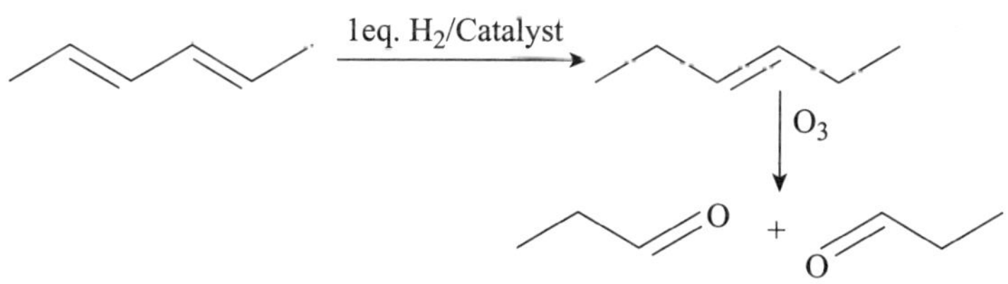

Ans: (d)

85. $$CH_3—\overset{O}{\overset{\|}{C}}—C_2H_5 \xrightarrow{\overset{\oplus\ \ominus}{NaCN}} CH_3—\underset{CN}{\overset{ONa}{C}}—C_2H_5 \xrightarrow[NH_4OH]{NH_4Cl} CH_3—\underset{CN}{\overset{NH_2}{C}}—C_2H_5$$

Ans: (c)

86.

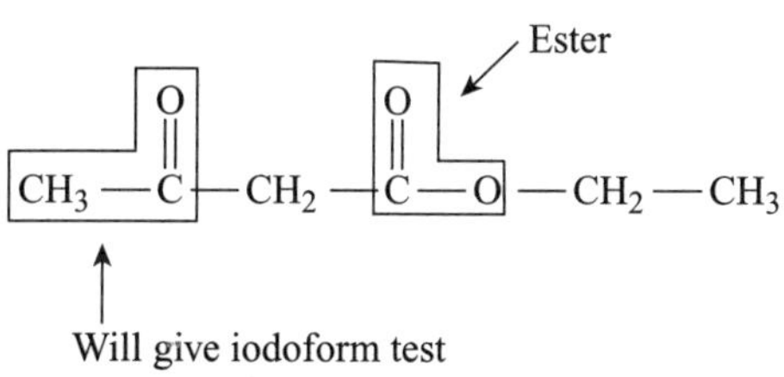

$$CH_3—\overset{O}{\overset{\|}{C}}—O—CH_2—CH_3 + H_2N—Z \rightleftharpoons CH_3—\overset{O}{\overset{\|}{C}}—NH—Z$$

Amide is formed

Ans: (b)

91.

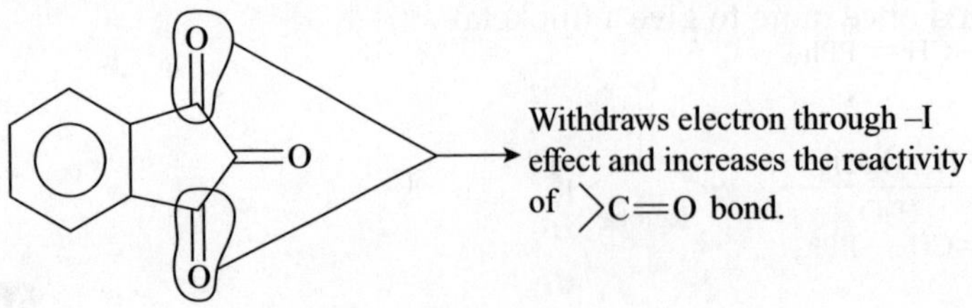

Ans: (a)

92.

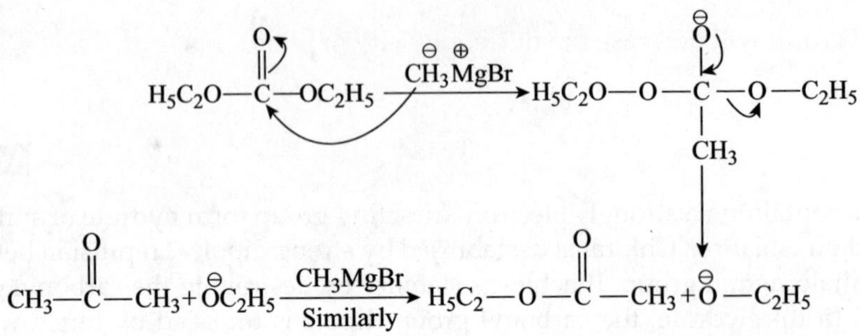

Ans: (d)

93.

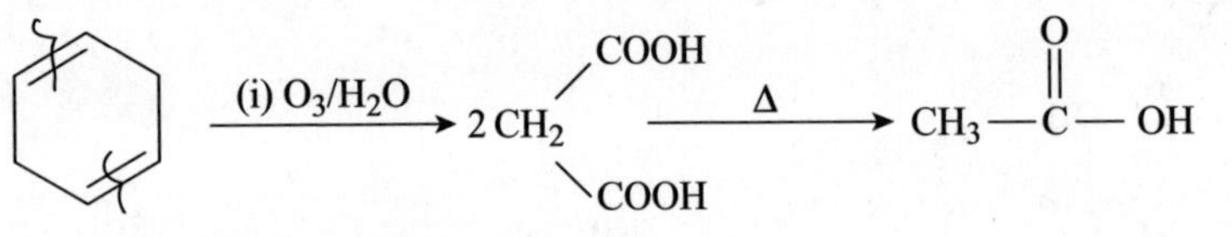

Ans: (b)

102. (See the solution of Question Number 109).

Ans: (a)

103. $Al(CH_2{-}CH{-}Me_2)_2H$ and $Li^+[Al(OCMe_3)_3H]^\ominus$ are weak reducing agents.

Ans: (a)

104.

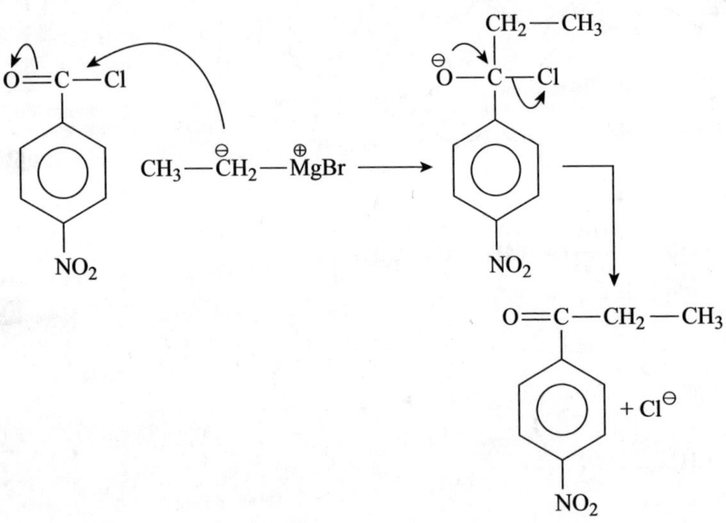

Ans: (c)

105. Paraldehyde is polymer of acetaldehyde while others are polymer of formaldehyde.

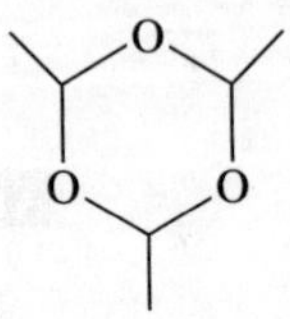

Paraldehyde

Ans: (b)

106. 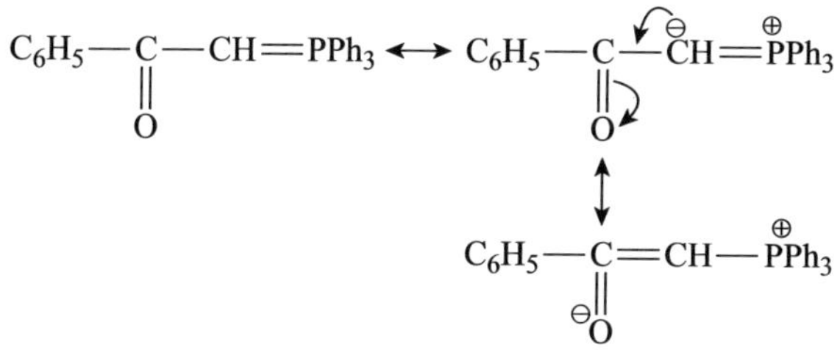

Hence, the presence of —C(=O)— group will decrease the nucleophilicity of ylide. **Ans:** (d)

108. 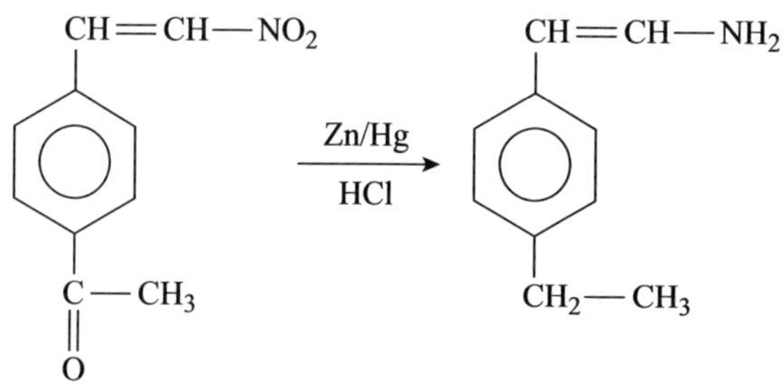

$$R-NO_2 \xrightarrow{Zn/H^+} R-NH_2$$

$$R-NO_2 \xrightarrow[\text{Active metal} = \text{Fe, Zn, Sn}]{\text{Active metal and H+}} R-NH_2$$

Ans: (b)

109. 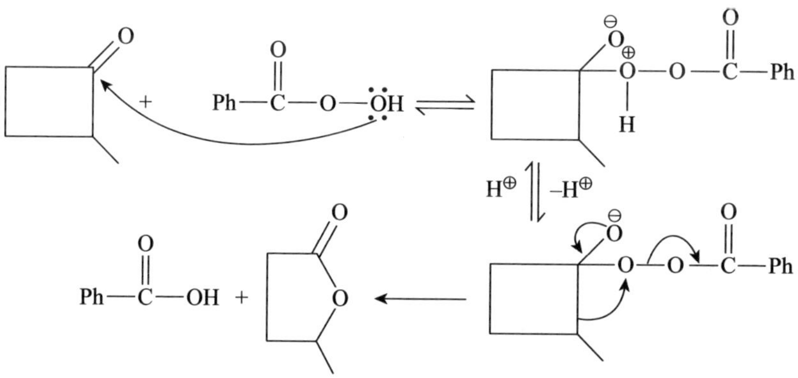

Relative migration tendencies H > Tert-alkyl > Sec-alkyl ≥ Phenyl > Primary alkyl > Methyl **Ans:** (a)

110. 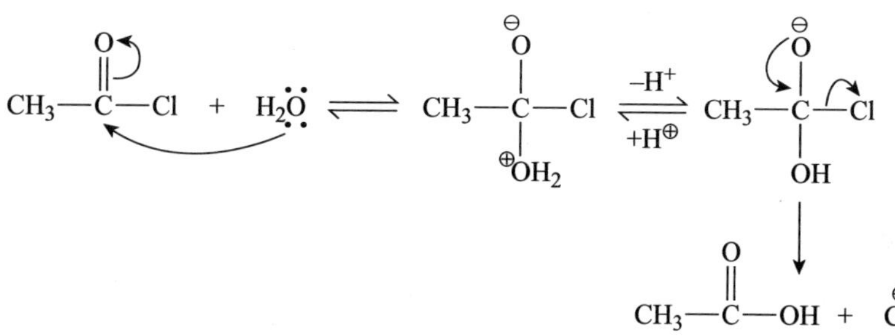

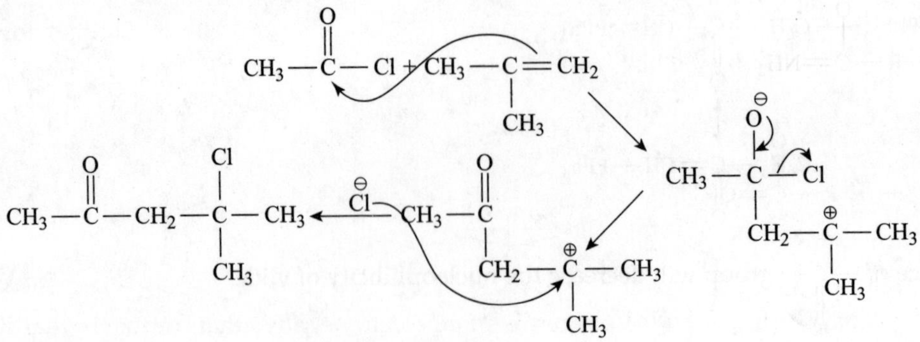

Ans: (d)

112. Relative reactivities of carboxylic acid derivatives.

$R-\overset{O}{\overset{\|}{C}}-Cl > R-\overset{O}{\overset{\|}{C}}-O-\overset{O}{\overset{\|}{C}}-R > R-\overset{O}{\overset{\|}{C}}-OR > R-\overset{O}{\overset{\|}{C}}-NH_2$ it simply tells that acid chloride can be converted into anhydride but reverse is not possible.

(i)

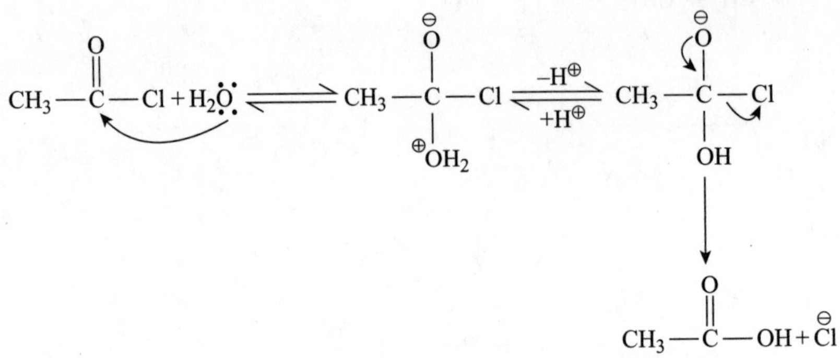

Acid chloride is going to carboxylic acid which is possible because $CH_3-\overset{O}{\overset{\|}{C}}-Cl$ is more reactive than $CH_3-\underset{O}{\underset{\|}{C}}-OH$

(ii) $CH_3-\overset{O}{\overset{\|}{C}}-O-CH_3 + HBr \rightleftharpoons CH_3-\overset{O}{\overset{\|}{C}}-Br + CH_3OH$ is not possible because $CH_3-\overset{O}{\overset{\|}{C}}-X$ is more reactive than ester, similarly (III) is also not possible. Make it a concept

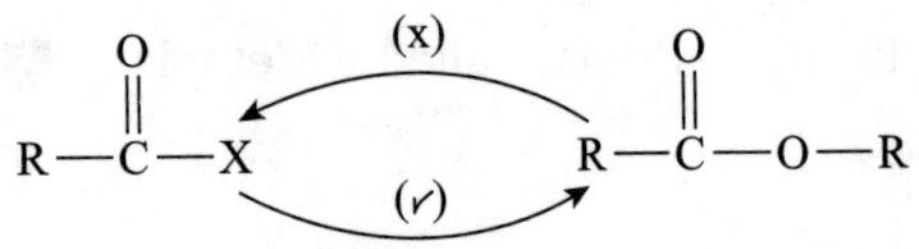

Converting into (ok)

(iv) $CH_3-\overset{O}{\overset{\|}{C}}-O-\underset{O}{\underset{\|}{C}}-CH_3 + H_2O \rightleftharpoons 2CH_3-\overset{O}{\overset{\|}{C}}-OH$

Ester is converting into acid which is possible but reverse is not possible.

Ans: (a)

114.

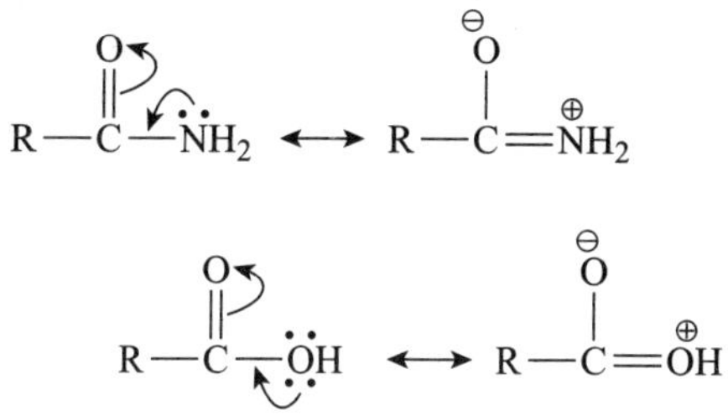

Reactivity order $R-\overset{\overset{O}{\|}}{C}-OH > R-\overset{\overset{O}{\|}}{C}-NH_2$ oxygen is more electronegative than nitrogen, that means lone pair of nitrogen will easily participate in resonance.

Ans: (d)

116. Electron withdrawing group increases the reactivity of acid derivatives towards H_2O (or towards nucleophilic addition).

Ans: (d)

118.

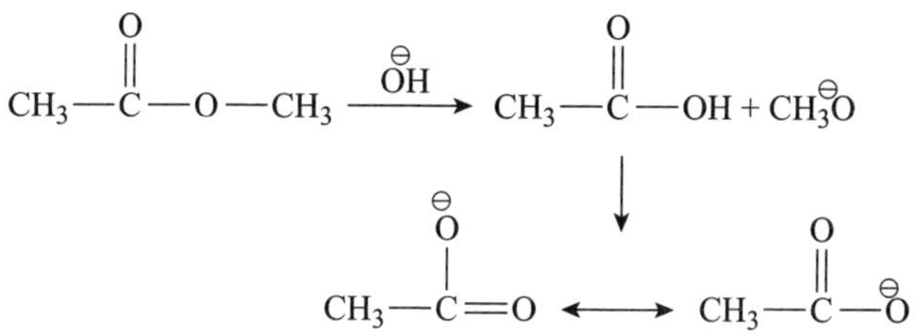

For backward reaction acid needs to be protonated. The basic form of carboxylic acid will not go into nucleophilic alyl substitution reaction because the negatively charged carboxylate ion is resistant to nucleophilic attack. So carboxylic acids can undergo nucleophilic alyl substitution reactions only if they are in acidic medium.

Ans: (c)

119.

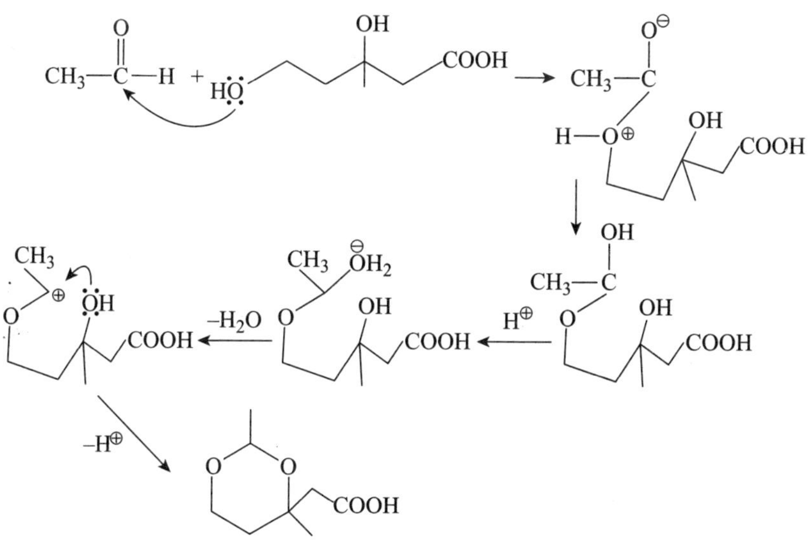

Ans: (d)

121. Just apply Le-Chatelier's principle. **Ans:** (d)

123. Intermediate carbocation is formed, i.e., relative stability of the intermediates is checked. **Ans:** (b)

124.

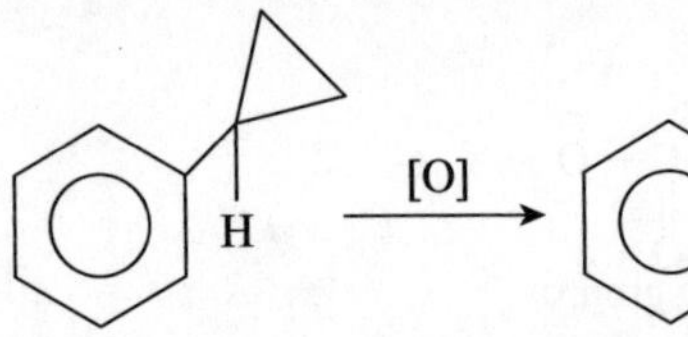

Ans: (d)

125.

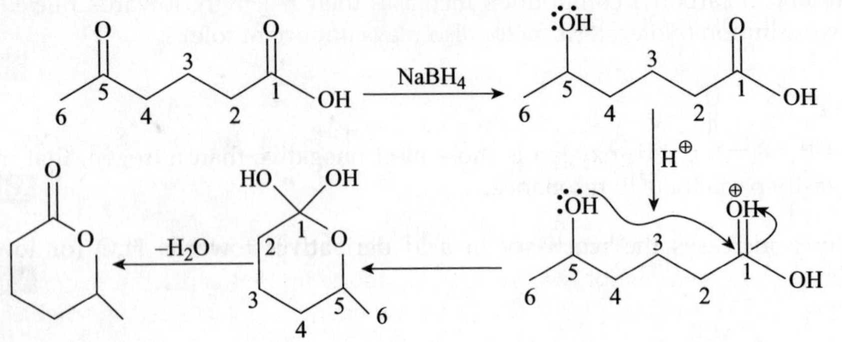

Ans: (b)

126.

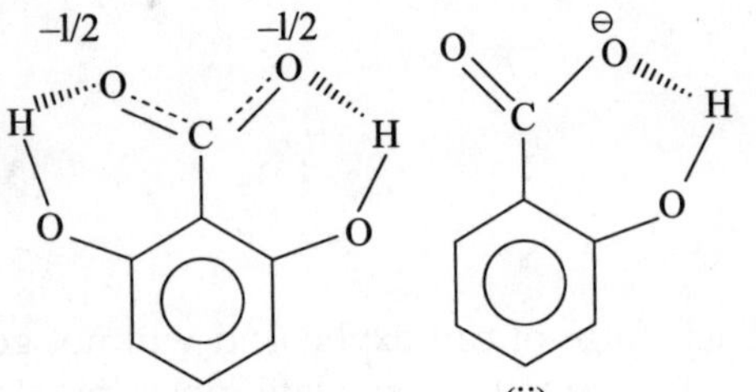

(v) (ii)

In both the cases H bonding is stabilising the conjugate base but stabilisation is more in case of (v) than (iii).

Ans: (d)

128.

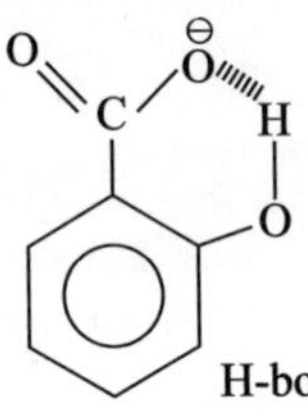

H-bonding increasing the acidic strength.

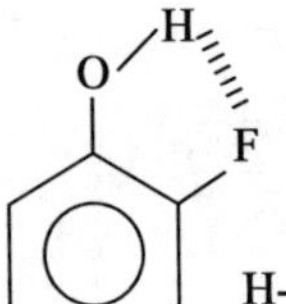

H-bonding decreasing the acidic strength.

If H-bond is formed (stronger) after removal of H$^{\oplus}$ ion then it will support acidic strength.

Ans: (c)

130.

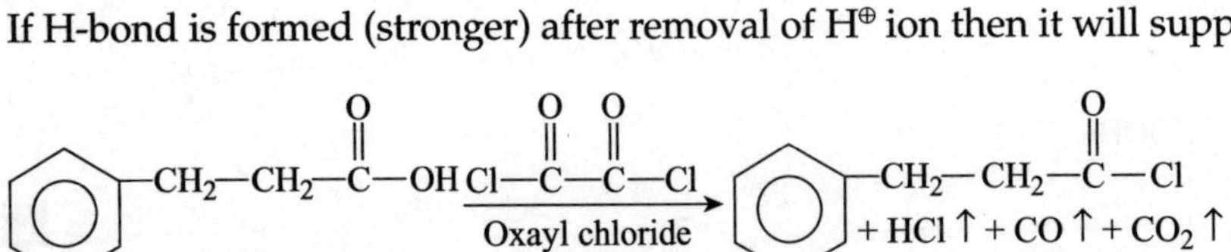

The best reagents for converting carboxylic acids to acid chlorides are thionyl chloride ($SOCl_2$) and oxalyl chloride [$(COCl)_2$] because they form gaseous by-products that does not contaminate the product. Oxalyl chloride is particularly easy to use because it boils at 62°C and any excess is easily evaporated from the reaction mixture.

Ans: (d)

137.

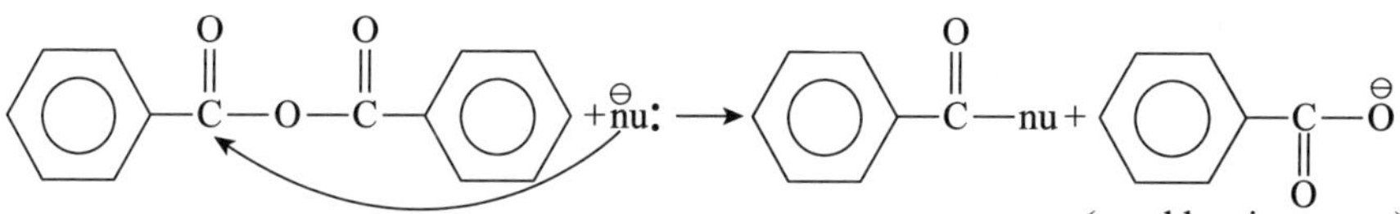

(good leaving group)

Electron withdrawing present in carbonyl compounds increases their reactivity towards nucleophilic addition reaction or acyl substitution (Note: steric factor also plays important role).

Ans: (c)

138.

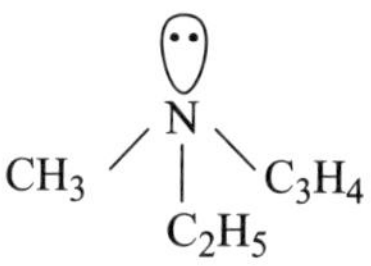

A tetrahedral amine with three different substituents (and a lone pair) is non-superimpossible on its mirror image (chiral). In most cases we cannot resolve such an amine into two entantiomers, however, because the two enantiomers interconvert rapidly. This interconversion takes place by nitrogen inversion, in which the lone pair moves from one face of the molecule to the other.

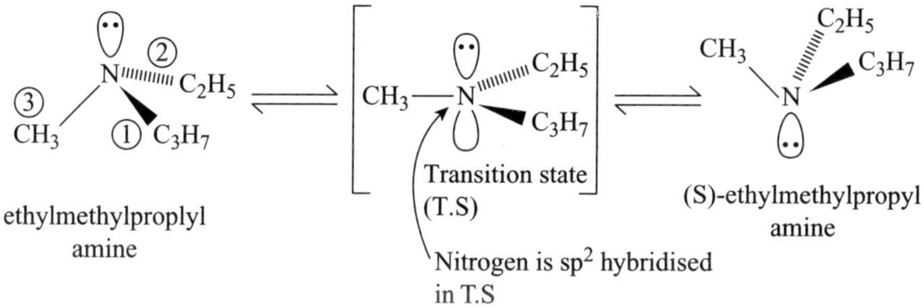

Ans: (a)

140. Chlorobenzene is a non-polar solvent, hence solvation effect is not observed which is observed in case of aqueous solution.

Ans: (a)

141.

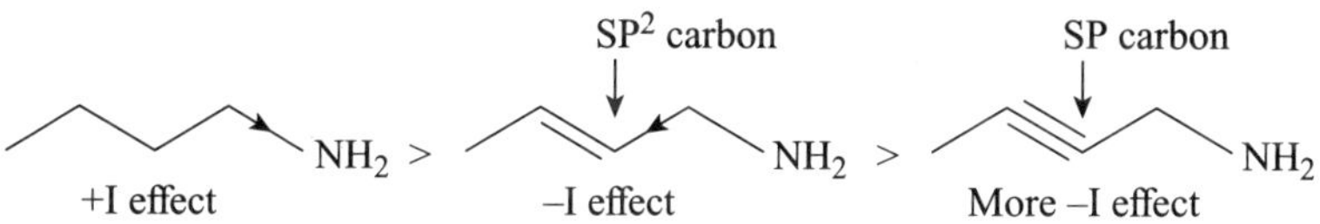

Ans: (a)

144.

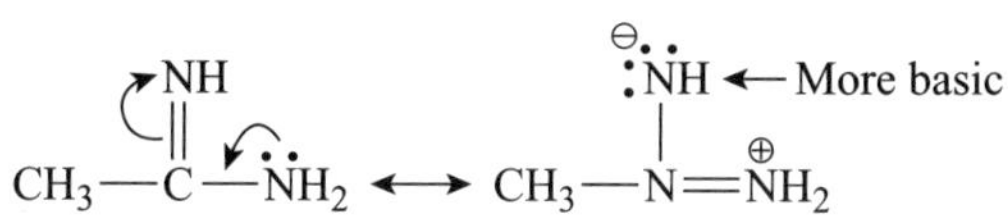

Ans: (d)

147.

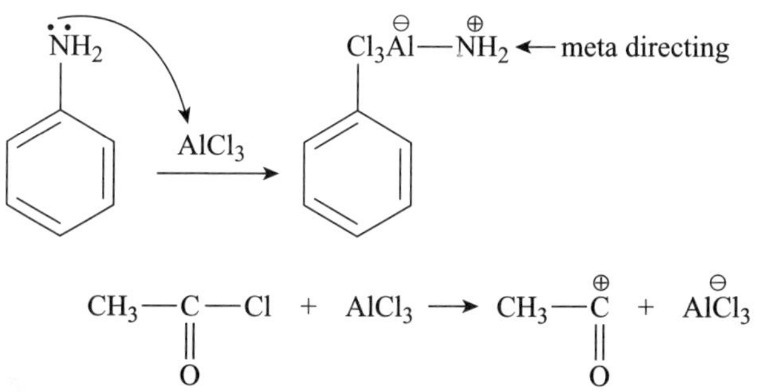

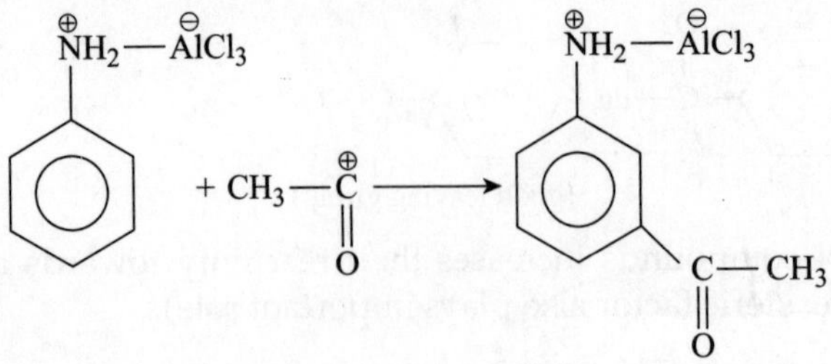

Ans: (d)

148. $CH_3-CH_2-CH_2-NH_2 \xrightarrow{NaNO_2 + HCl}$

$$\overset{+}{Na}\,\overset{\ominus}{O}-N=O + \overset{\oplus\,\ominus}{HCl} \rightleftharpoons \underset{\text{Nitrous acid}}{H-O-N=O} + NaCl$$

$$H-\ddot{\underset{..}{O}}-\ddot{N}=\ddot{\underset{..}{O}}: + \overset{\oplus}{H} \rightleftharpoons H-\overset{\oplus}{\underset{\underset{H}{|}}{O}}-N=O \rightleftharpoons H_2O + \underset{\text{Nitrosoniumion}}{[\overset{\oplus}{:N}=\ddot{O} \leftrightarrow :N\equiv\overset{\oplus}{O}]}$$

$$CH_3-CH_2-CH_2-NH_2 + NaNO_2 + 2HCl \longrightarrow CH_3-CH_2-CH_2-N\equiv NCl^- + 2H_2O + NaCl$$

$$CH_3-CH_2-CH_2-\overset{\oplus}{N}\equiv N\overset{\ominus}{Cl} \longrightarrow CH_3-CH_2-\overset{\oplus}{CH_2} + \underset{\text{Nitrogen}}{:N\equiv N:}$$

The driving force for this reaction is formation of N_2, an exceptionally stable molecule.

$$CH_3-CH_2-\overset{\oplus}{CH_2} + H_2O \longrightarrow CH_3-CH_2-CH_2-OH$$

$$\longrightarrow CH_3-\underset{\underset{OH}{|}}{CH}-CH_3$$

$$\xrightarrow{NaCl} CH_3-\underset{\underset{Cl}{|}}{CH}-CH_3$$

Ans: (c)

150.

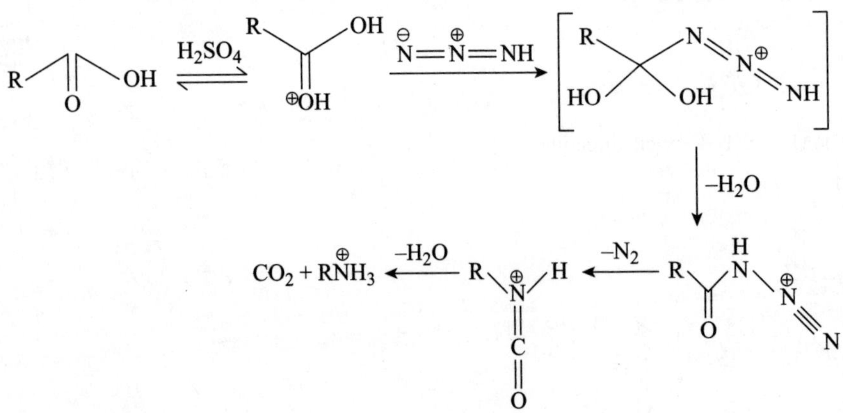

(Schmidt reaction)

Ans: (b)

151.

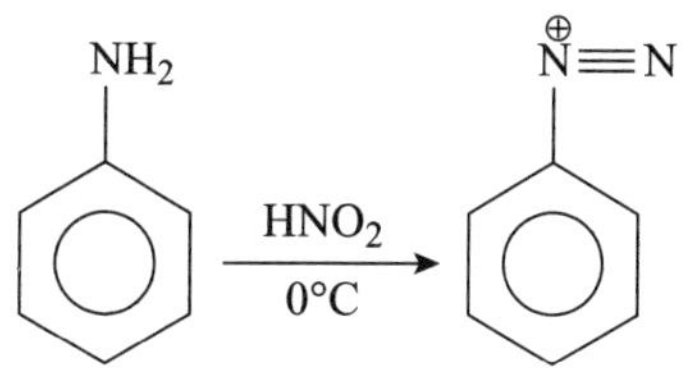

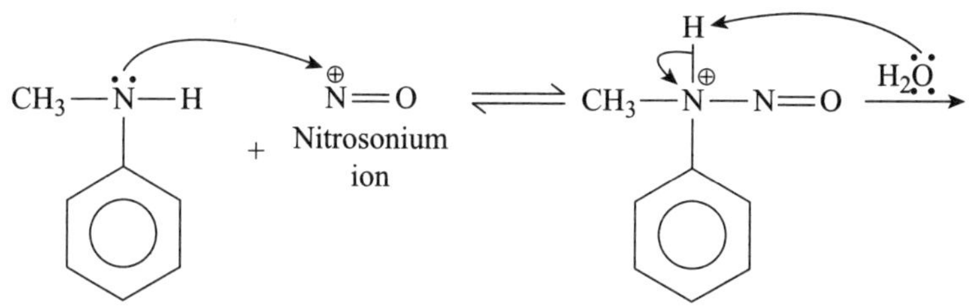

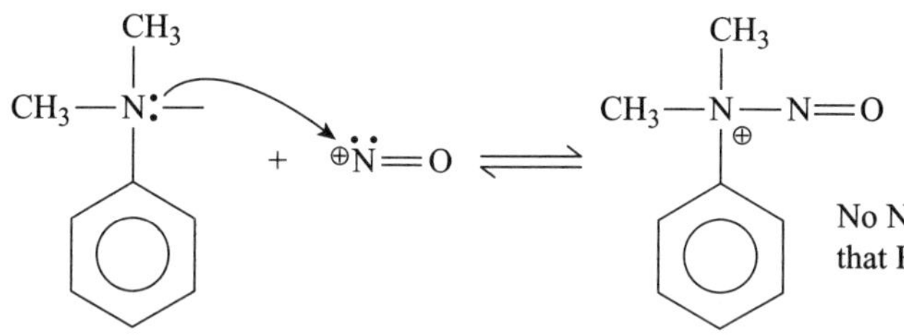

No N — H bond present so that H_2O could abstract $-H^{\oplus}$

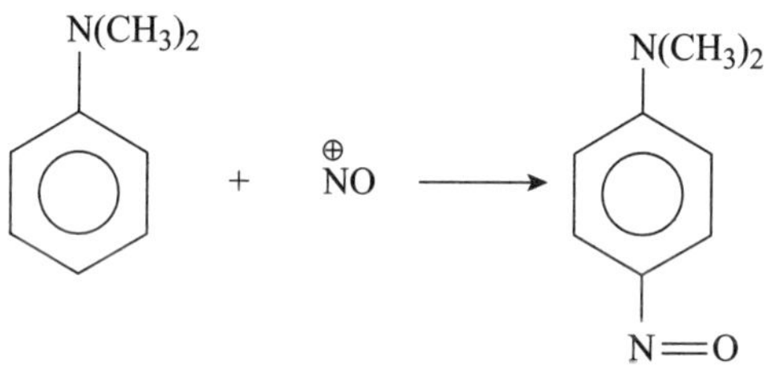

Ans: (c)

152. No isotopic effect is observed. **Ans:** (c)

154. (iii)

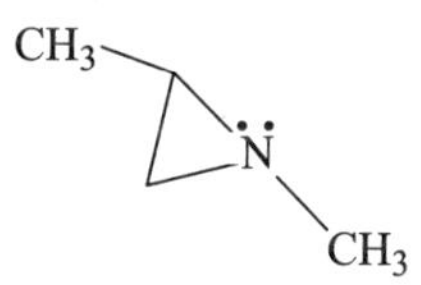

Chiral and resolvable.

If the nitrogen atom is contained in a small ring, it is prevented from attaining the 120° bond angles that facilitate inversion. Such a compound has a higher activation energy for inversion, the inversion is slow and the enantiomer may be resolved.

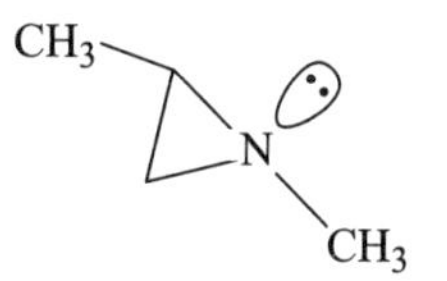

(R)-1, 2-dimethylaziridine

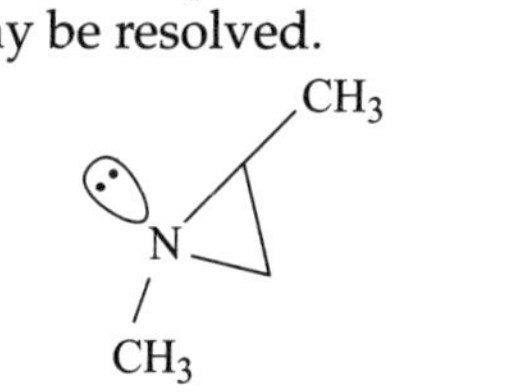

(S)-1, 2-dimethylaziridine

(iv)

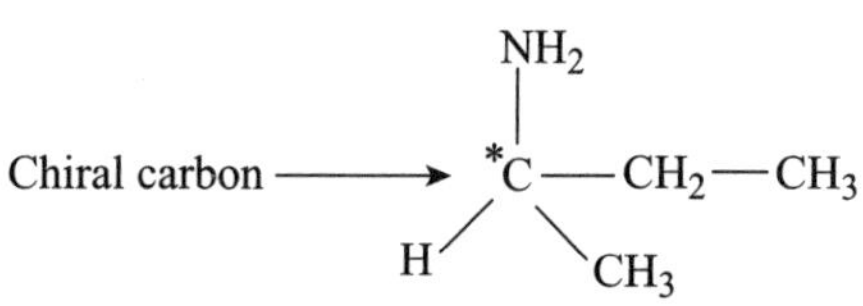

(For more details see the solution of Question Number 138 of this exercise) Aziridines = (Three-membered rings containing a nitrogen atom)

Ans: (d)

155. 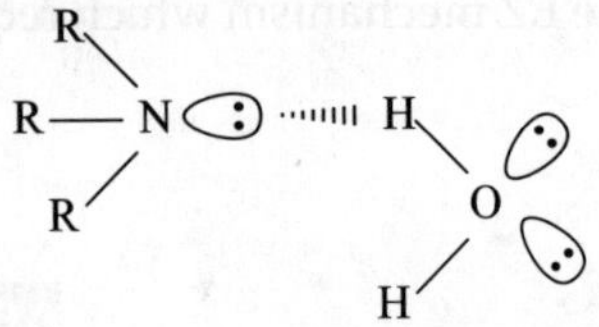

Hydrogen bond acceptor only.

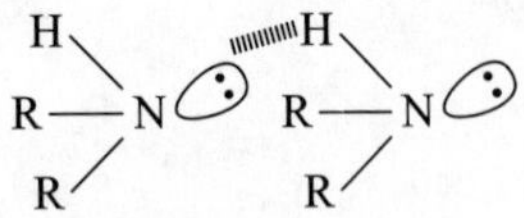

1° or 2° amines

Hydrogen bond donor and acceptor primary and secondary amines have N-H bonds, allowing them to form hydrogen bonds. Having no N-H bonds, pure tertiary amines can't engage in H-bonding. They can, however, accept hydrogen bonds from molecules having O-H or N-H bonds.

Ans: (b)

156.

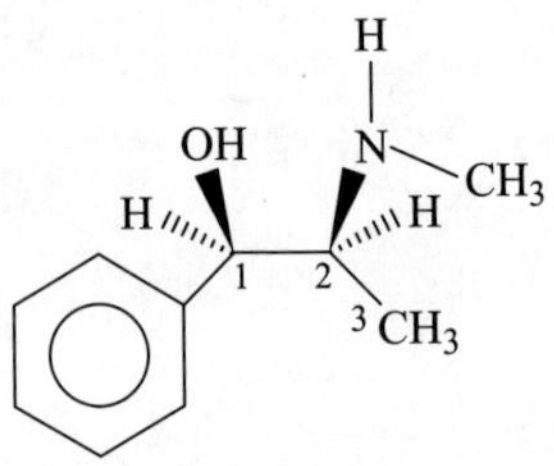

(1R, 2S) – (–) Ephedrine

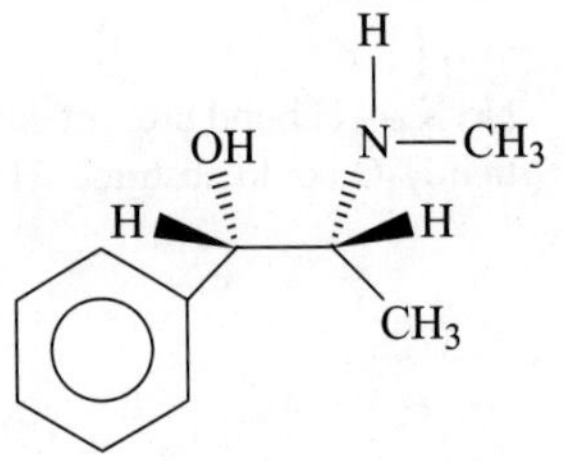

(1S, 2R) – (+) Ephedrine

The IUPAC names of the two enantiomers are (1R, 2S) respectively (1S, 2R) –2, methylamino-1-Phenylpropan-1-0l.

Ans: (d)

160. 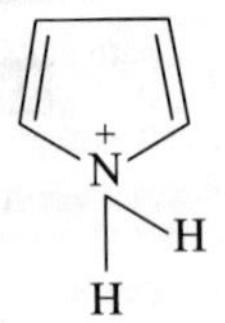 ⟶ non-aromatic (because no cyclic resonance)

Ans: (b)

161.

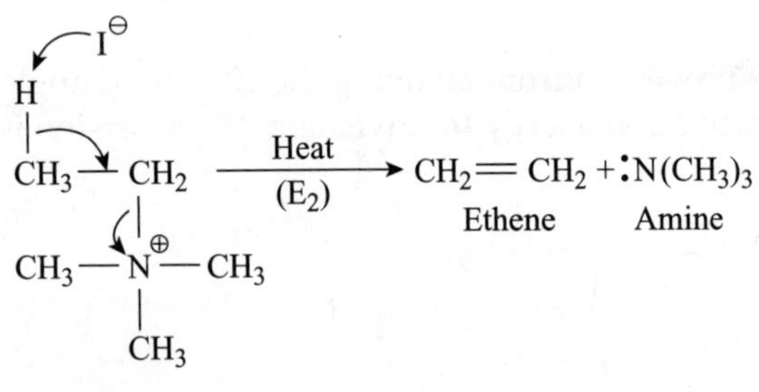

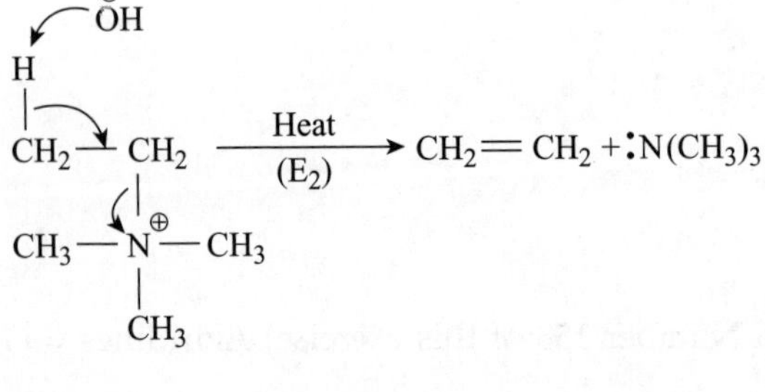

Elimination of the quaternary ammonium salt generally takes place by the EZ mechanism which requires a strong base.

$HI > H_2O$ Acidic order

$\overset{\ominus}{O}H > I^{\ominus}$ Basic order

Ans: (b)

162.

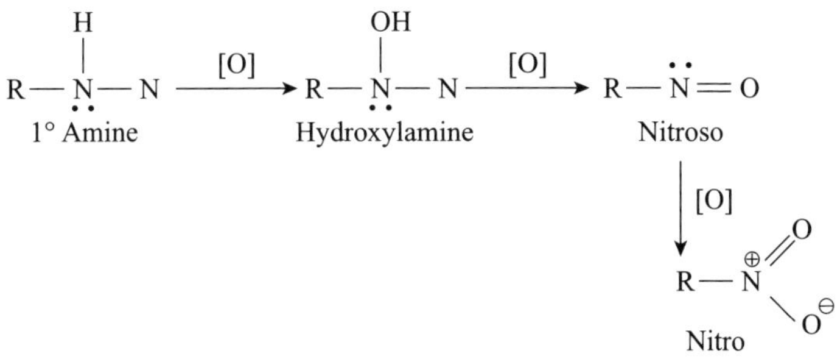

The symbol [0] is used for a generic oxidising agents, which may be H_2O_2, permanganate and peroxyacids.

Ans: (d)

165. The reaction of an aldehyde or a ketone with a secondary amine produces an enamine.

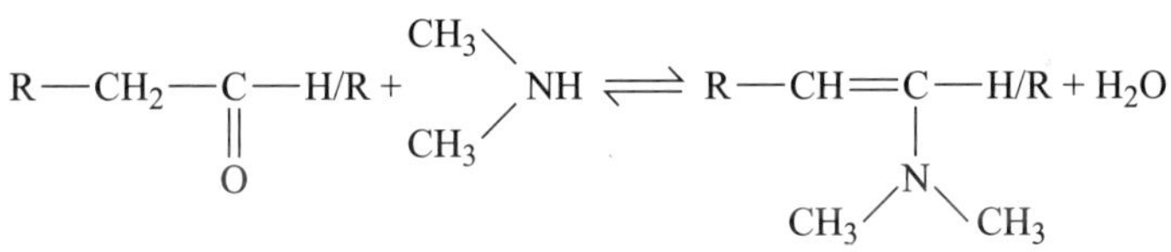

Ans: (d)

166. Pyridine resembles a strongly deactivated benzene ring. Friedel-Crafts reactions fails completely. Deactivation results from the electron withdrawing effect of the electronegative nitrogen atom. When pyridine does react, it gives substitution at the 3-position, analogous to the meta substitution shown by deactivated benzene derivatives.

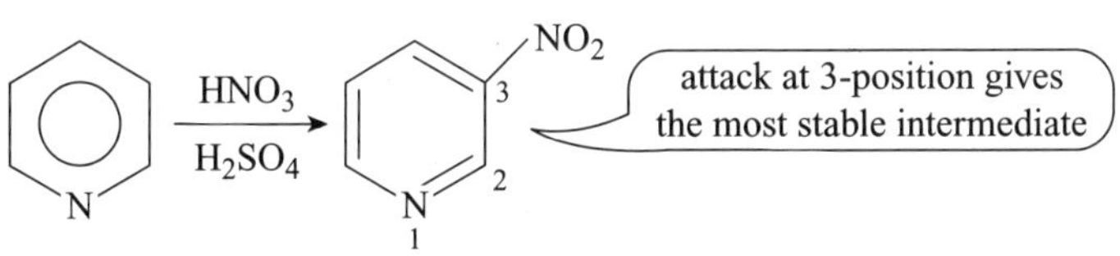

Relative reactivity towards electrophilic aromatic substitution.

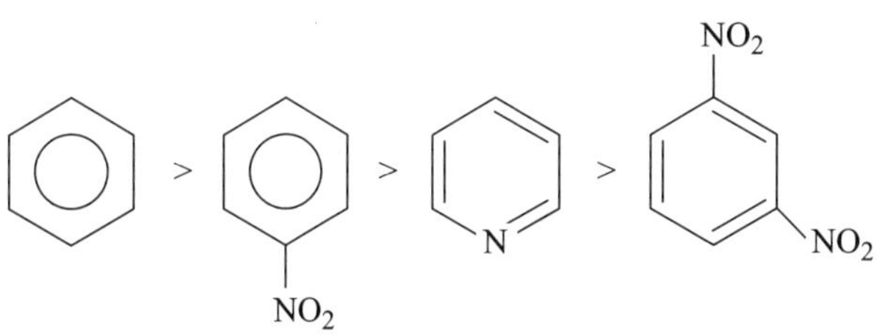

Ans: (c)

167. Relative reactivity towards ERS

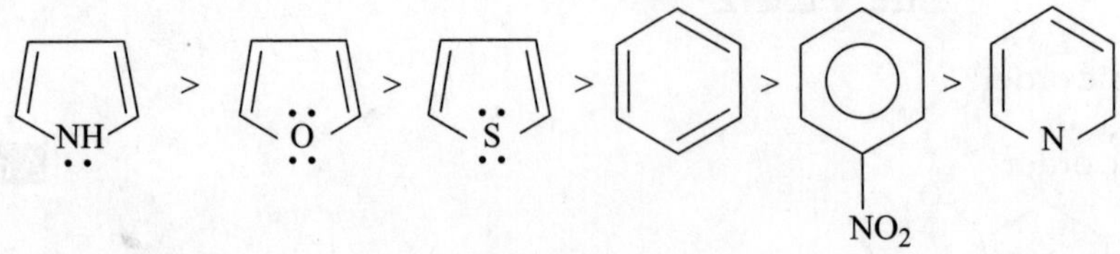

Ans: (d)

168.

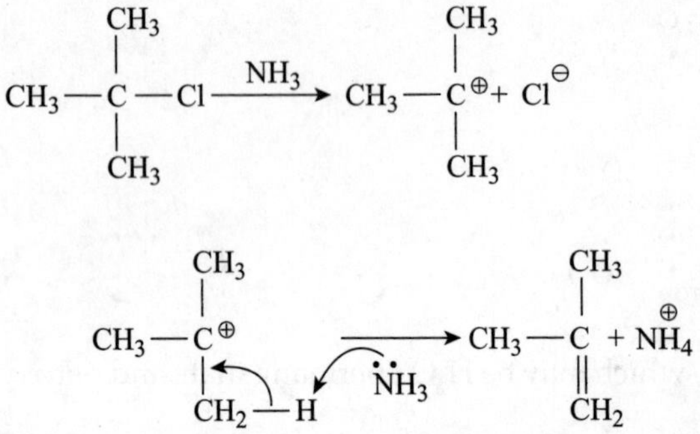

Major product

Ans: (a)

LEVEL 2

31.

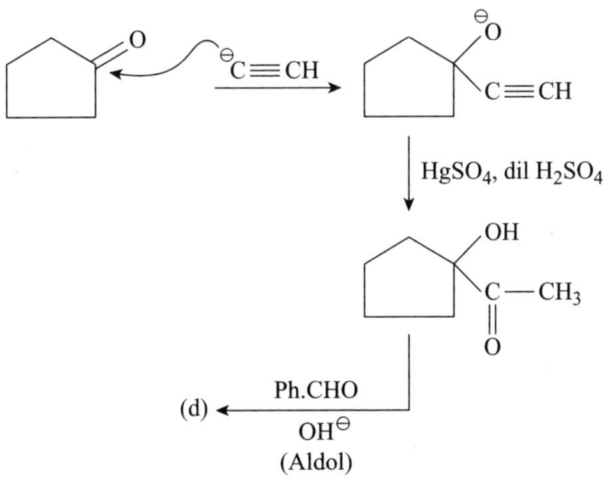

Ans: (d)

32.

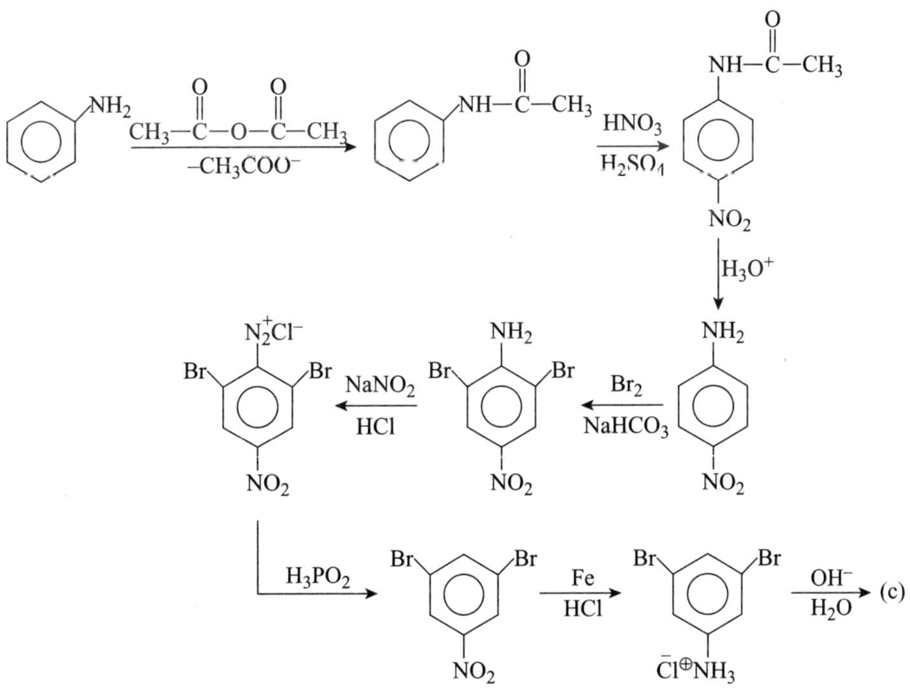

Ans: (c)

33.

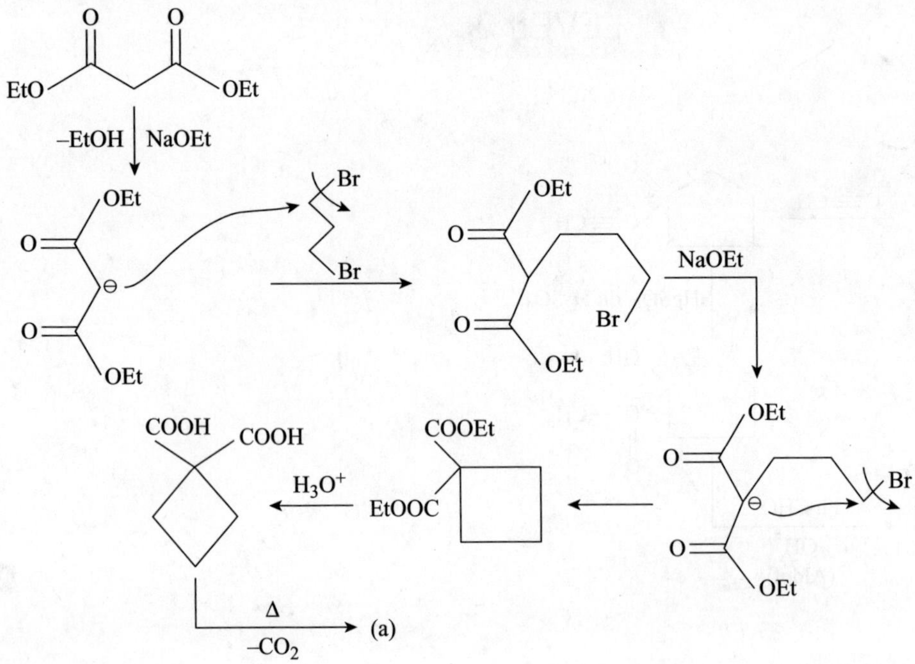

Ans: (a)

34.

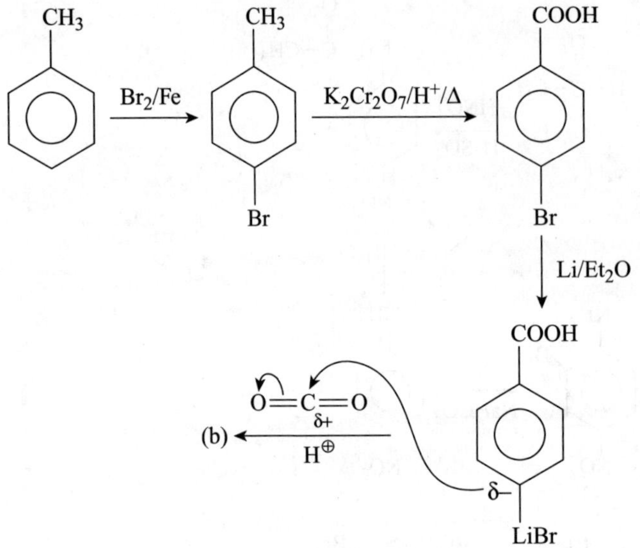

Ans: (b)

35.

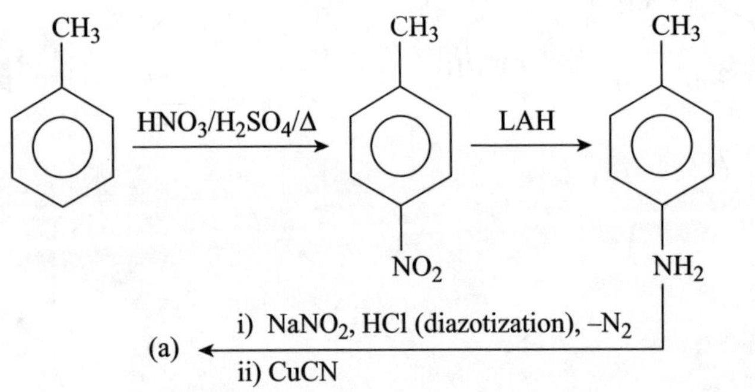

Ans: (a)

36. 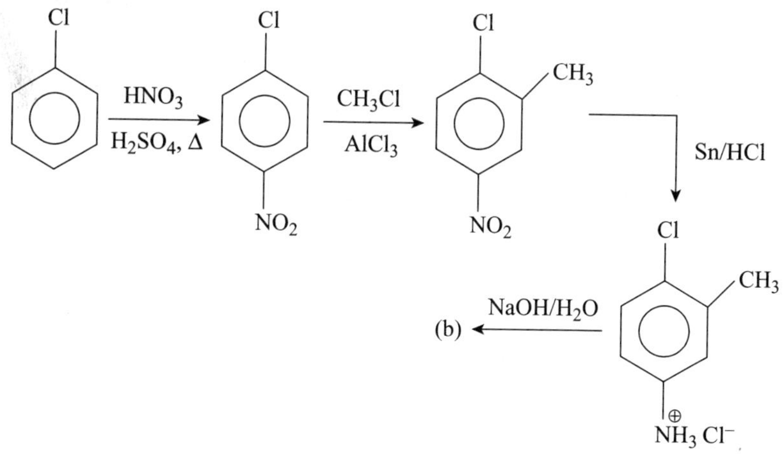

Ans: (b)

37. 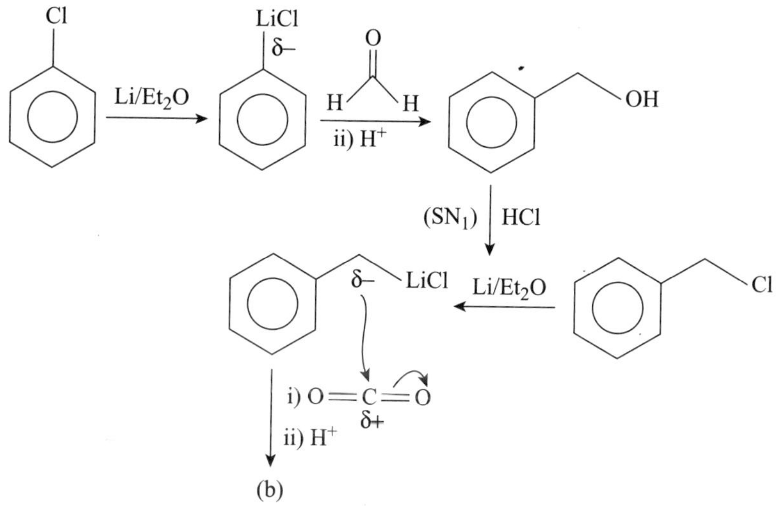

Ans: (b)

38. Hofmann's bromamide degradation.

Ans: (c)

39. 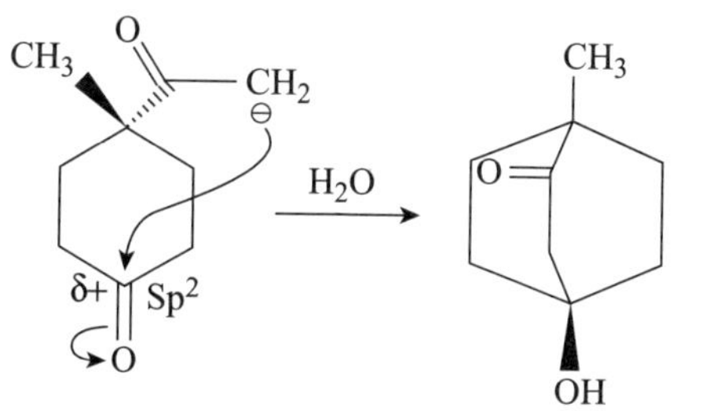

(c)
Rigid structure
(Bicyclo compound)

Ans: (c)

40. 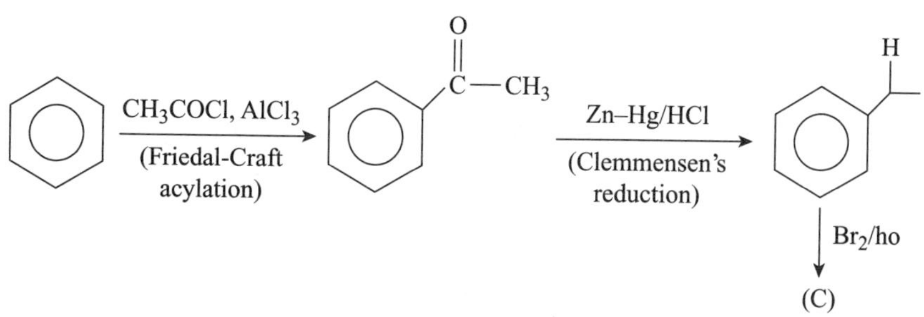

Ans: (c)

41.

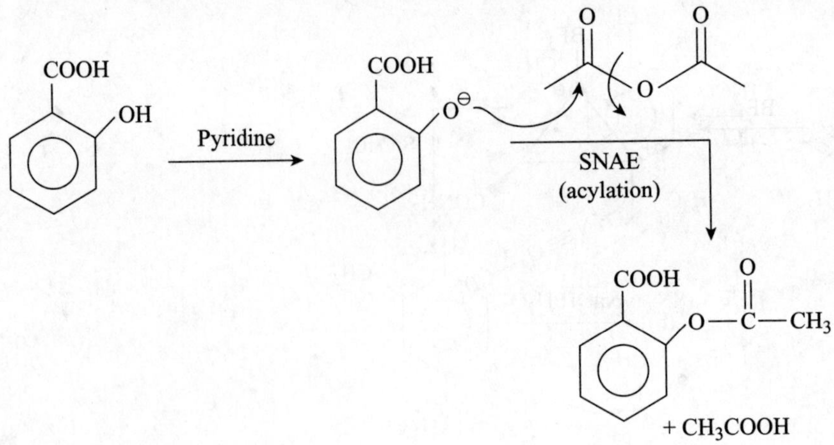

Ans: (a)

42.

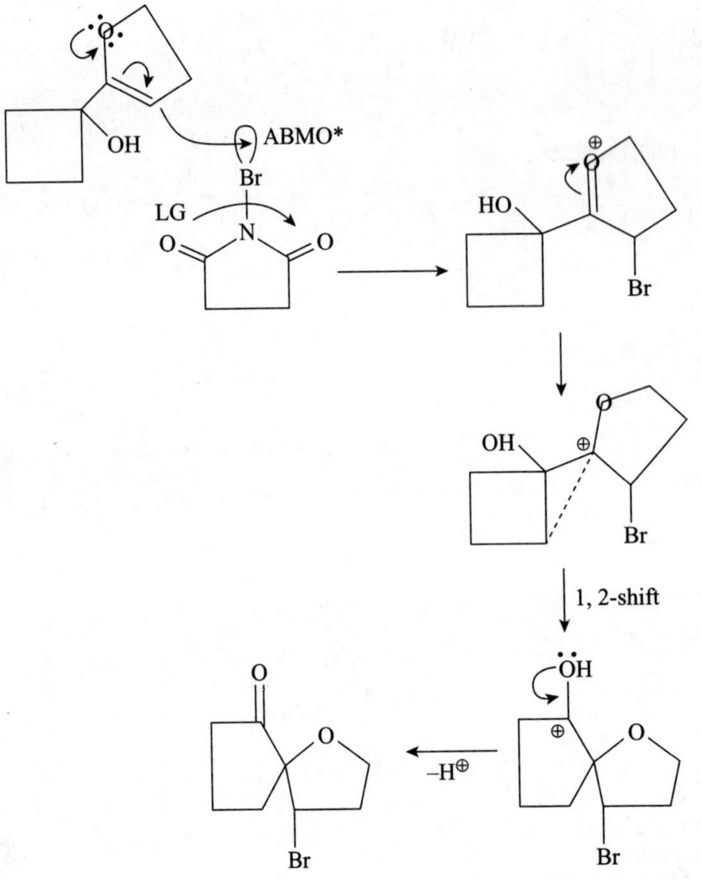

Ans: (b)

43.

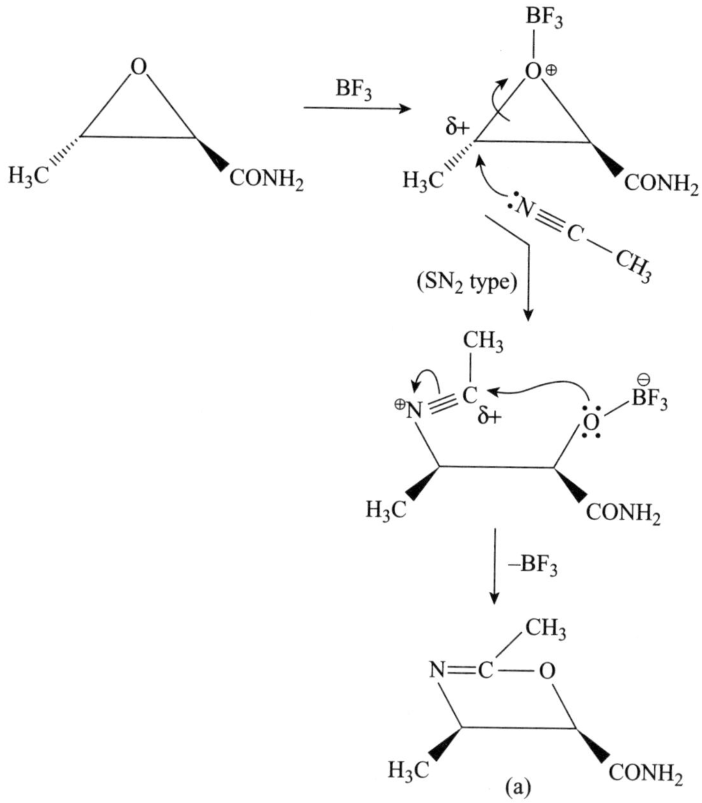

Ans: (a)

44.

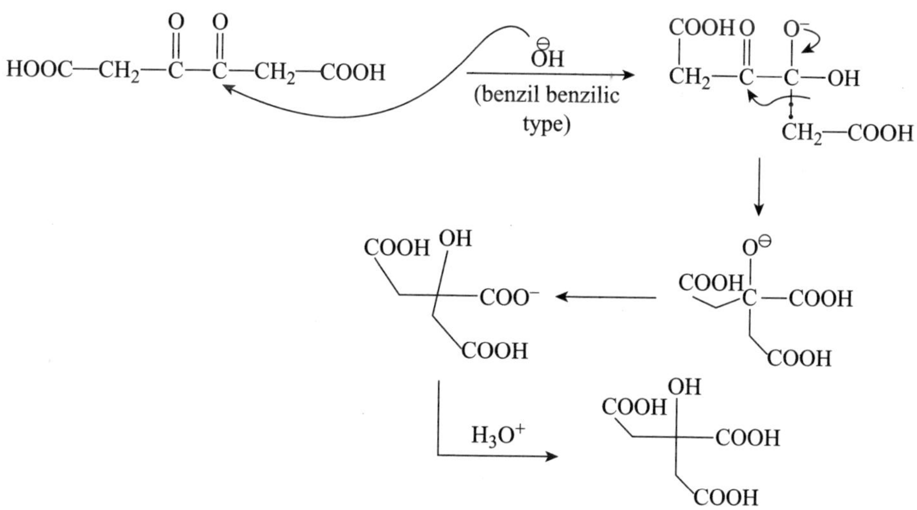

Ans: (b)

45.

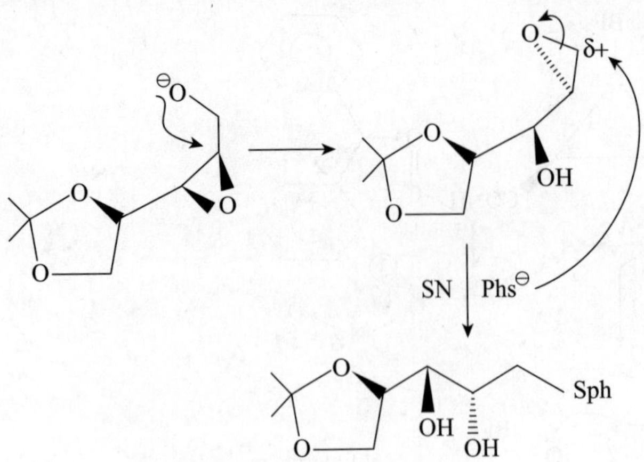

Ans: (b)

46.

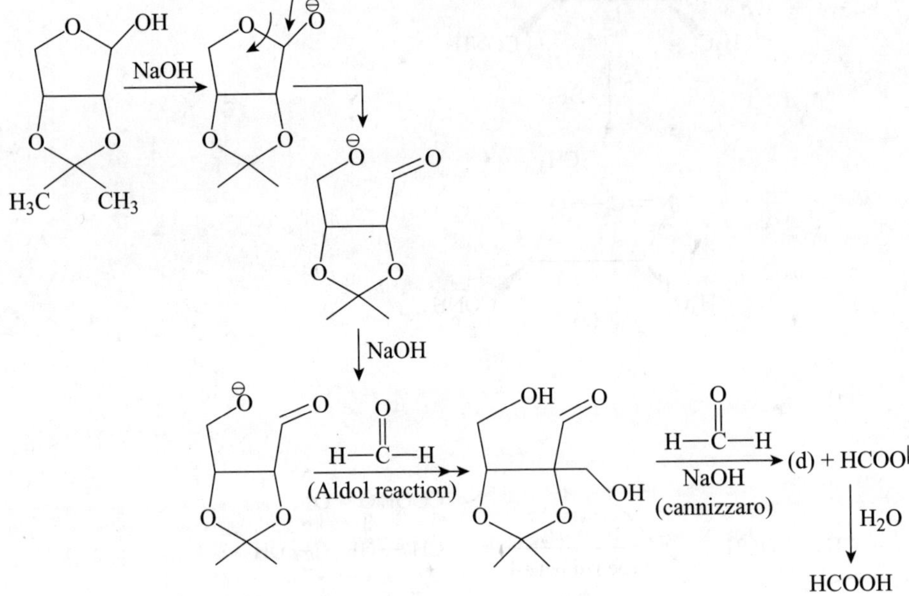

Ans: (d)

47.

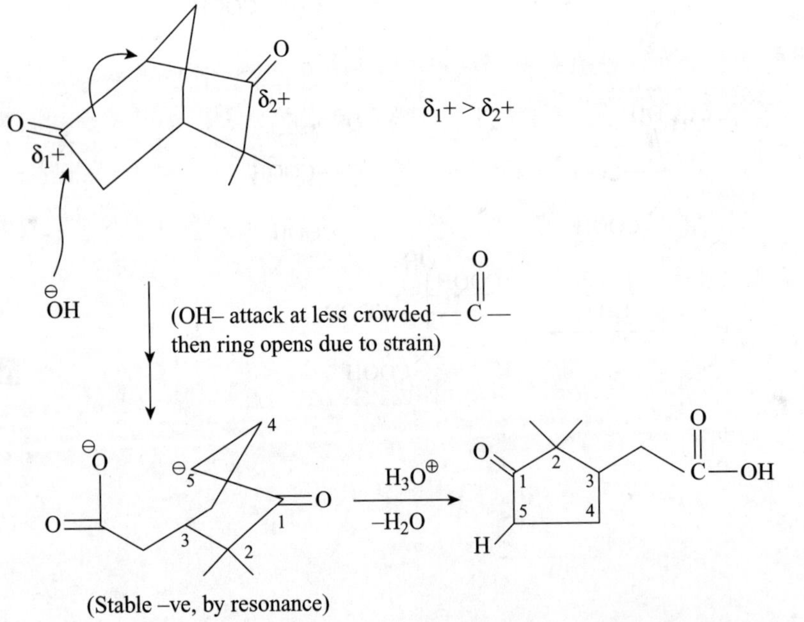

Ans: (b)

48.

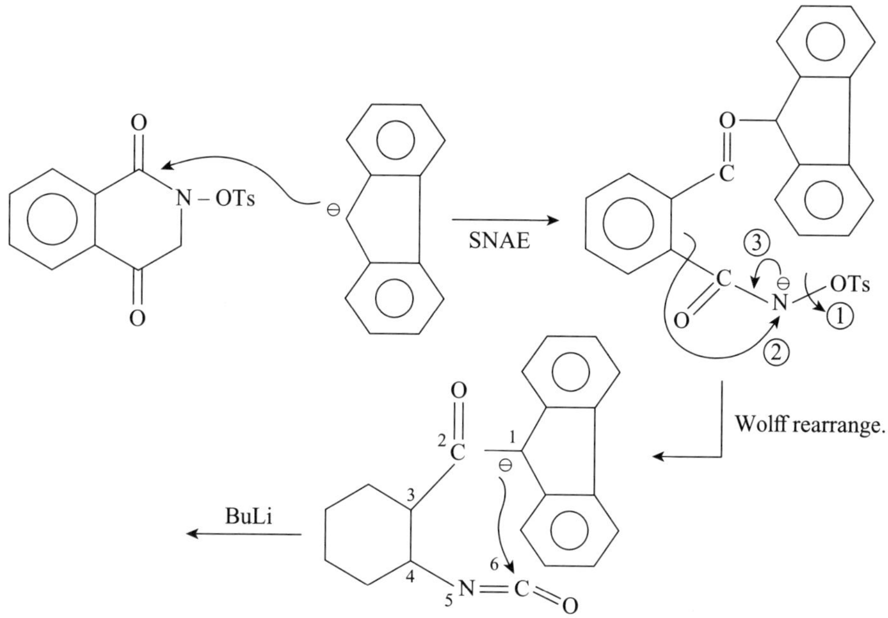

Ans: (b)

49.

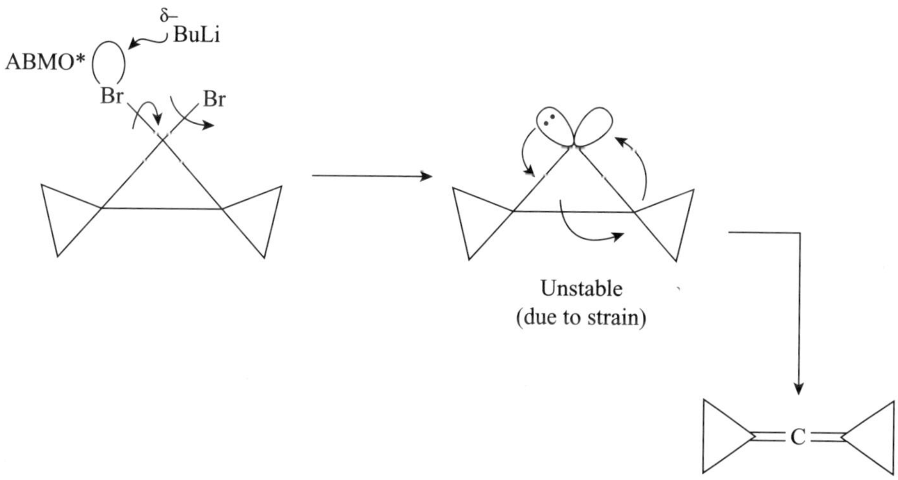

Ans: (c)

50.

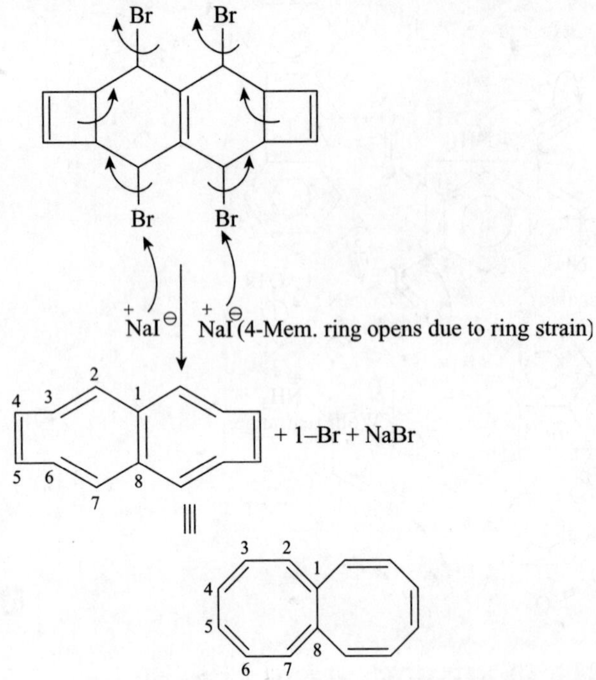

(d)

Ans: (d)

51.

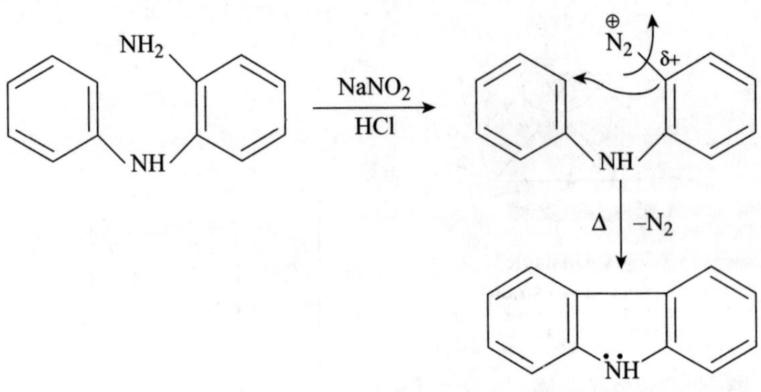

(Aromatic compound)

Ans: (c)

52.

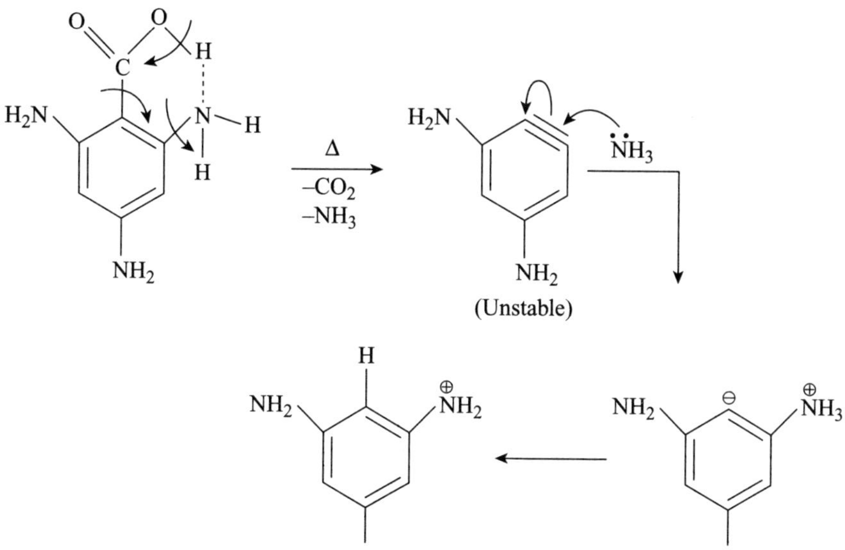

Ans: (a)

53. to 65. Refer solution of Question Number 27 to 38 respectively at level 2 (Carbonyl compound and acid-derivative).

Ans: (a)

67.

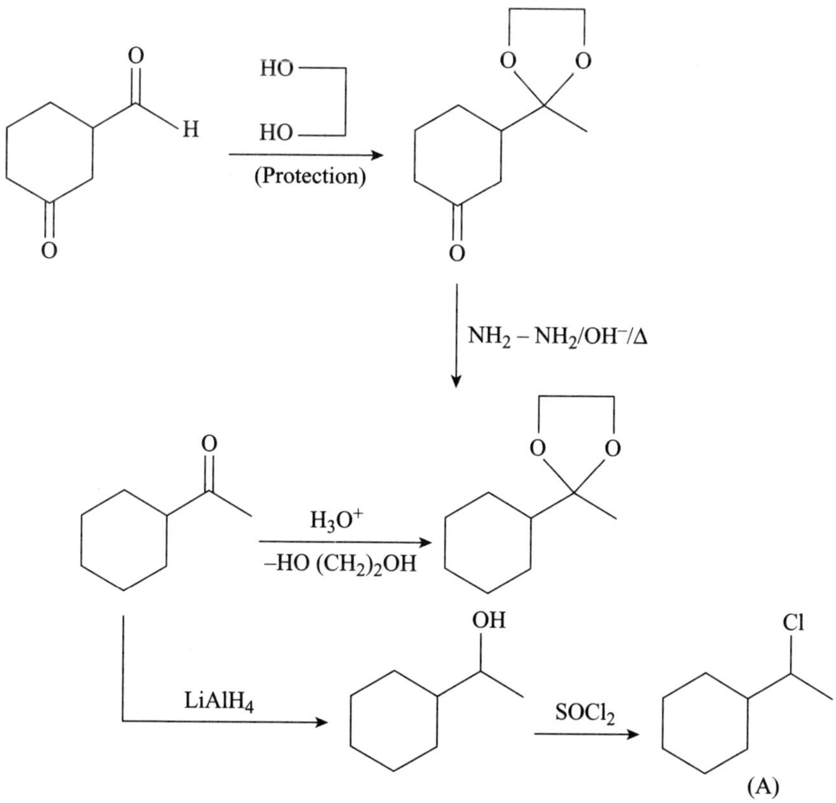

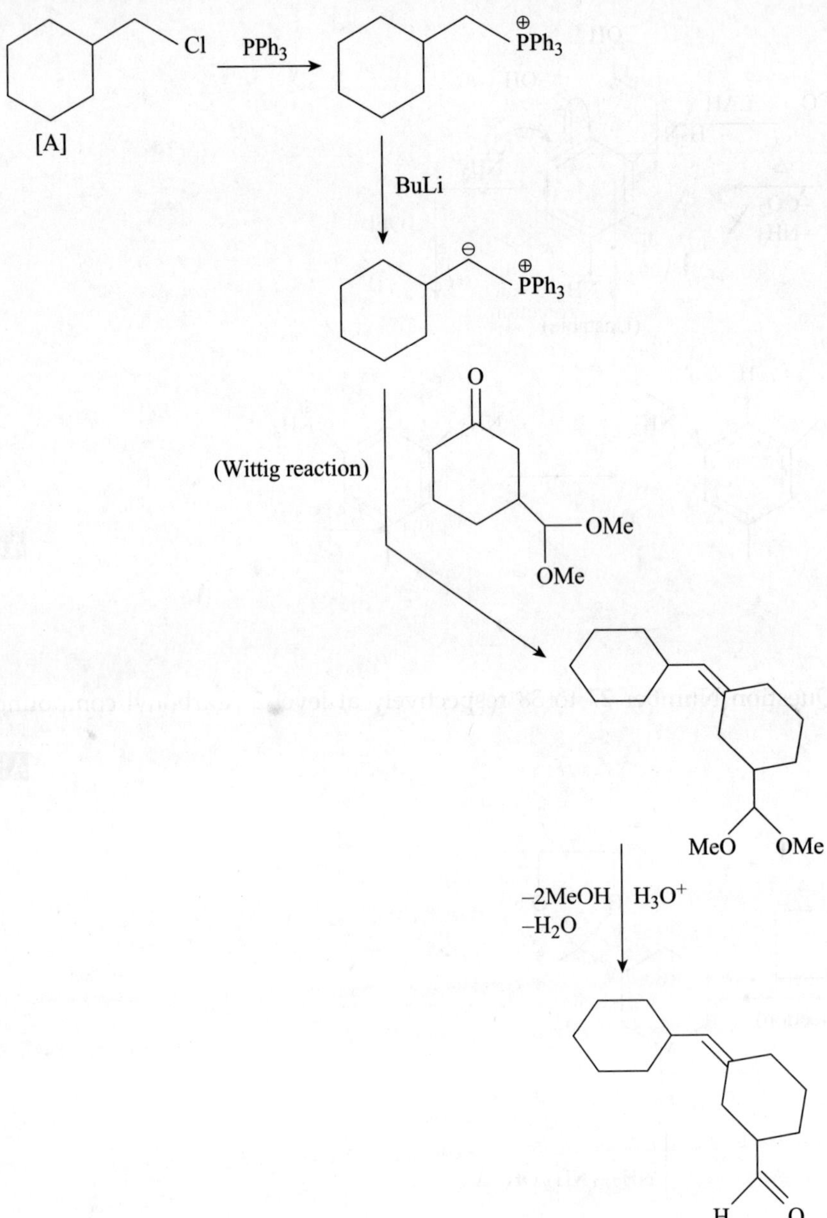

(B)

Ans: (a) and (d)

68.

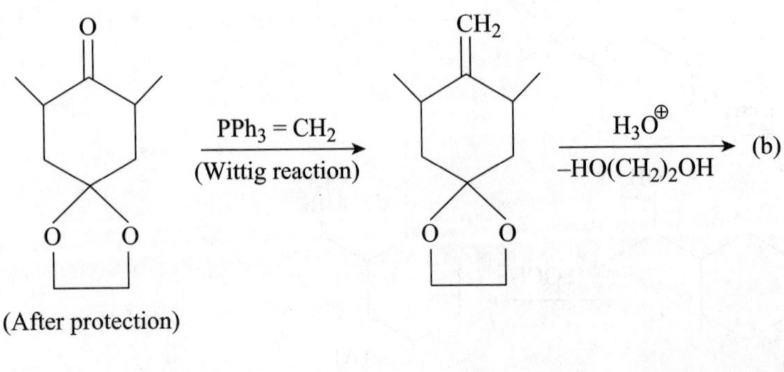

(After protection)

Ans: (b)

69.

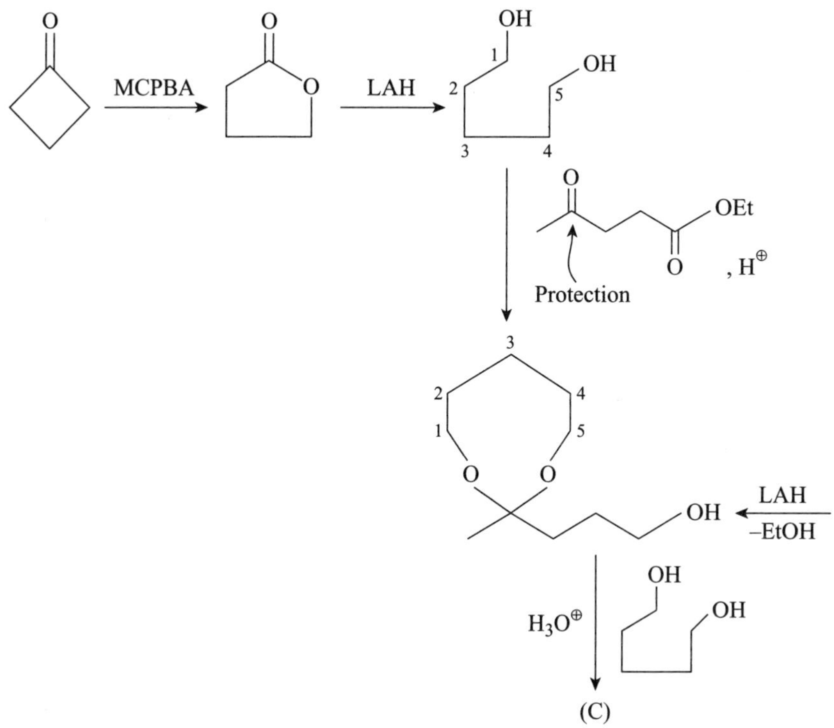

Ans: (c)

71.

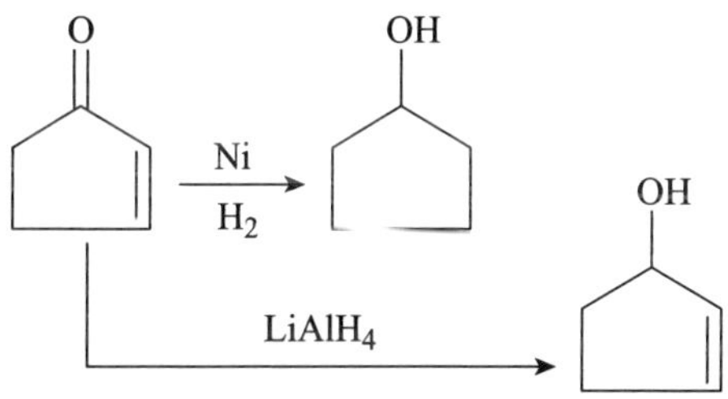

Ans: (a) and (d)

74.

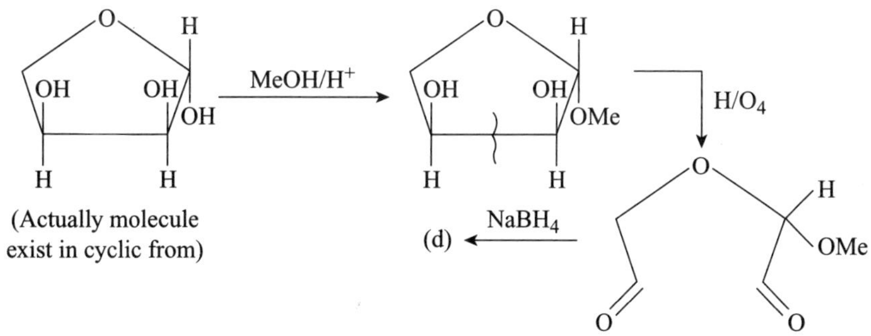

Ans: (d)

75.

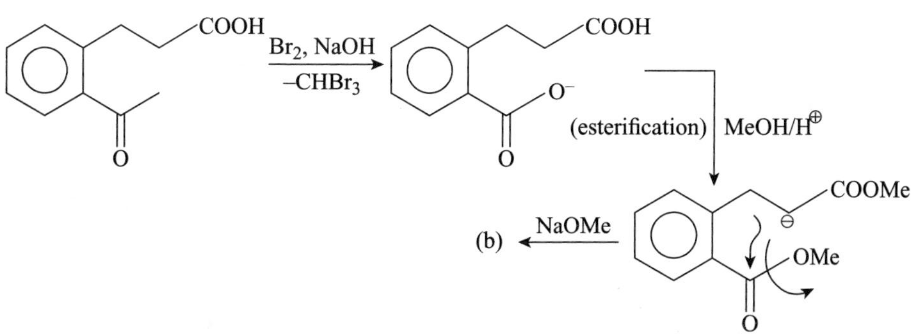

Ans: (b)

76.

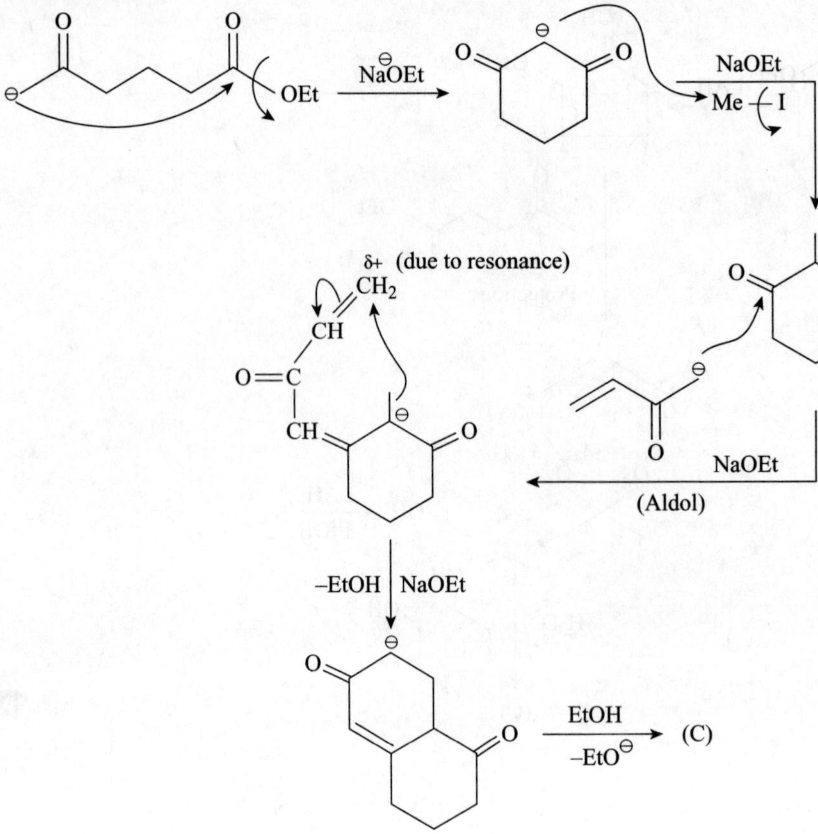

Ans: (c)

77.

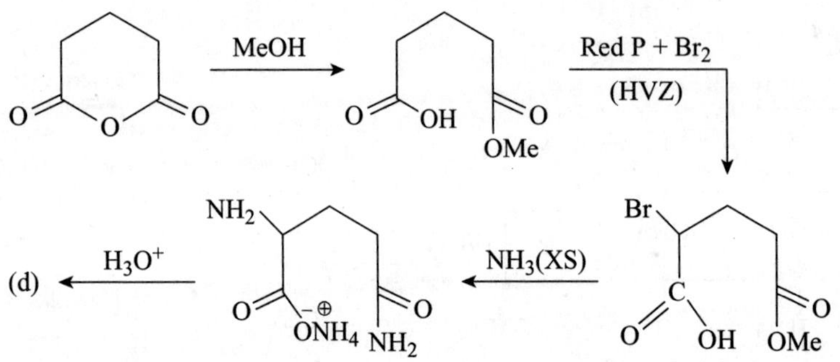

Ans: (d)

78.

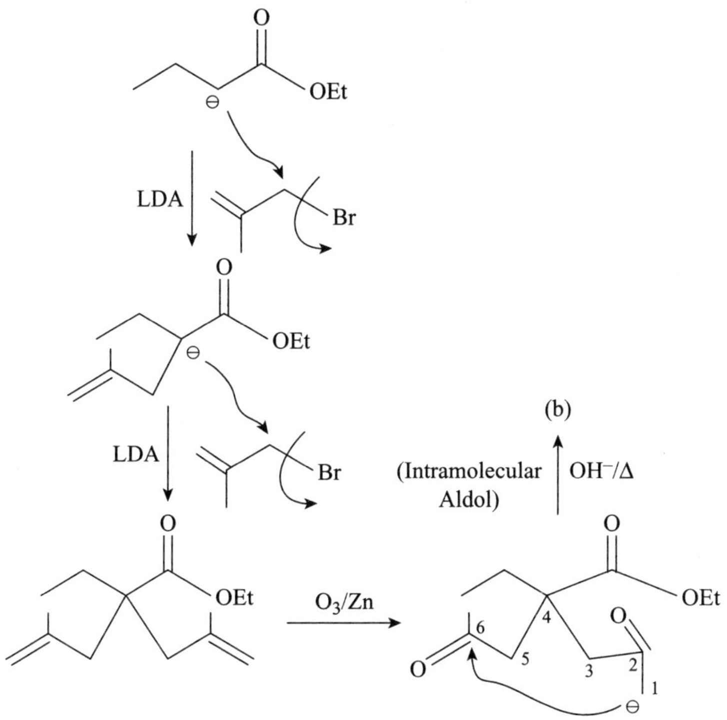

Ans: (b)

79.

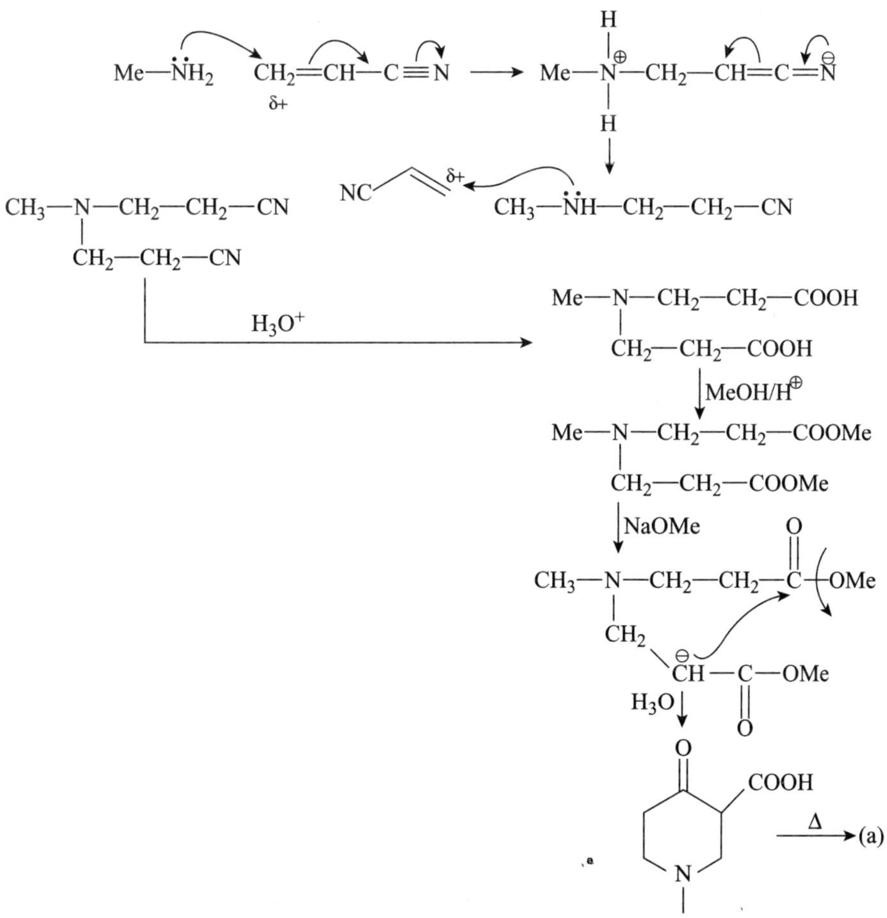

Ans: (a)

80.

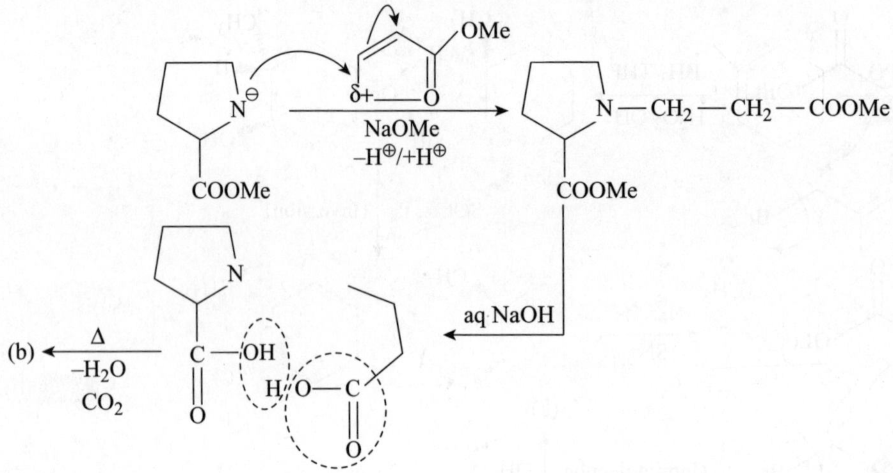

Ans: (b)

81.

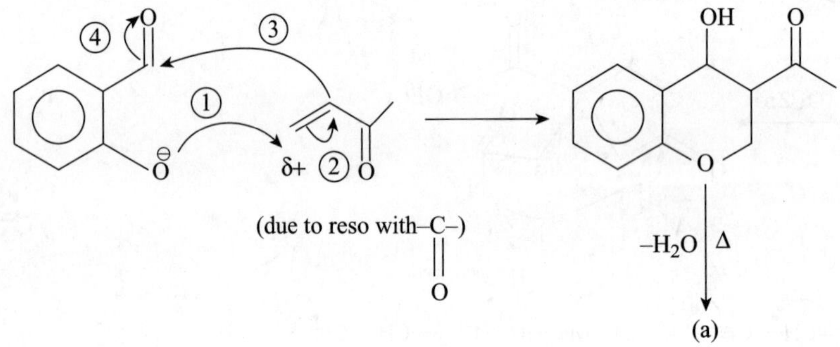

Ans: (a)

82.

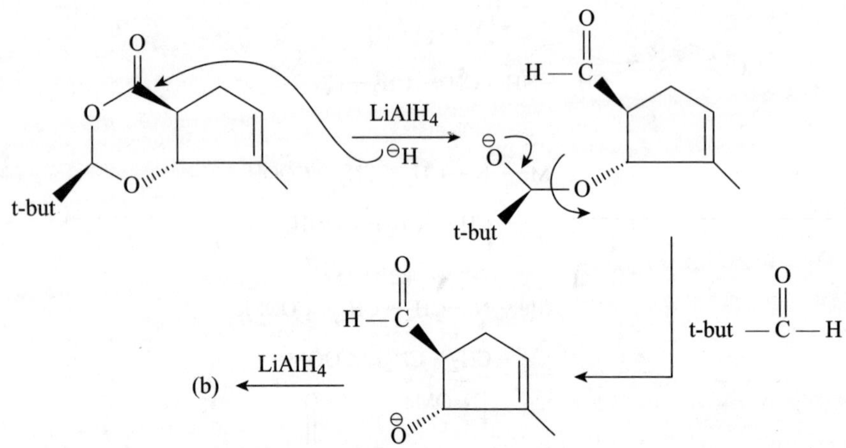

Ans: (b)

83.

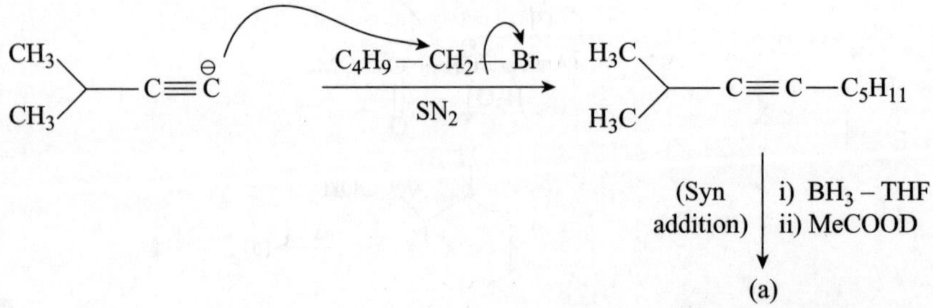

Ans: (a)

84.

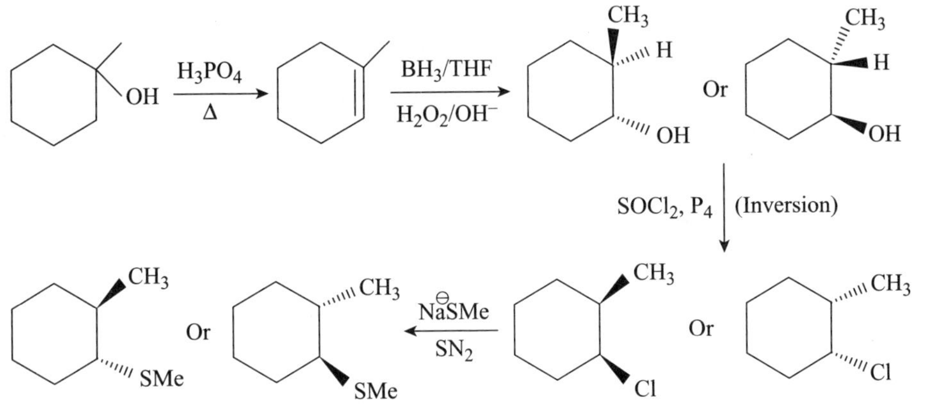

Ans: (c) and (d)

85.

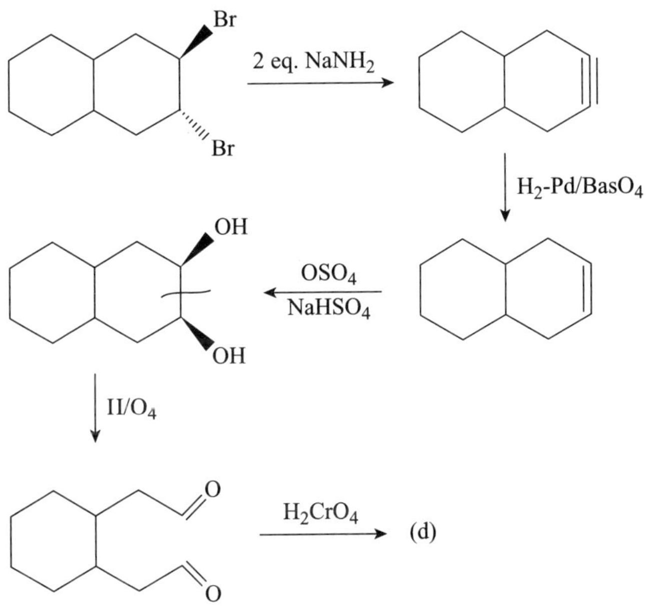

Ans: (d)

86.

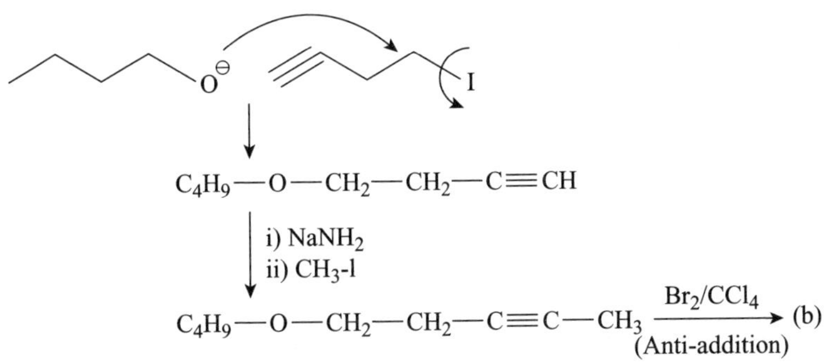

Ans: (b)

87.

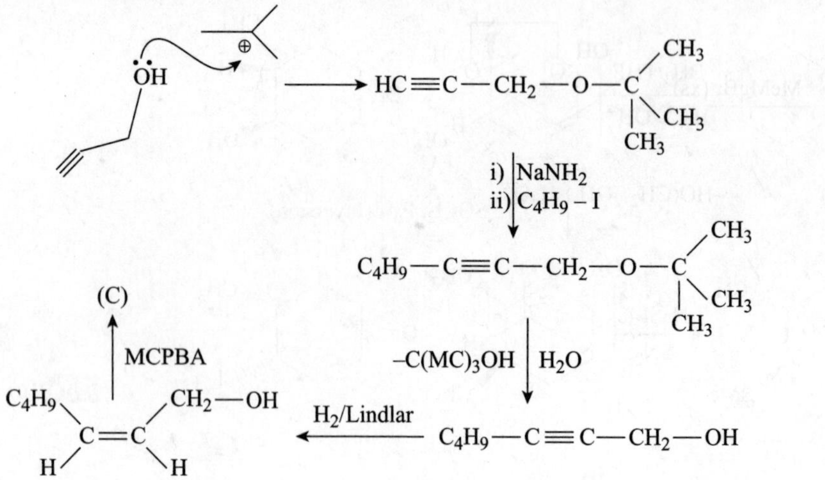

Ans: (c)

88.

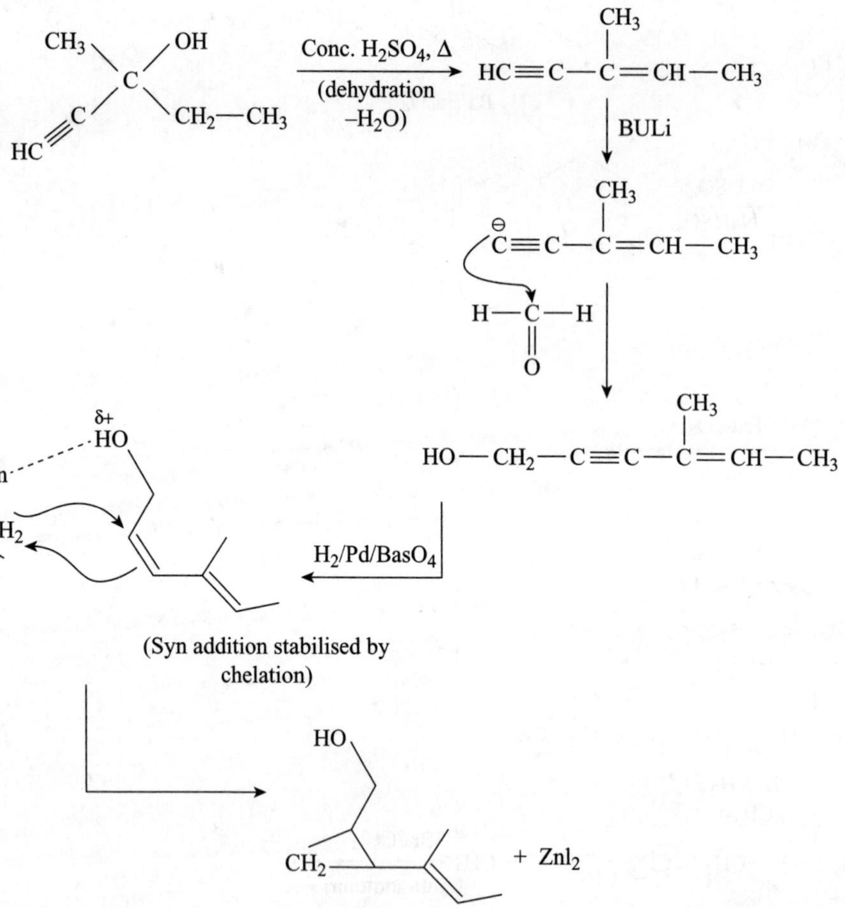

Ans: (a)

89. Aldol Condensation. **Ans:** (b)

90.

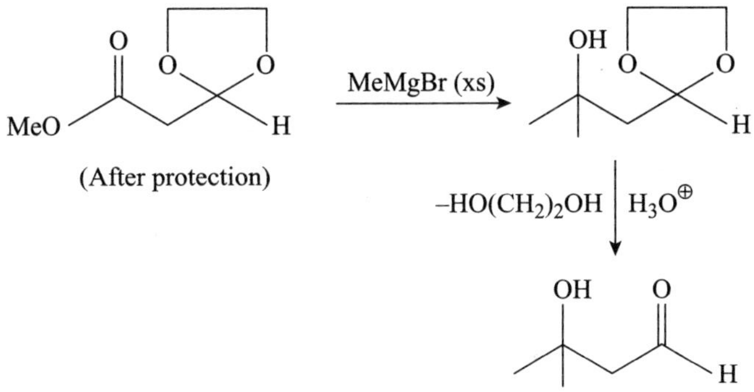

Ans: (c)

91.

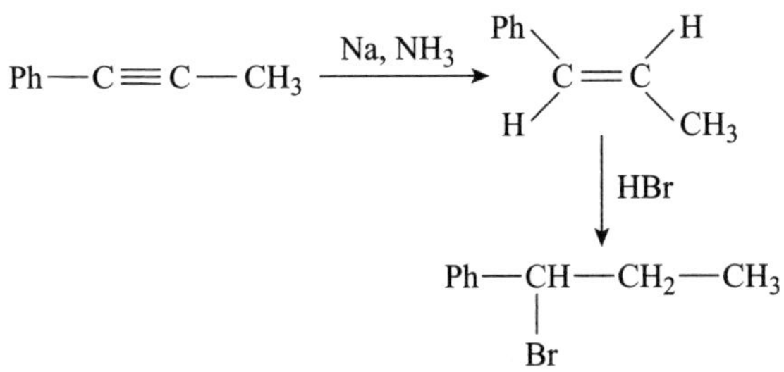

Ans: (c)

92.

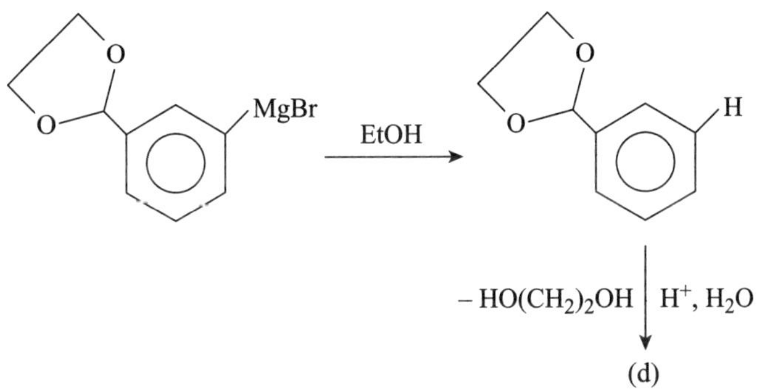

Ans: (d)

93. To avoid self-condensation of ketone, it is added slowly in test-tube containing aldehyde and base; and for same reason aldehyde is used in excess. Ans: (a), (b) and (d)

94.

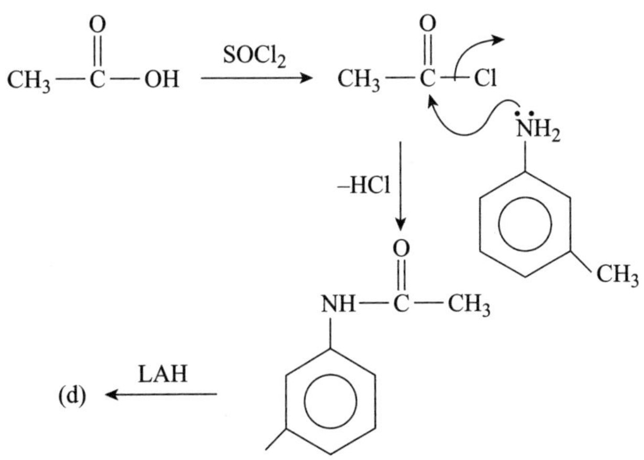

Ans: (d)

95. 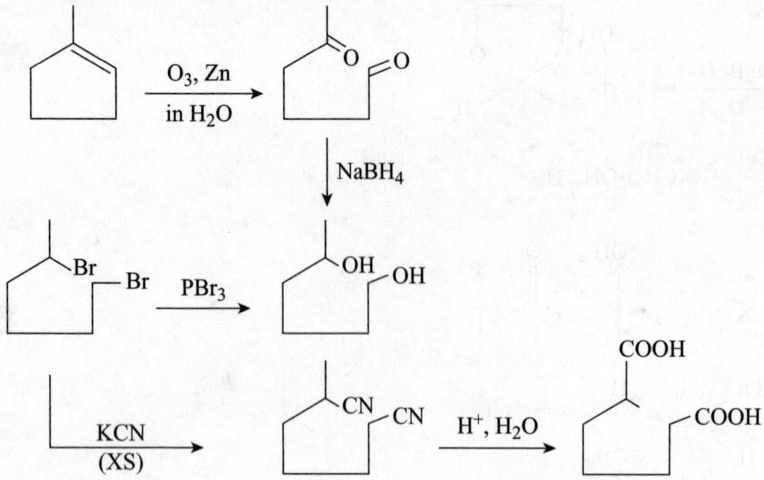

Ans: (b)

96. 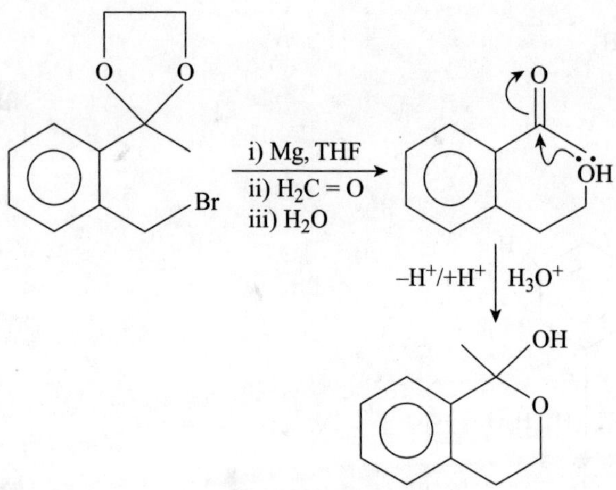

Ans: (a)

97. 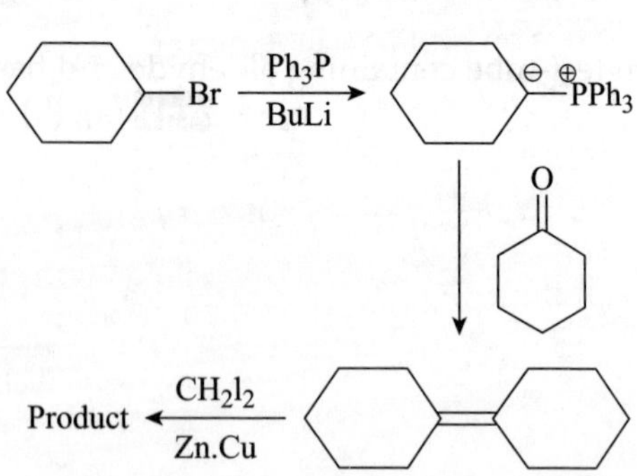

Ans: (c)

98. 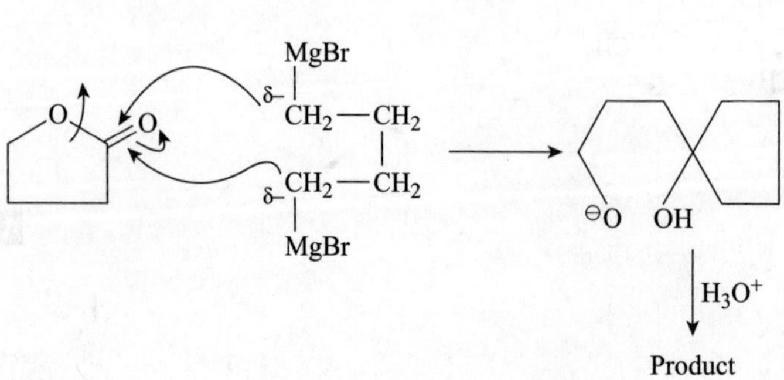

Ans: (d)

99. Trans $\xrightarrow[\text{addition}]{\text{Anti}}$ Erythro products

Ans: (d)

100.

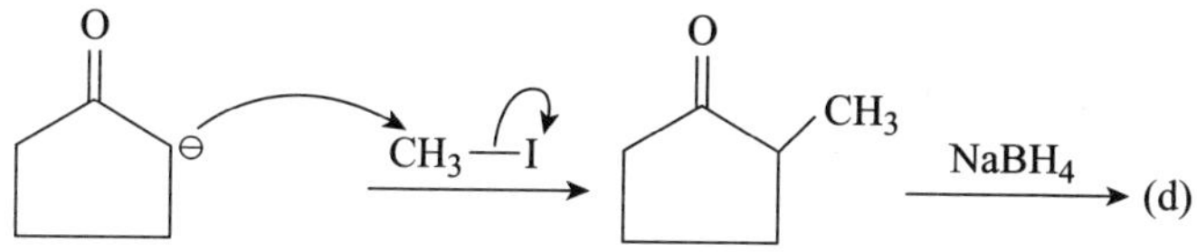

Ans: (d)

101.

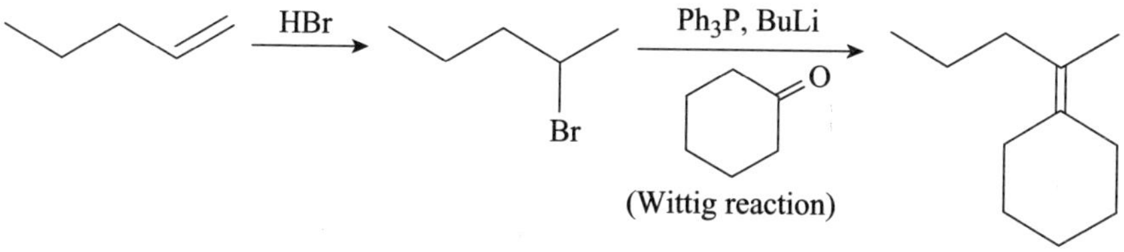

Ans: (a)

102.

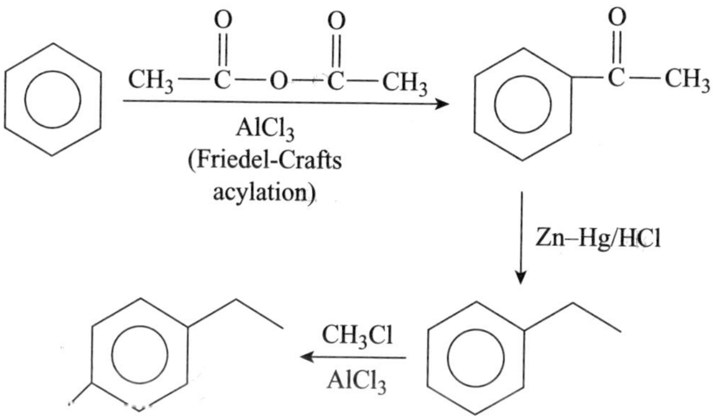

Ans: (c)

103.

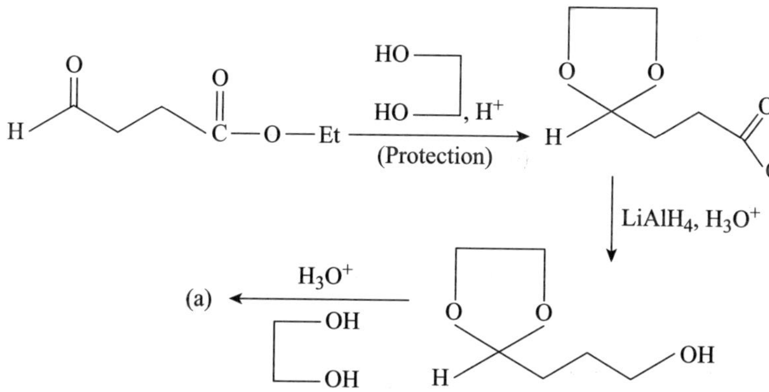

Ans: (a)

108. $RMgx + H_2O \rightleftharpoons R—H + Mg(OH)x$

Acid 1   Acid 2

A.S. : Acid1 > Acid2, So equillibrium goes forward.

Ans: (b)

109.

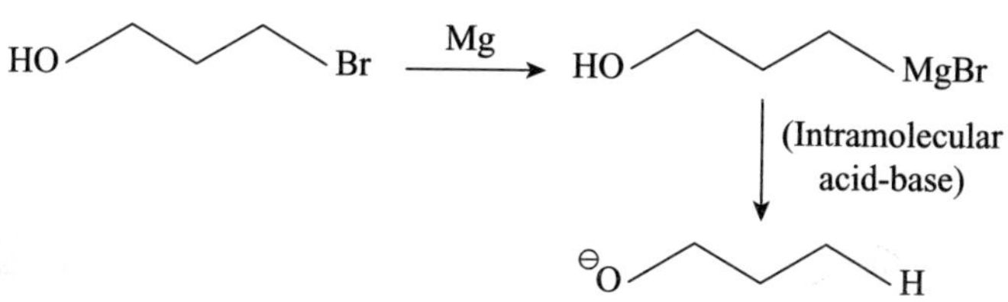

Ans: (a)

110.

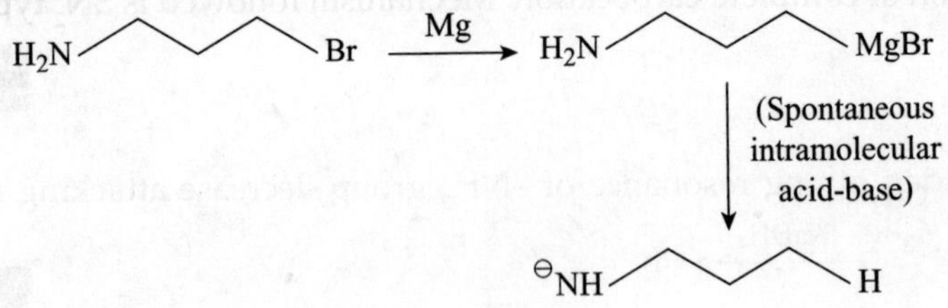

Ans: (c)

112.

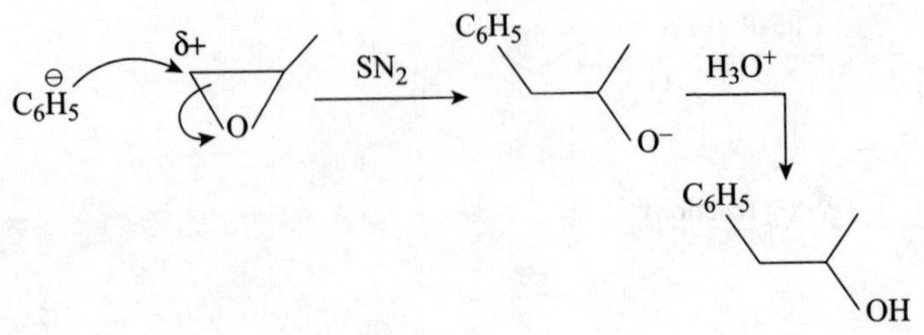

Ans: (a)

113.

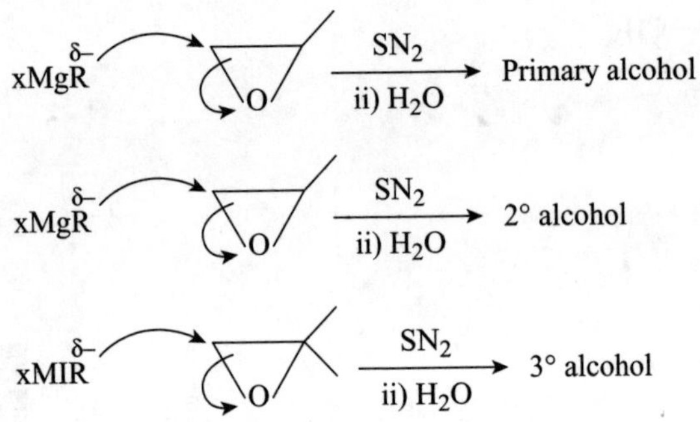

Ans: (d)

114.

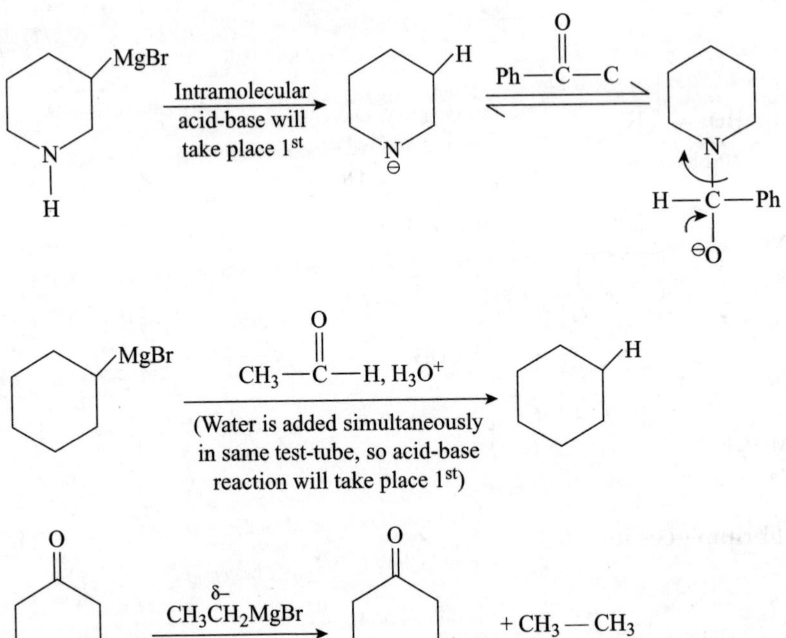

Ans: (d)

116. More polar protic solvent favours better solvation.

Ans: (a)

117. $nu^{\ominus}$ attack to $R-O^{\oplus}H_2$ before formation of complete carbocation. Mechanism followed is SN_2 type.

Ans: (d)

126. During acidic hydrolysis of acid amides, strong resonance of $-NH_2$ group decrease attacking rate of nucleophile.

Ans: (b)

127.

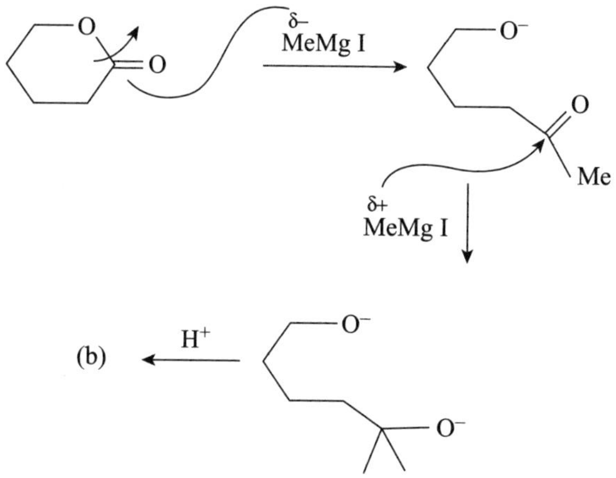

Ans: (b)

128. $R-X + NaCN \longrightarrow RCN + RNC$

Major (May also form in minor form)

Ans: (b)

131.

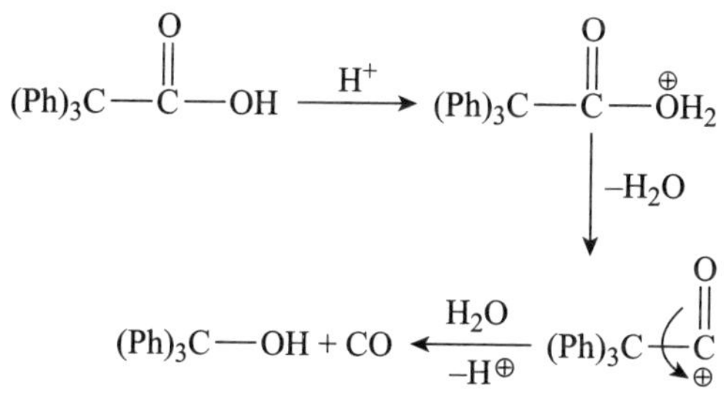

Ans: (c)

132.

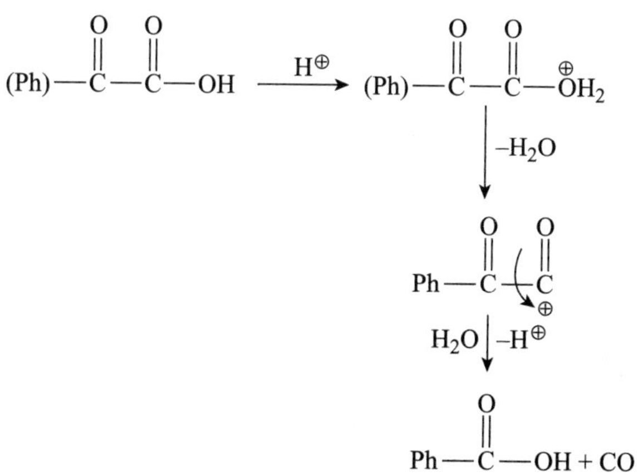

Ans: (c)

CHAPTER 9

Practical Organic Chemistry

LEVEL 1

1. $Na_2S + Na_2[Fe(CN)_5NO] \longrightarrow \underset{\text{Deep violet}}{Na_4[Fe(CN)_5NO]}$

Ans: (a)

2. $3Na_4[Fe(CN)_6] + 4FeCl_3 \xrightarrow{HCl} Fe_4[Fe(CN)_6]_3 + 12\ NaCl$

Ans: (b)

3. $Na + S + C + N \xrightarrow{\Delta} \underset{\text{Sodium extract}}{NaSCN}$

Ans: (b)

4. $Na + C + N \xrightarrow{\Delta} \underset{\text{Sodium extract (S. E.)}}{NaCN}$ For preparation of S.E. along with N, C should also present.

Ans: (c)

5. $NaSCN + FeCl_3 \longrightarrow \underset{\text{Blood red colour}}{[Fe(SCN)]Cl_2 + NaCl}$

Ans: (b)

10. $RCOOH + Ag^+ \longrightarrow RCOOAg + H^+$

Silver salt method : $\underset{W(gm)}{2RCOOAg} \xrightarrow{\Delta} \underset{W_1(gm)}{2Ag} + CO_2 + R - R$

$$\text{EQ mass of acid} = \left(\frac{W}{W_1}\times 108\right) - 108 + 1$$

Ans: (a)

12.

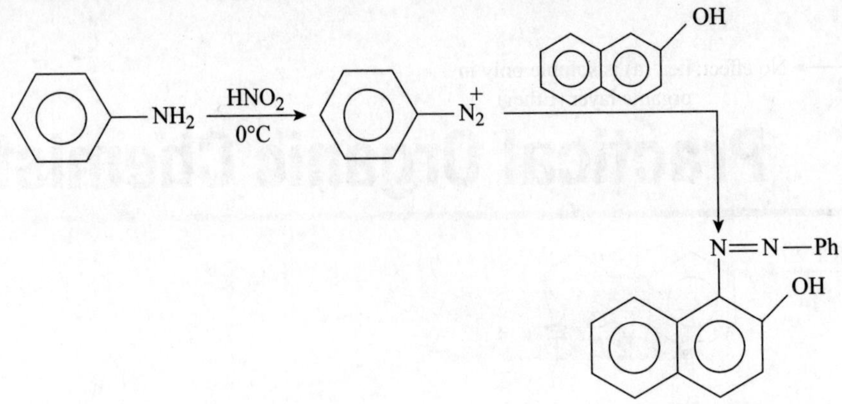

Red dye

Ans: (c)

16.

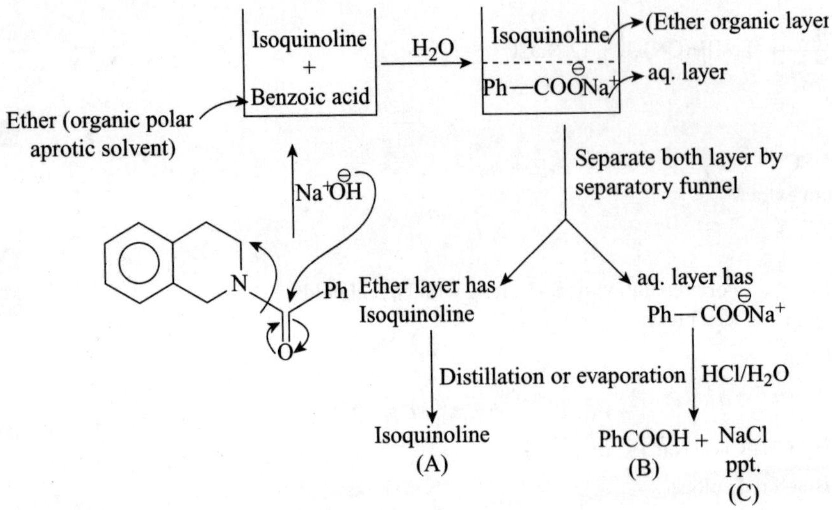

Benzoic acid is soluble in organic layer in undissociated form, after dissociation to Ph - COO– it will be soluble in aq. layer (polar protic solvent).

Ans: (b)

17.

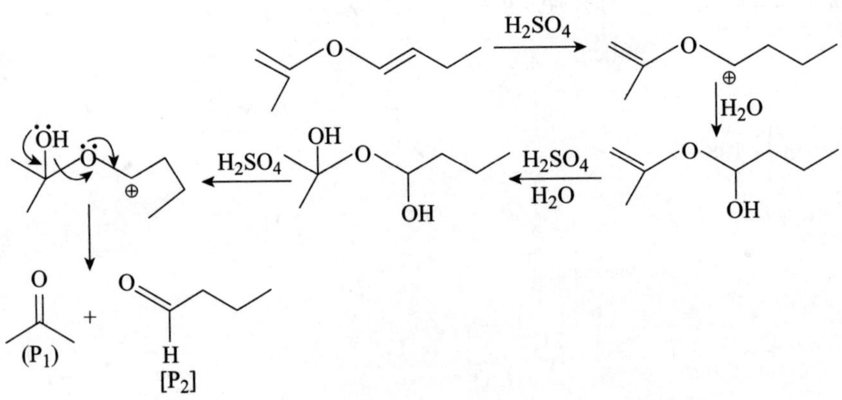

Ans: (a) and (b)

18.

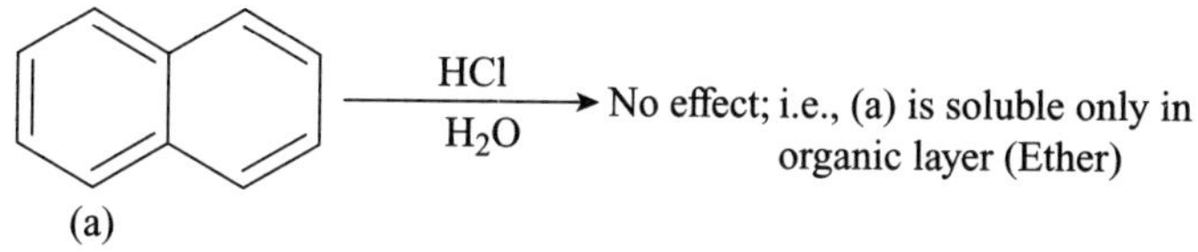
HCl
H2O
No effect; i.e., (a) is soluble only in
organic layer (Ether)
(a)

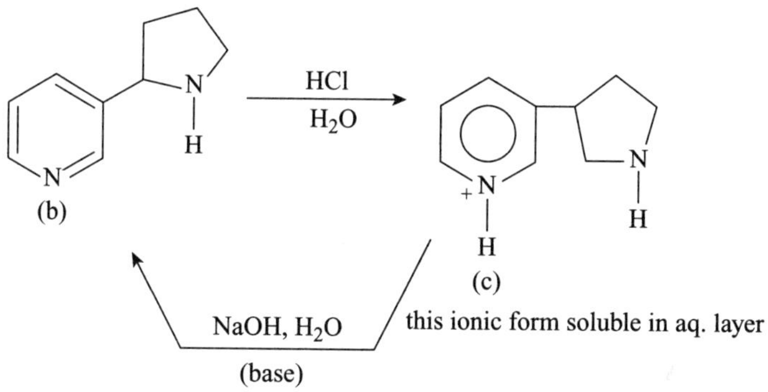
HCl
H2O
(b)
(c)
this ionic form soluble in aq. layer
NaOH, H2O
(base)

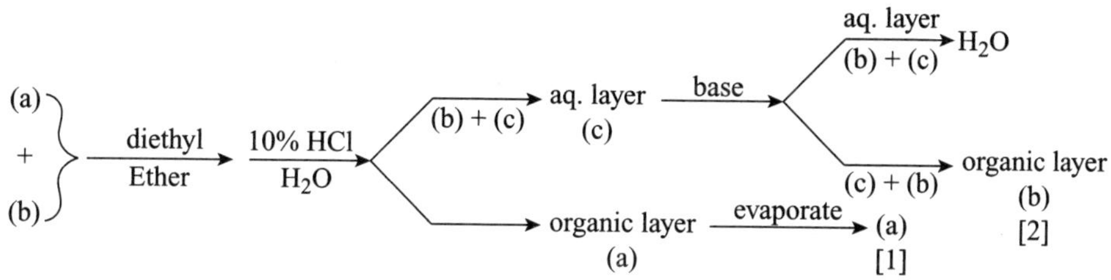
(a)
+
(b)
diethyl
Ether
10% HCl
H2O
(b) + (c)
aq. layer
(c)
base
aq. layer
(b) + (c)
H2O
(c) + (b)
organic layer
(b)
[2]
organic layer
(a)
evaporate
(a)
[1]

Ans: (c)

19.

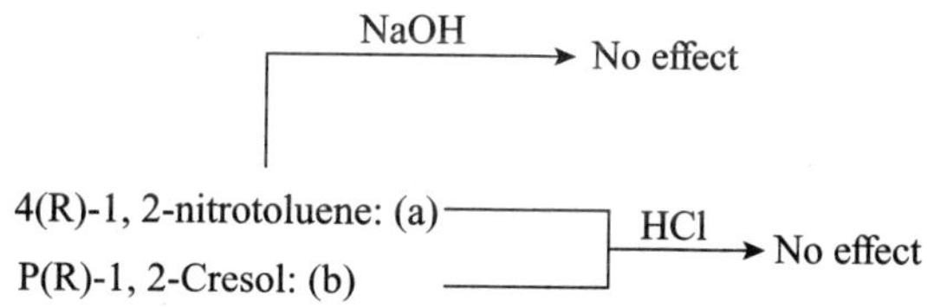
NaOH
No effect
4(R)-1, 2-nitrotoluene: (a)
P(R)-1, 2-Cresol: (b)
HCl
No effect

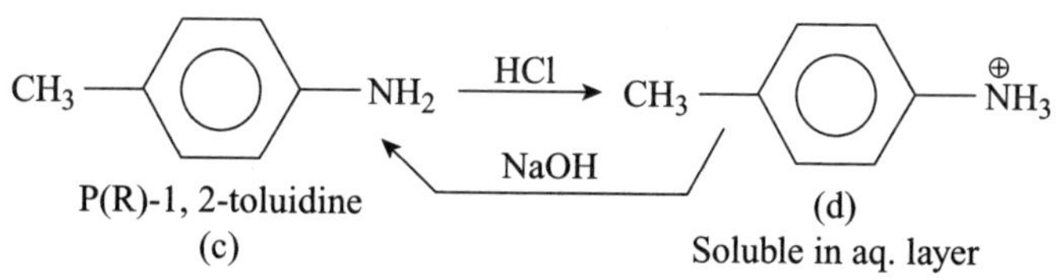
CH3
NH2
HCl
NaOH
CH3
NH3
P(R)-1, 2-toluidine
(c)
(d)
Soluble in aq. layer

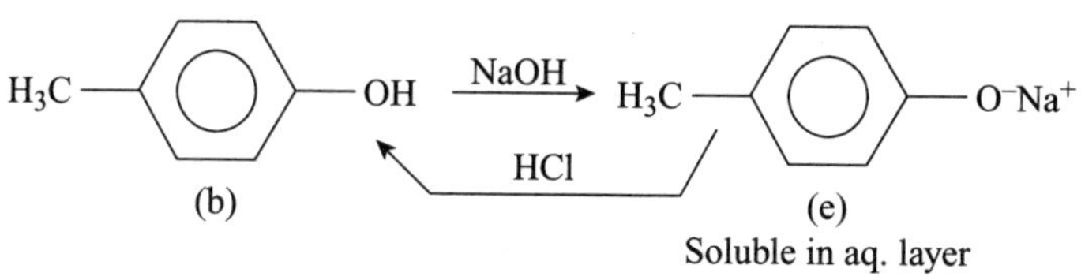
H3C
OH
NaOH
HCl
H3C
O-Na+
(b)
(e)
Soluble in aq. layer

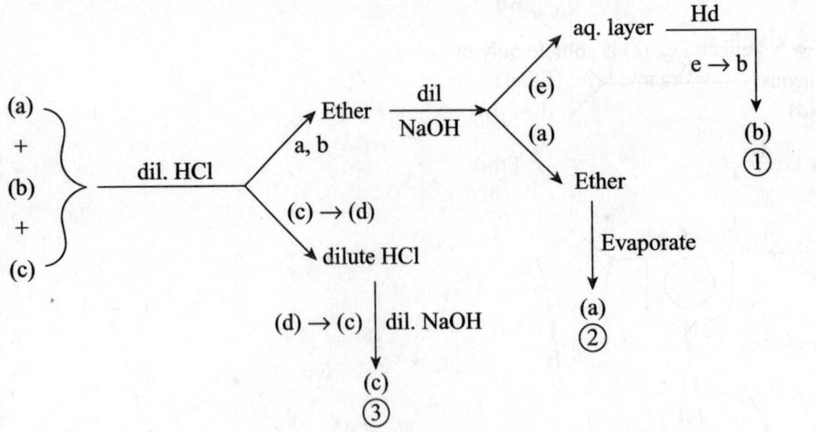

Ans: (d)

20. (a)

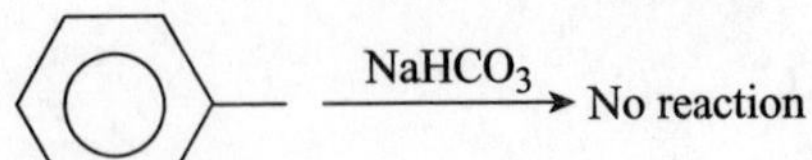

(b)

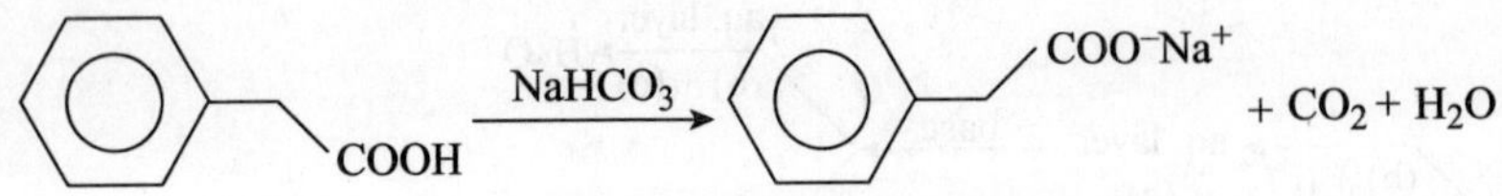

Soluble in aq. layer
(d)

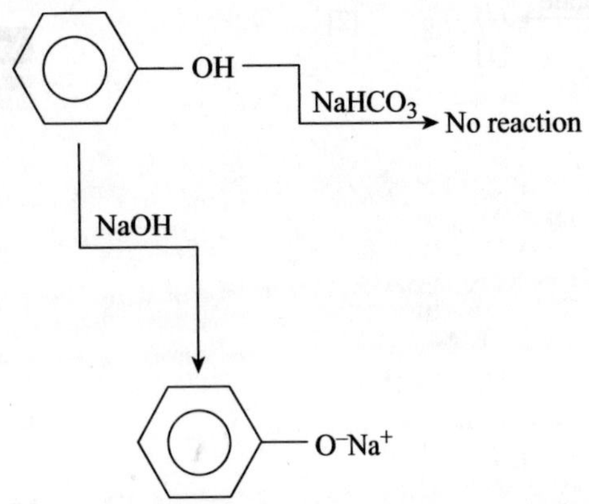

(e)
Soluble in aq. layer

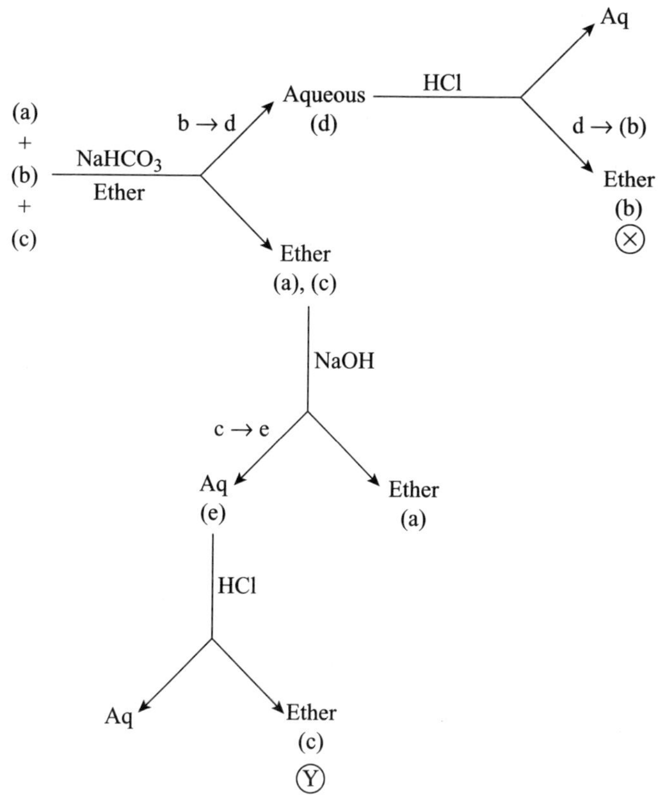

Ans: (d)

21.

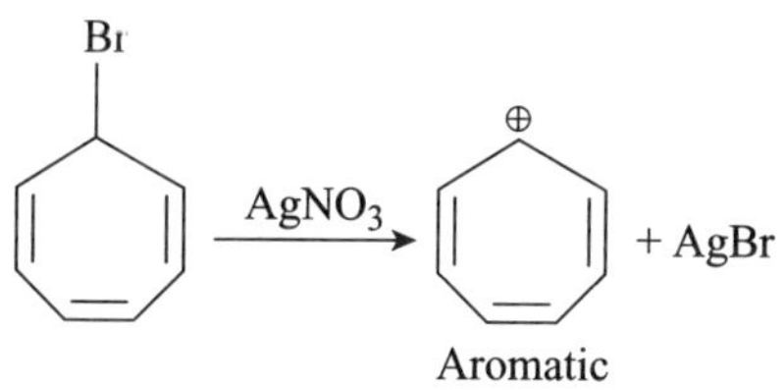

In (a), (c) C – Br bond does not break due to partial double bond character. **Ans:** (b)

22. $Na_2S + Na_2[Fe(CN)_5NO] \longrightarrow Na_4[Fe(CN)_5NOS]$ **Ans:** (d)

23. A. S. : HCOOH > CH_3COOH > Picric acid > H_2CO_3 > Phenol

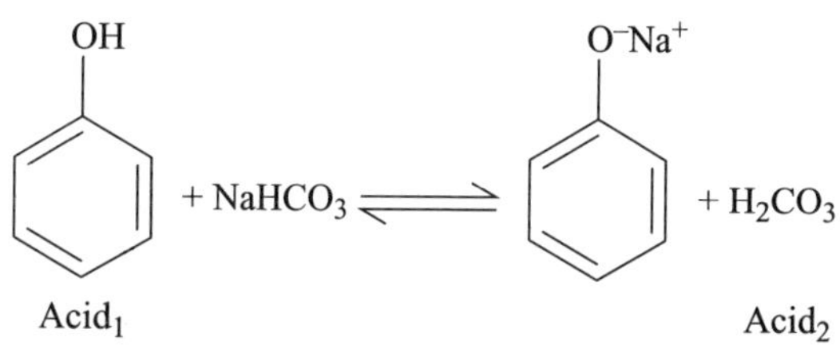

A.S. : $Acid_1 < Acid_2$. Therefore, reaction goes backward. **Ans:** (a) ,(b) and (c)

24.

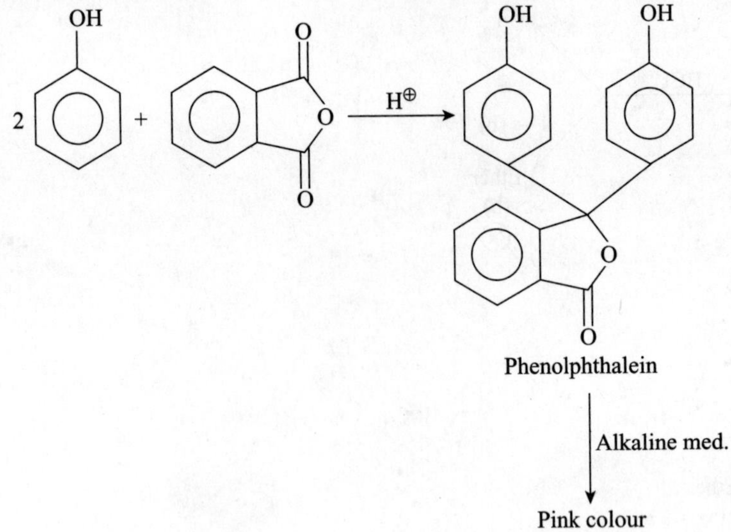

Ans: (c)

25.

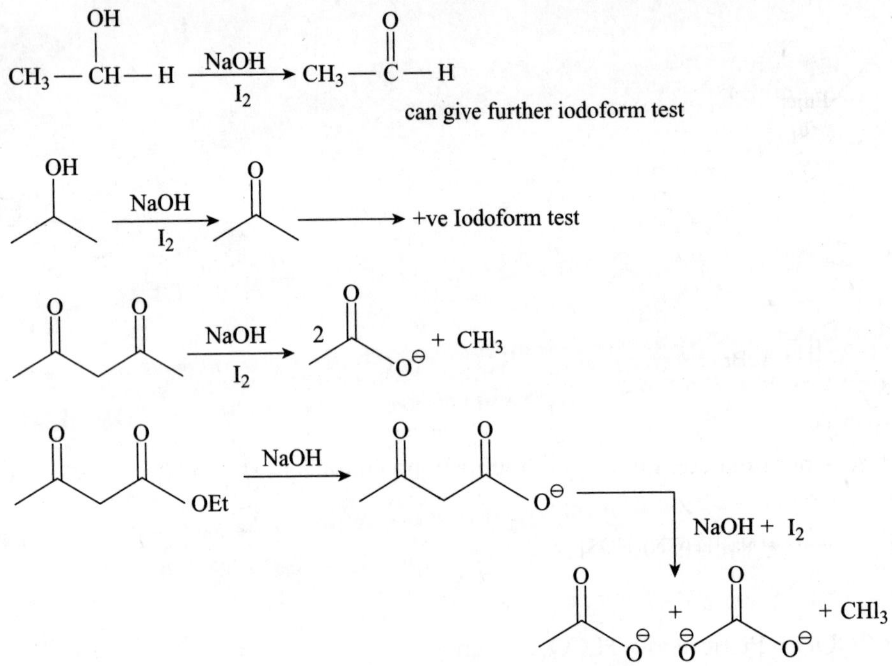

Ans: (a), (b), (c) and (d)

28.

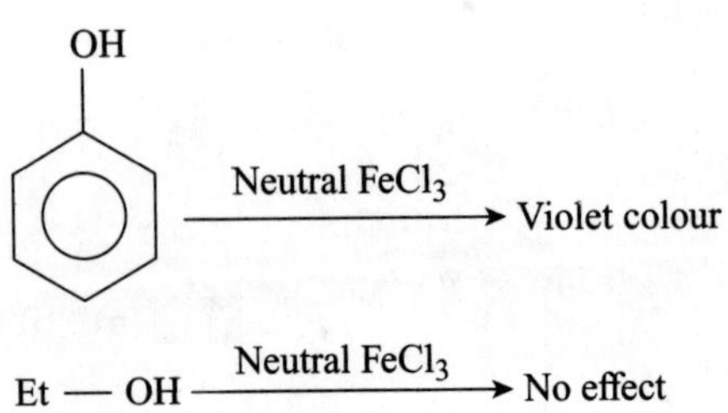

Ans: (c)

29.

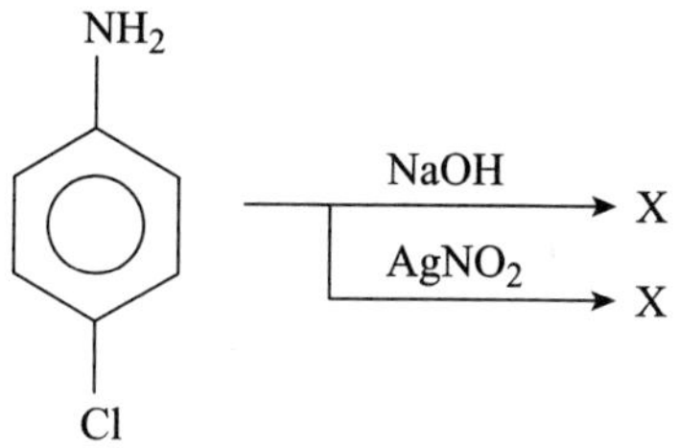

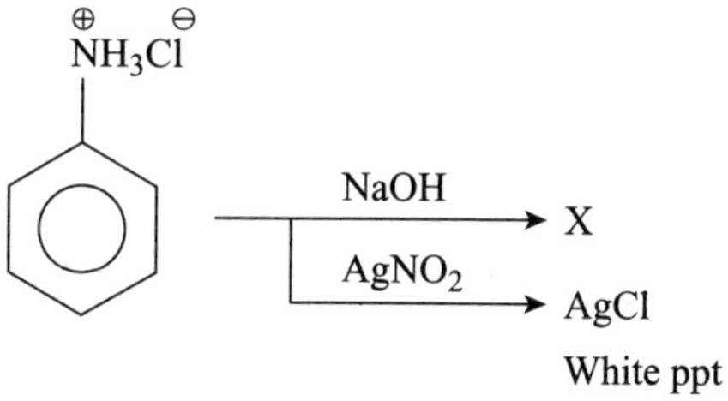

A.S. : $H_2O > R-NH_3 > R-NH_2$

Ans: (b)

LEVEL 2

1. The element present in compound are converted from covalent form into ionic form by fusing compound with Na metal.

Ans: (a), (b) and (c)

3. $$\text{Alcohol} \xrightarrow[\text{(CAN)}]{\text{Ceric ammonium nitrate}} \text{Red colour}$$

$$\text{Aldehyde} \xrightarrow{\text{CAN}} X$$

Ans: (a), (b) and (d)

6. $$CH_3-C(=O)-OH \xrightarrow{\text{NaOH}} CH_3-C(=O)-O^{\ominus} \xrightarrow[I_2]{\text{NaOH}} X$$

$$CH_3-C(=O)-Cl \xrightarrow[SN_{AE}]{\text{NaOH}} CH_3-C(=O)-O^{\ominus}$$

$$CH_3-C(=O)-NH_2 \xrightarrow[SN_{AE}]{\text{NaOH}} CH_3-C(=O)-O^{\ominus}$$

Ans: (b), (c) and (d)

7. $H-C(=O)-OH$

$$\xrightarrow{\text{T. R.}} +ve$$
$$\xrightarrow{\text{Fehling}} +ve$$
$$\xrightarrow{KMnO_4} HO-C(=O)-OH$$

$CH_3-C(=O)-OH$

$$\xrightarrow{\text{Tollen}} -ve$$
$$\xrightarrow{\text{Fehling}} -ve$$
$$\xrightarrow{KMnO_4} X$$

Ans: (a), (b) and (c)

8. $Ph-C(=O)-H$

$$\xrightarrow{\text{Tollen's reagent}} +ve\ \text{Test}$$
$$\xrightarrow{\text{Fehling's solution}} -ve\ \text{Test}$$

The mass of H_2O produced is determined by passing mixture through a weighed U-tube containing anhydrous $CaCl_2$ or $Mg(ClO4)_2$.

Ans: (a) and (c)

9. $$\text{Phenol/phenol derivative} \xrightarrow[\substack{(NH_4)_2Ce(NO_3)_6 \\ \{\text{Yellow}\}}]{\text{Ceric ammonium nitrate}} \text{Red colour}$$

$$RCOOH + NaHCO_3 \longrightarrow RCO\overset{-}{O}\overset{+}{Na} + H_2O + CO_2$$

Ans: (a), (b) and (c)

13. $HCOOH \xrightarrow{NaHCO_3} CO_2\uparrow + H_2O + HCOO^-Na^+$

$HCHO \xrightarrow{NaHCO_3} X$

Ans: (b) and (c)

14.

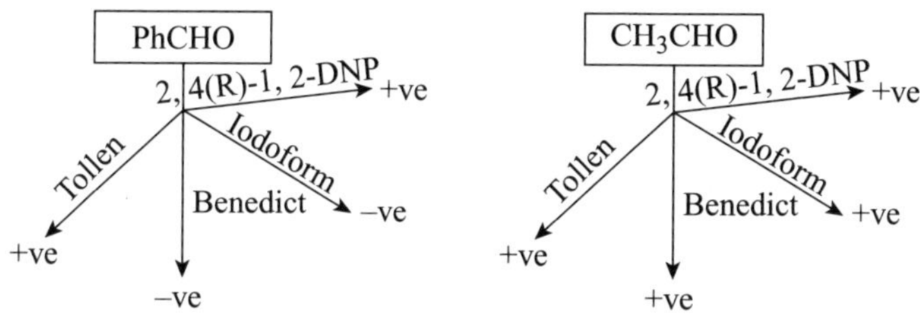

Ans: (b) and (c)

15.

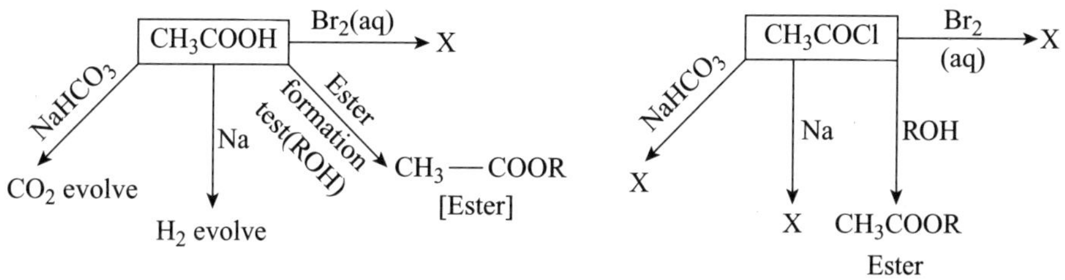

Ans: (a)and (b)

18. $\dfrac{100}{x} = \dfrac{18\times 220}{83\times 100} \Rightarrow x \simeq 258\,\text{g}$

Ans: (b)

Passage 2

Test (Q) : $Na + C + S + N \xrightarrow{\Delta} NaSCN$

{Compound X} Sodium extract

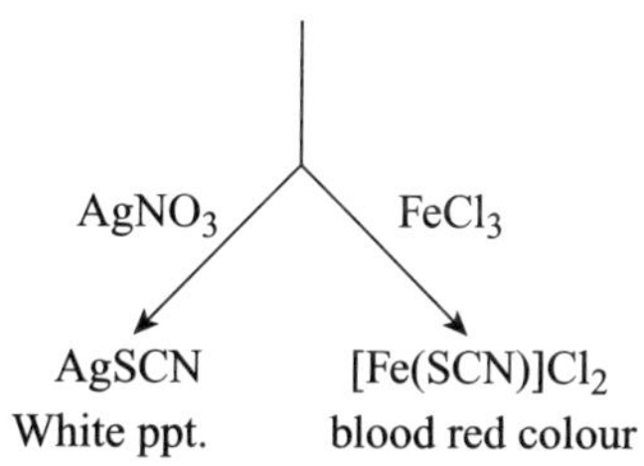

$Na + Cl \xrightarrow{\Delta} NaCl \xrightarrow{AgNO_3} AgCl$

{may present in compound} X; NaCl: Sodium extract; AgCl: white ppt.

Test (R) : $Na + C + N \longrightarrow NaCN \xrightarrow[\text{ii) } FeCl_3,\ HCl]{\text{i) } FeSO_4,\ NaOH} \text{Prussian blue}$

NaCN: gives characteristic test

$Na + X \xrightarrow{\Delta} NaX \xrightarrow{AgNO_3} AgX$

NaX: gives characteristic test; AgX: Coloured ppt.

$2Na + S \xrightarrow{\Delta} Na_2S \xrightarrow{Pb^{+2}} \text{Pbs}$

Na_2S: gives characteristic test; Pbs: black ppt.

$$Na + P + HO \longrightarrow Na_3PO_4$$

No characteristic test available

Test (S) :

$$3Na_2CO_3 + 2P + (O) \longrightarrow 2Na_3PO_4 + 3CO_2$$

(Y)

$$Na_3PO_4 + 3HNO_3 \longrightarrow H_3PO_4 + 3NaNO_3$$

$$\underset{\text{Ammonium molybdate}}{H_3PO_4 + 12\,(NH_4)_2MoO_4} + 21HNO_3 \longrightarrow \underset{\text{Ammonium phosphomolybdate}}{(NH_4)_3PO_4\cdot 12\,MoO_3} + 21\,NH_4NO_3 + 12H_2O \quad \text{(yellow ppt.)}$$

Passage 4

(a)

$$CH_3 - Ph - COOH \underset{HCl}{\overset{\text{aq. NaOH}}{\rightleftharpoons}} CH_3 - Ph - COO^{\ominus}$$

(d) Soluble in aq. layer

(b)

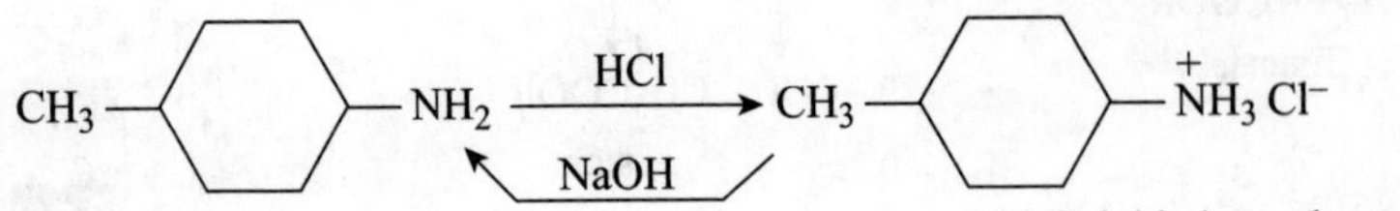

(e) Soluble in aq. layer

(c)

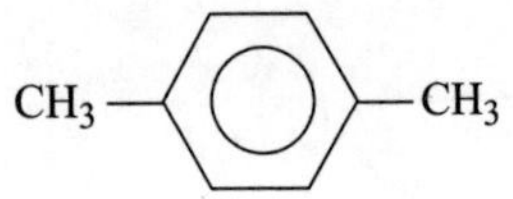

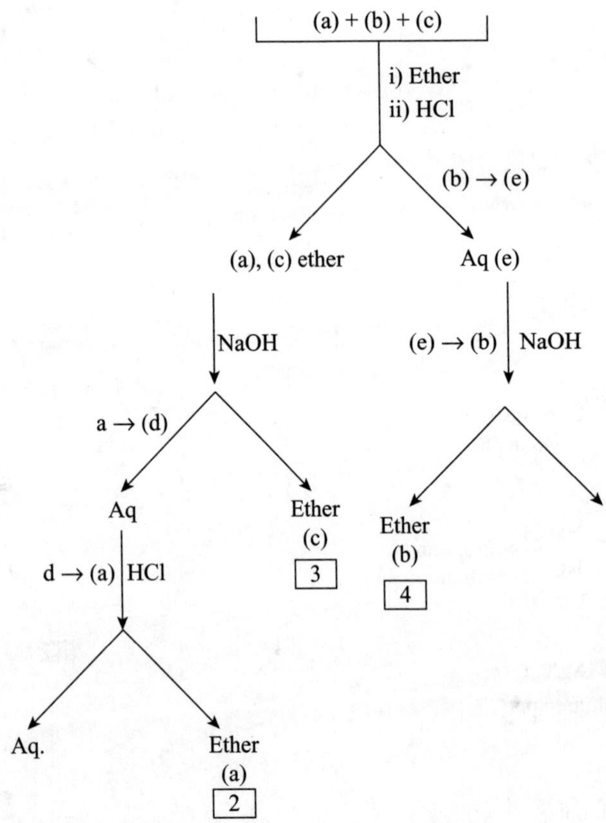

34. (a)

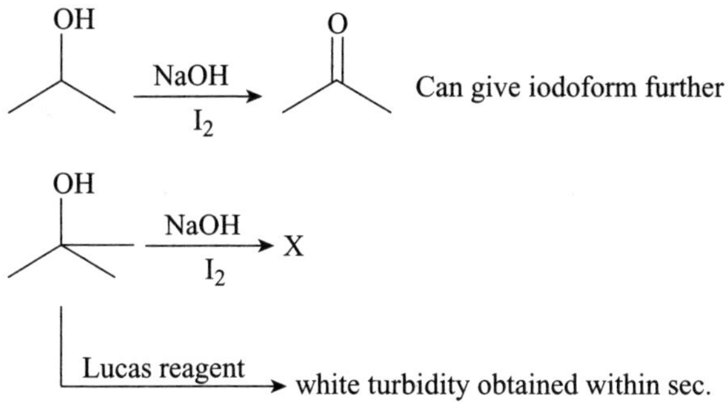

Ans: (r) and (s)

(b)

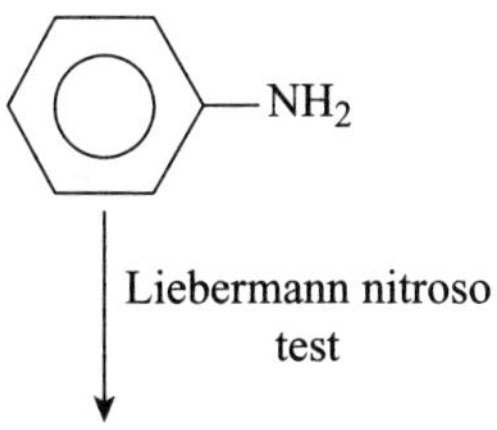

+ve

–aniline and phenol gives dye at different PH

Ans: (p) and (q)

(c) $R—NH_2 \xrightarrow[\text{Liebermann's test}]{HNO_2}$ Evolve N_2 gas (seen as bubbles)

$R_2NH \xrightarrow[\text{Liebermann's test}]{HNO_2}$ Yellow oily layer ($R_2N — NO$)

Ans: (p)

(d)

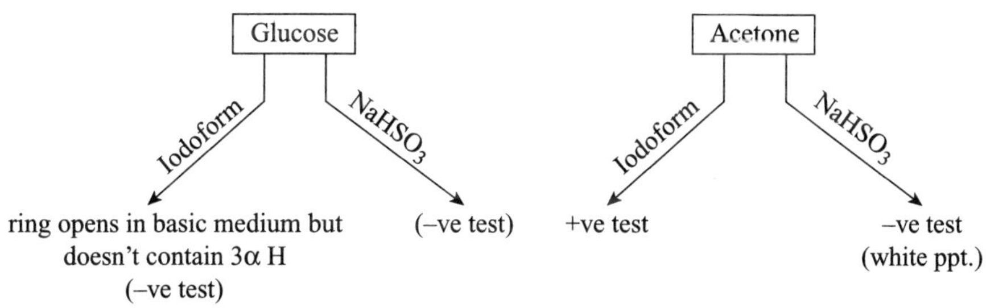

Ans: (r) and (t)

35. (a) $CH_3—\overset{O}{\overset{||}{C}}—OH \xrightarrow{NaOH} CH_3—\overset{O}{\overset{||}{C}}—O^{\ominus} –$

Ans: (q)

(b) $CH_3—\overset{O}{\overset{||}{C}}— Cl \xrightarrow[SN_{AE}]{NaOH} CH_3—\overset{O}{\overset{||}{C}}—O^{\ominus}$

Ans: (r) and (s)

(c) $CH_3—\overset{O}{\overset{||}{C}}— NH_2 \xrightarrow[SN_{AE}]{NaOH} CH_3—\overset{O}{\overset{||}{C}}—O^{\ominus}$

Ans: (p)

36. (a)

$$CH_3—C\equiv CH \xrightarrow[\text{dil } H_2SO_4]{\% \ HgSO_4} CH_3—\overset{\overset{O}{\|}}{C}—CH_3$$

Ans: (t)

(b) Bromine water test is given by alkene and alkynes. **Ans:** (p), (q), (r) and (s)

37. (a)

$$Ph—\overset{\overset{O}{\|}}{C}—H \xrightarrow{\text{Tollen's reagent}} +\text{ve Test}$$

$$Ph—\overset{\overset{O}{\|}}{C}—H \xrightarrow{\text{Fehling's solution}} -\text{ve Test}$$

Ans: (p) and (s)

38. (a) $\overset{\ominus}{Cl}\ \overset{\oplus}{N}H_3—NH_2 \xrightarrow[\text{Tollen's reag.}]{[Ag(NH_3)_2]OH} \underset{\text{White ppt.}}{AgCl \downarrow}$ **Ans:** (q) and (r)

39. (c) **Ans:** (r), (s) and (t)

$$2° \text{ alcohol} \xrightarrow{\text{Victor Meyer's test}} \text{blue colour}$$

$$3° \text{ alcohol} \xrightarrow{\text{Victor Meyer's test}} \text{colourless}$$